Sequence Stratigraphy of the Mid-Continent

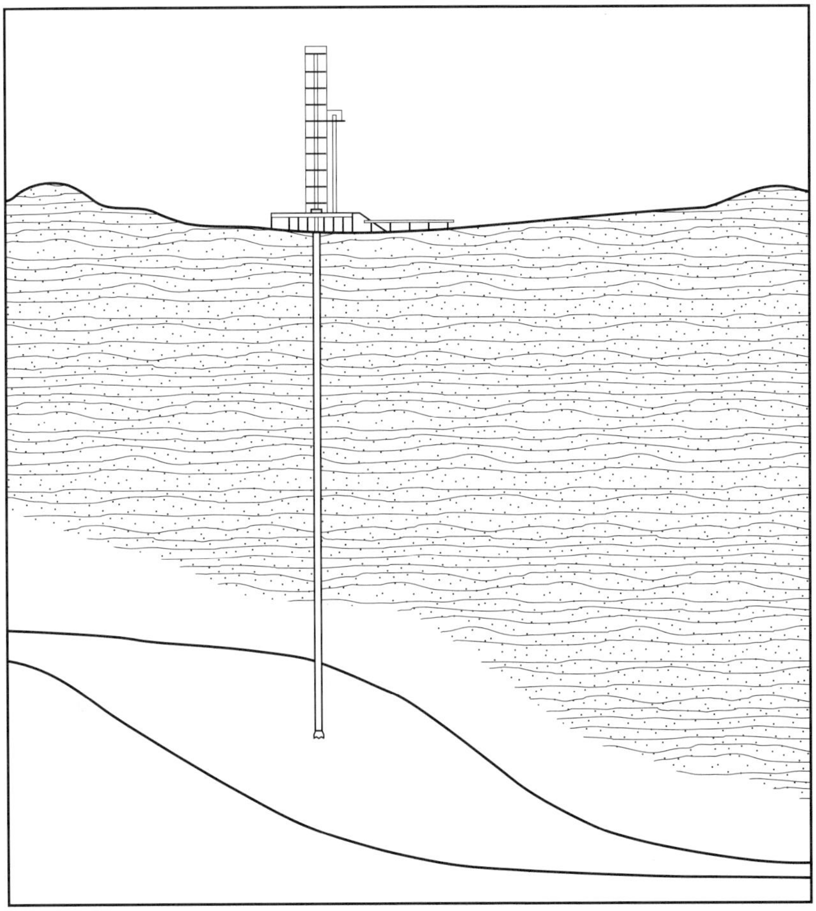

Editor: Norman J. Hyne
Professor of Petroleum Geology
Continuing Education
The University of Tulsa
Tulsa, OK 74104

ISBN: 0-945087-01-2

Printed in the United States of America
Copyright and Publication, 1995

Tulsa Geological Society acknowledges Judith R. Barton as the copy editor and proofreader for this book, and Brenda Howell for designing the art for the cover and title page. The Mid-Continent Section of the AAPG made a generous loan that enabled the Tulsa Geological Society to print this book. The Kansas Geological Survey gave financial support for color printing in this book.

CONTENTS

Tulsa Geological Society

75th Anniversary

1920-1995

SEQUENCE STRATIGRAPHY
A NEW LOOK AT OLD ROCKS

Norman J. Hyne
Professor of Petroleum Geology
Continuing Education
The University of Tulsa
Tulsa, OK 74104

ABSTRACT

Sequence stratigraphy is the use of time-stratigraphic units to classify, correlate, and map sedimentary rocks. Time-stratigraphic mapping in the field was first done using unconformities and marker beds. It did not become popular until the late 1970s when seismic reflection was applied to basin fill on a regional basis *(seismic stratigraphy).* Outcrop and well data extended seismic stratigraphy to a finer resolution during the 1980s (*sequence stratigraphy*). During a fall and rise of sea level, a characteristic package of sedimentary rocks called a sequence is deposited. The sequence is bounded on the top and bottom by major unconformities and often contains lowstand, transgressive, and highstand systems tracts. These are assemblages of sedimentary rocks that were deposited during specific stages of a regression and transgression cycle. Systems tracts can be identified on outcrops, cores, and well logs. The vertical and lateral succession of facies, along with source and reservoir rock locations, can be predicted within each systems tract. Time horizons are formed by the (1) lowstand surface of erosion, (2) transgressive surface, and (3) maximum flooding surface. These either define the border of a sequence or the borders of systems tracts in a sequence. Seismic-sequence stratigraphy is being used to reinterpret many sedimentary rocks and to identify stratigraphic plays.

ROCK-STRATIGRAPHIC UNITS

Stratigraphy began during coal mining and canal digging operations in England during the 18th and 19th centuries. In 1719, John Strachey was one of the first to use a geological cross section to show dipping coal seams and their intervening beds in Sommerset, south of Bristow, England. Later, William Smith, a coal mine surveyor who became involved in canal building, published the first comprehensive geological map of England and Wales in 1815. They used rock-stratigraphic units composed of relatively uniform lithologies because it was important to project the rock layers out ahead of the coal mining and canal dig-

FIGURE 1 — The angular unconformity at Siccar Point, Scotland (one of the locations where geologists first recognized the enormous time span represented by unconformities).

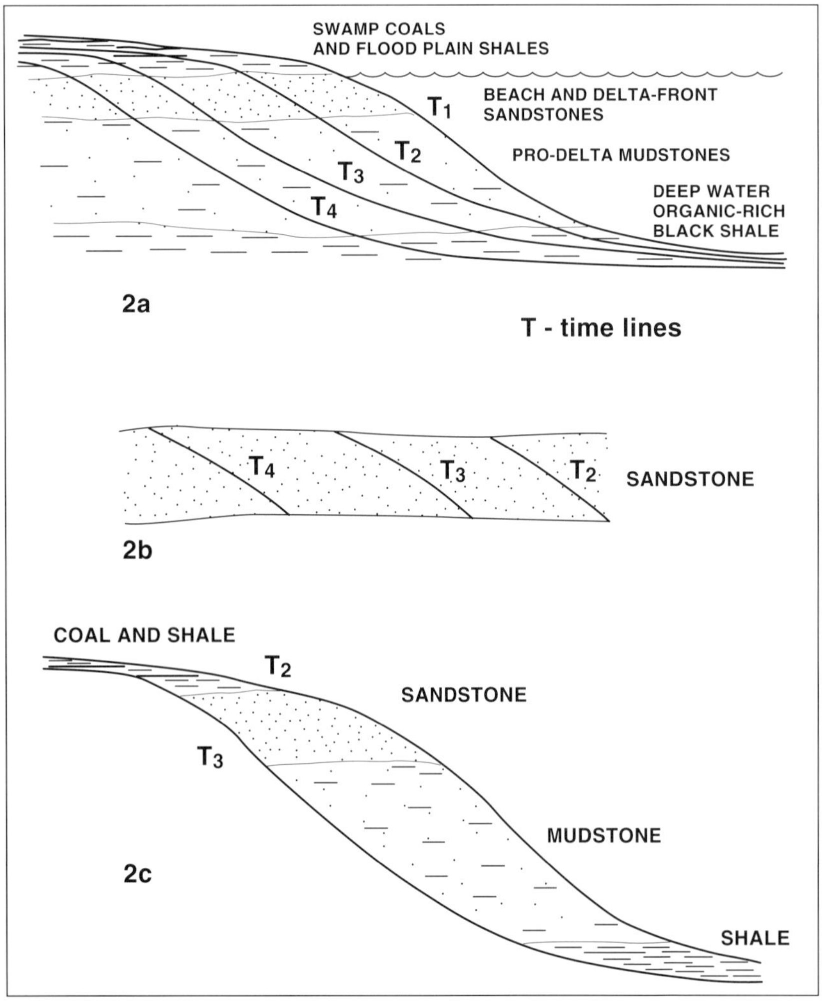

FIGURE 2 — A cross section showing (2a) the depositional environments and sedimentary facies on a prograding delta; (2b) rock-stratigraphic unit mapping of deltaic sedimentary rocks; and (2c) time-stratigraphic unit mapping of deltaic sedimentary rocks.

ging. The fundamental rock-stratigraphic unit used for mapping is the formation. Similar adjacent formations can be combined to form a group and, on a finer scale, formations can be subdivided into members.

At the time stratigraphy was first being developed, a great debate raged as to the age of the earth. In 1788, James Hutton, James Hall, and John Playfair rowed a boat to Siccar Point on the coast of southern Scotland. At Siccar Point, an obvious angular unconformity (Fig. 1) separates the underlying graywacke sandstones of Silurian and Ordovician ages from the overlying fluvial red beds of the Upper Old Red Sandstone of Devonian age. It was here, and at several other unconformity locations, that geologists recognized the enormous gap in the record of time missing between the rocks above and below an unconformity and, thus, the great antiquity of the earth.

In 1838, Amanaz Gressly developed the concept of facies. A sedimentary facies is a rock body that differs in composition from adjacent sedimentary rocks that were deposited at the same time. A facies is the sedimentary rock record of a specific depositional environment. A formation is often composed of a single facies, but can also be composed of two or more facies. This has been apparent to any petroleum geologist who has drilled updip on a sandstone formation only to find that it shaled out in a facies change. In 1894, Johannes Walther recognized the relationship between facies in Walther's law of facies. The law states that

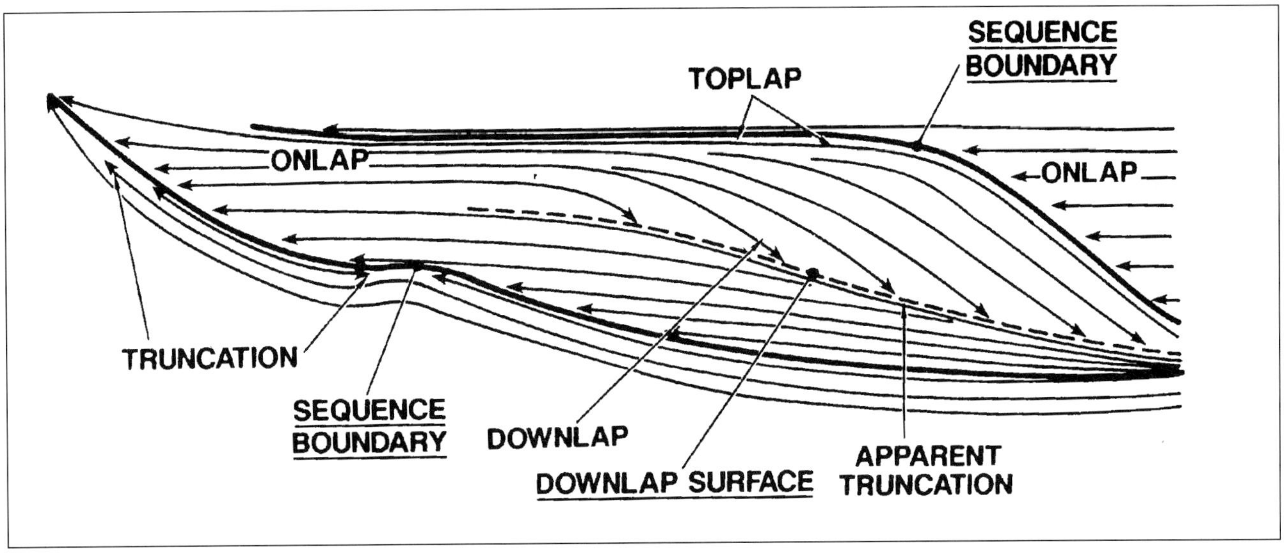

FIGURE 3 — A cross section model of primary seismic reflection terminations shown as arrows against unconformities. Disconformity names are underlined (after Vail, 1987).

only those facies that occur beside each other can be superimposed (Fig. 2a). The same lateral variation in facies is seen vertically. The relationship occurs only if there are no major time breaks in the sedimentary rocks, such as unconformities.

TIME-STRATIGRAPHIC UNITS

During recent decades, exploration for petroleum in sedimentary basins intensified and it became clear that formations were not the same age everywhere. Time lines cut across formations because a formation is not deposited during the same time interval along the extent of the formation (Fig. 2b). For this reason, rock-stratigraphic units are not very useful in predicting the location of various facies and, specifically, source and reservoir rocks. Time-stratigraphic units are strata deposited during specific intervals of time. Their boundaries are the same age everywhere. Time-stratigraphic units display the succession of facies deposited during that time interval (Fig. 2c) allowing for a paleogeographic reconstruction of the basin. Paleogeographic maps used for petroleum exploration (Adler et al., 1971) can be made only from time-stratigraphic units. "Our final interpretations of paleogeography, sediment dispersal patterns, and basin history can be no better than the stratigraphic foundation upon which they are based (Galloway, 1989a)." However, locating precise time lines in outcrop and well data has been difficult.

Sloss (1988) recognized that unconformities separated younger from older rocks. He identified six major unconformities on the North American craton to separate mappable, time-stratigraphic units (Sloss, 1963). These unconformities were later correlated worldwide (Sloss and Speed, 1974). The time-stratigraphic units between the unconformities were termed sequences.

Using marker beds, unconformities, and marine to nonmarine facies changes to locate time lines (Busch and Link, 1985), genetic units were defined to map Mid-Continent sedimentary rocks. These time-stratigraphic units were used with some success in locating petroleum prospects. [*In this book, Rascoe shows the geological background used to define time-stratigraphic units by geologists in the Mid-Continent before seismic and sequence stratigraphy were ever introduced.*]

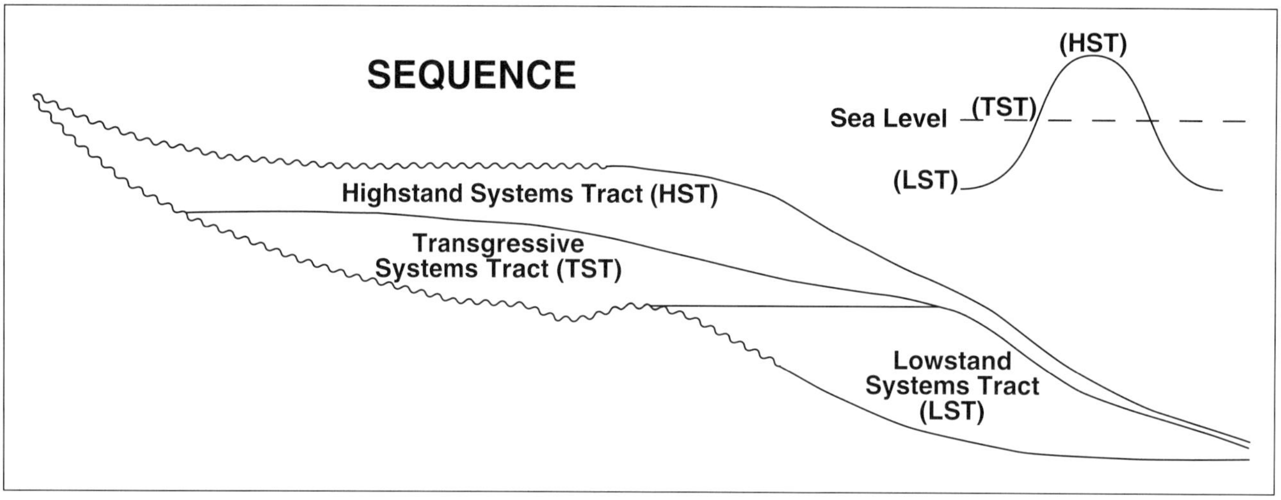

FIGURE 4 — A cross section of the systems tracts in a type 1 sequence. The general stage at which each systems tract is deposited is shown on the eustatic sea-level cycle.

SEISMIC STRATIGRAPHY

During the 1960s, digital recording and computer processing had significantly improved seismic reflection data. Researchers at Exxon, many of whom were former students of Sloss, recognized during the exploration of the North Sea that not only stratal surfaces but unconformities can also be seen as primary seismic reflectors. Unconformities have chronostratigraphic significance, because all rocks above the unconformity are younger than any rocks below the unconformity and, therefore, they must contain time lines. However, the gap in time between the upper and lower rocks can vary laterally along an unconformity. They determined that the different types of seismic reflections at the unconformities (Fig. 3) could record eustatic sea-level changes. Established terms such as onlap and erosional truncation were used, and the terms toplap and downlap were introduced for description of the reflectors at the unconformities. Seismic reflection patterns were also used to define several different seismic facies related to generalized depositional environments (e.g., shelf, reef, delta platform, slope, and basinal plain). The time-stratigraphic unit of basin fill defined by major unconformities (Fig. 4) is called a depositional sequence (Mitchum et al., 1977). It is defined as a relatively conformable succession of genetically related strata bounded at its top and base by unconformities and their correlative conformities (Vail et al., 1977). The use of seismic reflection records to delineate unconformities, sequences, and facies is called seismic stratigraphy. The principles of seismic stratigraphy were first published in *Seismic Stratigraphy — Application to Hydrocarbon Exploration* (Payton, 1977). Seismic stratigraphy was originally applied to siliciclastic sediments, but can also be applied to carbonates (Sarg, 1988; Hunt and Tucker, 1993).

Four factors were identified as influencing the sediments deposited in a siliciclastic sequence: eustasy, subsidence, sediment supply, and climate. Eustasy, the worldwide rise and fall of sea level, controls the highest elevation to which the marine sediments can be deposited. Subsidence of the basin and eustasy both control accommodation space, the space for potential sediment accumulation between sea level and the sediment/water interface (basin floor). Accommodation space determines the shape of the sequence and how thick or thin it is. Seismic stratigraphy was originally developed for basins with slow, steady subsidence with maximum subsidence in the basin center. Sediment supply controls the rate of sedimentation and facies development. Climate controls the type and volume of sediments deposited: siliciclastic, carbonate, or mixed. Eustasy was originally identified as the primary control on basin fill in most basins (Vail et al., 1977). Eustatic control was controversial and was a major factor in the lack of early acceptance of seismic stratigraphy in areas that had been influenced by tectonics such as the Mid-Con-

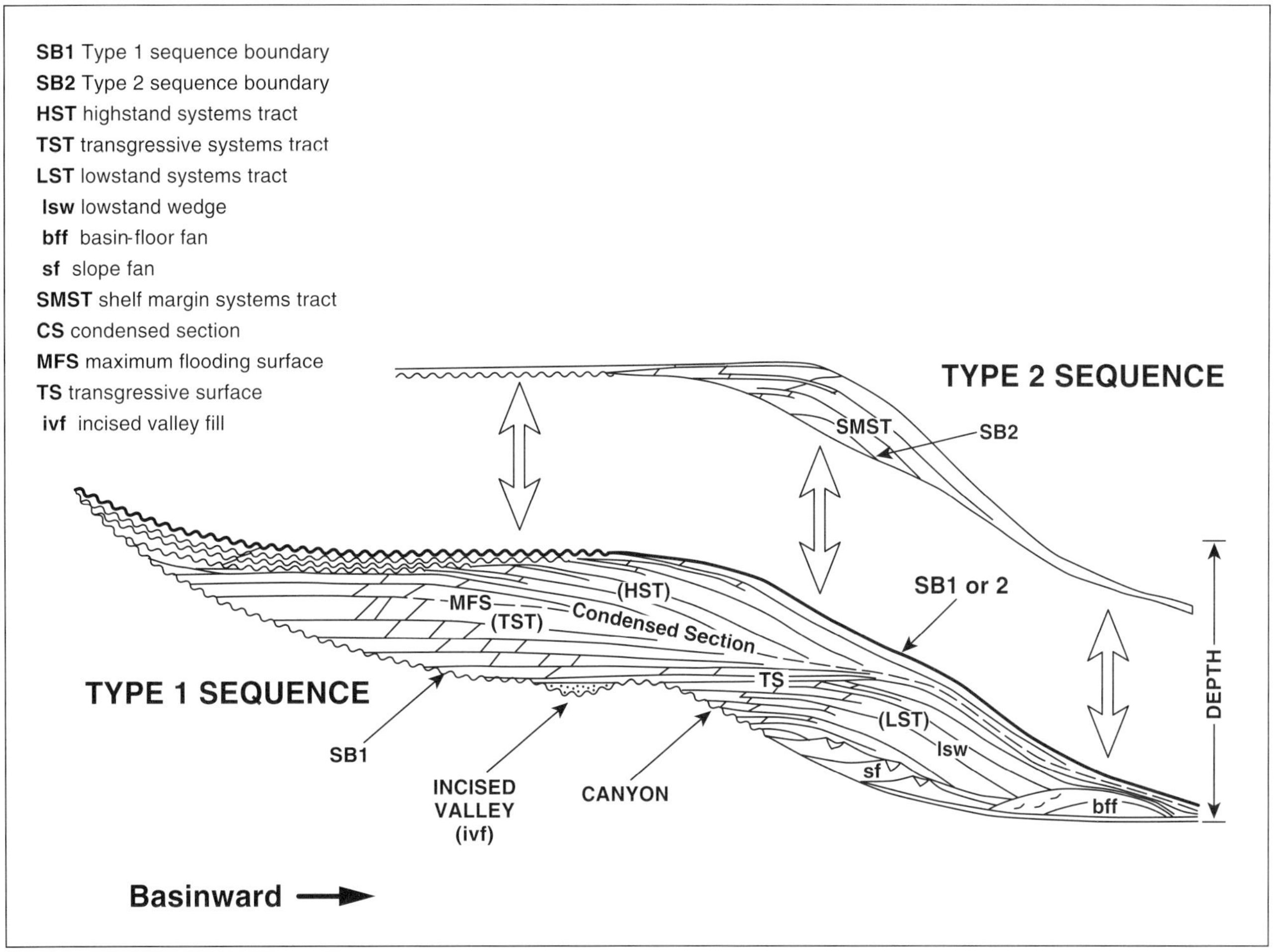

FIGURE 5 — The surfaces and parts of a sequence on a cross section oriented parallel to the depositional dip of the basin (modified from Vail, 1987). The slug diagram shows a type 1 sequence. A type 2 sequence occurs above it. The sequence thickness is in the order of tens to thousands of feet thick.

tinent. In a carbonate sequence, a significant difference is that carbonate sediments are generated within the depocenter. Because of this, carbonate depositional rates are more a function of intrabasinal conditions than external source area conditions that are more important for siliciclastics (Loucks and Sarg, 1993).

The sequence oriented parallel to the depositional dip of the basin (perpendicular to the shoreline) is often thickest in the middle and outer shelf and upper and middle slope environments and thins toward either end (Figs. 4 and 5). It is called a slug diagram. The upper and lower bounding surfaces of a sequence are major unconformities (Mitchum et al., 1977). When the sea level falls below the depositional shelf break of the sequence below it, the shelf surface is eroded and the sedimentary facies are shifted seaward. This is a type 1 sequence boundary. The eustatic fall of the sea level is on the order of 100 m, and valleys are incised on the exposed shelf. The unconformity can be traced out into the basin where it becomes a conformable bedding surface with no evidence of erosion. Sequences are composed of systems tracts, related sedimentary facies that were deposited adjacent to each other during the same segment of the sea-level fall and rise cycle. The most common sequence, a type 1 sequence which lies on a type 1 sequence boundary, contains lowstand, transgressive, and highstand systems tracts, each deposited during a different stage on the sea-level curve (Fig. 4).

The lowermost systems tract in the sequence is the lowstand systems tract (LST) which lies on the type 1 sequence boundary (Fig. 5). These facies were deposited during the rapid sea-level fall, the late fall, lowstand, and early rise of the sea level. Deposition of the LST is a time of active submarine canyon erosion and submarine fan deposition. It is often composed of a basin-floor fan (bff), slope fan (sf), and lowstand wedge (lsw). The basin-floor fan is primarily sand and is usually located at the mouth of a submarine canyon. Its formation is associated with the erosion of the submarine canyon and incision of fluvial valleys on the shelf. The slope fan that overlies it is formed by turbidites and debris flows and is composed of turbidite-leveed channel and overbank deposits (Van Wagoner et al., 1990). The fans are sometimes illustrated as a single lowstand fan (lsf) (Posamentier and Vail, 1988). The lowstand wedge consists of coastal, shoreline, and marine sedimentary facies.

The transgressive systems tract (TST) is the middle systems tract that was deposited during the rapid rise of sea level (Fig. 5) and consists of coastal, shoreline, and marine sedimentary facies. Its base is the transgressive surface on the LST known as the first flooding surface or transgressive surface (TS), depending on its characteristics. The TS is the first significant marine-flooding surface across the shelf in the sequence. Along the top of the TST is the condensed section (CS) deposited during the maximum extent of shoreline transgression. The CS is composed of thin marine beds deposited very slowly on the middle and outer shelf, slope, and basin floor. It is often an organic-rich, phosphatic, radioactive, black shale (hot shale on natural gamma ray logs). Calcareous nannofossils and planktonic foraminifera are most abundant in the offshore portion of the CS because of the low terrigenous sedimentation rate. The CS in many sequences yields abundant microfossils to date the sequence (Loutit et al., 1988). A second abundance peak of microfossils may occur at the top of the slope fan. [*In this book, Scott explains how biostratigraphy is used in the dating and correlation of sequences, and Boardman et al. apply microfossil analysis to sequences in the Mid-Continent.*] On a seismic record, the top of the TST may be a downlap surface. It is a maximum flooding surface (MFS) on which the base (toes) of the prograding clinoforms of the overlying highstand systems tract downlap (Figs. 3 and 5). The MFS marks the peak of sea-level rise and is usually located in the condensed section.

The highstand systems tract (HST) is located in the upper part of a sequence (Fig. 5) and consists of coastal, shoreline, and marine sedimentary facies. It was deposited during the late sea-level rise, high stillstand, and early sea-level fall. It is overlain by a major unconformity or a regressive systems tract. The eustatic fall of sea level causes a forced regression which is characterized by a seaward shift of the sedimentary facies and a shoreline regression (Posamentier et al., 1992a). It is associated with a zone of sediment bypass, subaerial exposure, and possible fluvial erosion. In some instances, the sediments are reworked and redeposited forming a regressive systems tract (RST). An RST can also form if sea-level fall is relatively slow. It will be located on the HST and below the sequence boundary.

A type 2 sequence, in contrast, is formed when relative sea level does not fall below the depositional shelf break of the underlying sequence. The unconformity below a type 2 sequence does not show the deep erosion and valley incision characteristic of a type 1 sequence boundary. Also, the shelf facies do not shift seaward beyond the old shelf edge. It is called a type 2 sequence boundary. The sequence has a shelf margin systems tract (SMST) lying on the type 2 sequence boundary (Fig. 5). The SMST, consisting of coastal, shoreline, and marine sedimentary facies, can be overlain by transgressive and highstand systems tracts. A type 1 or 2 sequence boundary marks the top of the sequence.

In a basin with relatively little tectonic influence, eustasy is often the most important control on the deposition of a sequence and sediment fill is episodic. The amount of relative sea-level fall and rise in each sequence can be estimated (Vail et al., 1977). These cycles have been age-dated using biostratigraphy, and a hypothetical, eustatic sea-level curve was constructed and published (Vail et al., 1977; Haq et al., 1988). There are short period cycles within longer cycles with at least five different orders of sea-level fall and rise cycles. The first-order cycle has a period greater than 50 m.y. and is thought to be related to plate tectonics. As continents came together, the volume of the ocean basins increased and sea level fell. After the continents broke up, the volume of the ocean basins decreased and sea level rose. The second-order cycle with a period of 3 to 50+ m.y. is a combination of five to seven third-order cycles with shorter peri-

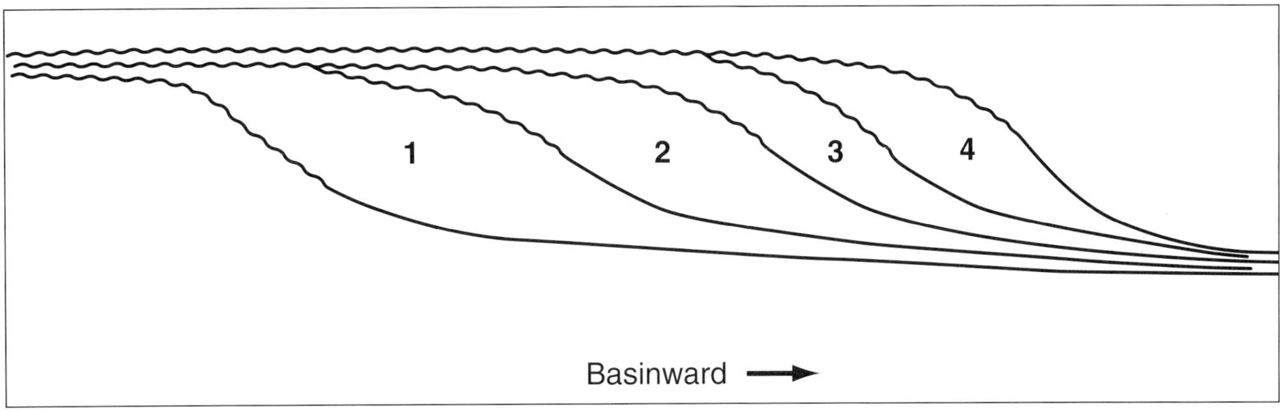

FIGURE 6 — A cross section of a series of sequences that have prograded out into a basin. Notice how unconformities on the shelf become conformable surfaces in the basin.

ods. Second-order cycles correspond in duration to the sequences of Sloss (1963) which are now called supersequences. Third-order cycles have periods of 0.5 to 3 m.y. and correlate with the formation of sequences (Vail, 1987). The cause of third-order cycles is not known. Fourth-order cycles have a duration of 0.2 to 0.5 m.y., whereas fifth-order cycles are 0.01 to 0.2 m.y. long. These cycles are thought to be related to cyclic changes in the earth's tilt and orbit that are known as the Milankovitch cycles. A special issue of the Journal of Sedimentary Petrology: *Orbital Forcing and Sedimentary Sequences*, edited by Fischer and Bottjer (1991), discussed these cycles and their effects. Sequences, formed by third-order cycles, are at the limit of traditional biostratigraphic resolution. The magnitude of the sea-level rise and fall is difficult to determine from the seismic record because of sediment compaction, crustal subsidence during the rise, and the extent of alluvial plain onlap during the subsequent sea-level fall. Because of this, coastal onlap and offlap curves have been published with a relative change of sea level in the order of several hundred meters for the long period cycles (Haq et al., 1988). These curves have not been completely accepted (Miall, 1992). Originally, the relative rise of the sea level was thought to be gradual compared to the rapid fall of the sea level, and the curves had a sawtooth shape. A more equal rate of rise and fall similar to a sine curve has been proposed for some geological times (Kerr, 1980).

Theoretically, sequences driven or forced by eustatic cycles should correlate throughout the basin and worldwide. This is used as a test to differentiate tectonic from eustatic cycles. A tectonically forced sequence will be bounded by regional unconformities that will not necessarily correlate across the basin. [*In this book, Bennison discusses the modification of sequences in a tectonically active basin.*] Sequences caused by eustatic sea-level changes should correlate across the basin. Attempts at correlating sequences between basins have had mixed results (Ross and Ross, 1981; Scott et al., 1988; Boardman and Heckle, 1989; Heckle and Weible, 1991).

As a basin is filled with sediments, a series of sequences will prograde out into the basin (Fig. 6). With seismic stratigraphy, a geologist can use seismic records to delineate the depositional packages in a basin and predict the facies distribution before a well is drilled. Once a well is drilled and the vertical succession of facies is determined, Walther's law can be applied to predict the lateral succession of depositional environments and facies with time. A major advantage of seismic stratigraphy is that it can be applied to basins with little or no well control.

Vertical seismic resolution is usually no more than 20 m. Because of this, only the largest aspects of a sequence can be resolved with seismic data.

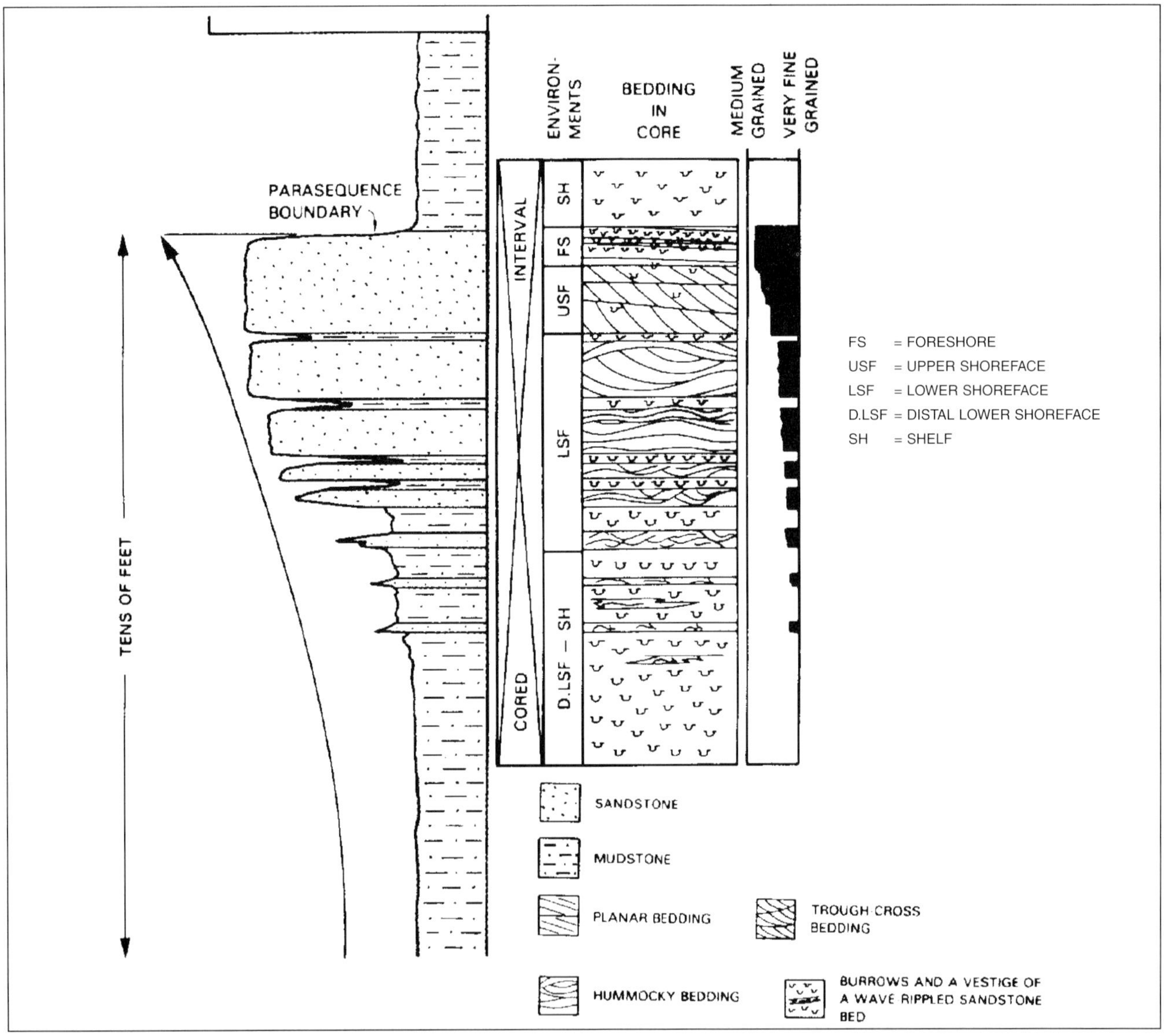

FIGURE 7 — An example of a parasequence from a beach depositional environment (from Van Wagoner et al., 1990).

SEQUENCE STRATIGRAPHY

In the 1980s, the principles of seismic stratigraphy were applied to outcrops, cores, and well logs, and were called sequence stratigraphy. The higher resolution showed that sequences are made of parasequences and parasequence sets (Table 1).

A parasequence (Fig. 7) is a relatively conformable succession of genetically related beds or bed sets bounded by marine-flooding surfaces (Van Wagoner, 1985). A marine-flooding surface separates younger from older sedimentary rocks, and the rocks above the surface show evidence of a sudden increase in water depth. The marine-flooding surface is often characterized by minor submarine erosion or nondeposition. This is in contrast to a sequence boundary which results from a sudden decrease in water depth. Parasequences may be composed of bed sets, beds, lamina sets, and laminae in descending order of scale (Van Wagoner et al., 1990) (Table 1). A parasequence and a genetic increment of strata (Busch, 1971) are apparently similar in scale and concept (Walker, 1990).

A parasequence set is a relatively conformable succession of genetically related parasequences. It

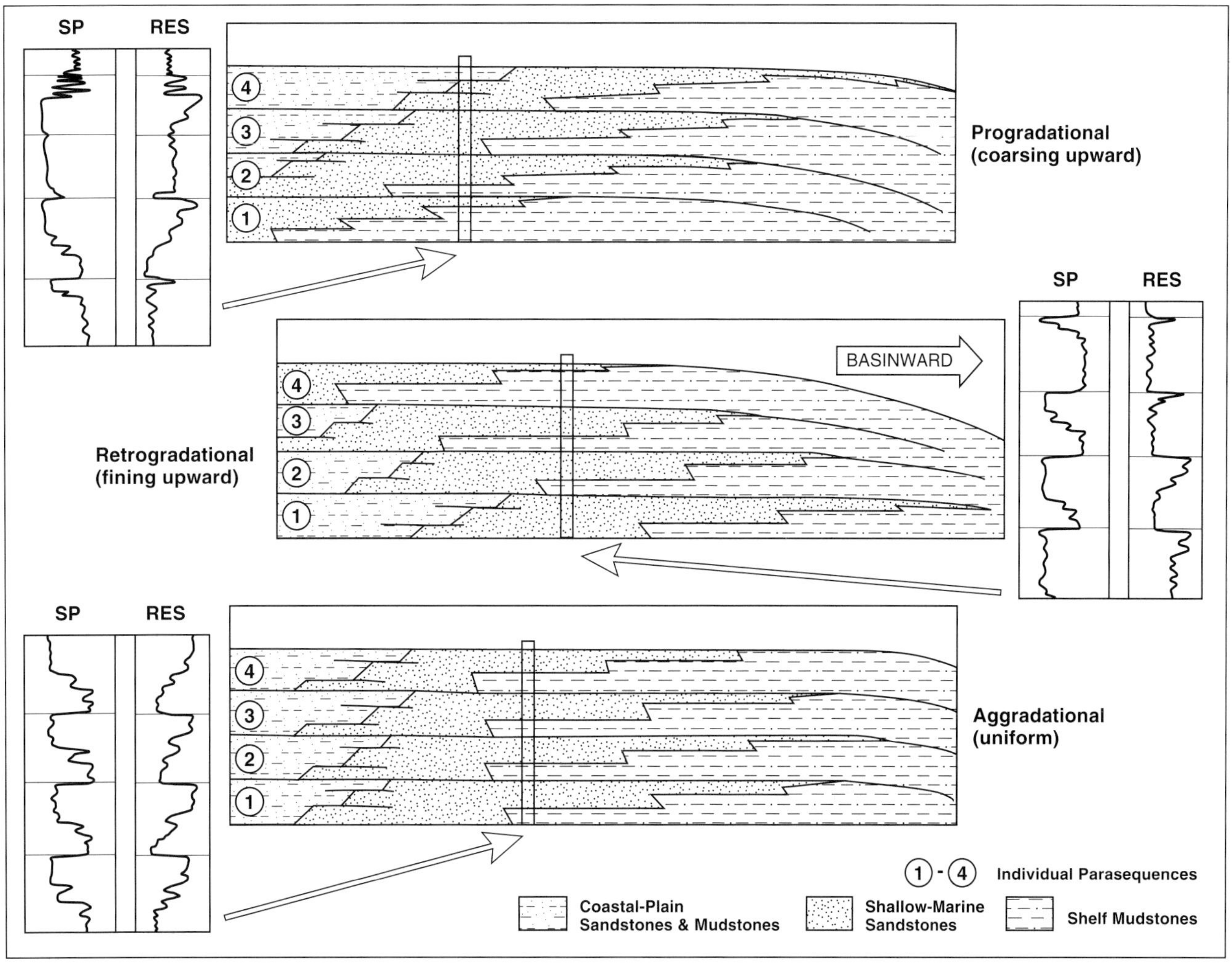

FIGURE 8 — Parasequence set stacking patterns. The well log stacking patterns are (a) prograding (upward coarsening), (b) retrograding (upward fining), and (c) aggrading (uniform) (modified from Van Wagoner et al., 1990).

forms a distinctive stacking pattern and in many places is bounded by major marine-flooding surfaces. Parasequence sets and parasequences are related to fourth- and fifth-order eustatic cycles.

Parasequence sets can show progradational, retrogradational, or aggradational stacking patterns (Fig. 8). In a progradational parasequence set, progressively younger parasequences are deposited further basinward. In siliciclastics, this parasequence set usually will coarsen upward in grain size and can be identified on well logs (Fig. 8a). In a retrogradational parasequence set, progressively younger parasequences are deposited further landward. The grain size in this parasequence set will usually fine upward (Fig. 8b). In an aggradational parasequence set, progressively younger parasequences do not shift either landward or basinward in position. This parasequence set will be uniform in upward grain size (Fig. 8c).

Systems tracts in a sequence show characteristic stacking patterns (Fig. 9). The stacking patterns of the basin-floor (bff) and slope fan (sf) in an LST are difficult to recognize, because upward-shallowing criteria (such as grain size) are lacking in those environments. However, the lowstand wedge (lsw) is composed of one or more progradational parasequence sets. The TST may be characterized by a retrogradational parasequence set or sets. The early HST is characterized by an aggradational parasequence set, whereas the late HST has one or more progradational parasequence sets. In a type 2 sequence, the SMST consists of one or more slightly progradational or aggradational parasequence sets. Sequence boundaries may be located by the change in stacking pattern from the underlying parasequence set to

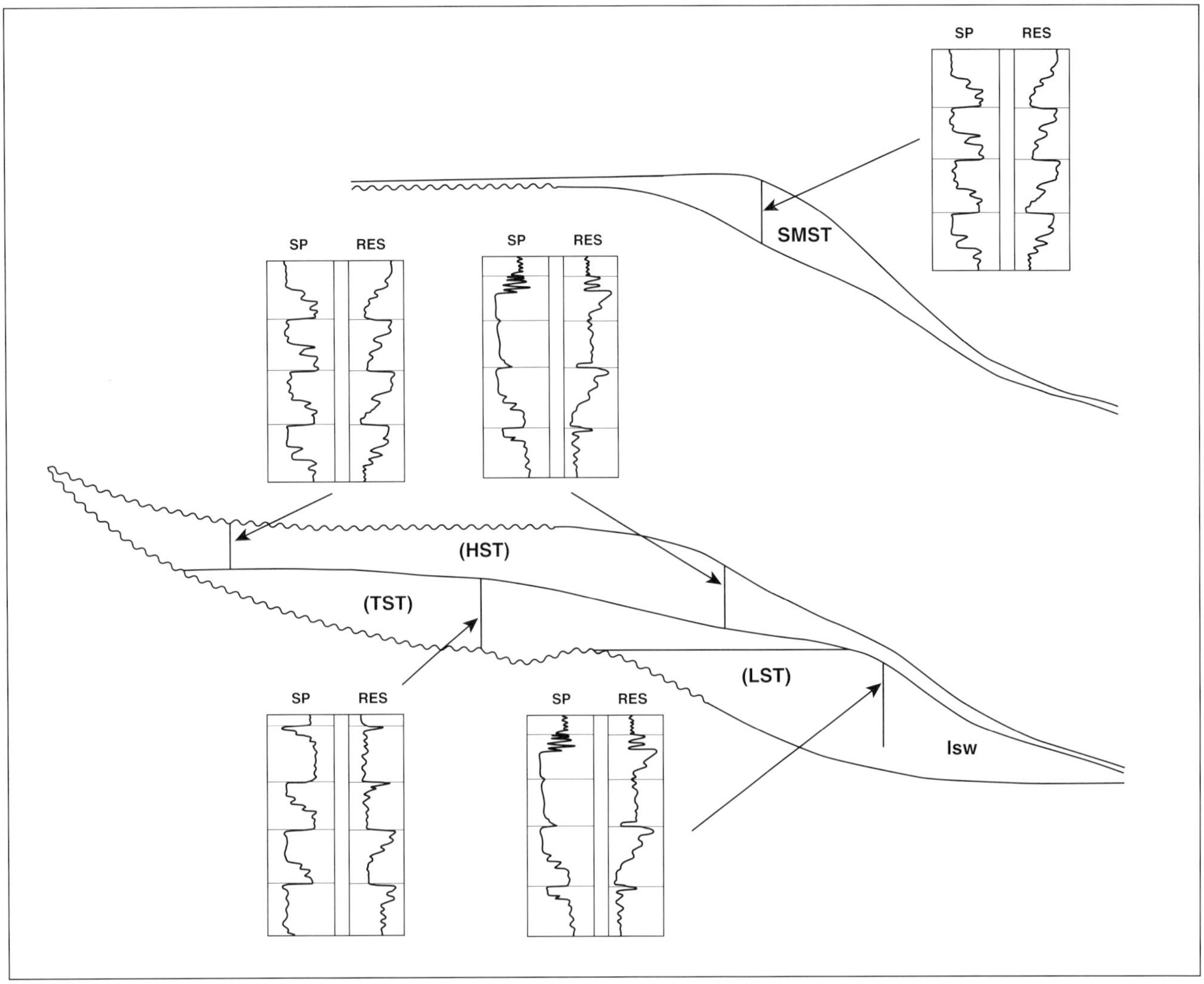

FIGURE 9 — Characteristic well logs of parasequence set stacking patterns in the different systems tracts of a sequence.

the overlying parasequence set.

The sequence and parasequence set boundaries are used in sequence stratigraphy for stratigraphic subdivision, correlation, and mapping. Three important surfaces are the (1) lowstand surface of erosion (lse) or sequence boundary (SB1 or SB2), (2) transgressive surface (TS) or first flooding surface, and (3) maximum flooding surface (MFS) in the condensed section (CS). [*In this book, Shelton and Gerkin show how to recognize sequence surfaces in siliciclastics, and Fritz and Medlock describe carbonate surfaces.*]

TERMINOLOGY

Seismic-sequence stratigraphy has introduced several unique terms. These specialized terms (i.e., sequence boundary, systems tract, parasequence set, and parasequence) are defined in papers by Van Wagoner et al. (1988), Loutit et al. (1988), and Van Wagoner et al. (1990). However, some geologists (Walker, 1990; Wiemer, 1992) argue that not all of these terms are necessary and that some of the concepts are already named in the North American Stratigraphic Code (1983). In the code, an allostratigraphic unit is a "mappable stratiform body of sedimentary rock that is defined and identified on the basis

STRATAL UNITS	DEFINITIONS	RANGE OF THICKNESSES (FEET)	RANGE OF LATERAL EXTENTS (SQ. MILES)	RANGE OF TIMES FOR FORMATION (YEARS)	TOOL RESOLUTION
		1000 100 10 1 INCHES	10,000 1000 100 10 1	10^6 10^5 10^4 10^3 10^2 10 1	PALEO; EXPLORATION SEISMIC; WELL LOG; CORE AND OUTCROP
SEQUENCE	A RELATIVELY CONFORMABLE SUCCESSION OF GENETICALLY RELATED STRATA BOUNDED BY UNCONFORMITIES AND THEIR CORRELATIVE CONFORMITIES (MITCHUM AND OTHERS, 1977)				
PARASEQUENCE SET	A SUCCESSION OF GENETICALLY RELATED PARASEQUENCES FORMING A DISTINCTIVE STACKING PATTERN AND COMMONLY BOUNDED BY MAJOR MARINE-FLOODING SURFACES AND THEIR CORRELATIVE SURFACES				
PARASEQUENCE	A RELATIVELY CONFORMABLE SUCCESSION OF GENETICALLY RELATED BEDS OR BEDSETS BOUNDED BY MARINE-FLOODING SURFACES AND THEIR CORRELATIVE SURFACES				
BEDSET					
BED					
LAMINA SET					
LAMINA					

TABLE 1 — Characteristics of sequences and their components (after Van Wagoner et al., 1990).

of its bounding disconformities." The units are allogroups, alloformations, and allomembers. Provisions are made for allostratigraphic type sections similar to those for lithostratigraphic units. Formations and sequences span different stratigraphic intervals, although they may be given the same names. There is no convention for naming sequences. They have been named after the condensed section, the most prominent marine lithostratigraphic unit, the initial and terminal lithostratigraphic units, and other methods. *[In this book, Watney et al. suggest a letter and number method based on the age of the condensed section and the sequence's stratigraphic position. They also suggest criteria for location of a reference section in a sequence.]*

When the sequence is caused by a change that is external to the sedimentary body such as eustasy, it is called an allocyclic sequence. However, every sharp base of a sandstone need not mark the base of an allocyclic sequence. Some cycles are caused by changes that are internal to the sequence (such as channel or bar migration) and are called autocyclic sequences.

APPLICATION

Seismic-sequence stratigraphic models are tools to analyze basin fills rather than a rigid template for constructing stratigraphic cross sections. The mechanics and some rules of thumb for sequence stratigraphic interpretation of well logs and cross sections were outlined by Mulholland (1994). *[Studies in this book that apply sequence stratigraphy include those on the Cherokee Group (Brenner), Chase Group (Mazzulo et al.), Simpson Group (Candelaria and Handford), Douglas Group (Archer and Feldman), upper Morrowan (Al-Shaieb et al.), Council Grove Group (Puckette et al.), and the Middle and Upper Pennsylvanian (Bennison and Watney et al.).]*

The type 1 and type 2 sequences that are bounded by unconformities and were originally described by Vail et al. (1977) are called Exxon sequences or models. In a variation, Galloway (1989a) recognized the influence of sediment supply and basin subsidence and uplift in the Gulf Coast. He uses genetic stratigraphic sequences, regressive depositional units bounded by transgressive surfaces, to correlate and map the northwest Gulf of Mexico Cenozoic basin (Galloway, 1989b). Fourteen major periods of continental margin outbuilding are identified. A genetic stratigraphic sequence is bounded by maximum flooding surfaces usually located in condensed sections. The sequence may contain highstand, lowstand, and transgressive systems tracts in that order upward. It also contains an lse (sequence boundary). The condensed section with the enclosed MFS boundaries of genetic stratigraphic units are much easier to identify on wireline well logs than the erosional surface and paleosol boundaries of sequences. A genetic stratigraphic sequence is also useful where the sequence boundaries are separated by very thin sequences and cannot be distinguished on the seismic record. However, using unconformities as sequence boundaries has the advantage that the unconformities are often overlain by potential sandstone reservoirs of deep-sea submarine fan, beach, and incised valley-fill origins.

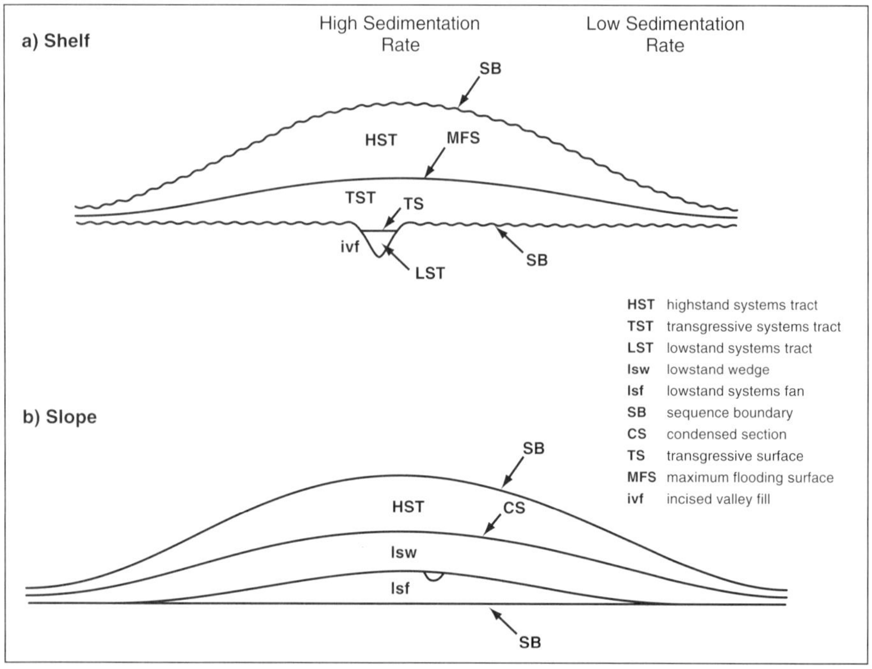

FIGURE 10 — Depositional strike cross sections of a sequence. This is in contrast to a depositional dip cross section (Fig. 5); (a) a cross section on the shelf and (b) a cross section on the slope. Note the thin or absent TST on the slope.

Seismic-sequence stratigraphy is best applied to profiles oriented parallel to the depositional dip of the basin. Out in the basin, the sequence units become thin and are bounded by subparallel reflectors ("railroad tracks") that are still useful in correlation (Fig. 6). Eventually, the sequence boundaries can amalgamate. Along the depositional strike of the basin, which is parallel to the shoreline, the sequence boundaries also form railroad tracks away from depocenters (Fig. 10) and can amalgamate.

The seismic-sequence stratigraphy concept is independent of scale and time. Neither the size and depth of the basin nor the period of the sea-level fall and rise should affect the model. Posamentier et al. (1992) applied the principles of sequence stratigraphy to a fan delta deposited in a roadside drainage ditch during a rise and fall of the water level that lasted several days. Seismic-sequence stratigraphy has been applied to modern sedimentary environments (Boyd et al., 1989) and extended to continental sedimentary rocks (Shanley and McCabe, 1994). This new method of studying sedimentary rocks could revive exploration in mature petroleum provinces, such as the Mid-Continent.

PETROLEUM GEOLOGY

Source rocks can occur in any systems tract but the richest source rocks have been shown to be condensed sections (Creaney and Passey, 1993). The total organic carbon can be very high and more marine in origin (oil prone) in the CS. The CS can often be identified on a natural gamma ray log by its high radioactivity. Meyer and Nederlof (1984) showed how source rocks can be identified on other wireline logs by using crossplots. Gamma ray/sonic and gamma ray/resistivity pairs were used by Loutit et al. (1988) to identify condensed sections.

In a review of global basins, Klemme and Ulmishek (1991) identified six, large-scale stratigraphic time intervals which contain the source rocks that have generated more than 90% of the world's oil reserves. Although they were not able to identify any cyclicity in the temporal distribution of the source rocks, they noted that black shale deposition was common during transgressive periods.

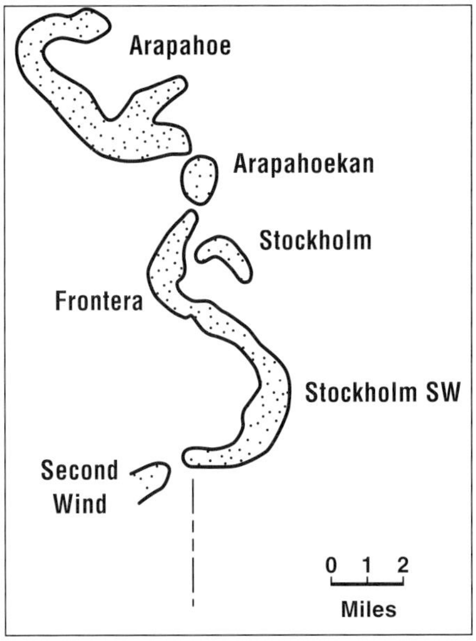

FIGURE 11 — Incised valley fill, Morrowan sandstone oil and gas fields trend along the Kansas-Colorado border (modified from Shumard, 1991).

Siliciclastic reservoir rocks may be widely distributed within a sequence. Vail (1987) identified five potential siliciclastic reservoirs in the LST. These are basin-floor fans, slope-leveed channel fills, prograding turbidite wedges, prograding/aggrading coastal sands, and incised valley fills. Sangree et al. (1990) contended that the LST has the greatest potential for stratigraphic plays. Many of the deepwater Gulf of Mexico reservoirs being developed today are in turbidite sandstones located in LSTs. Fluvial, deltaic, and coastal sandstones are potential reservoirs in the TST. The highstand and shelf margin systems tracts may have point bar, distributary channel fill, delta front, and barrier island sandstones (Sangree et al., 1988). Many of these sandstones are encased in shale and are potential stratigraphic traps.

In a carbonate sequence (Sarg, 1988), a freshwater lens underlies the shelf during deposition of the LST. Significant dissolution creating or enhancing reservoir rock can occur on the shelf during this time. A karst limestone can occur just below the sequence boundary. The effect is less extensive with a type 2 sequence boundary. In contrast, mixing and hypersaline dolomitization can be important processes during the highstand. Reservoirs also may develop in carbonate grainstones and other rock types in transgressive and highstand systems tracts (Scott et al., 1993).

A successful application of sequence stratigraphy to petroleum exploration has been in incised valley fills of the lowstand and trangressive systems tracts of the middle United States and Alberta, Canada. Incised valley plays occur in the Pennsylvanian Morrow Formation of eastern Colorado and western Kansas, Lower Cretaceous Muddy Formation of Colorado and Wyoming, and Lower Cretaceous Viking Formation of Alberta. The Morrow Formation primarily consists of marine shale with sandstones in long, narrow, sinuous trends averaging 2 km wide and 3 to 10 m thick (Weimer, 1992). Krystinik and Blakeney (1990) identified seven subaerial exposure surfaces in the Morrow Formation. A sequence overlies each of these surfaces which has incised valleys cut into the underlying marine shale, and possibly an older surface (Wheeler et al., 1990). During lowstand and rising sea level, the incised valleys were filled with fluvial lag, fluvial sands, estuarine sands, and then estuarine muds. The TS was formed as the sea level rose over the deposits. Black, anoxic muds were then deposited on the surface. Over 150 million BOE have been found in this Morrow Formation trend (Fig. 11). The Stockholm Southwest field on this trend in Kansas was discovered in 1979 and will produce between 11 and 15 million barrels of oil. [*In this book, Archer and Feldman describe different valley-fill facies, and Al-Shaieb et al. describe the Morrowan valley-fill sandstones.*]

SEQUENCE STRATIGRAPHY IN THE MID-CONTINENT
A HISTORICAL REVIEW

Bailey Rascoe, Jr.
Consulting Geologist
Bartlesville, OK 74006

INTRODUCTION

In 1977 the American Association of Petroleum Geologists (AAPG) published the first of three works on seismic stratigraphy: *Memoir* 26, "Seismic Stratigraphy — Application to Hydrocarbon Exploration," 1977, C.E. Payton, editor; *Memoir* 39, "Seismic Stratigraphy — An Integrated Approach to Hydrocarbon Exploration," 1985, O.R. Berg and D.G. Woolverton, editors; and *Studies in Geology* 27, v. 1, 2, and 3, "Atlas of Seismic Stratigraphy," 1987, A.W. Bally, editor.

These volumes conclusively demonstrate that seismic data are new, enlightening, and valid sources of stratigraphic information, and it is from these works that the principles of modern *sequence stratigraphy* have evolved. Sequence stratigraphy is defined by Vail (1987, p. 1) as the "application of seismic stratigraphic interpretation techniques to sedimentary basin analysis . . . to subdivide, correlate, and map sedimentary rocks. The application of this procedure to a grid of seismic data groups seismic reflections into packages that correspond to chronostratigraphically constrained genetic depositional intervals. These intervals are called *depositional sequences* and *systems tracts.* They have predictable stratal patterns and lithofacies; thus they provide a new way to establish a chronostratigraphic correlation framework based on physical criteria." In essence, as Weimer (1992, p. 3) stated, sequence stratigraphy is the study of depositional sequences. A depositional sequence is "a stratigraphic unit composed of a relatively conformable succession of genetically related strata and bounded at its tops and base by unconformities of their correlative conformities" (Mitchum et al., 1977, p. 53) (Fig. 1). Sequence stratigraphy has evolved since 1977 to also include analysis which may be based solely on wireline logs.

For more than 60 years, many workers on the Mid-Continent have defined and mapped stratigraphic units that are very similar to depositional sequences. This study focuses on the concepts on which this early work was based in order to demonstrate that these efforts were the forerunners of modern sequence stratigraphy.

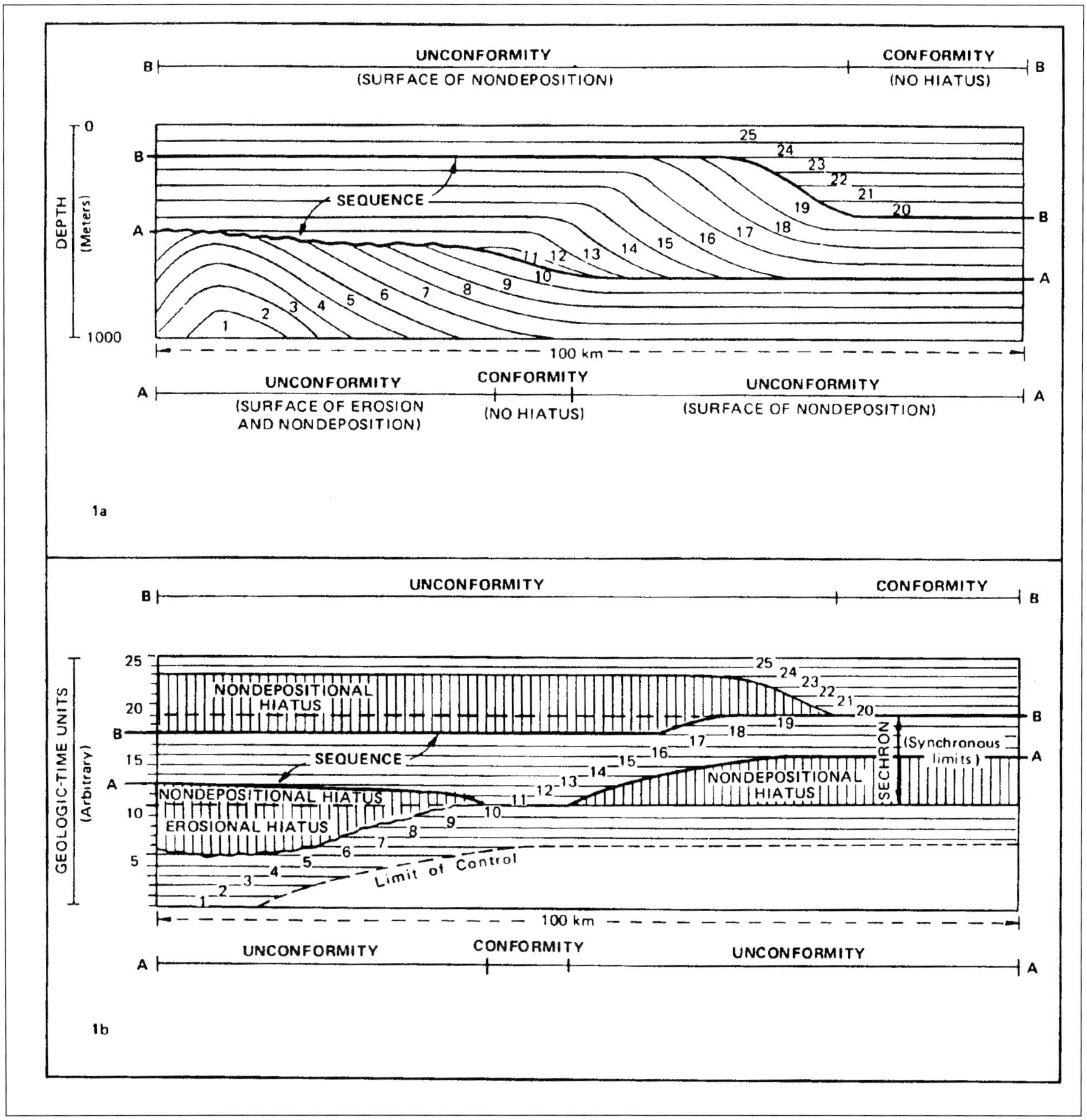

FIGURE 1 — Basic concepts of a depositional sequence. A depositional sequence is a stratigraphic unit composed of relatively conformable successions of genetically related strata and bounded at its top and base by unconformities or their correlative conformities (Mitchum et al., 1977).

a. Generalized stratigraphic section of a sequence. Boundaries are defined by surfaces A and B which pass laterally from unconformities to correlative conformities. Individual units of strata 1 through 25 are traced by following stratification surfaces, and are assumed conformable where successive strata are present. Where units of strata are missing, hiatuses are evident.

b. Generalized chronostratigraphic section of a sequence. Stratigraphic relations shown in A are replotted here in chronostratigraphic section (geologic time is the ordinate). Geologic-time ranges of all individual units of strata are given as equal. The geologic-time range of the sequence between surfaces A and B varies from place to place, but variation is confined within synchronous limits. These limits are determined by those parts of sequence boundaries which are conformities. Here, limits occur at beginning of unit 11 and end of unit 19. A sechron is defined as maximum geologic-time range of a sequence.

CYCLOTHEMS

The term "cyclothem" was applied by Wanless and Weller (1932, p. 1003) to the deposits of Pennsylvanian age that form a sedimentary cycle. Historical accounts of the early development of cyclothem concepts can be found in papers by Weller (1930, 1931), Wanless and Weller (1932), and Moore (1936, 1957).

Significantly — from the standpoint of sequence stratigraphy — Moore (1936, p. 21) emphasized that ". . . a cyclothem expresses an exceedingly important genetic relationship, and this grouping is of decided value in stratigraphic study." Of equal significance, geologists of the Illinois Geological Survey began mapping cyclothems in the 1930s, and they recognized that each cyclothem is separated from adjacent ones by disconformities. Thus, two important aspects of the cyclothem model — genetic relationships and the placement of boundaries at breaks in sedimentation — are cornerstones of modern sequence stratigraphy.

In recent years significant investigations have clarified, and in a sense simplified, the understanding of Pennsylvanian cyclothems in the Mid-Continent. Numerous workers have recognized that the so-called "black shale member" of the cyclothem sequence was deposited in the deepest water when marine transgression reached its apex. Thus, Heckel (1977) proposed the simplified "Basic Cyclothem" (formerly the megacyclothem of Moore, 1936), which recognizes a single transgressive-regressive (T-R) event. In the older megacyclothem model of Moore (1936), two episodes of transgression, represented by the limestones, alternated with three episodes of regression, in the form of the shales. The two interpretations are illustrated in Fig. 2.

Watney (1984) recognized 12 cyclothems in the Lansing-Kansas City Group (Upper Pennsylvanian) of western Kansas. His analysis of the lithofacies components which comprise the cyclothem members is extremely useful to the exploration for petroleum accumulations in the Lansing-Kansas City reservoirs. His work demonstrates the continuing validity and practical usefulness of cyclothem concepts.

UNITS

Weirich (1953) published the results of his many years of subsurface study of the Middle Pennsylvanian (Atokan and Desmoinesian) sandstones, which are productive over eastern Oklahoma and eastern Kansas. In his study, Weirich (1953) mapped the thickness and extent of Atoka and Lower, Middle, and Upper Cherokee "units" (Fig. 3), and the regional extent of the Bartlesville Sandstone. Weirich did not discuss the stratigraphic implications of units, other than to state that they were assemblages of regionally correlative formations.

Today, more than 40 years later, it is apparent that the units which Weirich mapped are similar to, or identical with, the stratigraphic subdivisions proposed by subsequent workers such as the "format" of Forgotson (1957), and the "genetic increment of strata" of Busch (1971).

SURFACES OF DEPOSITION

Rich (1951) defined "three critical environments of deposition," and described criteria for the recognition of rocks deposited in each of them; the relation to wave base was judged to be a critical factor in the kind of sediments that were deposited in each environment. Rich (1951, p. 2) further proposed that in each environment, the sediments which accumulated there produced a subaqueous

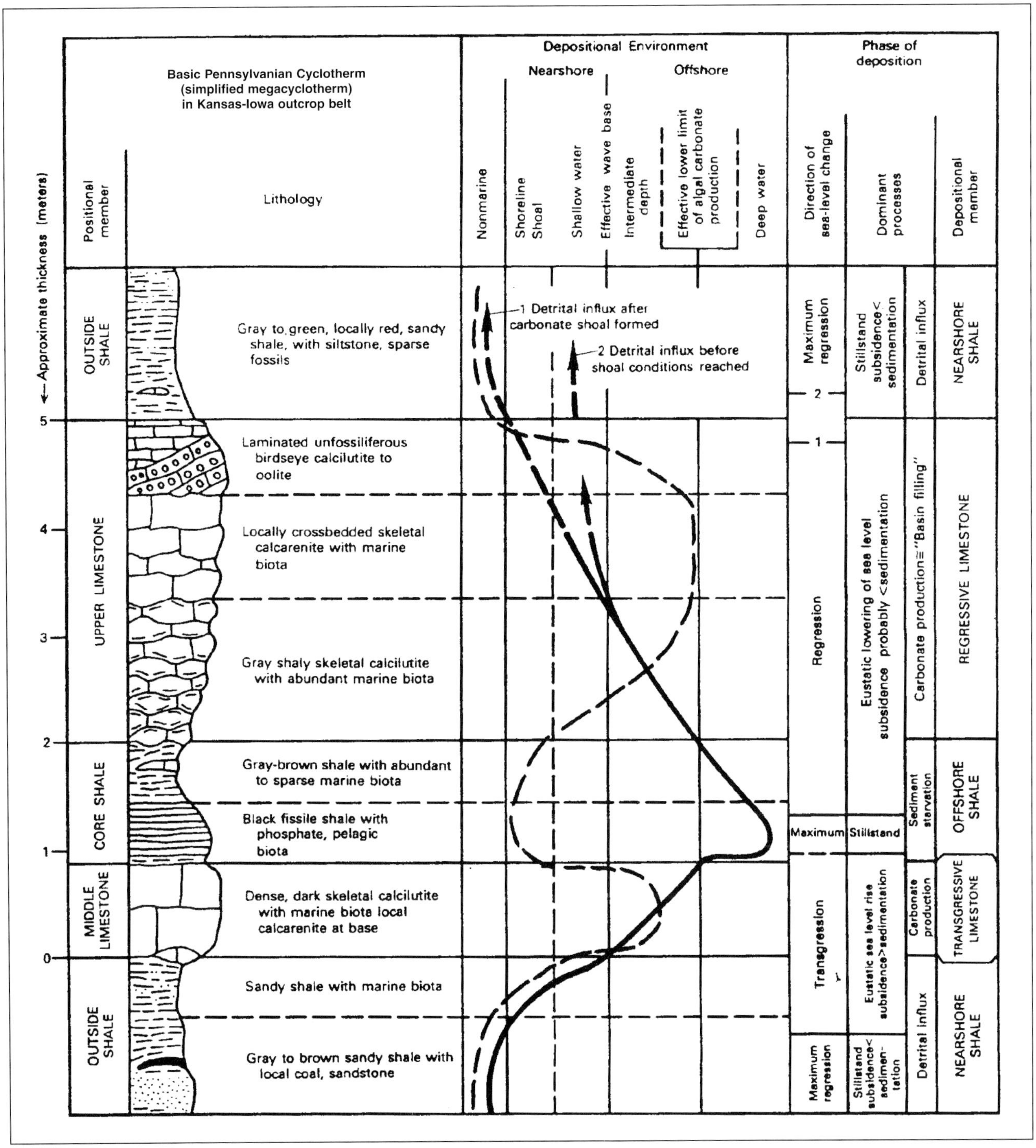

FIGURE 2 — Basic vertical sequence of an individual Kansas Pennsylvanian cyclothem (modified from Heckel, 1977, Fig. 2). The heavy line indicates long-term sea-level oscillation with black shale differentiated from other shale members. The dashed line shows the short-term sea-level oscillation which differentiated shale members from limestone members (modified from Heckel and Baeseman, 1975, Fig. 1, and Ebanks et al., 1979, Fig. 6). Terms on left for cyclothem members describe position in cyclothem; terms on right for cyclothem members describe phase of deposition and are preferable when environments are reasonaby well established.

land form which is characteristic of that environment. For example, sediments deposited above the wave base were deposited on the *undaform*; the sediments extending from the wave base to the floor of the water body were laid down on the *clinoform*, an ". . . inclined subaqueous depositional plane . . ." (Rich, 1951, p. 3); and the strata deposited on that floor accumulated on the *fondoform*. These land forms and the rock units which accumulated on these surfaces are shown in Fig. 4.

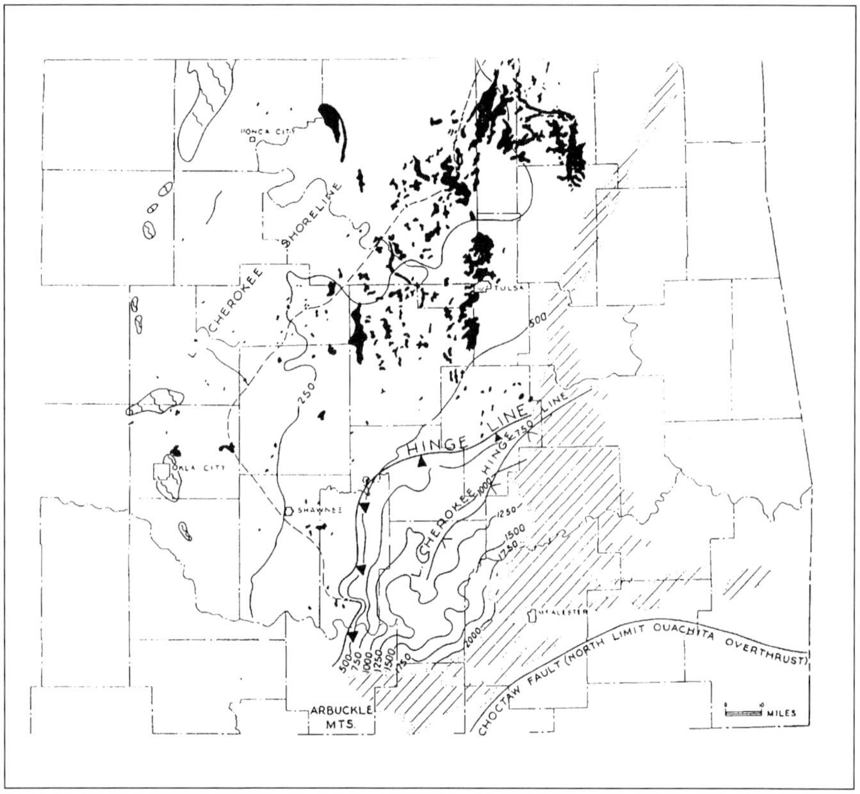

FIGURE 3 — Map of middle Cherokee unit (modified from Weirich, 1953, Fig. 4). Isopach interval is 250 ft. The positions of lower Cherokee shoreline and hinge line (separating shelf from basin) are shown. The middle Cherokee sea advanced 40 miles west in Oklahoma, and the hinge line moved 25 miles west of its lower Cherokee site. Oil fields producing from Red Fork and Bartlesville sandstones are confined to the shelf, which is 100 miles (160 km) wide. Most fields are stratigraphic traps.

Van Siclen (1958, p. 1897) noted that sediments deposited by the Pennsylvanian and Permian seas in western Texas assumed topographic forms which resemble those formed by the topset, foreset, and bottomset beds of the classic delta, and he termed these forms "depositional topography." Van Siclen (1958, p. 1901; Figs. 2, 3, and 8) equated depositional topography with the "subaqueous land form" surfaces of Rich (1951) as described above; he also constructed cross sections which identified undaform, clinoform, and

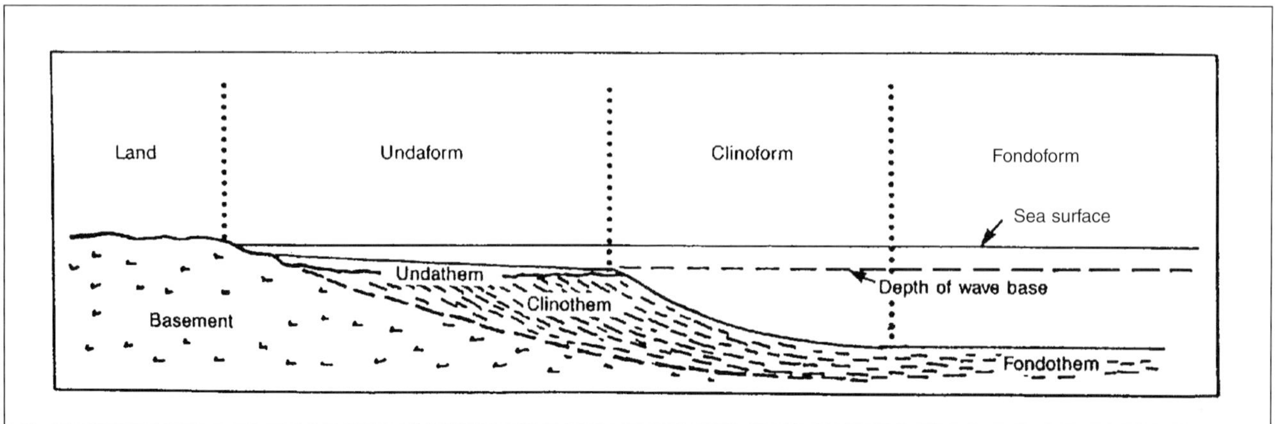

FIGURE 4 — Undathem, clinothem, and fondathem designate the rock units formed in the three environments related to the wave base. Undaform, clinoform, and fondoform indicate the topographic expression or land form represented by each type of rock unit (modified from Rich, 1951, Fig. 1).

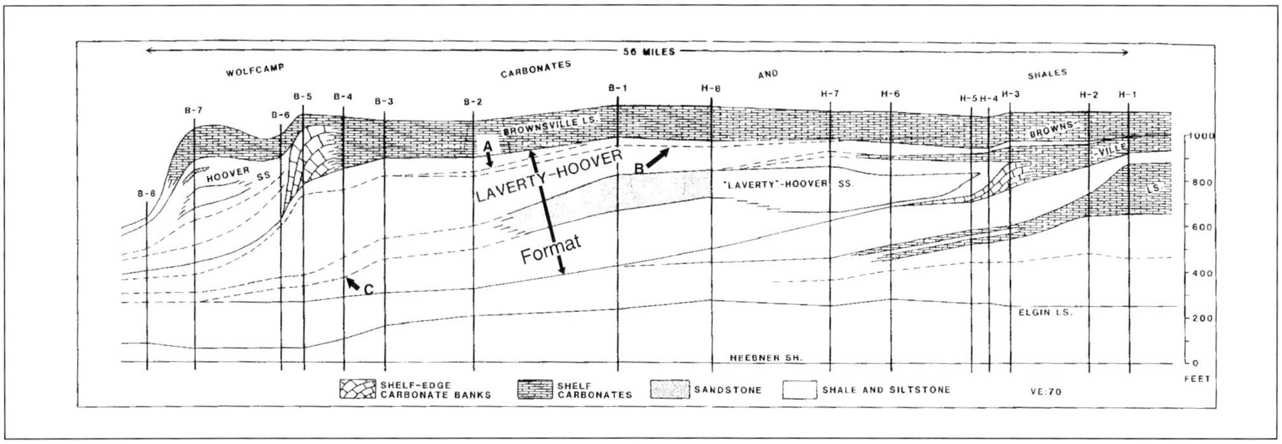

FIGURE 5 — Log section across "Laverty" Hoover format, Harper and Beaver counties, Oklahoma. Wells labeled "H" and "B" are in Harper and Beaver counties, respectively. Dashed lines indicate thin "marker" beds. "A" and "B" markers terminate or "toplap" against the upper surface of the format; marker "C" terminates or "downlaps" against the surface on which this unit was deposited.

undaform surfaces, and isopach intervals that reflected original sea-floor topography.

Although disconformities and unconformities had long been used as boundaries of stratigraphic units, the work by Rich (1951) and Van Siclen (1958) demonstrated that depositional surfaces are appropriate boundaries of stratigraphic units. The boundaries of depositional sequences — the *sine qua non* of sequence stratigraphy — are recognized as unconformities *and* their correlative conformities.

FORMATS

Forgotson (1957, p. 2109) expressed the need for a suitable terminology to denote the stratigraphic units used in regional studies, particularly those involving subsurface data. He stated, "For detailed regional work, it is necessary to map laterally continuous units that are lithologically different. The tops and bottoms of these units are usually recognized in the subsurface by characteristic electrical log markers, which may or may not coincide with accepted formation boundaries. These units are extended laterally beyond the limits of recognition of their original defining markers, by interval correlation and position with respect to additional markers which appear in the section. A unit of this type is not a formation, it is not a lithosome, nor is it a facies. Yet it is a rock unit that is defined on the basis of observable criteria, and it is the most suitable kind of rock unit for regional mapping." To designate such units, Forgotson (1957, p. 2110) proposed the term *format* which he states is ". . .a rock unit. . .which crosses facies boundaries. . .but is defined by electrical log markers or other criteria which can be extended across facies boundaries."

Although a format is identified as a rock unit (i.e., one identified on the basis of lithology alone), Forgotson (1957, p. 2110) notes that "in many areas, evidence suggests that certain electrical log markers are time-parallel over long distances. The isopach and lithofacies pattern shown by this type of marker-defined unit is one of the most significant factors to be considered in the interpretation of the paleogeography, the tectonic framework, and the regional environmental pattern which existed during the deposition of the unit." It must be remembered that the validity of regional paleogeographic maps requires that the units be time-stratigraphic.

In the opinion of many workers in the Mid-Continent, the cyclic Pennsylvanian and Permian marker-defined units are indeed time-stratigraphic in nature. Visher et al. (1971, p. 1209) considered the format of Forgotson (1957) to be a time-stratigraphic concept, and they proposed that ". . .it be used as a formal time-stratigraphic subdivision of series." In their analysis of Pennsylvanian delta patterns in Oklahoma, Visher et al. (1971, p. 1212) identified nine formats in the Meramecian and Chesterian series of the Mississippian, and the Lower and Middle series of the Pennsylvanian. They recognized a deltaic model which consists of six subdivisions: (1) lower alluvial plain; (2) upper deltaic plain; (3) lower deltaic plain; (4) subaqueous sand sheets; (5) marginal basin; and (6) marginal depositional plain. This model was applied to the Bartlesville Sandstone (Visher et al., 1971, Fig. 5, p. 1217).

Rascoe (1978) subdivided the Virgilian Series (Upper Pennsylvanian) of the Anadarko basin into 13 formats, and these were regarded as time-stratigraphic units. Each Virgilian format in this study encompassed one complete T-R cycle: each cycle began with the transgressive phase in which a thin veneer of shallow marine carbonates was deposited over the shelf area and on the topset surface of the preceding deltaic wedge; along the shelf-slope break of the underlying deltaic wedge, a shelf-edge carbonate bank of oolites and fossil "hash" accumulated. Basinward the carbonate bank is contiguous with, but was deposited prior to, the deltaic sediments of the regressive phase. The regressive deltaic wedges prograded westward into the Anadarko basin in response to episodes of lowered sea level which allowed clastic material from the Ouachita Mountains to be transported across the shelf area and into the basin. The boundaries of these Virgilian formats are the land forms or depositional sur-

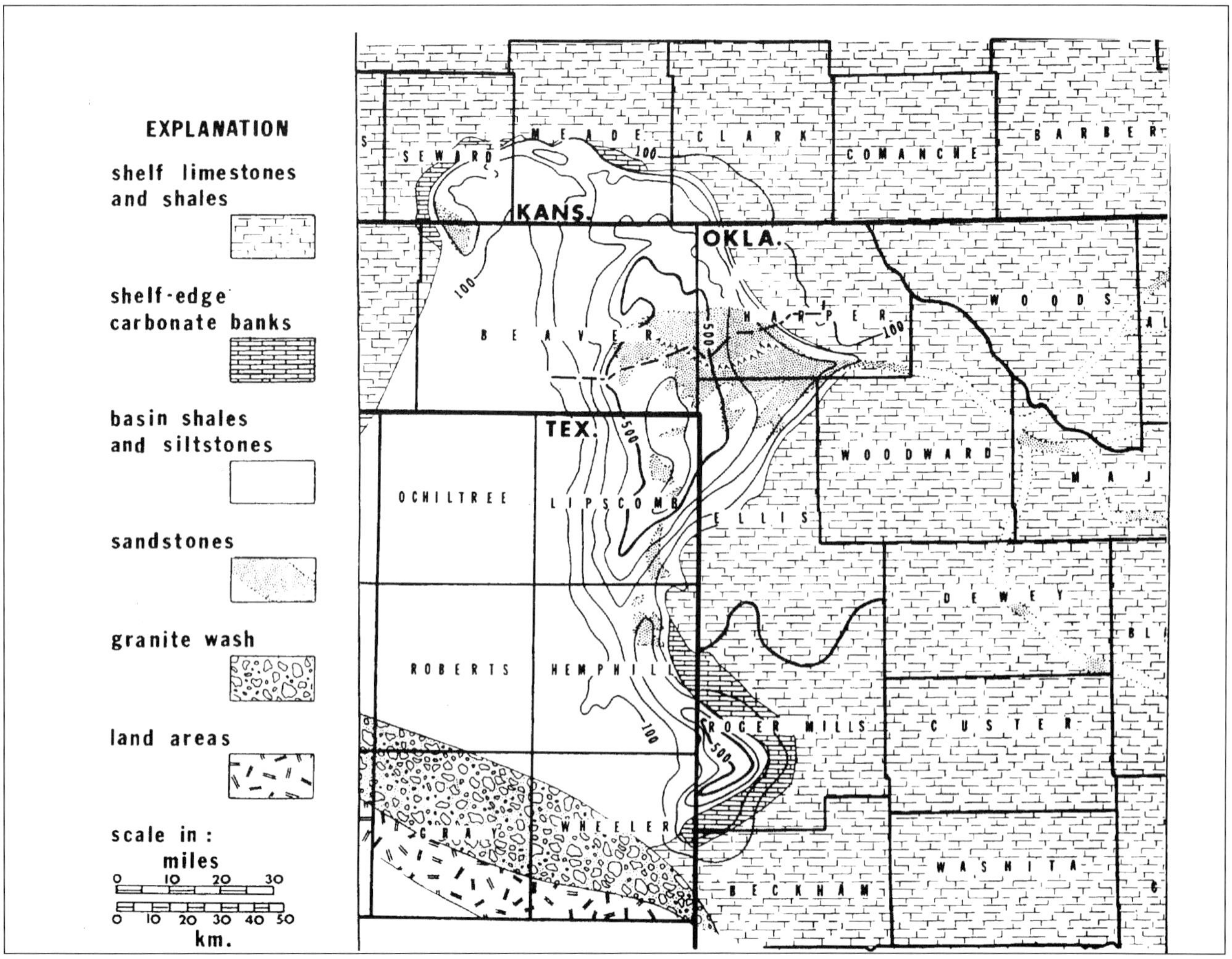

FIGURE 6 — Thickness and depositional systems of the Laverty-Hoover format. Contour interval is 100 ft. The dashed line is the location of the cross section shown in Fig. 5.

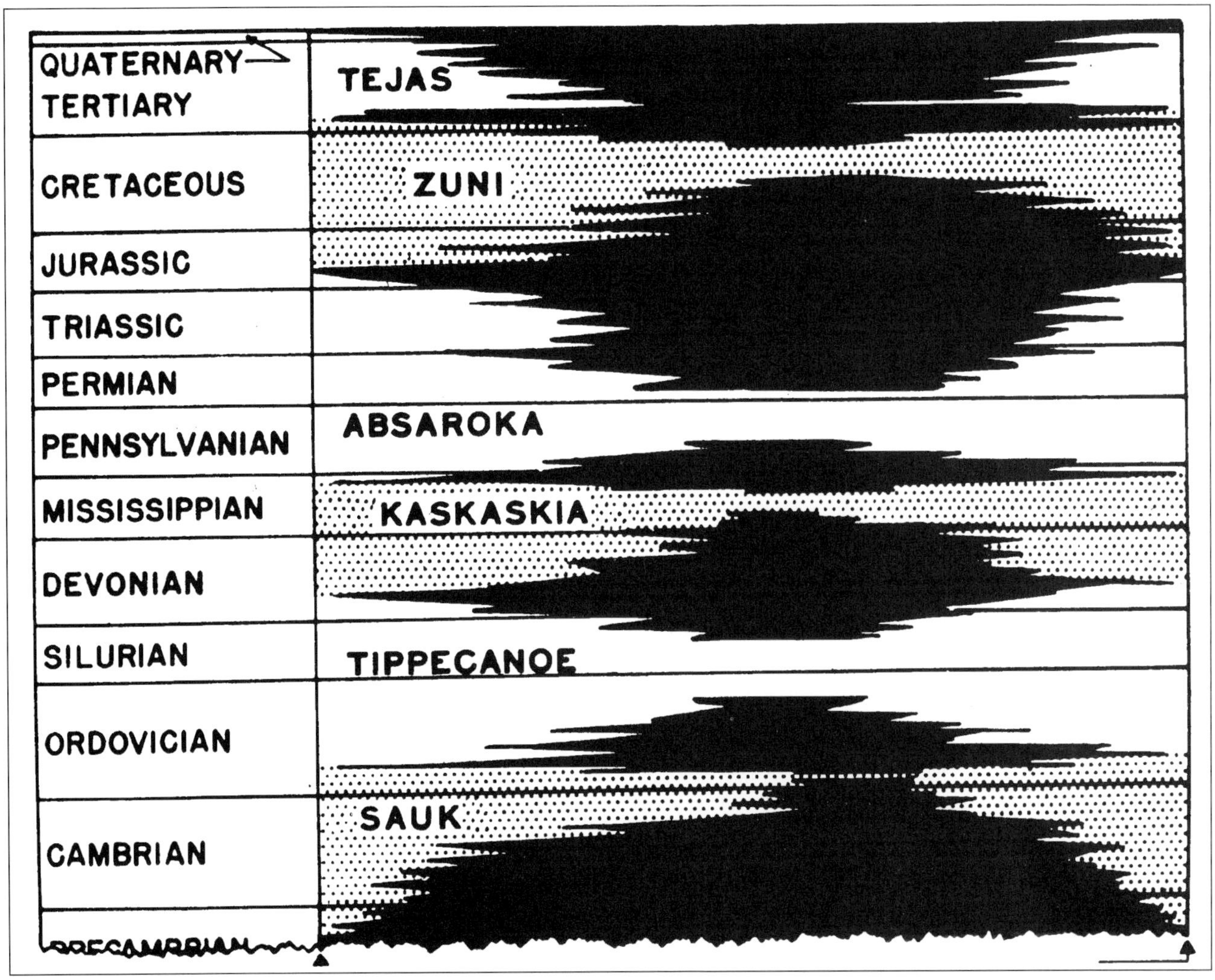

FIGURE 7 — Time-stratigraphic relationships of the sequences in the North American craton (Sloss, 1963). Black areas represent nondepositional hiatuses; white and stippled areas represent deposition. (Stippling introduced only to differentiate successive depositional episodes.)

faces defined by Rich (1951) and applied by Van Siclen (1958). Each passes laterally from undaform (above wave base, or shelf) to clinoform (slope surface from undaform to basin floor), to fondoform (basin floor). These 13 formats record the progressive filling in from east to west of most of the Anadarko basin during Virgilian time.

Figure 5 identifies the boundaries of the Laverty-Hoover format, and the relation of this unit to older and younger formats. Figure 6 illustrates the thickness and major depositional systems of the Laverty- Hoover format.

STRATIGRAPHIC SEQUENCES

The evidence of major, interregional unconformities has long been recognized over the continental interior of North America, and Sloss (1963) subdivided the cratonic stratigraphic column into six stratigraphic sequences which are bounded by seven major unconformities. Sloss (1963, p. 93) noted that "although the time values of the unconformities vary widely because of differences in degree of

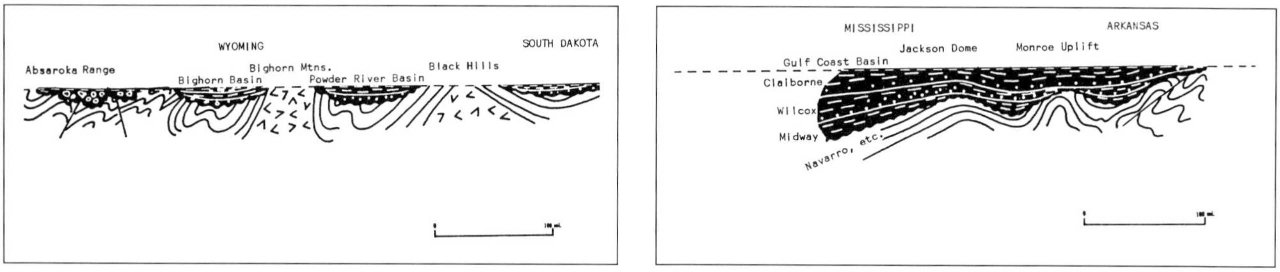

FIGURE 8E — Diagrammatic cross sections showing relationships of lower part of Tejas sequence (black) to older units (white).

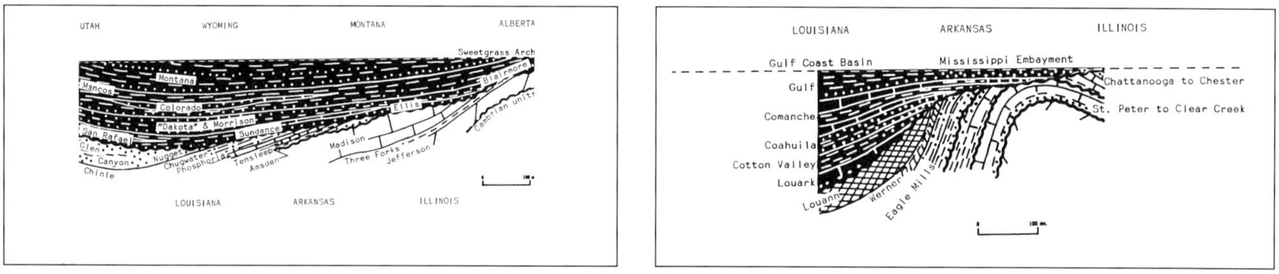

FIGURE 8D — Diagrammatic cross sections showing relationships of lower part of Zuni sequence (black) to older units (white). Sequence boundaries are shown by heavy wavy lines.

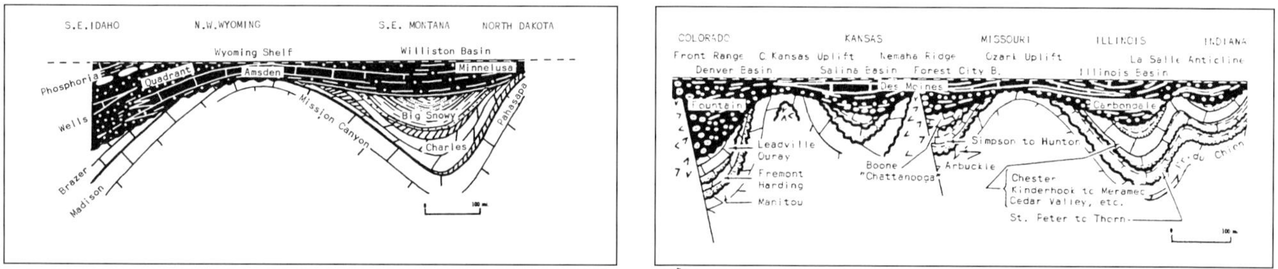

FIGURE 8C — Diagrammatic cross sections showing relationships of lower part of Absaroka sequence (black) to older units (white). Sequence boundaries are shown by heavy wavy lines.

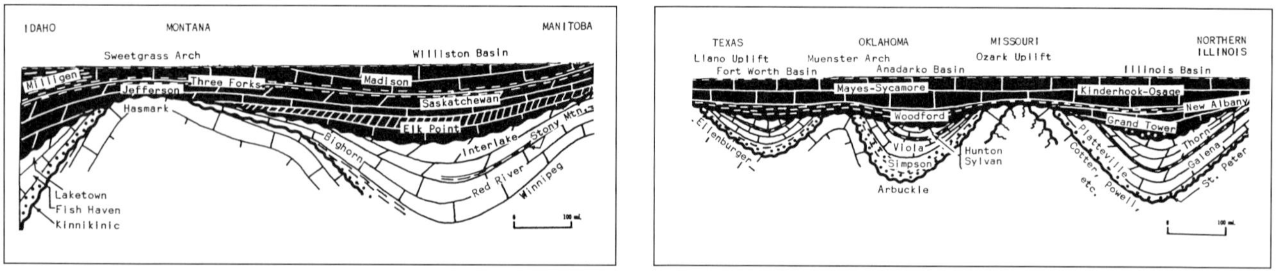

FIGURE 8B — Diagrammatic cross sections of the Sauk and Tippecanoe sequences (white) and the basal part of the Kaskaskia sequence (black). Sequence boundaries are indicated by heavy wavy lines.

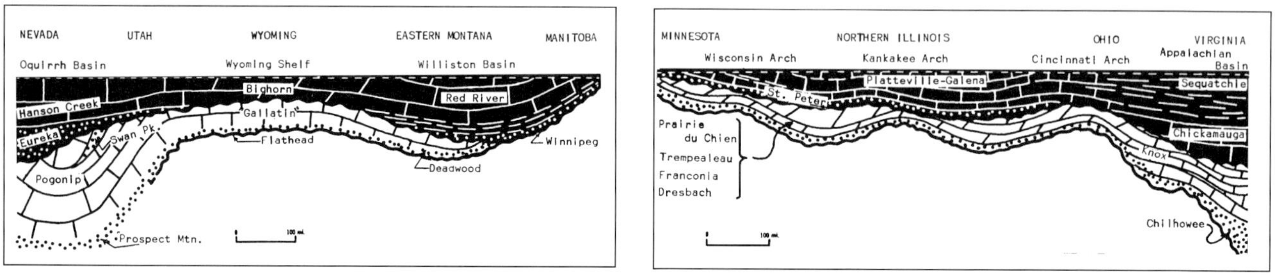

FIGURE 8A — Diagrammatic cross sections of the Sauk sequence (white) and the lower part of the Tippecanoe sequence (black).

nondeposition and amount of erosion, the approximate dates of the regressional maxima represented are (1) very late Precambrian, (2) early Middle Ordovician, (3) early Middle Devonian, (4) 'post-Elvira' Mississippian, (5) early Middle Jurassic, and (6) late Paleocene. A seventh major regression is now in progress." Figure 7 illustrates the time-stratigraphic relationships of the stratigraphic sequences which are enclosed by these unconformities; it should be noted, however, that Sloss (1963, p. 93) regarded these sequences as rock-stratigraphic units. Figures 8A through 8E show the stratigraphic units which comprise these sequences.

Adler et al. (1971) used the Sloss concept of stratigraphic sequences in their analysis of the "Future Petroleum Provinces of the Mid-Continent," and they presented four subcrop maps showing the distribution of the various sequences lying on or beneath adjacent sequences (Adler et al., 1971; Figs. 5, 11, 17, and 30). The distribution of sequences underlying the Absaroka sequence is shown in Fig. 9.

As stated by Adler et al. (1971, p. 1004), "The sequence concept is based on the belief that the times of widespread epeirogeny, or nondeposition and erosion, in the North American interior properly constitute the first-order points of subdivision of the cratonic sedimentary record. Regional stratigraphic analysis strongly affirms the practical utility of this approach. For example, the major T-R cycles of deposition on the North American craton are made readily apparent by mapping of the major units of the sequences, and geographic distribution maps of the various sequences reveal the paleotectonic evolution of the craton since Precambrian time. In brief, the sequence concept provides an epeirogenic, in addition to the traditional biostratigraphic, basis for the long-range time-stratigraphic correlation of major rock units across very large areas and between basins separated by post-depositional erosion. Besides having general utility to stratigraphers in general, sequence mapping seems especially well suited to the petroleum geologists, who long have recognized the control and influence that unconformity surfaces — and the geomorphic and sedimentary processes whereby they were formed and then buried — exert on a variety of geomorphic, stratigraphic, and structural trapping situations where hydrocarbons have accumulated."

In their paper in *Memoir* 26, "The Depositional Sequence as a Basic Unit for Stratigraphic Analysis," Mitchum et al. (1977, p. 55) stated that their depositional sequence concept is modified from the stratigraphic sequences which Sloss (1963) identified and defined. It may be that the Sloss stratigraphic sequences shaped and influenced the concepts of modern sequence stratigraphy more than any other work.

GENETIC UNITS

Busch (1953) recognized the economic significance of deltaic deposits in petroleum exploration, and he studied the lithologic relationships within the Holocene and ancient delta complexes. Busch (1971) concluded that all the lithologic components of a delta are genetically related to each other, and accordingly he proposed two stratigraphic units to analyze these relationships. The smaller and more fundamental of these units was designated the Genetic Increment of Strata (GIS). It consists of a vertical sequence of strata in which each lithologic component is genetically related to all others. The GIS is defined at the top by a time-lithologic marker bed (such as a thin limestone or bentonite) and at the base by a time-lithologic marker bed, an unconformity, or a facies change from marine to nonmarine beds (Fig. 10). The larger unit was called the Genetic Sequence of Strata (GSS). It includes two or more continuous Genetic Increments of Strata and represents more or less continuous sedimentation (Fig. 11). The stratigraphic interval must be devoid of angular unconformities, but disconformities of limited extent may be present. An isopach map of the GSS defines the shelf, hinge

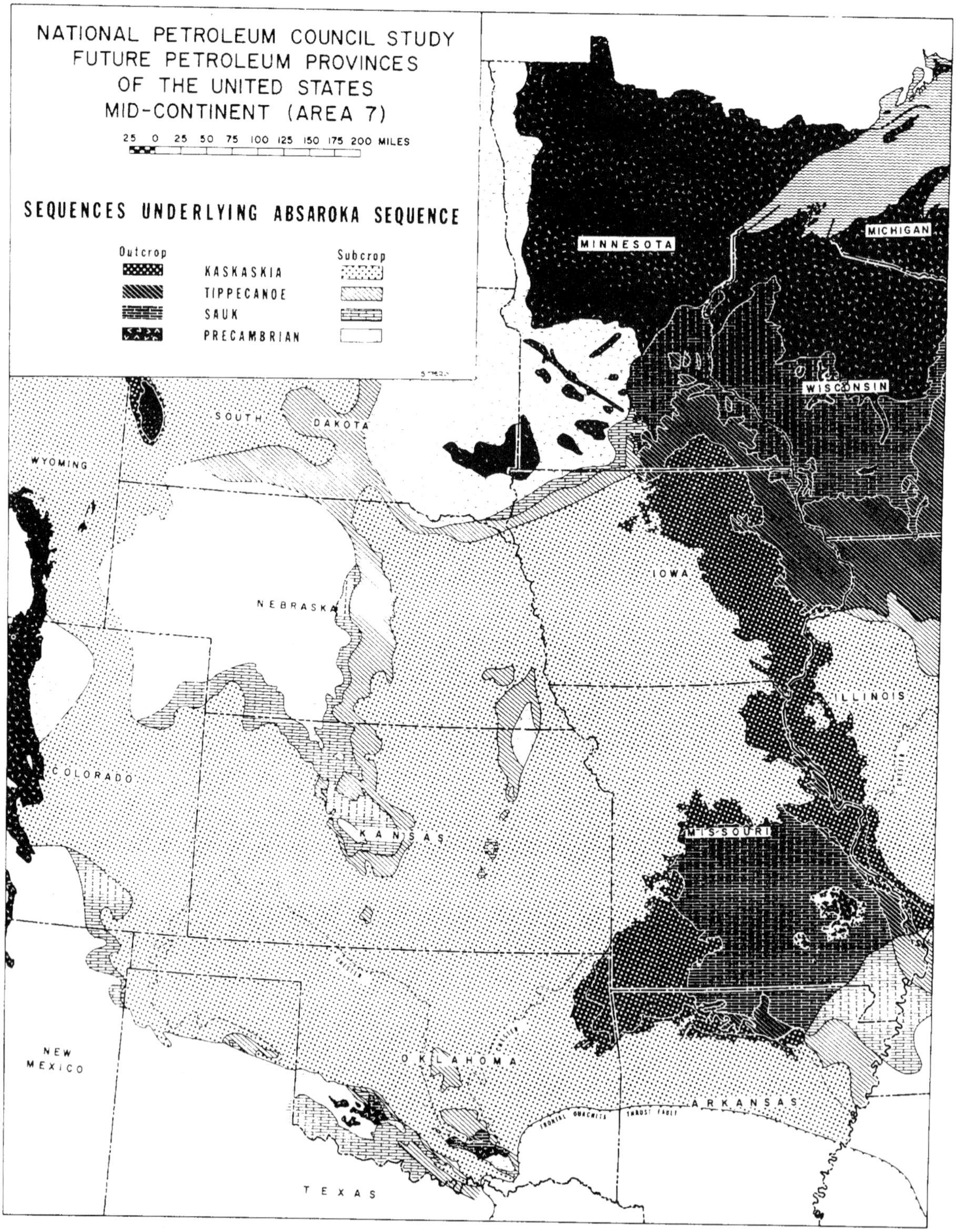

FIGURE 9 — Subcrop-outcrop map shows distribution of sequences underlying Absaroka sequence in Mid-Continent (Adler et al., 1971, Fig. 17).

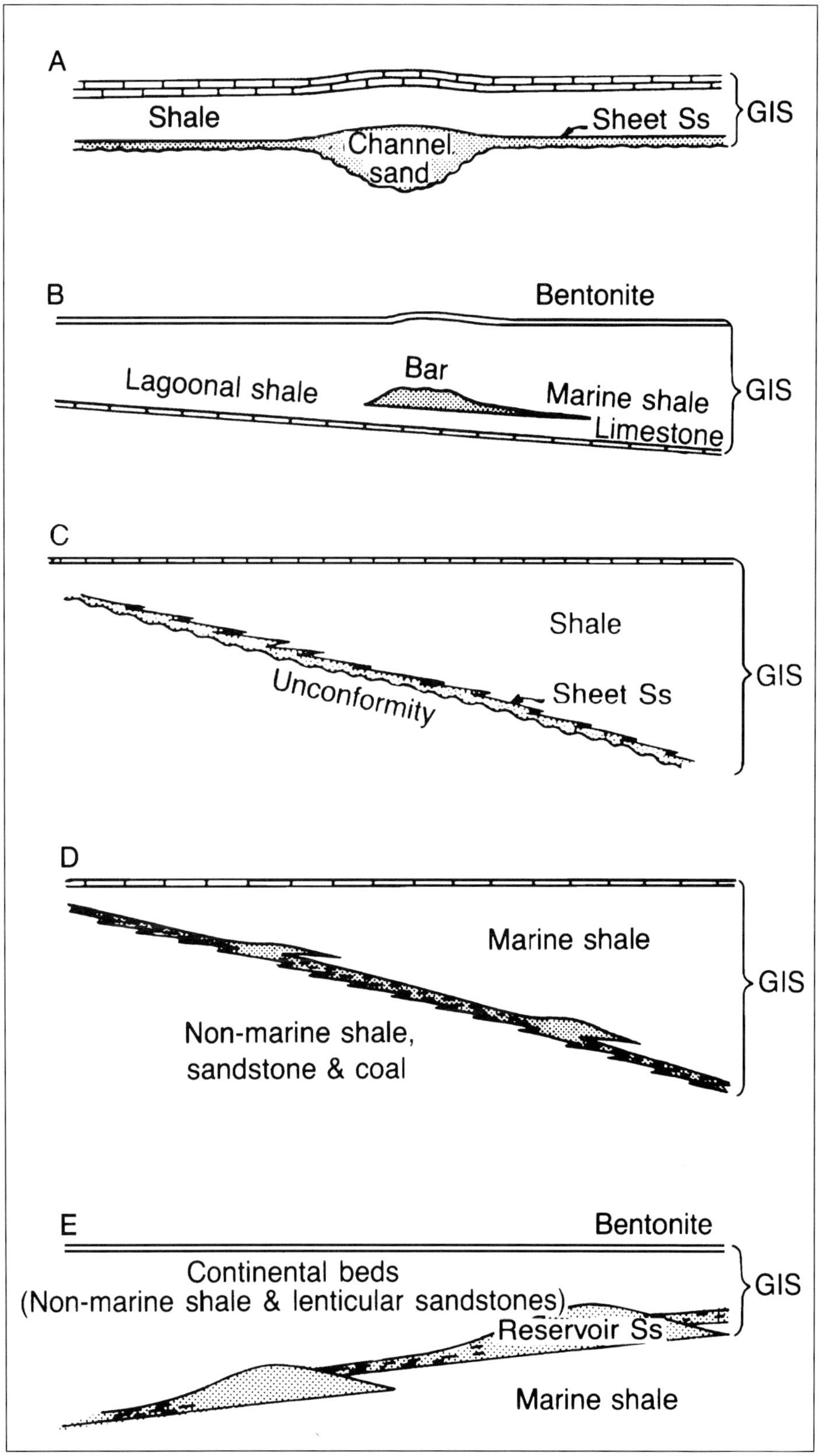

FIGURE 10 — Diagrammatic illustrations of genetic increments of strata (GIS).

(A) Channel sandstone grading laterally into a sheet sandstone with overlying shale and limestone; upper limit of GIS is defined by lithologic-time marker (limestone), base defined by an unconformity.
(B) Barrier-bar sandstone bordered by lagoonal shale on one side and marine shale on the other; upper and lower limits of GIS are defined by lithologic-time marker beds (bentonite and limestone).
(C) Sheet sandstone, resulting from either cyclic subsidence of land or cyclic emergence of sea level, overlain by basinward thickening of marine shale; upper limit of GIS is defined by lithologic-time marker bed (limestone), base by an unconformity.
(D) Sheet sandstone resulting from either cyclic subsidence of land or cyclic emergence of sea level, showing two areas of locally thicker (and better-sorted) sandstone at former stillstand positions of shoreline; upper boundary is defined by lithologic-time marker bed (limestone), base by facies change from continental to marine beds.
(E) Sheet sandstone resulting from either cyclic emergence of land or cyclic subsidence of sea level, showing two areas of locally thicker (and better-sorted) sandstone at former stillstand positions of shoreline; upper boundary is defined by lithologic-time marker bed (bentonite), base by abrupt facies change from marginal-marine sandstone to marine shale.

In C and D the en echelon sandstone wedges interfinger with the overlying marine shale, whereas in E they interfinger with the underlying marine shale. (Busch and Link, 1985)

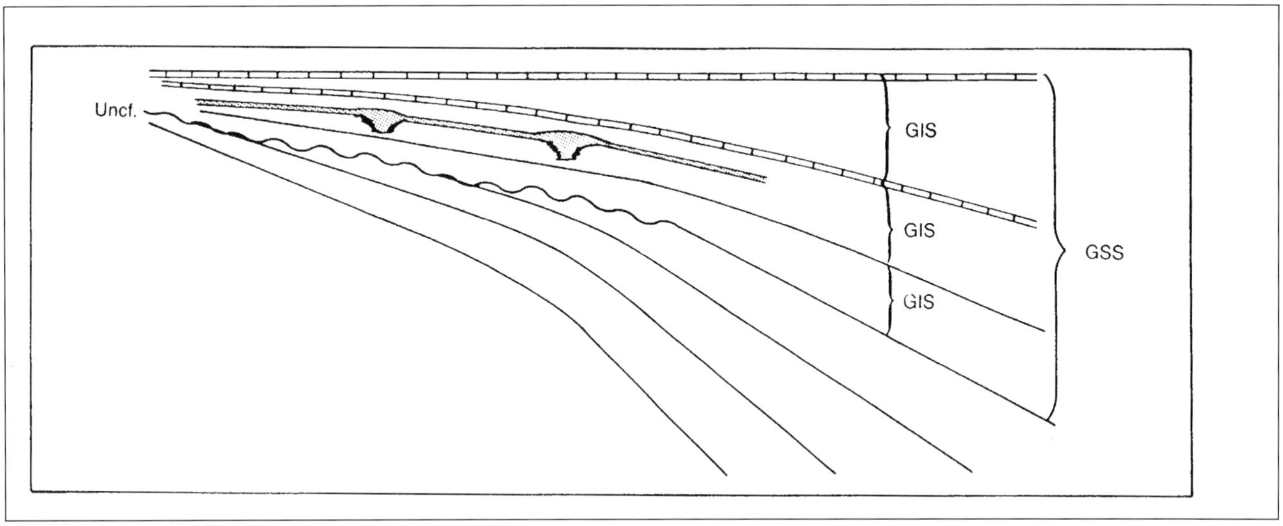

FIGURE 11 — Relationship of a Genetic Sequence of Strata (GSS) to several Genetic Increments of Strata (GIS). The upper boundary of the GSS is defined by a thin limestone; the base is marked by an unconformity which disappears basinward as it grades into a marker horizon (Busch and Link, 1985).

line, and less stable part of a depositional basin; the principal purpose of this type of map is the reconstruction of the paleodepositional configuration of a basin.

CONCLUSION

Cyclothems, formats, genetic units, and depositional sequences share a common trait that is the central reason for their existence: each has been defined or described as an assemblage of strata which is genetically related. The use of these various units for more than 60 years is a measure of the importance of subdividing the stratigraphic section into operational units, the members of which were deposited by a variety of genetically related depositional processes.

Sequence stratigraphy has been practiced in the Mid-Continent for over 60 years and, undoubtedly, it continues to be used today in the search for accumulations of petroleum in stratigraphic traps. Concepts regarding depositional sequences which have emerged through the definition of seismically defined sequences can be usefully applied to log-defined depositional sequences in the Mid-Continent. Specifically, the upper and lower contacts of marker-bounded units can be regarded as ancient depositional surfaces which, given the cyclic nature of Pennsylvanian and Permian sediments, would not seem an unreasonable interpretation. Such a conclusion would denote the chronostratigraphic significance of depositional sequences in the Mid-Continent.

RECOGNIZING SEQUENCE SURFACES IN MID-CONTINENT SILICICLASTIC STRATA

John W. Shelton and Larry D. Gerken
Masera Corporation
Tulsa, OK 74136

ABSTRACT

Sequences in the Mid-Continent tend to be abbreviated because of limited accommodation space. Some sequence components (e.g., systems tracts) present on continental margins are either absent or merged with other elements. Nevertheless, sequence boundaries are rather easily distinguished because of the associated unconformity. The boundary commonly corresponds to the base of channelized sandstones or to regional unconformities, where the difference in ages of the strata above and below it is readily apparent. Some regional unconformities are recognizable because of convergent strata, but they are actually less than spectacular on a local scale. Other less prominent sequence boundaries are associated with paleosols. Transgressive units have been recognized and have been effectively used in dividing strata into mappable units for delineation, not only of local structure but also delineation of sandstone reservoirs. These units not only include transgressive surfaces, but they may coincide with or approximate marine-flooding surfaces, condensed sections, or parasequence boundaries, or a combination.

INTRODUCTION

Concepts

Sequence stratigraphic concepts were first developed by workers studying cratonic strata — in particular by Sloss (1963), who recognized major stratal packages (sequences) bounded by unconformities, and by Wanless and Weller (1932), who earlier recognized smaller packages (cyclothems) also bounded by unconformities. The concept of Sloss was greatly expanded and revised through the work of Exxon and ex-Exxon workers, led by P.R. Vail (Fig. 1) [as reported in AAPG Memoirs 26 (1977) and 36 (1984), Studies in Geology 27 (1987), Methods in Exploration Series No. 7 (1990), and SEPM Spec. Pub. 42 (1988), for example]. The concepts and methodology, now known as sequence stratigraphy, have received widespread, enthusiastic acceptance.

Accompanying the sequence stratigraphy (of Exxon) is a rather elaborate terminology. The sequences of Sloss are now supersequences; the Exxon sequence is "a stratigraphic unit composed of a relatively conformable succession of genetically related strata and bounded at its top and base by unconformities or their correlative conformities" (Mitchum (1977). Recognized within sequences are stratal subdivisions (systems tracts of Brown and Fisher, 1977); two main types of assemblages of systems tracts compose two types of sequences (e.g., Baum and Vail, 1988; Posamentier and Vail, 1988). Of particular importance in the study of the mature Mid-Continent hydrocarbon provinces are genetically related strata within sequences that are bounded by units representing marine inundation or deepening.

Van Wagoner et al. (1987, 1990) defined this bundle of strata as a parasequence ("a relatively conformable succession of genetically related beds or bed sets bounded by marine-flooding surfaces or their correlative surfaces"). According to the definition of Van Wagoner et al. (1990), some parasequences are bounded either above or below by sequence boundaries. A marine-flooding surface is seemingly equivalent to a transgressive surface of previous workers.

The ultimate cause for the development of these stratal packages of various scales is continuous eustatic sea-level changes of corresponding scales (Figs. 1 and 2) (Vail et al., 1977). Indeed the basic premise of Exxon sequence stratigraphy is that a suite of eustatic cycles is correlatable on a global scale.

It should be noted that there are some dissenters; probably the most ardent is Miall (1986, 1992), who objects to the correlation detail proposed by the Exxon workers and to "circularity" of reasoning inherent in utility of the method. There are also some nonconformists, such as Sloss (1988), who object to the exclusion of tectonics as a major factor in explaining the supersequences. Galloway (1989a, b) prefers to use a "genetic stratigraphic sequence," bounded by hiatal surfaces that reflect maximum marine flooding, as the basic strata package and to consider sediment supply and subsidence as two additional variables (to eustatic change) in controlling basinal sedimentary development.

Objective

The purpose here is not to become entangled in any dispute concerning the terminology, accuracy of the Exxon global cycle chart(s), utility of the various methods in mapping (in particular, facies and trends), or cause, or causes of the cyclicity, which itself is essentially universally accepted. Rather this study is intended to assist explorationists and productionists in mapping reservoir trends by recognizing meaningful stratal units and surfaces within a context of sequence stratigraphy, used in a broad, or generic, sense.

FEATURES OF SEQUENCES IN THE MID-CONTINENT

The six Phanerozoic sequences of Sloss (1963), now supersequences (second order or supercycles) of Exxon, are unconformity-bounded stratal packages representing approximately 60 to 140 m.y. (Fig. 1). The bounding unconformities are interregional, and recognized by their geographic extent and persistence in basins. In general, the unconformities do not correspond to systemic boundaries. Almost all Mid-Continent strata belong to the oldest four sequences which formed during the Paleozoic; the youngest of the four (Absaroka) incorporates basal Pennsylvanian to Lower Jurassic strata.

The Exxon sequences, as subdivisions of the supersequences and as third-order cycles, each represent approximately 0.5 to 7 m.y. Some workers doubt that in the Mid-Continent the shortest of these time intervals can be recognized, or confirmed, because of the limitations of age determination

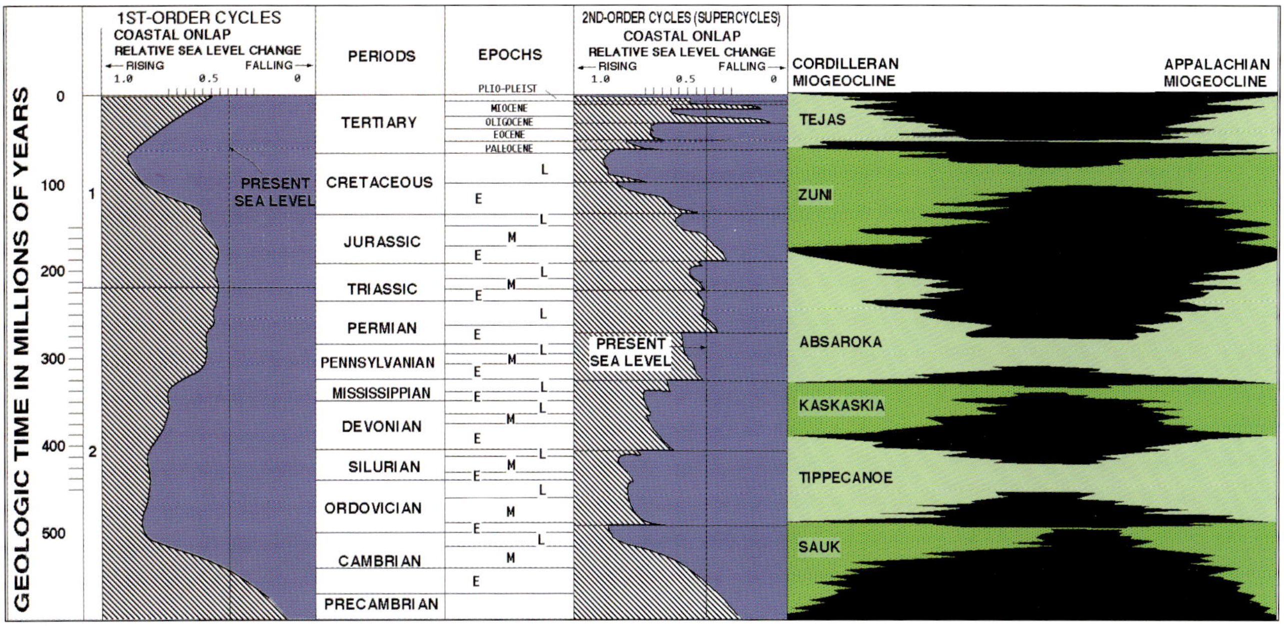

FIGURE 1 — Exxon's first-order and second-order cycles (eustatic sea-level changes) and sequences of Sloss (supersequences of Exxon) (after Sloss, 1963; Vail et al., 1977).

of Paleozoic strata. Nevertheless, Heckel (1986) considered the time represented by cyclothems (fourth-order sequences or "major cycles") to range from 0.2 to 0.4 m.y., and the time represented by "minor cycles" (small reversals of marine inundation and retreat within major cycles) to range from 0.04 to 0.12 m.y. The parasequence, as a fundamental element of the sequence, probably corresponds to the time of either type of cycle.

Lower Paleozoic sequences in the Mid-Continent (Fig. 3) generally are dominated by marine strata (Fig. 4), although considerable erosional relief may exist at some sequence boundaries (Fig. 5). On the other hand, Upper Paleozoic sequences are characterized by cyclothems, with nonmarine, coastal, and marine sediments. In the Mid-Continent the parasequence is commonly the basic stratal increment used in subsurface mapping (Figs. 4 and 6).

Cratonic Sequences

Cratonic areas, with very gently sloping surfaces and limited storage capacity (accommodation space), are characterized by limitations in the types of depositional processes, patterns (facies architecture), and regimes that were present even during highstands. During lowstands there were more restrictions (e.g., common dominance of local fluvial deposition).

Intracratonic Basinal Sequences

Intracratonic basins are characterized by considerable storage capacity, such that the range of sequence components is greater than that in the other cratonic areas. Accompanying the additional capacity for stratal volume is an increase in the slope of the earth's surface resulting from differential subsidence and deposition. Recognizing the sloping depositional surface is important in reconstructing the depositional history, as it influences depositional processes, patterns, and regimes under a rather wide range of sea-level positions, especially lowstands (Fig. 7).

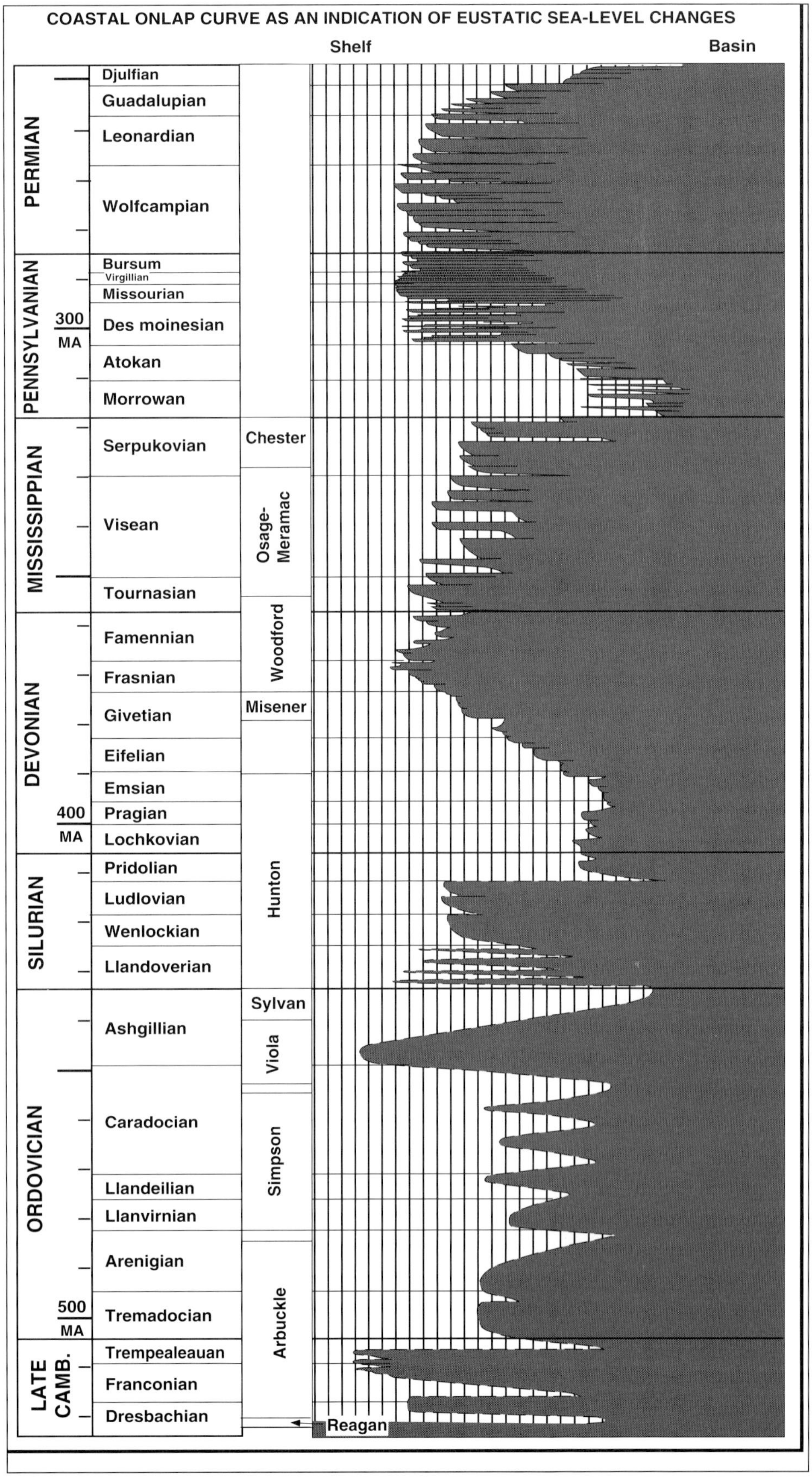

FIGURE 2 — Coastal onlap curve for Paleozoic: Late Paleozoic, based in large measure on the repetitious nature of strata (mainly cyclothems) after Ross and Ross (1988); curve for Devonian after Johnson et al. (1985) and Ross and Ross (1988); curve for Early Paleozoic after Ross and Ross (1988).

SEQUENCE SURFACES IN THE MID-CONTINENT: RECOGNIZABLE AND DISTINGUISHING CHARACTERISTICS

Sequence Boundaries

General Description. By definition of the Exxon workers, sequence boundaries are represented by unconformities (or correlative conformable surfaces). In the Mid-Continent, the most prominent type is probably the basal contact of lenticular, channelized sandstone. Another type of sequence boundary is the top of a paleosol, with some indication of soil profile, root casts/rhizoconcretions, caliche/concretionary cornstone/hydromorphic soil, or vadose karstic features, or a combination. A third type is bounded by transgressive surfaces. A fourth type of unconformity, commonly with little local relief, is recognized by onlap of the overlying unit(s), convergence/thinning of the sequence, and sharp base of the overlying unit(s). The last two surfaces may be of regional extent.

Examples. Mid-Continent examples of the bounding surface of a sequence presented herein include prominent unconformities, ones regionally developed and the more common disconformities or local unconformities, and less prominent surfaces with or without accompanying paleosols.

Bases of Channelized Sandstones. The base of channelized deposits of coarse clastics, where developed, is probably the most easily recognized sequence surface; this type has been documented on outcrops, in cores, on logs, and on some seismic profiles.

Where it is well developed, the Desmoinesian Bartlesville (Bluejacket) Sandstone in northeastern Oklahoma shows an erosional base. This is demonstrated on an outcrop and in a core (Fig. 8A). The disconformity (sequence boundary) may be inferred by the bell-shaped log (Fig. 8B). These channelized sandstone bodies overlying erosional bases are thought to reflect alluvial (fluvial) deposits as valley fills. Some channelized sandstones in the Bartlesville succession may represent deltaic distributary deposits that formed during highstands, and by Exxon definition they occur within sequences.

The base of the Virgilian Kisinger Sandstone of north-central Texas is erosional, as documented from

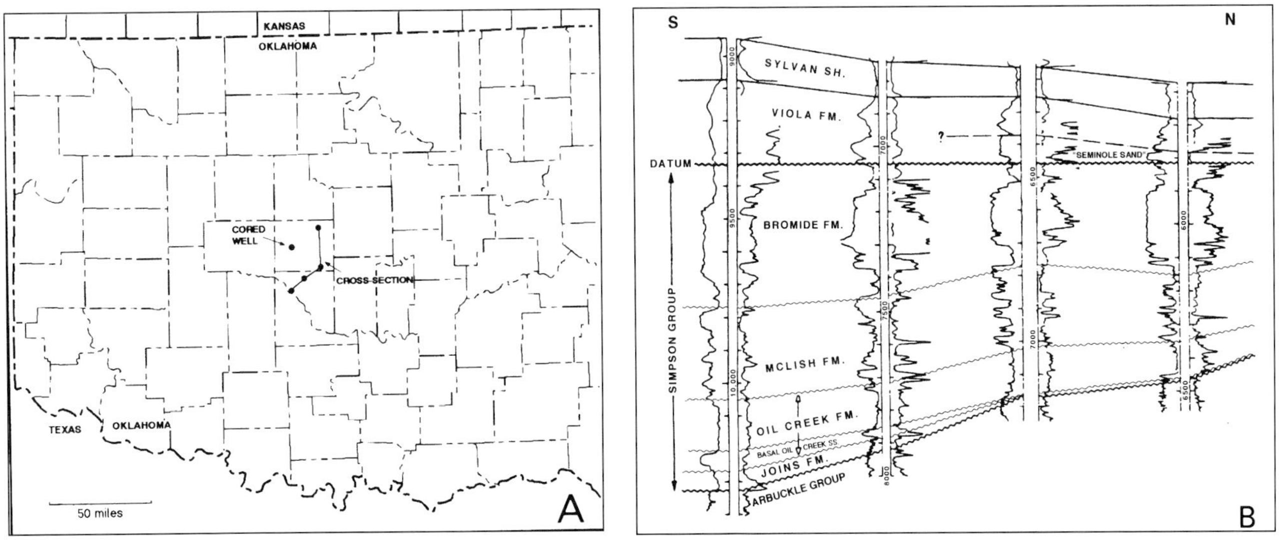

FIGURE 3 — (A) Index map for cross section and core of Ordovician Simpson Group. (B) Stratigraphic cross section of Simpson Group in central Oklahoma, composed of the Joins, Oil Creek, McLish, and Bromide formations, showing northward thinning and bounding unconformities (after Statler, 1965).

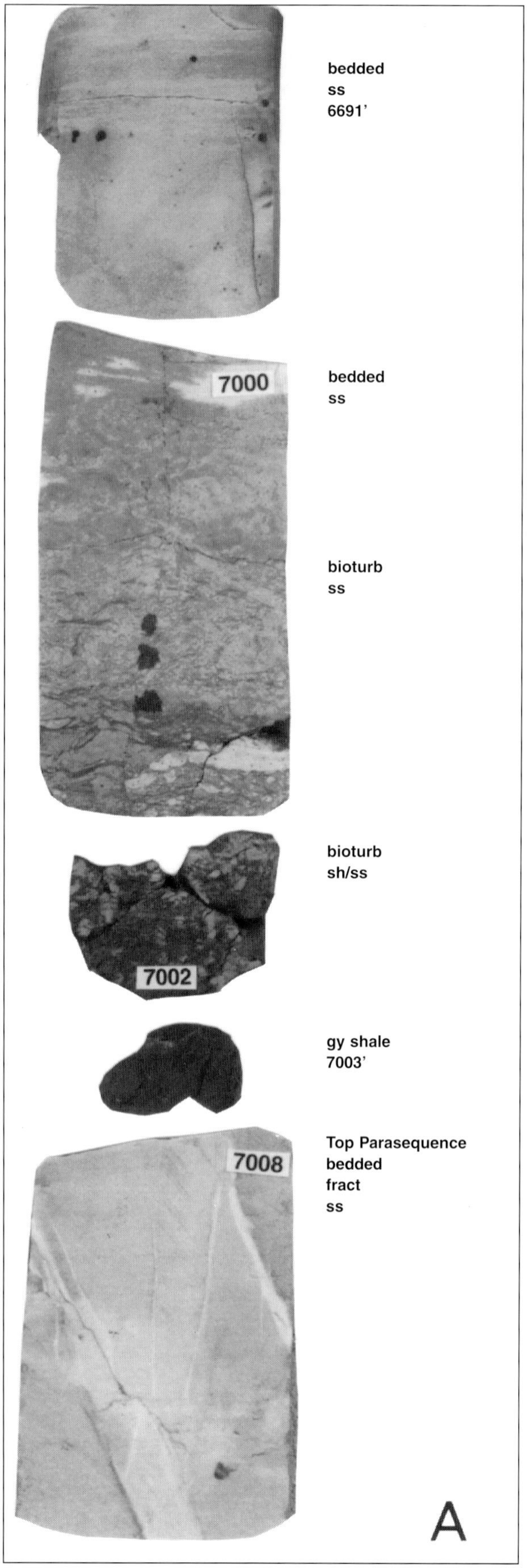

FIGURE 4 — (A) Uppermost part of a parasequence and the lowermost part of an overlying parasequence (upward-coarsening section with bioturbated beds below bedded units) in that part of the Bromide Formation informally designated as "2nd Wilcox," Oklahoma County, Oklahoma. (B) Well log of interval showing parasequences.

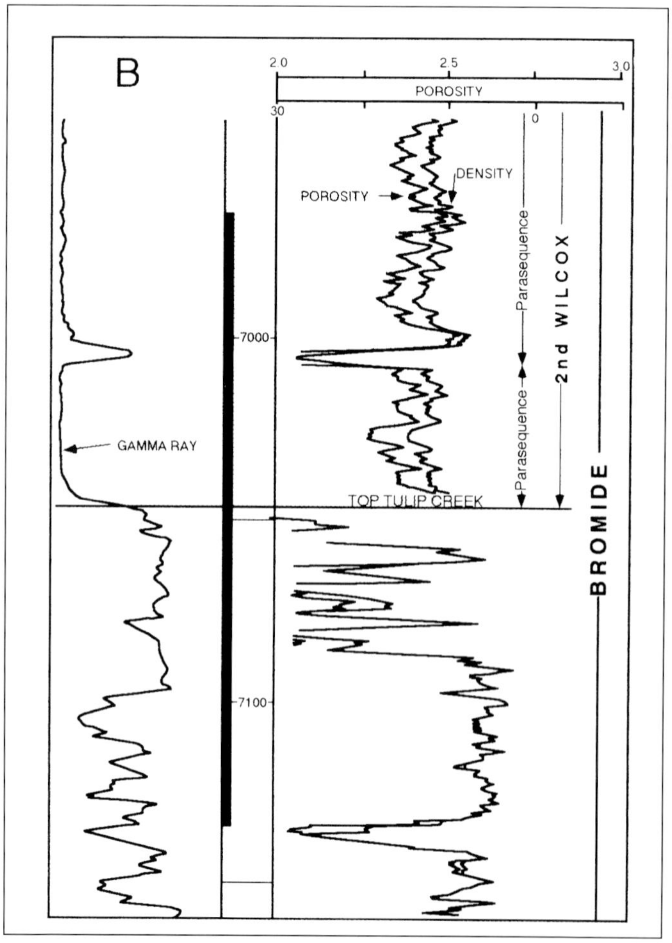

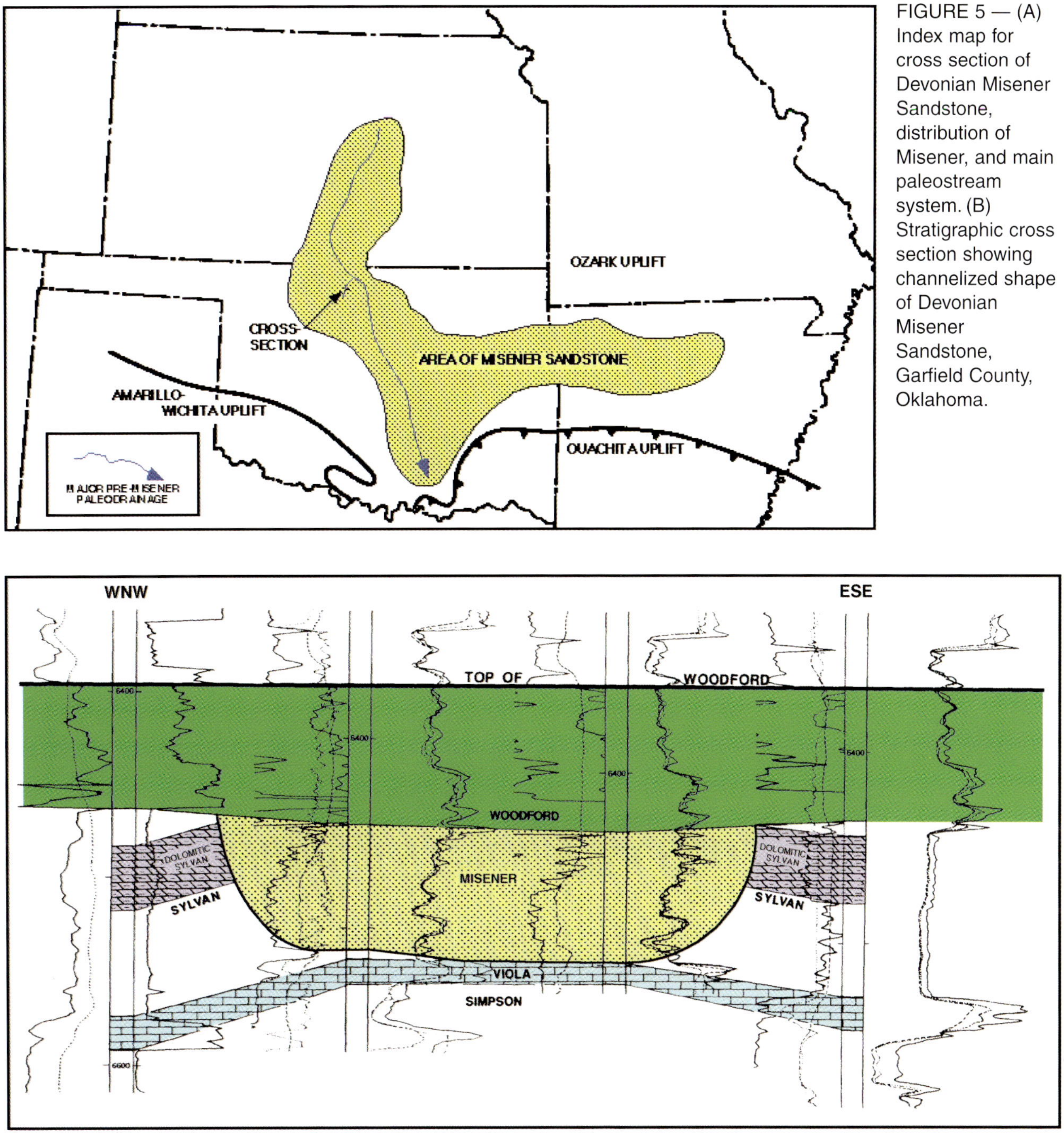

FIGURE 5 — (A) Index map for cross section of Devonian Misener Sandstone, distribution of Misener, and main paleostream system. (B) Stratigraphic cross section showing channelized shape of Devonian Misener Sandstone, Garfield County, Oklahoma.

outcrop and shallow subsurface data (Fig. 9). Downcutting up to 150 ft locally removed one limestone and eroded the uppermost part of an underlying limestone. The sandstone interval is dominated by alluvial valley deposits above the sequence boundary corresponding to the basal contact, but estuarine sandstone and shale characterize the uppermost 30 ft.

In the Illinois basin, the Desmoinesian Anvil Rock Sandstone shows a well-documented stream system, based on log shapes and thick sandstone intervals (Hopkins, 1958) (Fig. 10). Two types of alluvial/fluvial sandstones are recognized by differences in thickness and by inferred depths of erosion, responsible for the unconformable basal contact (or sequence boundary). It is noteworthy that the decidedly channelized Anvil rock is younger than the Bankston Fork Limestone, which over-

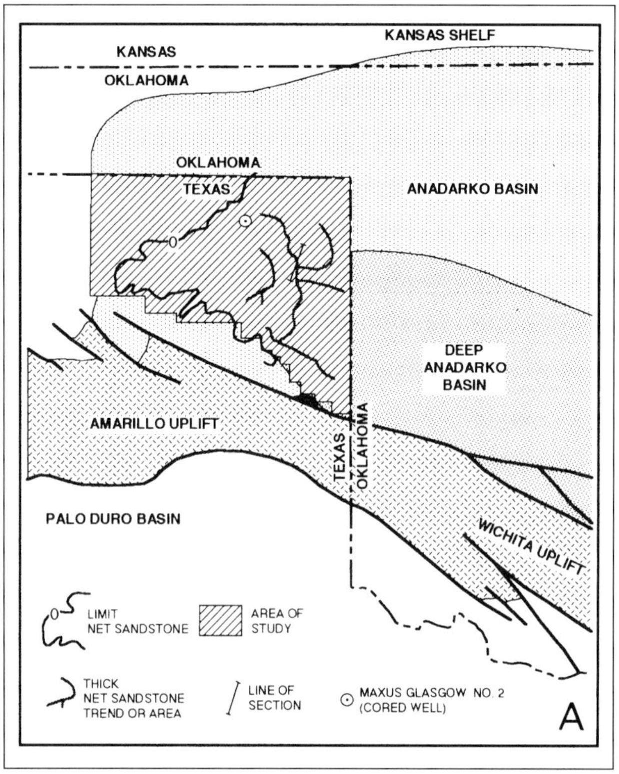

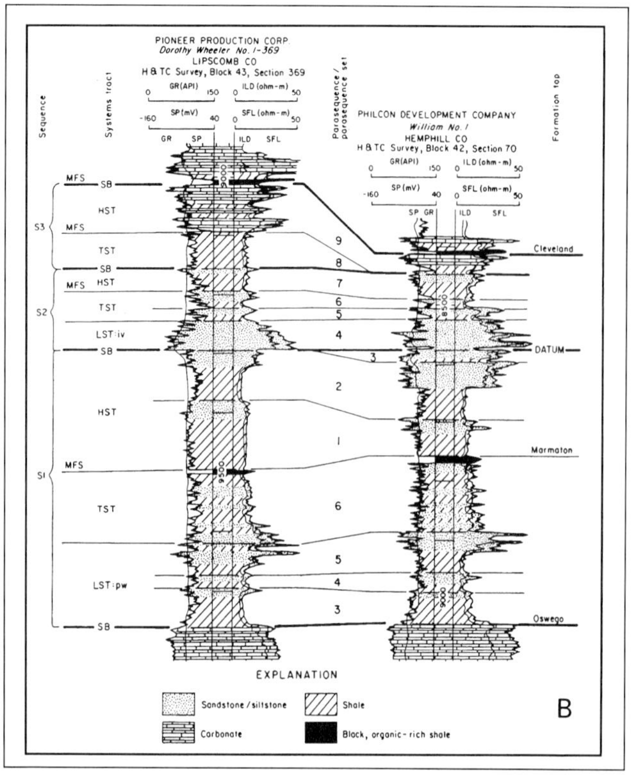

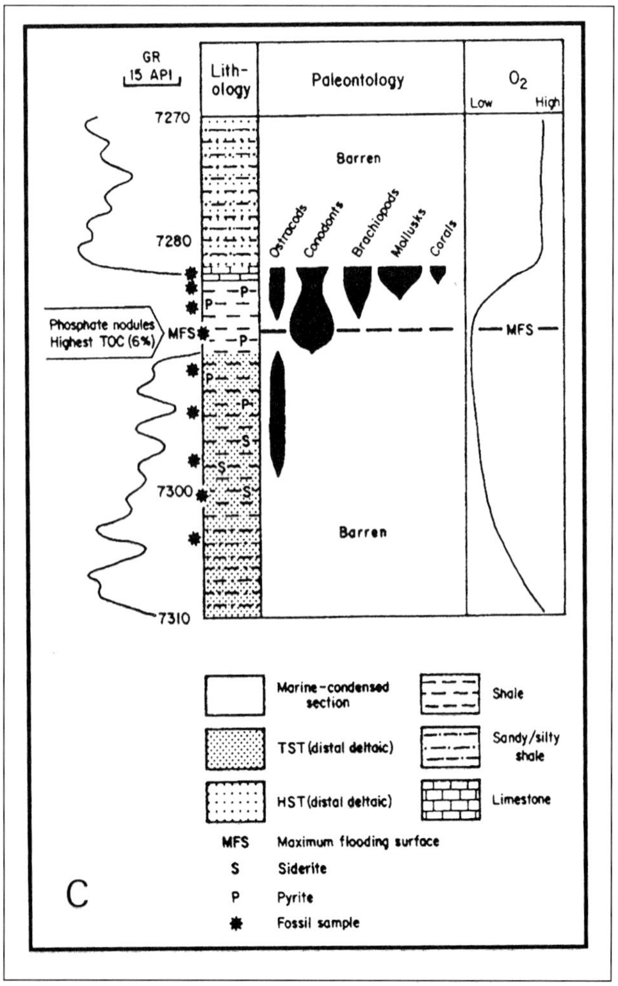

FIGURE 6 — (A) Index map for logs of lower Missourian (Pennsylvanian) Cleveland Sandstone along with its distribution and thick trends. (B) Stratigraphic cross section using representative well logs of upper Desmoinesian Marmaton Group and lower Missourian Cleveland Formation, showing subdivision of succession primarily by recognition of organic-rich, black shale and secondarily by parasequences and systems tracts (Hentz, 1994). (C) Wireline log and constituents of interval containing organic-rich shale, representing condensed section and containing transgressive and maximum flooding surfaces in uppermost part (top) of Marmaton, in Maxus Glasgow No. 2 (after Hentz, 1994).

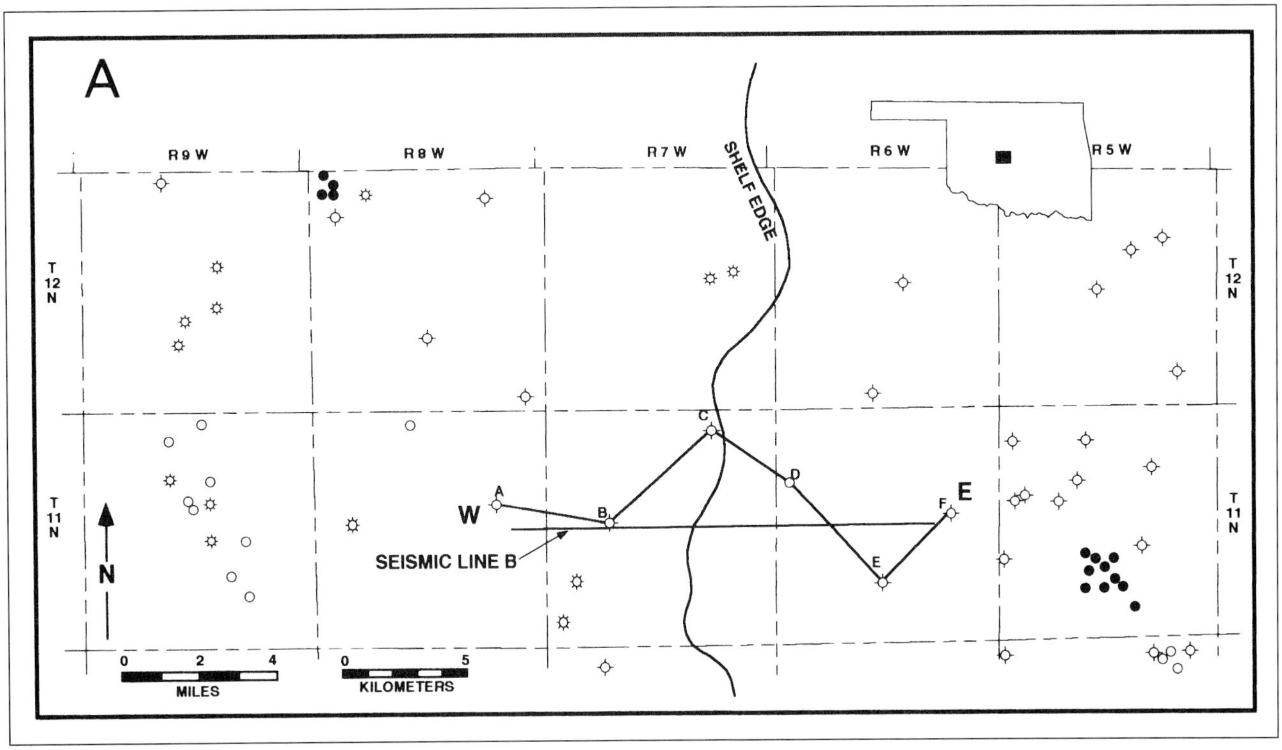

FIGURE 7 — (A) Index map for cross section and seismic line showing Missourian shelf edge. (B) Seismic section and corresponding section of Upper Pennsylvanian (Missourian) strata in Anadarko basin, Oklahoma, with stratal dip that reflects depositional slope, which in turn reflects storage capacity (or accommodation space) (from Galloway et al., 1977). The lower boundary of a sequence is thought to coincide with the top of the Belle City Limestone and with the equivalent base of the "offshelf" Medrano Sandstone.

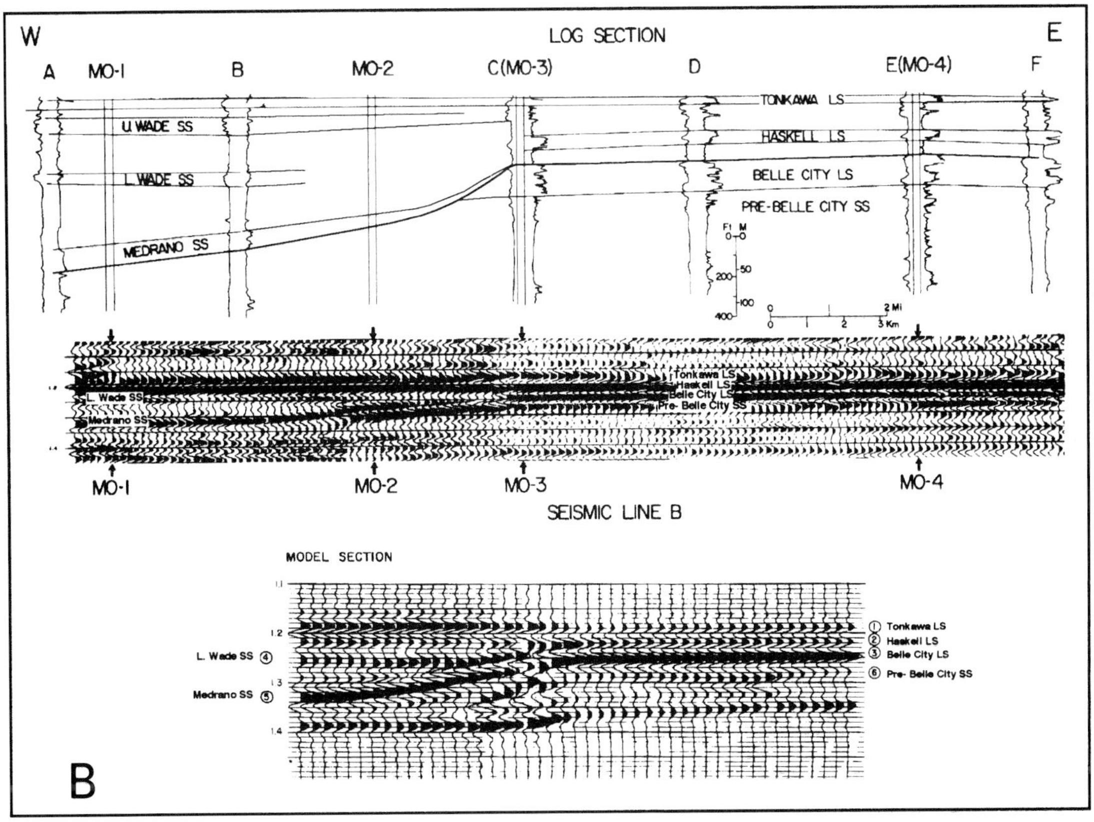

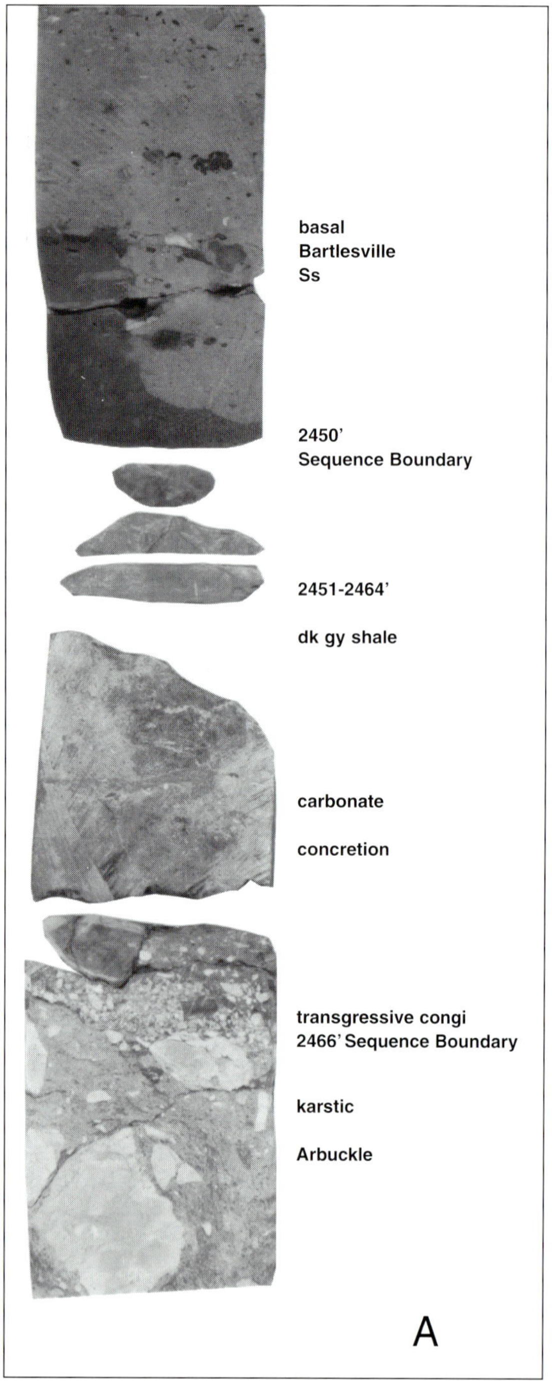

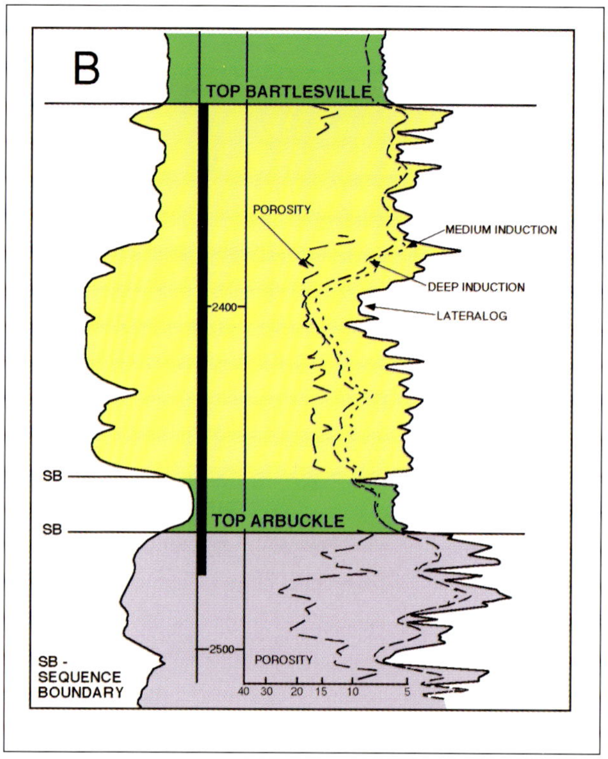

FIGURE 8 *(above right and left)* — (A) Basal conglomerate of Pennsylvanian Bartlesville Sandstone, representing unconformable boundary, Cushing field, Creek County, Oklahoma, overlies shale, with concretion in lowermost part, which in turn overlies basal Pennsylvanian conglomerate. Conglomerate overlies karstic Cambro-Ordovician Arbuckle strata. Conglomerate represents both the transgressive surface and the sequence boundary overlying a major unconformity. (B) Well log of cored interval, with electric-log response representing sharp boundary at base of Bartlesville.

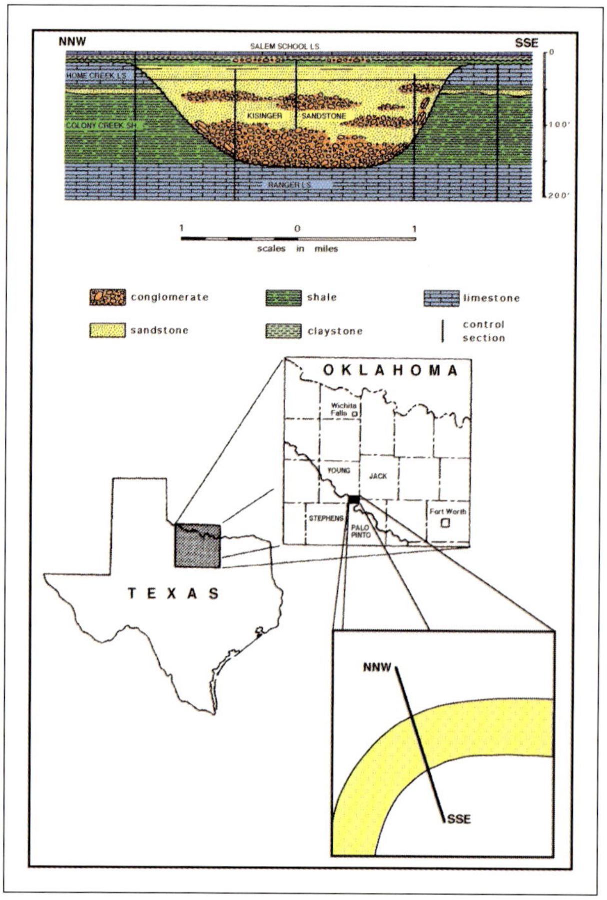

FIGURE 9 — Cross section of Pennsylvanian (Virgilian) Kisinger Sandstone, north-central Texas, showing unconformable basal boundary, transgressive conglomerate, and overlying Salem School Limestone, a marine-flooding unit (after Shelton, 1973).

lies a thin sandstone that is commonly regarded as widespread "sheet" Anvil rock.

In the northeastern part of the Anadarko basin, the basal sandstone of the Mississippian-Pennsylvanian Springeran Series overlies an unconformity marking the base of the section even where the sandstone is not developed (Clement, 1977) (Fig. 11). The basal contact of the sandstone (with underlying mudstones) is sharp, whereas the upper contact is apparently

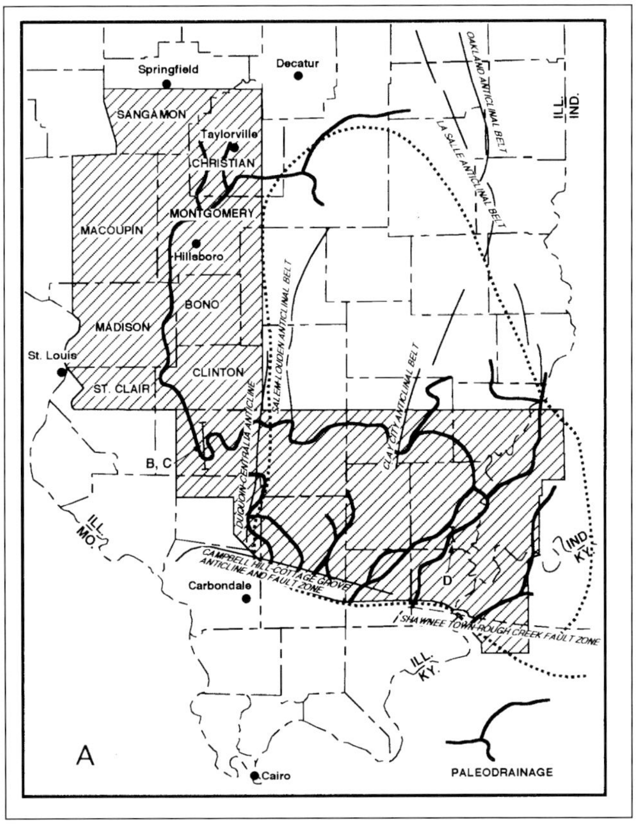

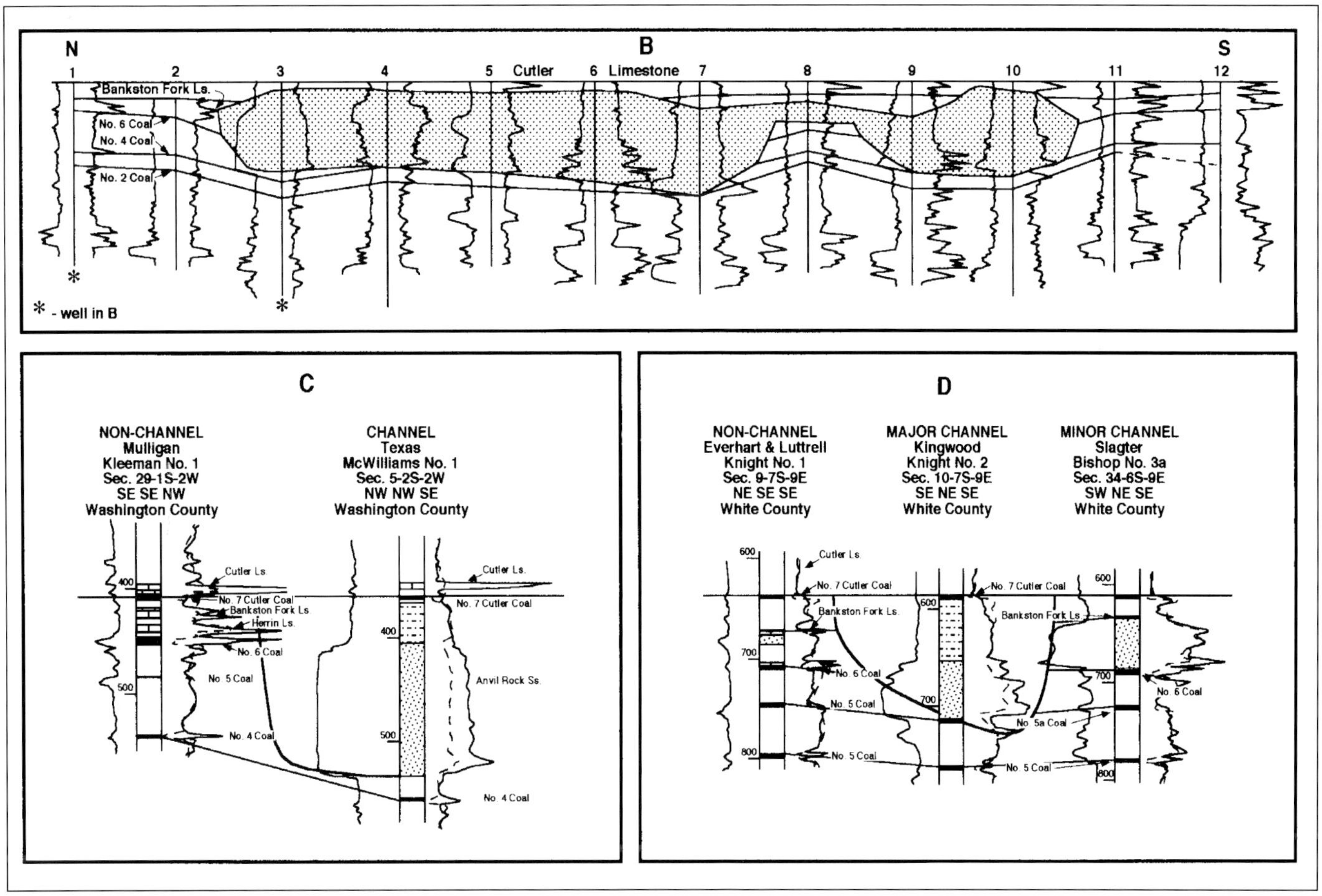

FIGURE 10 — (A) Index map for Desmoinesian (Pennsylvanian) Anvil Rock Sandstone, Illinois, with major paleodrainage system represented by the sandstone. (B) Correlation section of Anvil Rock Sandstone, showing unconformable basal boundary. (C and D) Representative electric logs of stratigraphic interval containing Anvil Rock Sandstone; prominent base of thick sandstone is lower boundary of sequence that is younger than the sequence that includes the thinner Anvil Rock sandstones.

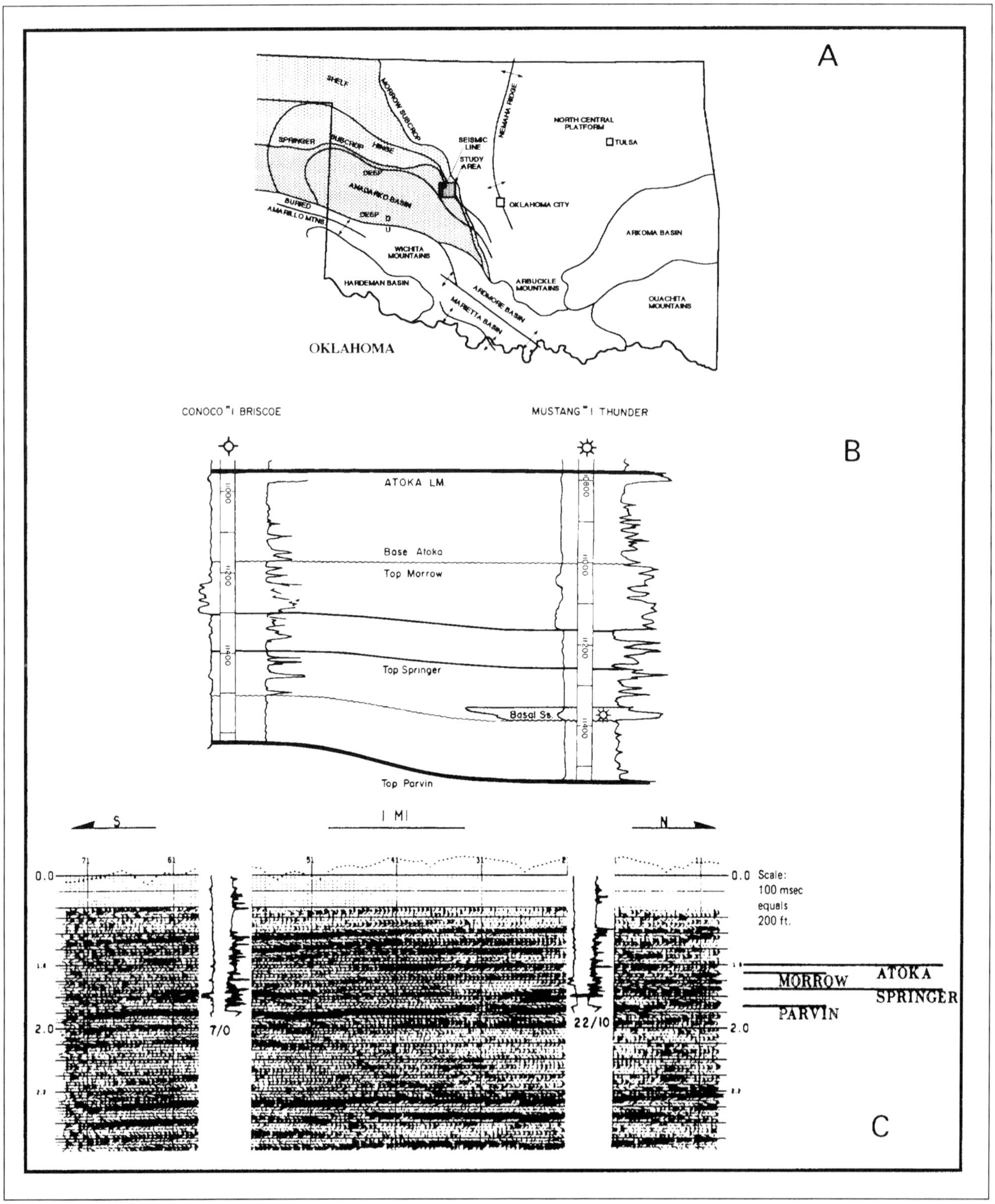

FIGURE 11 — (A) Index map for cross section and seismic line of basal Springer (Pennsylvanian-Mississippian) Sandstone. Stratigraphic cross section (B) and seismic line (C) showing unconformity at base of Springer Group, Anadarko basin, Oklahoma, with unconformity at base of locally developed basal Springer Sandstone (from Clement, 1977).

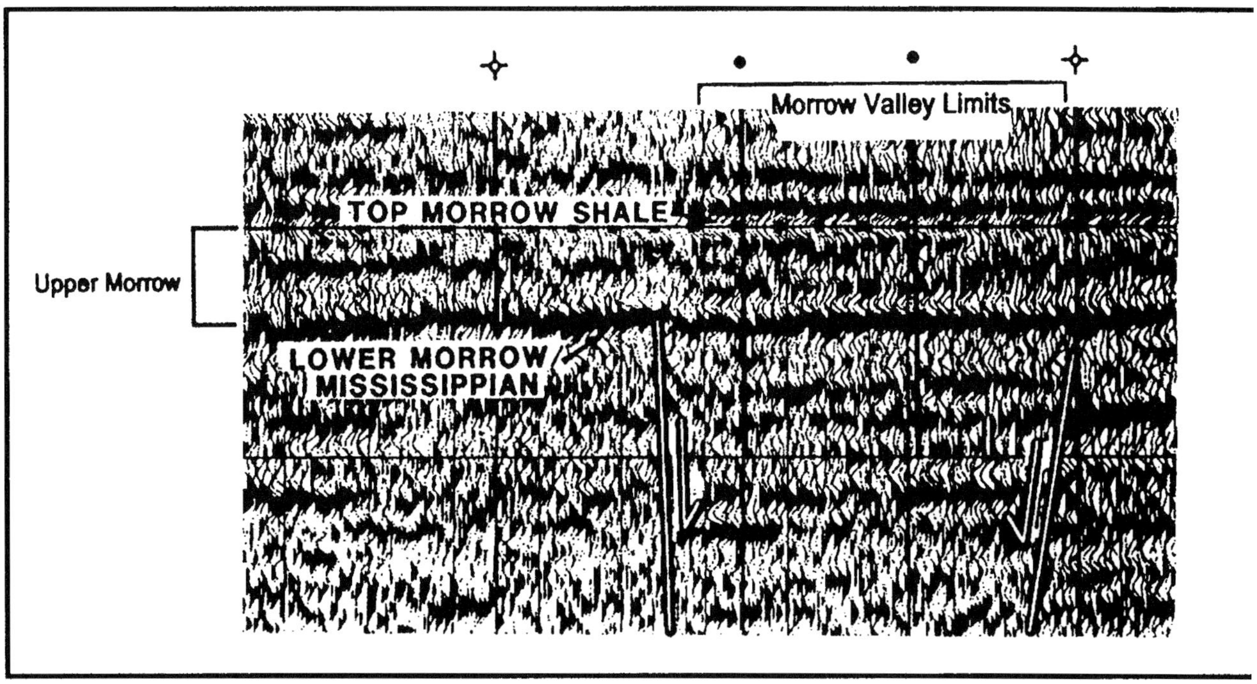

FIGURE 12 — Seismic profile across Pennsylvanian (Morrowan) valley fills in eastern Colorado, distinguishable by change in seismic characteristics, Mt. Pearl field (from Bowen et al., 1993).

gradational. The basal sandstone, with the associated unconformity, is recognizable on seismic profiles (Fig. 11C), after customized processing.

Upper Morrow (Pennsylvanian) shallow-marine shale in eastern Colorado was incised by southeast-flowing streams during lowstand periods. The valleys that formed were subsequently filled with sediments that include "amalgamated" fluvial sandstone bodies (Bowen et al., 1993). In some cases the valley is recognizable on seismic by corresponding changes in seismic reflection features (Fig. 12).

Not all channelized sandstones in the Mid-Continent represent fluvial deposits, although the prior erosion was by streams. In Oklahoma, the Devonian Misener Sandstone, which is locally developed in stable shelf areas of Oklahoma below the extremely widespread Woodford Shale, shows on logs and in a core the sharp, channelized basal contact and lateral juxtaposition to Ordovician strata (Fig. 5). Although the Misener occupies incised valleys, the fill is of marine (and possibly coastal) sediments, based on fauna, sedimentary structures (burrows), and mineralogy (glauconite and phosphate). It is likely that deposition of the Misener represents the earliest phase of deposition of the Woodford euxinic facies — in estuaries and embayments.

Paleosols. Paleosols in the Virgilian Lawrence Formation and in the Snyderville Shale Member of the Oread Formation are recognized over a large area in Nebraska, Iowa, Missouri, and Kansas (Joeckel, 1994). Features include slickensides and associated lenticular units, brecciated fabric, carbonate dikes and nodules, and clay coatings (Fig. 13).

Paleosols in the Permian Council Grove Group in Texas County, Oklahoma, are parts of sequences consisting of marine limestone, above a transgressive surface, and red beds, the latter above an exposure surface (Puckette et al., this book) (Fig. 14). The red beds represent hydromorphic soils (coniatolites), which formed on tidal flats with a variable water table but always at or near the surface. Workers place the Council Grove sequence boundaries at the top of the hydromorphic soils, corresponding to the

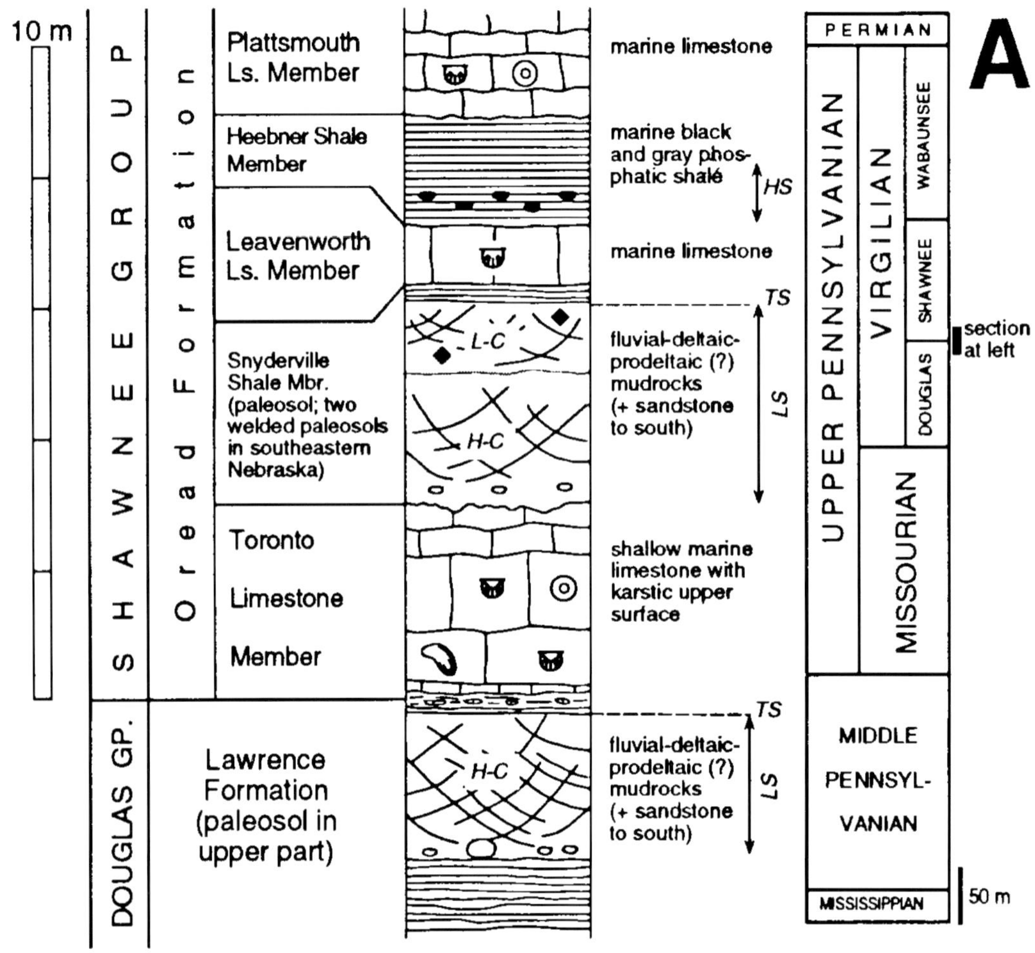

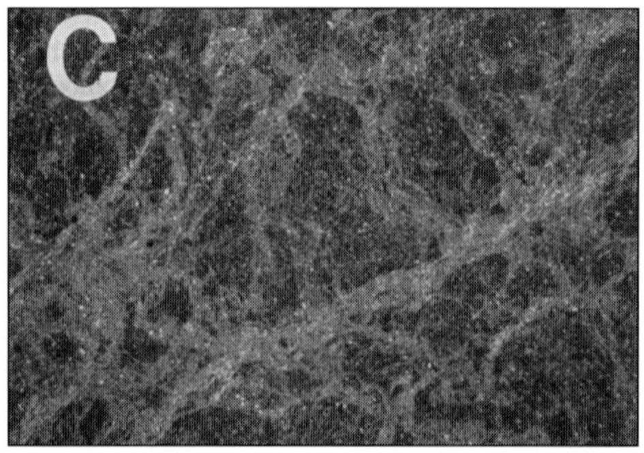

FIGURE 13 — Paleosols in Pennsylvanian (Virgilian) section in Nebraska and Missouri (from Joeckel, 1994, and with permission of Society for Sedimentary Geology). (A) Stratigraphic succession containing paleosols. (B) Lenticular bedding (l) with slickensides (s) in upper part of Lawrence paleosol. (C) Brecciated fabric in thin section in Lawrence paleosol, 2 m below top (11 mm across photomicrograph).

FIGURE 14 — Paleosols in core of fourth-order sequence that consists of (1) limestone (Crouse), with (1a) transgressive surface (lower TS at 3065 ft) and associated lag at base (above red beds, a hydromorphic soil), and (1b) exposure surface (ES) at top, and (2) the overlying Blue Rapids Shale, a red, hydromorphic soil (coniatolite) with vertically oriented carbonate nodules/masses. At top of soil is younger transgressive surface (TS at 3042 ft), defining the upper boundary of sequence in Permian Council Grove Group, Texas County, Oklahoma. (Photograph and information provided by J. Puckette; data and interpretation from Puckette et al., this book).

transgressive surfaces (Puckette et al., this book). Hydromorphic soils have been recognized at a number of stratigraphic positions in redbed successions in Wolfcampian and Leonardian strata in north-central Oklahoma.

Regional Unconformities. Four formations comprising the Middle Ordovician Simpson Group of Oklahoma, in ascending order, are Joins, Oil Creek, McLish, and Bromide (Statler, 1965) (Fig. 3). Although each formation is present over a rather large area, changes in the development of the basal sandstone may contribute to problems in correlating individual formations. The Joins is least extensive and most difficult to distinguish. Disconformities increase in magnitude toward the more stable part of the shelf. The Simpson Group, which thins from more than 2000 ft in southern Oklahoma toward the northeast, north, and northwest, contains sandstone, shale (much of which is green), and carbonate (both mud- and grain-supported limestones and some dolomite, which increases in occurrence toward the stable shelf). It has been long recognized that the Simpson reflects an overall cyclical pattern of its units (Decker, 1941). With respect to the sandstones, they are quartzose and well sorted. Basal formational sandstones are generally thick and extensive, whereas the other sandstone bodies are commonly thinner and less continuous, and they grade laterally into limestone.

Presumably each of the Simpson formations represents a sequence (Fig. 3), but it is conceivable that each is a set of sequences. The group and each formation within it are bounded by unconformities, recognized by the onlap of the lower units, shelfward convergence and thinning of the sequences, and sharp base of the basal sandstones. All units of the Simpson are thought to be shallow marine deposits, a feature that is remarkable considering the role of unconformities before and during the period of time that included Simpson deposition.

Less conspicuous unconformities are present on shelf and shelf-edge limestones; the basinal equivalent is the base of lowstand clastics. This type of setting existed during deposition of the Missourian shelf Belle City Limestone and of the offshelf Medrano Sandstone in the Anadarko basin (Galloway et al., 1977) (Fig. 7).

Transgressive Surfaces

In Exxon terminology, the "first" marine-flooding surface is the transgressive surface, which forms before the maximum flooding surface associated with a condensed section. Transgressive surfaces also are commonly associated with condensed sections; they may be associated with sequence boundaries. The surface may show the effects of burrowing, or in some cases borings; the associated depositional unit generally is thin, bioturbated, glauconitic, phosphatic, pyritic, fossiliferous, and conglomeratic. Compositionally this type of unit contains significant authigenic constituents, and the fauna show a distinct pelagic influence.

Another type of transgressive surface is the top of a widespread coal, which is overlain by marine strata. In the Eastern Interior basins, it is not uncommon for this type of coal to be directly overlain by black shale, which, in turn, is overlain by limestone.

These units are recognizable on an outcrop, in cores (Figs. 6 and 8), and, with proper calibration, on well logs (as thin markers without a unique log signature).

Maximum Flooding Surfaces

In cratonic areas, the interval (thickness) between the transgressive surface and maximum flooding surface is insignificant, and the latter commonly is considered, without flaws in interpretation, to be associated with, or a part of, the transgressive unit. In fact, much of the sedimentary history of the

Mid-Continent has been studied and resolved by persons using different marine-flooding surfaces in correlating and dividing the stratigraphic section into transgressive/regressive couplets.

Marine-flooding surfaces above transgressive surfaces in the Mid-Continent are most commonly associated with thin, widespread carbonates, or black shales (identical or analogous to the black shale overlying the widespread coal noted above), or a combination. In the Eastern Interior basin, the underlying coal may be as widespread as the carbonate. Each of these approximates the maximum flooding surface.

The limestones are commonly mud supported, with a rather wide range of invertebrate fossils. The black shale is fissile, organic-rich, phosphatic, and fossiliferous. Common fossils include conodonts, ostracodes, brachiopods, mollusks, and corals (Hentz, 1994). Where organic-carbon content is highest, the fauna are largely restricted to conodonts and phosphatic nodules (Hentz, 1994).

In north-central Texas, lower Virgilian strata contain the thin but widespread Salem School Limestone, which overlies a marine shale that contains at its base a very thin, fossiliferous conglomerate (Fig. 9). Because of its distribution, the limestone undoubtedly is, or approximates, a maximum flooding unit, which for practical purposes may be considered a surface. The fossiliferous conglomerate is the oldest marine unit and, as such, is the oldest transgressive unit. Also because of its thickness, it may be considered a surface. The shale undoubtedly is considered by some workers to contain the maximum flooding surface. In fact, the interval inclusive of the conglomerate, shale, and limestone (and possibly with three marine-flooding surfaces) is quite thin so that considering it a surface, as opposed to a section or interval, does not represent a significant error.

The Desmoinesian-Missourian boundary in the Anadarko basin is represented on logs by maximum gamma radiation within a "condensed" section of black shale, which contains phosphate nodules, pyrite, conodonts, ostracodes, and other invertebrates (Hentz, 1994) (Fig. 6). The middle of the shale, which is less than 10 ft thick, is considered as the maximum flooding surface. In reality the shale may be considered to be that surface and the transgressive surface as well. This shale is considered to be representative of the commonly occurring black shale in the Pennsylvanian cyclothems of the Mid-Continent.

Parasequence Boundaries

Some Pennsylvanian sequences, especially where the accommodation space is not restrictive, contain parasequences that are rather easily recognized by upward-coarsening log signatures as well as lithologic/petrographic data (e.g., dominant shale below dominant sandstone; bioturbation below current bedding). Each boundary corresponds to the base of shale transitionally below the bioturbated unit but sharply above the bedded sandstone of the underlying parasequence; this corresponds to the top of the "funnel" on upward-coarsening logs that have been properly calibrated to rock data.

Perhaps the most impressive parasequences are present in the Simpson Group. Recognized within each Simpson formation (sequence) are upward-coarsening units that consist of shale, bioturbated mudstone and sandstone, and well-bedded sandstone, in ascending order. Again the boundary is at the base of the shale, as illustrated in Fig. 4, which presents representative parts of a core from the Bromide in central Oklahoma.

CONCLUSIONS

Because limited accommodation space in Mid-Continent sedimentary provinces results in abbreviated sequences, some surfaces approximate or coincide with others, especially those associated with the transgressive part of the section. The following sequence surfaces are most commonly recognized in Mid-Continent siliciclastic strata.

1. Sequence boundaries at base of channelized sandstones.

2. Sequence boundaries associated with regional unconformities.

3. Sequence boundaries associated with paleosols.

4. Transgressive surfaces which commonly represent the first flooding surfaces but may approximate maximum flooding surfaces or condensed sections, or may coincide with other marine-flooding surfaces that correspond to parasequence boundaries. In some cases, transgressive surfaces are sequence boundaries.

5. Parasequence boundaries, which may approximate or coincide with transgressive surfaces, maximum flooding surfaces, condensed sections, other marine-flooding surfaces, or sequence boundaries, or a combination.

Transgressive surfaces have served workers well in dividing strata into units for structural mapping and sandstone-reservoir delineation.

ACKNOWLEDGMENTS

The writers wish to thank their colleagues and former colleagues at Masera Corporation, in particular R.D. Fritz, P. Medlock, and M. Kuydendall, for permitting use of some of the results of their studies in this paper, and to Valerie Lindsey and Sandra PaskVan for their help in preparation of the manuscript. Also appreciation is expressed to J. Puckette, D. Boardman, and Z. Al-Shaieb, Oklahoma State University, for freely permitting use of their data and interpretation of Permian red beds in the Oklahoma Panhandle and for providing useful comments on review of the manuscript. Gary Stewart, Oklahoma State University, and Dennis Kerr, The University of Tulsa, also provided very helpful critical reviews of the paper. The AAPG and SEPM have kindly permitted use of various illustrations from their publications in this manuscript, and Norman Hyne provided timely and appropriate editorial support and coercion.

RECOGNITION OF UNCONFORMITIES AND SEQUENCES IN MID-CONTINENT CARBONATES

Richard D. Fritz and Patrick L. Medlock
Masera Corporation
Tulsa, OK 74170

ABSTRACT

The Mid-Continent contains three primary carbonate sequences: (1) Sauk Cambro-Ordovician Arbuckle Group, (2) Tippecanoe Ordovician Viola Group and Siluro-Devonian Hunton Group, and (3) Kaskaskia Mississippian Osage-Meramec-Chester limestone sequence. The Pennsylvanian and Permian Absaroka sequence also contains significant carbonate cyclothems which are associated and subordinate to thick clastic deposits. Unconformities (and related conformities) are the boundaries that define carbonate depositional sequences and are the basis for the interpretations of sequence stratigraphy. A hierarchy of unconformities is worldwide, interregional, regional, subregional, and local according to their extent and hiatus or measurable time gap. Diagenesis associated with unconformity is both porosity destructive and constructive and, if developed on a large scale, can be recognized as karst terrain by large dissolution cavities, breccias, etc. The six most used methods for recognizing unconformity and related diagenesis are (1) outcrop and core analysis, (2) well log evaluation and correlation, (3) biostratigraphy, (4) seismic interpretation, (5) sequence evaluation, and (6) basin modeling. Carbonate reservoirs are often associated with unconformities worldwide and especially in the Mid-Continent. The recognition of key sequence boundaries, sequences, and associated diagenetic modifications can provide the basis for exploring any subtle traps.

INTRODUCTION

Most of the world's oil and gas reserves are located in carbonate reservoirs, many of which are associated with unconformities. It has been estimated that at least 20 to 30% of recoverable reserves are related to unconformities (Moody et al., 1970; Levorsen, 1954; Gibson, 1965).

Carbonate deposits are particularly susceptible to secondary processes associated with unconformities. Diagenesis can be reservoir constructive or reservoir destructive depending on the geochemistry of meteoric waters, geometry of the hydrologic system, and duration of exposure. Recognition of the diagenetic processes and the changes they impose on particular facies is instrumental in

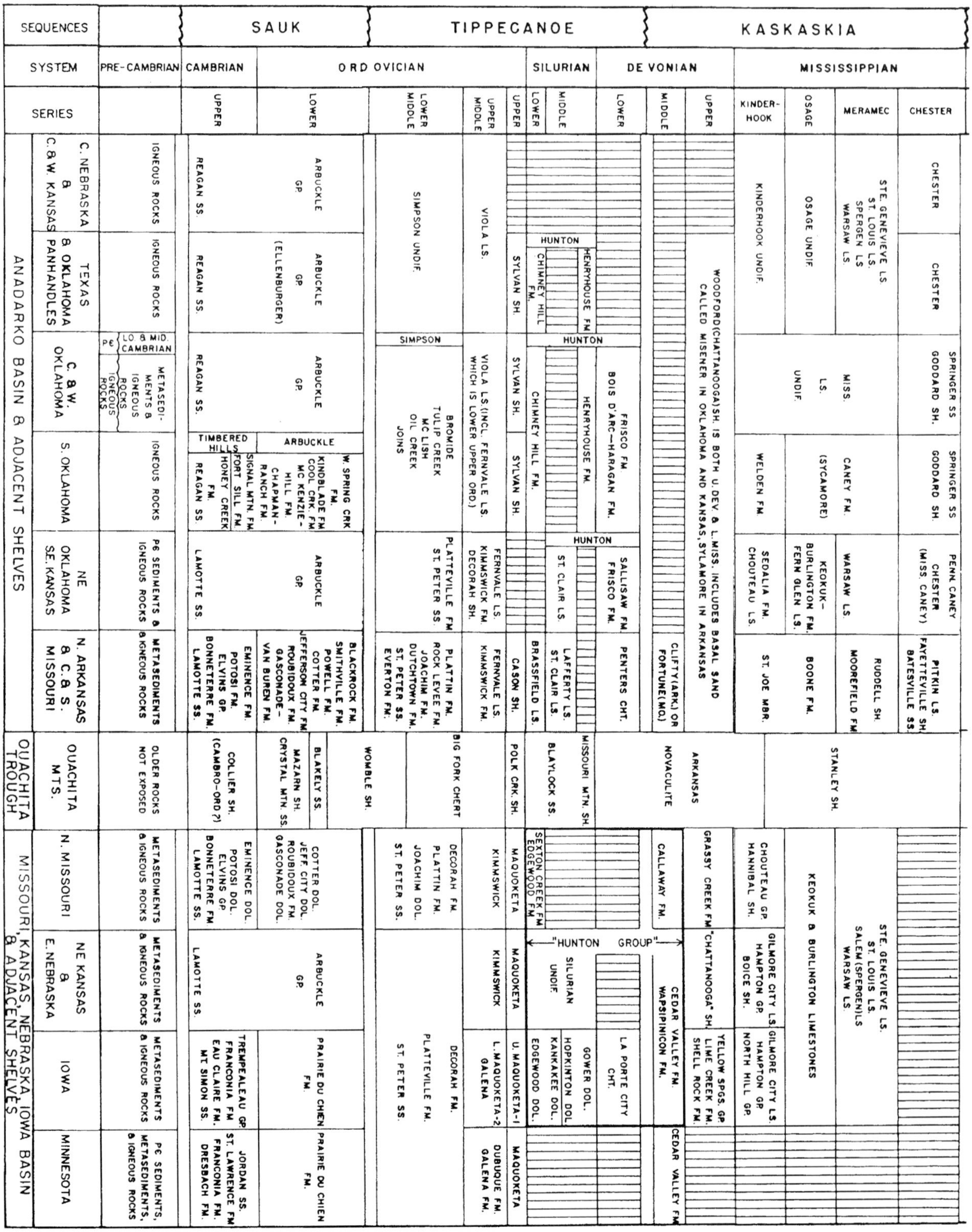

FIGURE 1 — Correlation chart for the Mid-Continent showing the sequences of Sloss (Rascoe and Hyne, 1987).

SEQUENCES	SYSTEM	SERIES	ANADARKO BASIN & ADJACENT SHELVES: C. NEBRASKA & C. & W. KANSAS	ANADARKO BASIN & ADJACENT SHELVES: TEXAS & OKLAHOMA PANHANDLES	ANADARKO BASIN & ADJACENT SHELVES: C. & W. OKLAHOMA	ANADARKO BASIN & ADJACENT SHELVES: S. OKLAHOMA	ANADARKO BASIN & ADJACENT SHELVES: NE. OKLAHOMA SE. KANSAS	ANADARKO BASIN & ADJACENT SHELVES: N. ARKANSAS & C. & S. MISSOURI	OUACHITA TROUGH: OUACHITA MTS.	MISSOURI KANSAS, NEBRASKA, IOWA BASIN & ADJACENT SHELVES: N. MISSOURI	MISSOURI KANSAS, NEBRASKA, IOWA BASIN & ADJACENT SHELVES: NE. KANSAS & E. NEBRASKA	MISSOURI KANSAS, NEBRASKA, IOWA BASIN & ADJACENT SHELVES: IOWA	MISSOURI KANSAS, NEBRASKA, IOWA BASIN & ADJACENT SHELVES: MINNESOTA
ZUNI	CRETACEOUS	UPPER	PIERRE SH. NIOBRARA FM. CARLISLE SH GREENHORN LS. GRANEROS SH. DAKOTA SS.	GREENHORN LS. GRANEROS SH. DAKOTA SS.		EAGLE FORD SH. WOODBINE SS.		NAVARRO GP. TAYLOR GP. AUSTIN GP.			NIOBRARA FM. CARLISLE SH. GREENHORN LS. GRANEROS SH. DAKOTA SS.	CARLISLE SH. GREENHORN LS. GRANEROS SH. DAKOTA SS.	PIERRE SH.(?) NIOBRARA FM. CARLISLE SH. GREENHORN FM. GRANEROS SH. DAKOTA SS.
		LOWER	KIOWA SH. CHEYENNE SS.	KIOWA SH. CHEYENNE SS.		WASHITA GP. FREDERICKS-BURG GP. TRINITY GP.							
	JURASSIC		MORRISON FM.	MORRISON FM.								FORT DODGE BEDS	
ABSAROKA	UPPER TRIASSIC			DOCKUM									
	PERMIAN	GUADA-LUPE	TALOGA FM. DAY CREEK DOL. WHITE HORSE SS.	QUARTERMASTER FM. WHITEHORSE GP.	QUARTERMASTER FM. CLOUD CHIEF FM. WHITEHORSE GP.								
		LEONARD	NIPPEWALLA GP. SUMNER GP.	BLAINE FM. GLORIETA SS. CLEARFORK GP. WICHITA FM.	EL RENO GP. HENNESSEY SH. WELLINGTON FM.						NIPPEWALLA GP. SUMNER GP.		
		WOLFCAMP	CHASE GP. COUNCIL GROVE GP. ADMIRE GP.	BROWN DOL.	CHASE GP. COUNCIL GROVE GP. ADMIRE GP.	PONTOTOC GP. UPPER PART					CHASE GP. COUNCIL GROVE GP. ADMIRE GP.		
	PENNSYLVANIAN	VIRGIL	WABAUNSEE GP. SHAWNEE GP. DOUGLAS GP.			CISCO GP.	WABAUNSEE GP. SHAWNEE GP. DOUGLAS GP.			WABAUNSEE GP. SHAWNEE GP. DOUGLAS GP.			
		MISSOURI	LANSING GP. KANSAS CITY GP. PLEASANTON GP.		OCHELETA GP. SKIATOOK GP.	HOXBAR (CANYON) GP.	LANSING GP. KANSAS CITY GP. PLEASANTON GP.			LANSING GP. KANSAS CITY GP. PLEASANTON GP.			
		DES MOINES	MARMATON GP. CHEROKEE GP.		BIG LS. OSWEGO LS.	DEESE (STRAWN) GP.	MARMATON GP. CHEROKEE GP.	BOGGY FM. SAVANNA FM. MC ALESTER FM. HARTSHORNE SS.		MARMATON GP. CHEROKEE GP.			
		ATOKA	13 FINGER LS.			UPPER DORNICK HILLS		ATOKA FM.	ATOKA FM.	ATOKAN			
		MORROW	KEARNY FM.	U. MORROW L. MORROW	MORROW	MORROW		BLOYD SH. HALE FM.	JOHNS VALLEY SH JACKFORK SS.				

the definition of reservoir characteristics and evaluation of reservoir potential.

In the Mid-Continent, Paleozoic carbonates often have had several episodes of diagenesis associated with unconformities. Carbonate reservoirs associated with unconformities often experienced rejuvenation by later processes associated with unconformities and have their own unique evolution. The purpose of this study is threefold: (1) to establish a comprehensive definition and classification of unconformities and related sequences which are applicable to Mid-Continent carbonates, (2) to discuss the processes associated with unconformities, and (3) to describe various methods of recognizing unconformable surfaces and associated diagenetic changes in Mid-Continent carbonates.

MID-CONTINENT CARBONATE STRATIGRAPHY

Sedimentation and stratigraphy in the Mid-Continent are best understood by application of the concepts described by Sloss (1963) who described six Phanerozoic sequences (supersequences in Exxon terminology), of which the lower four define the stratigraphy of the Mid-Continent (Fig. 1).

There are three primary carbonate sequences within the Mid-Continent: (1) Sauk Cambro-Ordovician Arbuckle Group, (2) Tippecanoe Ordovician Viola Group and Siluro-Devonian Hunton Group, and (3) Kaskaskia Mississippian Osage-Meramec-Chester limestone sequence. Numerous carbonates exist in the Absaroka Pennsylvanian and Permian as parts of cyclothems. In Oklahoma, many of the Pennsylvanian-Permian carbonates are thin limestones not typically associated with production, although significant production from carbonate reservoirs is found in Desmoinesian Big Lime- Oswego. In Kansas and the Oklahoma and Texas panhandles, thick productive carbonate sequences are found in the Missourian Kansas City-Lansing and Wolfcampian Brown Dolomite, Council Grove and Chase Group (Fig. 2). Many of the Absaroka carbonates, both productive and nonproductive, are important in that they are associated with marine-flooding surfaces and help define transgressive- regressive (T-R) couplets or parasequences (Shelton, 1973).

The Arbuckle Group of the Sauk sequence is the thickest carbonate section in the Mid-Continent, averaging more than 1000 ft in thickness, with a maximum thickness of 8000 ft in southern Oklahoma. The Arbuckle is composed of cyclical peritidal carbonates that were deposited on a broad platform which covered most of current North America.

The Upper-Middle Ordovician Viola Group of the Tippecanoe sequence is dominated by ramp carbonates composed of dense, fine-grained, cherty limestones. The upper part of the Viola is often fossiliferous, and the middle tends to be argillaceous (Gentile, 1982). The maximum thickness of the Viola is over 1000 ft.

The Hunton Group, also of the Tippecanoe sequence, ranges in age from uppermost Ordovician to Lower Devonian. The Hunton was deposited on a shallow carbonate ramp and is composed primarily of shallow-water mudstones and wackestones; grainstones are also common. Marlstones are developed in the distal part of the ramp.

Mississippian carbonates (Osage-Meramec-Chester) of the Kaskaskia sequence are typically composed of cherty limestones and dolomites interbedded with grainstones, quartz sandstones, and marls (Harris, 1975).

Pennsylvanian rock units in the Mid-Continent are divided into the Morrowan, Atokan, Desmoinesian, Missourian, and Virgilian series. Pennsylvanian rocks were deposited during the deformational stage of the Mid-Continent and are composed of dominantly clastic cyclothems. The Permian Wolfcampian is also composed of cyclic deposits.

There are multiple unconformities of varying extent and scale within the sequences described

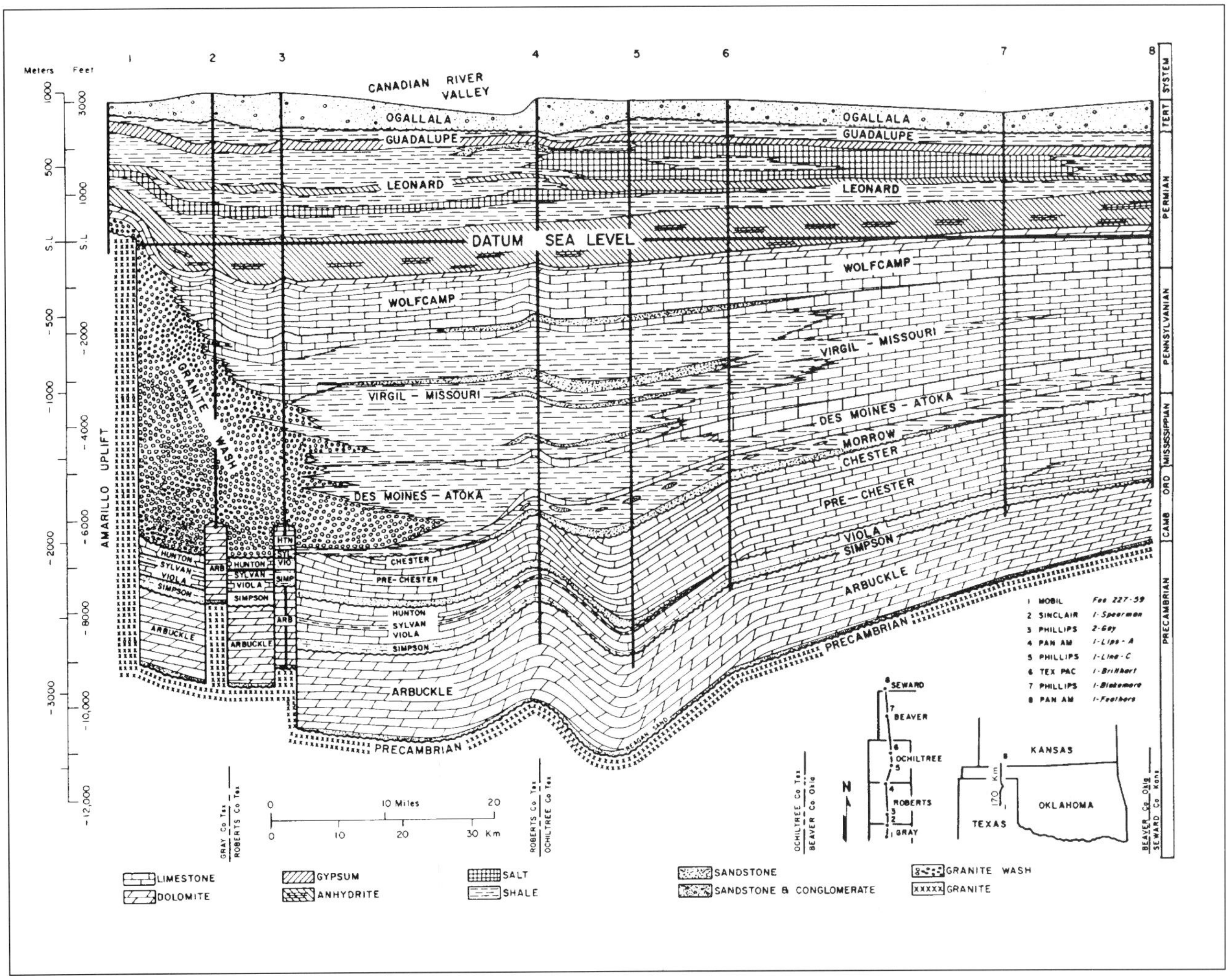

FIGURE 2 — North-south structural cross section in the western part of the Anadarko basin (Rascoe and Johnson, 1988).

above. The major unconformities separating the sequences of Sloss can be delineated in all but the deepest parts of the basins in Oklahoma (Fig. 2).

UNCONFORMITIES

Definition

A stratigraphic time gap, or *hiatus*, is defined as the total interval of geologic time not represented by strata at a specific position along a particular stratigraphic surface (Mitchum et al., 1977). A hiatus usually is caused by exposure processes at the Earth's surface, such as erosion and weathering, but can also be the result of nondeposition and/or condensed sedimentation.

An unconformity is defined as a buried erosional surface or surface of noncontinuity that separates an older rock unit from an overlying rock unit; at that surface, a significant time gap is present which is measurable in geologic time (Dunbar and Rodgers, 1957). Usually an unconformity also implies that a significant stratigraphic section is missing due to one of the processes discussed in the previous paragraph; however, the amount of section missing is not necessarily related to the magni-

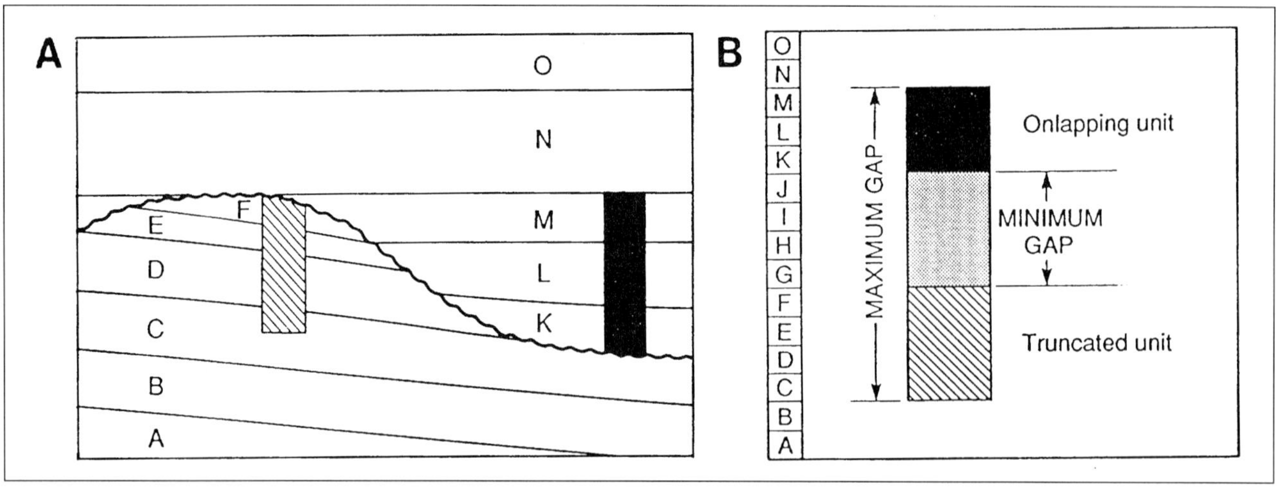

FIGURE 3 — Chronostratigraphic representation of a hiatus associated with an unconformity which shows (A) stratigraphic relationship of strata to unconformity, and (B) diagrammatic summary of time gap (Esteban, 1992).

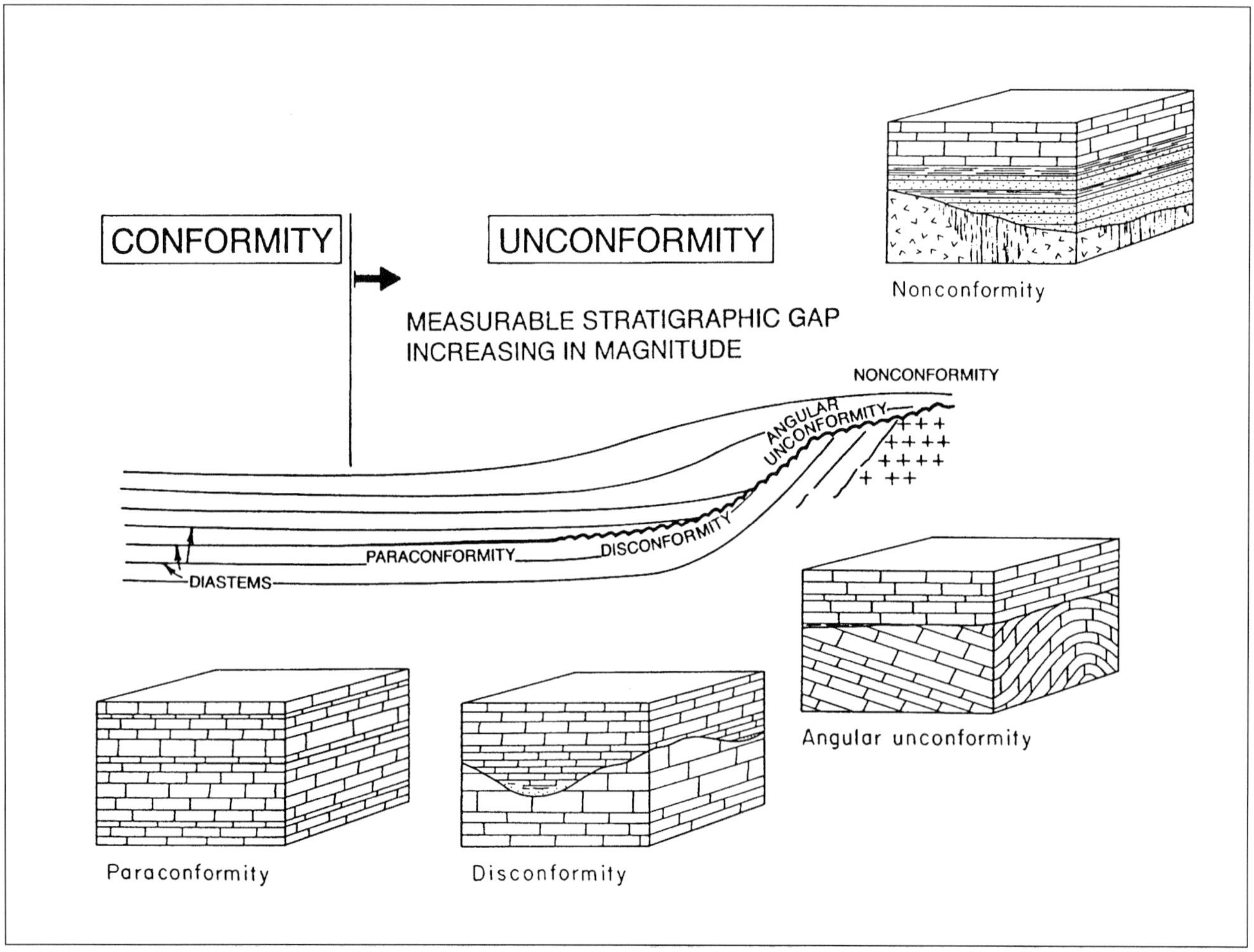

FIGURE 4 — Classification of unconformities and their geometry in cross section (modified from Dunbar and Rodgers, 1957).

tude of the time gap.

Chronostratigraphic representation of time gaps associated with unconformities can provide a better understanding of the evolution on an unconformable surface. The time gap and missing section along a particular unconformable surface can be described as minimum or maximum. For example, the minimum time gap can be defined as the actual age of the missing section, and the maximum time gap can be defined as the youngest section above the unconformity in contact with the oldest section below the unconformable surface (Esteban, 1992: Fig. 3). Although most carbonate reservoirs in the Mid-Continent are associated with major time gaps, there are many minor discontinuities which do not show a "measurable" stratigraphic gap between Mid-Continent strata; some of these also have associated reservoir development.

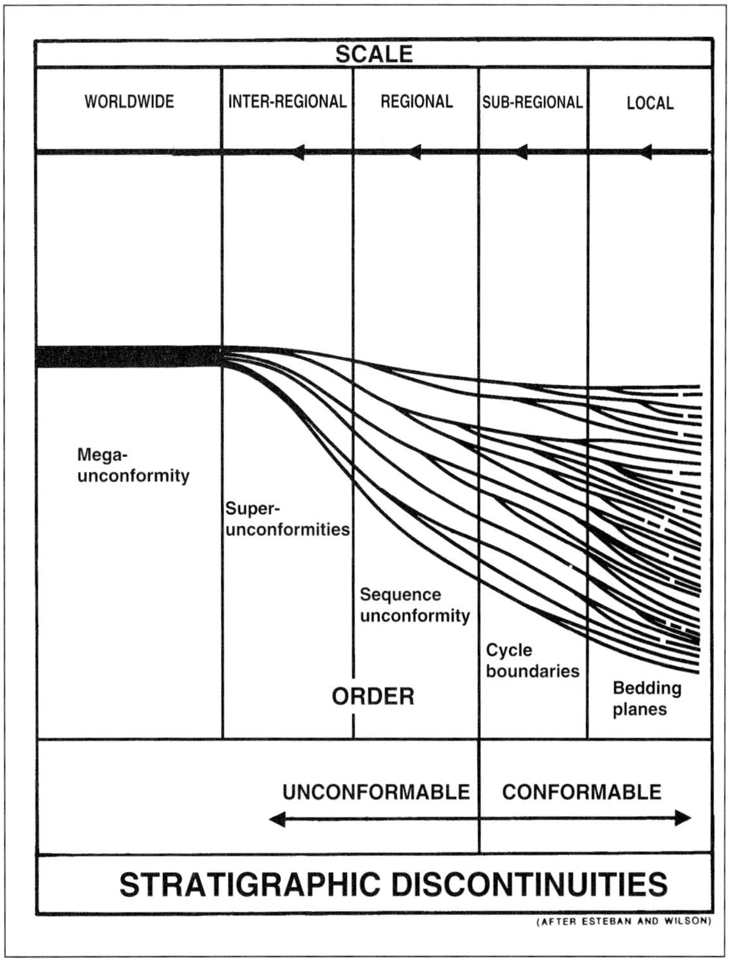

FIGURE 5 — Classification of stratigraphic discontinuities based on magnitude and extent (modified from Esteban, 1992).

Classification

Traditional textbook classification of unconformities is (1) nonconformity, (2) angular unconformity, (3) disconformity, and (4) paraconformity (Dunbar and Rodgers, 1957: Fig. 4). This conventional classification places emphasis on the angularity of the strata above and below the unconformity, and often describes the degree of folding prior to the hiatus (Mitchum et al., 1977). In face, the traditional classification is seldom used by recent workers in modern sequence stratigraphy who prefer to classify the relationship of the strata to the unconformity.

It is important to understand scale and order when evaluating the reservoir potential associated with an unconformity. Vail et al. (1984) and Van Wagoner et al. (1987) described two primary types of unconformities associated with sequence boundaries. *Type 1* is defined by subaerial to submarine erosion which is associated with stream rejuvenation, a basinward shift of facies, and a downdip shift of overlying strata. *Type 2* is also defined by subaerial erosion but without the associated stream rejuvenation and basinward shift of facies. Essentially, a type 1 unconformity occurs when the rate of sea-level fall exceeds the rate of basin subsidence at the depositional edge, e.g., the pre-Woodford unconformity. A type 2 unconformity occurs when the rate of eustatic fall is less than or equal to rate of subsidence at the depositional edge, e.g., the intraformation unconformities in the Arbuckle Group. The restrictive definition of unconformity by Vail, Van Wagoner and others excludes all surfaces which are not considered sequence boundaries, such as a parasequence boundary (e.g., top of an upward-shallowing cycle) and all submarine exposure surfaces without significant erosion. These surfaces, espe-

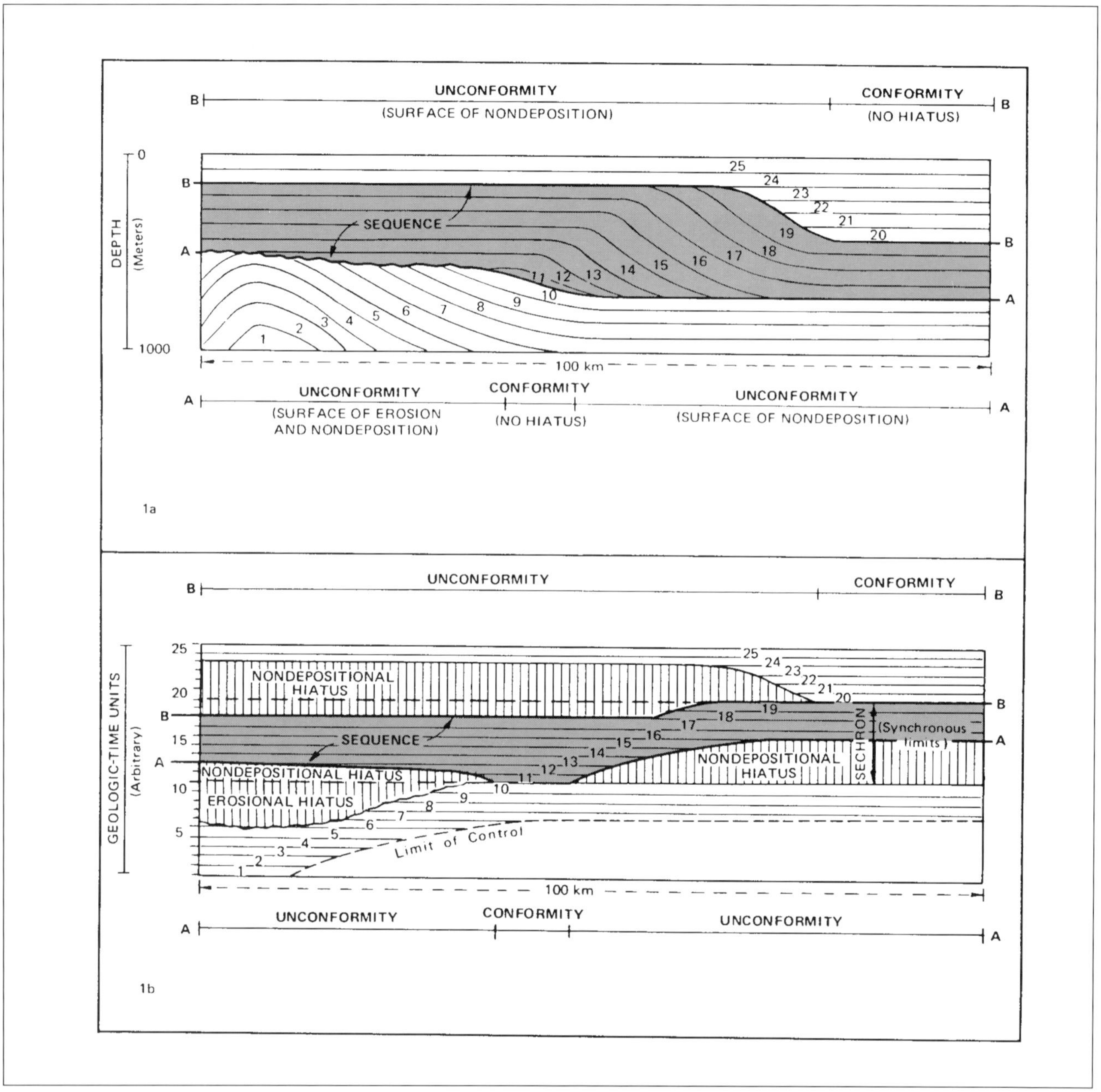

FIGURE 6 — Basic concepts of a depositional sequence: (a) generalized stratigraphic section — the sequence boundary is defined by surfaces A and B which pass laterally from unconformities to correlative conformities (individual units of strata, 1 through 25, are traced by following stratification surfaces); and (b) generalized chronostratigraphic section of sequence stratigraphic relations shown in 1a are replotted with geologic times as the ordinate (Mitchum et al., 1977).

cially parasequence boundaries, can be associated with important diagenetic events affecting carbonate reservoir development. It is for this reason that this study classified Mid-Continent unconformities by scale and order as proposed by Esteban (1992: Fig. 5).

Scale is defined by the lateral extent of an unconformable surface and may be classified as (1) worldwide, (2) interregional, (3) regional, (4) subregional, and (5) local. The order of a stratigraphic discontinuity refers to the time gap especially as defined by sequence stratigraphy such that the classification of order is (1) megaunconformity, (2) superunconformity, (3) sequence unconformity (Vail's

STRATIGRAPHIC SEQUENCES	ORDER	TIME SCALE: SEQUENCE	TIME SCALE: CHRONOSTRATIGRAPHY	TIME SCALE: HIATUS	STRATIGRAPHIC DISCONTINUITIES		
MEGASEQUENCE	1st	400	ERATHEM	200	UNCONFORMITIES	COMPOSITE	MEGA-UNCONFORMITY
SUPERSEQUENCE SET (Sloss Sequence)		200	SYSTEM	>60			
SUPERSEQUENCE (Synthem)	2nd	100	SERIES	30			SUPER-UNCONFORMITY
		20	STAGE	4-12			
DEPOSITIONAL SEQUENCE (Mesothem)	3rd	10 1	BIOZONE	~1		SINGLE	SEQUENCE UNCONFORMITY
PARASEQUENCE (Cyclothem)	4th	0.5 0.2	NOT APPLICABLE	0.01	CONFORMITIES		CYCLE BOUNDARY
BED	5th	0.01		0.001			BEDDING PLANE

FIGURE 7 — Relationship of stratigraphic sequences and sequence order to time gap produced by unconformities

SEQUENCE	FORMATION/GROUP AFFECTED BY UNCONFORMITY	DISCONTINUITY	ORDER	AREA/PRODUCING FIELDS
Absaroka	Intra-Pennsylvanian and Intra-Permian upward cycle	Regional and Boundary of shoaling	3rd and 4th order	Throughout Mid-Continent (Guymon-Hugoton Field)
Kaskaskia	Post-Mississippian	Super-unconformity	2nd to 3rd order	Southeastern Colorado and Southwest Kansas-
Kaskaskia	Pre-Chesterian	Regional to sub-regional unconformity	2nd to 3rd order	Central Oklahoma (Sooner Trend)
Kaskaskia	Post-Osage	Regional to - sub-regional unconformity	3rd order	Northeastern Oklahoma
Tippecanoe	Post-Hunton	Mega-unconformity/ Super-unconformity	1st to 2nd order	Forest City Basin (McClain Field) and Anadarko Basin Northern Shelf Arkoma Basin (Kinta Field)
Tippecanoe	Intra-Hunton	Regional to sub-regional	3rd and 4th order	Anadarko Basin: West Edmond Field and Cambell Field
Tippecanoe	Post-Viola	Regional unconformity	3rd???	???
Sauk	Post-Arbuckle	Mega-unconformity	1st to 2nd order	Central Kansas Uplift
Sauk	Intra-Arbuckle	Regional	3rd	Southern Oklahoma—Cottonwood Creek and Wilburton

TABLE 1 — Major stratigraphic discontinuities in the Mid-Continent.

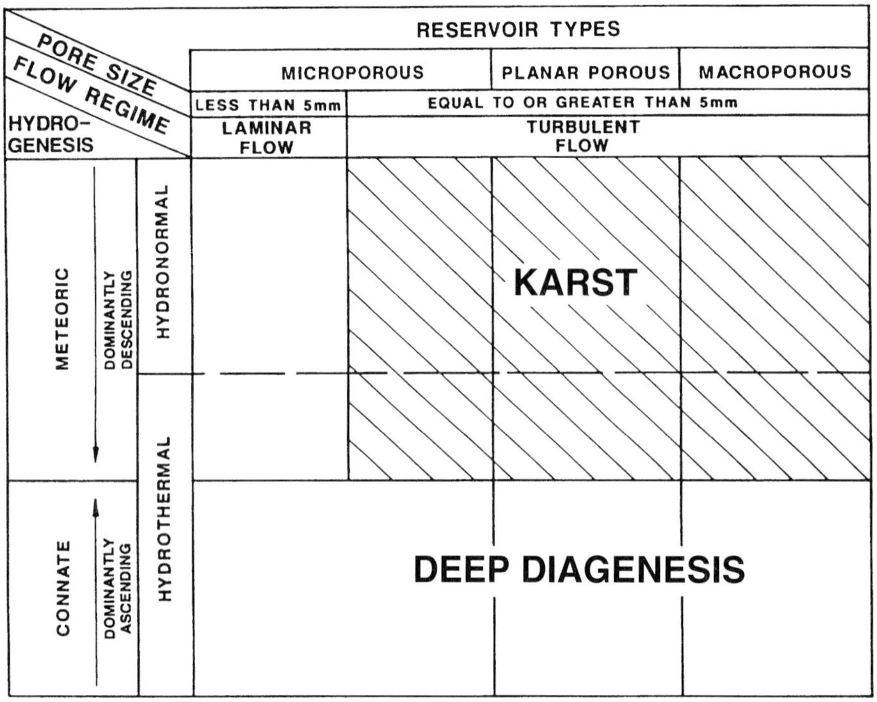

FIGURE 8 — Schematic diagram showing the parameters which define karst and the relationship of karst to meteoric and connate diagenesis.

type 1, type 2 sequence boundaries), (4) parasequence boundary, and (5) bedding plane (Fig. 5). The concept of the magnitude of a hiatus along a sequence boundary can be analogous to the concept of the magnitude of a sequence, or in this case a missing sequence (Mitchum et al., 1977).

Sequence Stratigraphy

Unconformity is one of the basic factors in stratigraphic evaluation and especially sequence stratigraphy. Sloss (1963) defined major stratal sequences bounded by unconformities on the craton, and Mitchum et al. (1977) defined a depositional sequence as "a stratigraphic unit composed of a relatively conformable succession of genetically related strata and bounded at its top and base by unconformities or their correlative conformities" (Fig. 6). Sequences as defined by Mitchum, Vail, and other Exxon and ex-Exxon workers are actually subdivisions of the Sloss sequences which they now define as "supersequences."

Exxon sequences or cycles were initially identified using the resolution of seismic data, and range in duration of time from 1 to 10 m.y. (Mitchum et al., 1977). Sequences are composed of meter to tens of meter-scale depositional cycles or parasequences. A parasequence is defined as a relatively conformable series of genetically related beds bounded by marine-flooding surfaces and their correlative surfaces; parasequences are progradational and the beds within the parasequence shoal upward (Van Wagoner et al., 1987). Sometimes recognizable are parasequence sets which are composed of "bundles" of parasequences which form distinct stacking patterns, such as those observed in the Arbuckle Group in southern Oklahoma.

According to Mitchum et al. (1977), three major orders of cycles can be recognized on global sea-level curves; based on work by Fischer (1964), Goldhammer (1987), Read (1989), and Van Wagoner

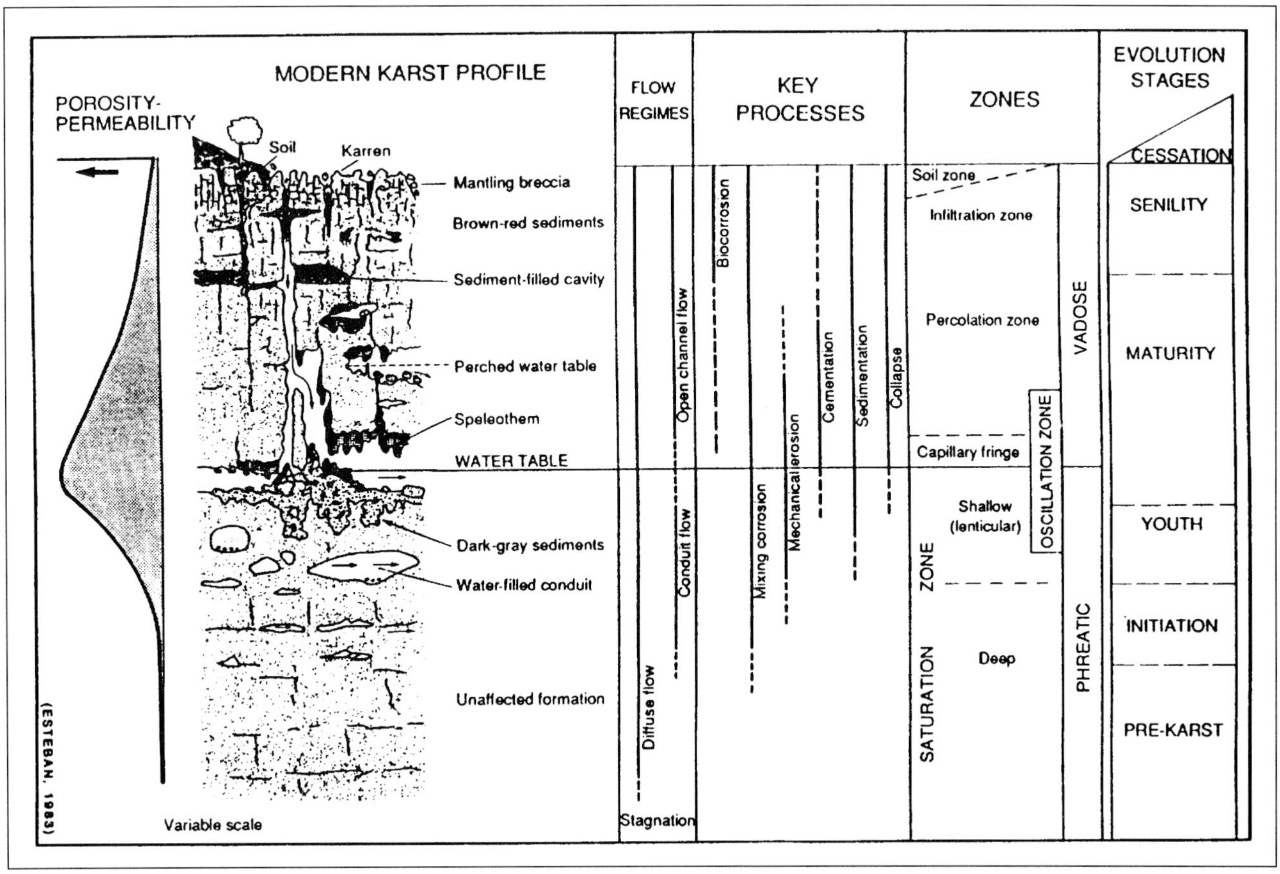

FIGURE 9 — Karst profile showing maturity hierarchy (Esteban and Klappa, 1983).

et al. (1990), at least two minor orders of cycles can be recognized. Comprehensively, these cycles can be referred to as first-, second-, third-, fourth-, and fifth-order cycles as shown in Fig. 7. The last two are considered parasequences and may be related to Milankovitch cyclicity, tectonic activity, magnitude and frequency of subsidence, sedimentation rate, depositional processes (e.g., progradation), and time and amplitude of each transgressive-regressive sequence (Read and Goldhammer, 1988; Hardie et al., 1991; Read et al., 1991). Table 1 shows the magnitude and related order of the most recognized sequence boundaries in the Mid-Continent.

It is important to understand that the concept of sequence stratigraphy was primarily derived from evaluation of siliciclastic deposits. Clearly, carbonate deposition is significantly different from siliciclastic deposition; i.e., the Exxon model derives most sediments from outside of a basin (extrabasinal), and most carbonate sediments are produced in a marine basin (intrabasinal) by organic and inorganic processes (Loucks and Sarg, 1993). Although the basic premises of the Exxon depositional sequence model will be the same, e.g., eustasy, unconformities, sequences, etc., modifications must be made in some of the interpretations due to the manner in which carbonates respond to processes such as eustasy. Handford and Loucks (1993) described a list of the infrastructual elements that must be considered when evaluating carbonate sequence stratigraphy. These elements are eustasy, subsidence, tectonics, continental linkage, margin type and orientation, climate, depositional system and related carbonate factory, morphology, width, circulation, and lithology.

FIGURE 10 — Hunton outcrop in north-central Arkansas along the Buffalo River. Arrow "a" points to the angular unconformity between the Hunton Group Henryhouse Formation/Chimneyhill Subgroup and Sylamore Sandstone. Arrow "b" points to the contact between the Sylamore Sandstone and overlying Mississippian carbonates.

FIGURE 11 — Collapse karst breccia in the uppermost Kindblade Formation along I-35 in the Arbuckle Mountains of southern Oklahoma (outcrop is approximately 25 ft high).

Diagenesis

The emphasis of this study is on unconformity associated with exposure surface and the relationship of the exposure surface to underlying strata. Exposure surfaces are typically the upper bounding surfaces of rock which show the results of exposure that is either terrestrial, coastal, or subaqueous, or combinations of the above. Diagenetic processes include corrosion, corrasion, cementation, sedimentation, and compaction (Esteban, 1992). Controls include fluid input and output, climate, permeability pathways, duration of exposure, hydrology, water chemistry, and geometry of exposure surface (Esteban, 1992; Saller and Yaremko, 1994).

Karst is one of the most often recognized diagenetic products associated with unconformities. Karst can be defined as subaerial terrain which results from macroscopic and microscopic diagenesis of carbonate strata in the surface or subsurface produced by chemical dissolution and associated physical modification as the result of turbulent flow of meteoric water in pores greater than 5 mm (modified from James and Choquette, 1987: Fig. 8). A karst profile is composed of two primary zones: (1) vadose with infiltration and percolation above the water table, and (2) phreatic with hydraulic dissolution and erosion below the water table (Esteban and Klappa, 1983: Fig. 9). Karst diagenesis is not restricted to the vadose zone and can extend deep into the phreatic zone. Modification of strata can reach a few inches to hundreds of feet below the surface. Karst can be porosity constructive or destructive, and young or juvenile karst terrains are often more porous and permeable than more mature karst terrains. This is because mature karst is more susceptible to cementation, sedimentation, collapse, and compaction.

Paleokarst is defined as ancient karst and can be classified as incipient, single event or composite. Incipient refers to exposure during deposition, while single event and composite refer to the number of episodes to which a carbonate has been subjected to karstification. In the Mid-Continent, most carbonates associated with exposure surfaces have been subjected to karst evolution and are composite in nature. Paleosoils, breccias, and dissolution features are typical karst products associated with unconformities.

Dolomitization may also be associated with exposure surfaces, diastems in deposition, and unconformities. Peritidal and ramp carbonates of the Arbuckle and Hunton groups are extensively dolomitized during early diagenesis as a result of subaerial conditions and hypersalinity. During interregional, regional, and subregional unconformities, additional modification by meteoric waters may have occurred (e.g., continued dolomitization in a mixed water lens, or dissolution [porosity formation to karstification], or cementation of the reservoir and occlusion of porosity)

RECOGNITION OF UNCONFORMITIES

There are six common methods that are used to identify unconformities: (1) outcrop and core analysis, (2) biostratigraphy, (3) well log analyses and correlation, (4) seismic interpretation, (5) sequence evaluation, and (6) basin modeling. Other methods include standard age dating techniques, stable isotope geochemistry, and paleomagnetic reversal correlations (Mitchum et al., 1977; Saller and Yaremko, 1994; Budd et al., 1995).

Outcrop and Core Analysis

The oldest method of identifying unconformities is outcrop analysis. Since James Hutton's description of the now famous unconformity at Siccar Point, Scotland, in 1788, workers have been studying outcrops to define stratigraphic discontinuities. Outcrop and core evaluations are two of the most reliable methods to recognize subaerial exposure in carbonates (Saller and Yaremko, 1994). Combined, outcrops and core studies can provide a good understanding of the geometry of an unconformity and related diagenetic events. Unfortunately, unconformities do not always leave recognizable diagenetic changes; in some cases, karst forms well below its associated unconformity surface. In addition, some diagenetic alterations, such as dissolution by brecciation and dolomitization, are related to connate water and are unrelated to subaerial exposure (Lynch and Al-Shaieb, 1991).

An example of the most obvious method of recognizing unconformities is found at the contact between the Hunton Group and overlying strata. In the Ozark uplift, the Woodford is thin or absent, so the pre-Woodford superunconformity is recognized by the angular relationship of the overlying Sylamore (Misener) Sandstone to the underlying Hunton Henryhouse Formation and Chimneyhill Subgroup (Fig. 10). Closer inspection reveals the presence of Sylamore cave fills in large Hunton solution cavities which developed prior to and during Sylamore deposition.

The Arbuckle and Wichita mountains are the best locales in Oklahoma to examine various types of unconformities in outcrops. Multiple unconformities, from interregional to subregional, are present, and the most studied and recognizable unconformity surfaces are those that formed during Pennsyl-

FIGURE 12 — Karst breccia and fill in Hunton Chimneyhill Subgroup for Sun Kilcrease No. 11 well, Section 36 T5N-R7E, Pontotoc County, Oklahoma.

FIGURE 13 — Porosity development in the Henryhouse Formation in core at 14,645 ft in the SunRay-DX Frans No. 1 well, Section 3 T15N-R16W, Custer County, Oklahoma.

vanian Wichita and Arbuckle orogenies (Table 1).

Recognition of unconformities associated with the major Pennsylvanian tectonic events is relatively easy due to the relationship of the underlying beds to overlying strata, i.e., angular unconformities. Karst terrains are often well developed and highly evolved; Fig. 11 shows a collapse megabreccia in the upper part of the Arbuckle Group. This example is probably a good case for karst evolution as this breccia lies immediately below the Early Orodovician regional intraformational unconformity between the Kindblade Formation and the overlying West Spring Creek Formation. This intraformational unconformity is associated with the karstic and dolomitic "Brown Zone" which is the primary Arbuckle reservoir. The collapse breccia of Fig. 11 also lies immediately below a Pennsylvanian unconformity that is interpreted to be the event during which most of the brecciation and solution cavities formed.

Unconformities are more difficult to evaluate in core than outcrop due to the lack of lateral geometry. Stratigraphic relationships and, of course, karst terrain are the two primary features to evaluate. For example, in the Hunton Group in the Anadarko and Arkoma basins, paleokarst is present in many cores immediately below the Hunton-Woodford contact (Fig. 12). In addition, and perhaps more important for reservoir development, well-developed porosity profiles are observed in cored intervals immediately below the Woodford unconformity that are associated with young or juvenile, phreatic karst development (Fig. 13).

Biostratigraphy

The second oldest method of recognizing unconformities is the use of the fossil record to identify time gaps. Often the reduction or extinction of a species can be evidence of a time gap. The problem with use of the fossil record exclusively is that there can be other explanations for a gap in the fossil record, such as a change in climate, geochemistry, faulting, etc. Also, the duration of many time gaps is less than the resolution of paleontological methods or even methods such as absolute age dating (Saller and Yaremko, 1994).

Early studies in the Mid-Continent used macrofossils to identify the relative age of rocks for correlation. For example, the unconformity between the Cambrian and the Ordovician can be identified by species extinction of common trilobites (Stitt, 1983: Fig. 14).

More recently, conodonts have been one of the most popular means of establishing sequences and time gaps. Sweet (1992) collected conodonts from cored wells across Kansas and outcrops in Oklahoma to study the biostratigraphy of the Upper Ordovician (Fig. 15a). Sweet (1984) used conodonts along with ostracodes and macrofossils to identify a major hiatus between the Oil Creek Formation and overlying McLish Formation (Fig. 15b). In 1991, Derby used conodonts to define uncon-

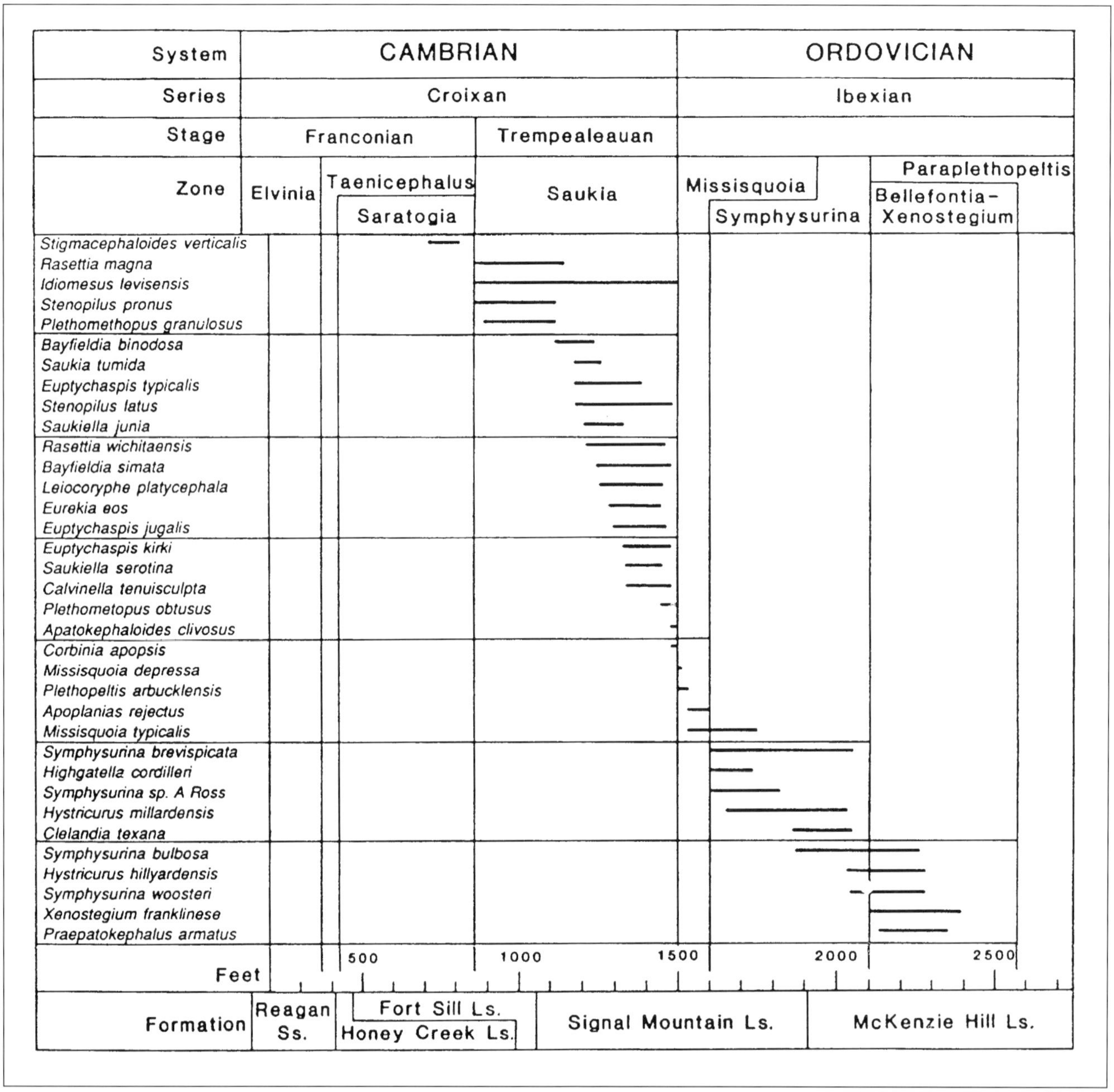

FIGURE 14 — Composite ranges of the most common trilobites along the Cambro-Ordovician boundary (Stitt, 1983).

formities in the West Spring Creek Formation below the contact of the Arbuckle Group with the overlying Joins Formation.

Regional to subregional stratigraphic discontinuities can also be identified, e.g., those at the top and middle of the Viola Group which are based on identification and correlation by various workers of graptolites, chitinozoa, and conodonts (Gentile, 1982).

Well Log Analysis and Correlation

Wireline data, especially gamma ray and porosity logs, can provide direct observation of diagenetic features associated with an unconformity. The gamma ray log, which measures shaliness of the

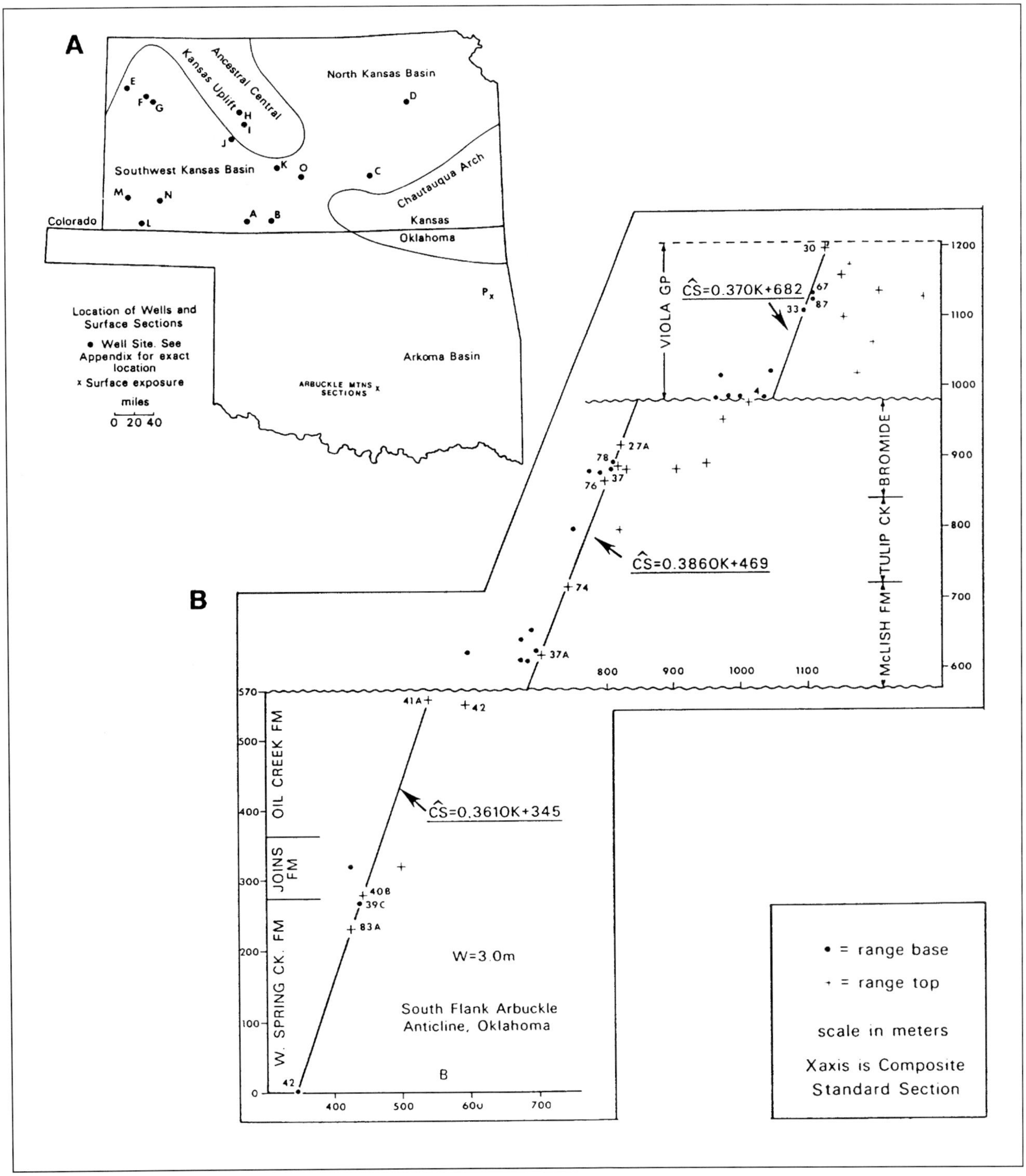

FIGURE 15 — Summary of work by Sweet (1984, 1992): (a) location of sampled sections and cored wells for conodonts study (Sweet, 1984), and (b) graphic correlation by Sweet (1984) with his composite standard section (CSS) showing the magnitude of unconformities in the Middle and Late Ordovician of Oklahoma. Individual species are indicated by two-digit numbers, some with alphabetic suffix (e.g., 37, 40B), and the pattern of tops are shown by plus signs (+).

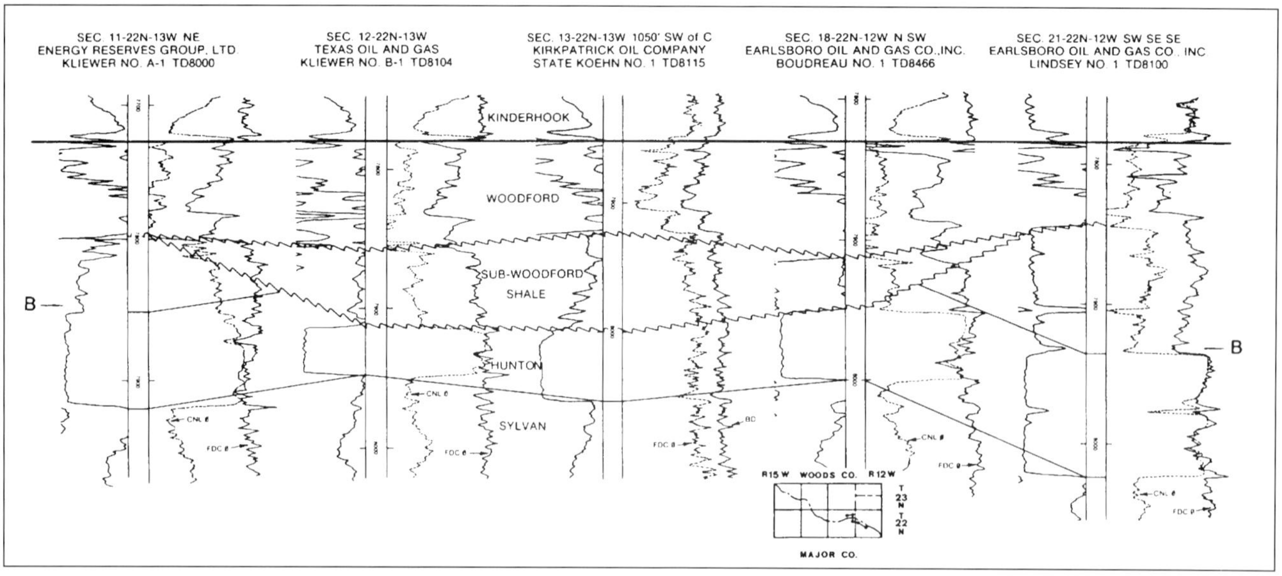

FIGURE 16 — Hunton/Woodford/Misener cross section in Major County, Oklahoma. Note the increase in porosity on logs adjacent to the pre-Woodford channel.

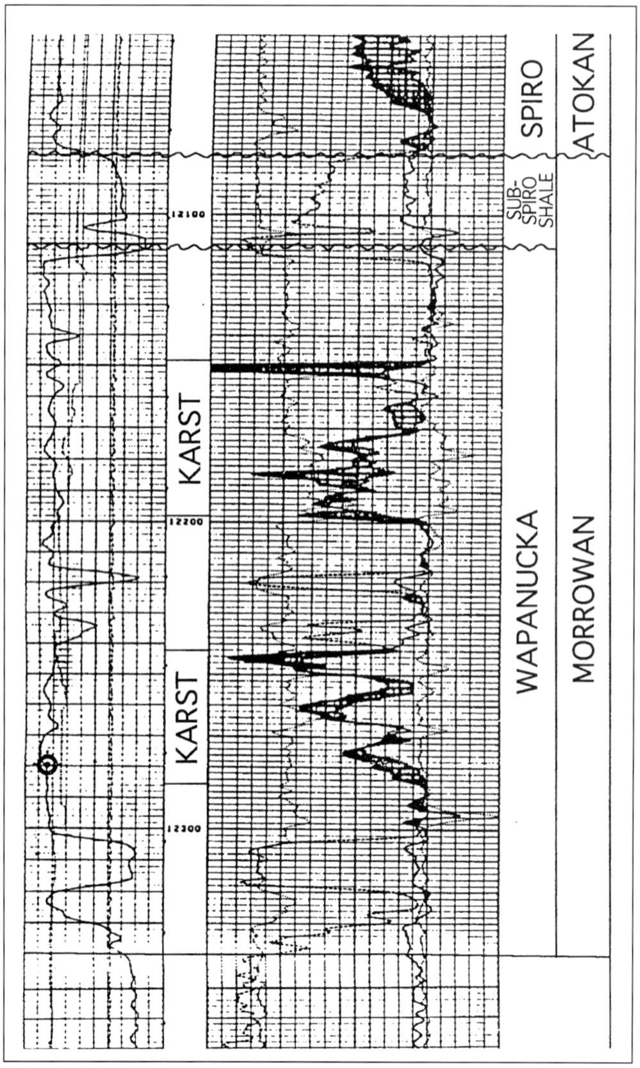

FIGURE 17 — Induction log of the Amoco Zipperer Unit No. 1 well showing karst development at depth in Section 32 T4N-R17E, Pittsburg County, Oklahoma.

rock, typically responds to collapse zones where internal sediments cause an increase in radioactivity. Most logs, including the formation density and neutron logs, show an irregular or "ratty" profile due to the chaotic nature of karst breccias. Formation density and neutron logs can also reveal a significant increase in porosity associated with the unconformity.

An example is the recognition of unconformable surfaces from the Hunton in north-central Oklahoma, where pre-Kinderhookian erosion developed valleys for later Misener Sandstone or equivalent "sub-Woodford Shale" deposition by nearshore marine processes. Figure 16 is a log cross section showing a valley fill of sub-Woodford Shale and the effect karst development has on gamma ray and porosity logs as described above. The gamma ray has a typical irregular profile from the top of the Hunton down to what is often designated the "B-marker," or the base of mature paleokarst. Also note the change in density-neutron response which shows a marked increase in porosity.

Figure 17 is a log of the Wapanucka Limestone in the Arkoma basin that has been interpreted as representing karst terrain. Dissolution associated with karst is represented by the areas of high porosity. This log and the identification of karst are unusual because the Wapanucka outcrop and core typically show very little effect from the overlying pre-Atokan unconformity.

Stratigraphic cross sections are arguably the best method for modeling the geometry of stratigraphic discontinuities. Consequently, the unconformable surface must be a correlatable event, i.e., recognizable on well logs. The unconformity can be relatively easy to recognize due to angularity or stratigraphic relationships with overlying strata, or it can be very subtle, requiring regional cross sections to identify truncation.

An example of an obvious unconformable relationship is the supersequence boundary between the Kaskaskia sequence (Mississippian) and the Absaroka sequence (Pennsylvanian) in the Sooner Trend of northwest Oklahoma (Fig. 18). Another example is the Middle Ordovician regional unconformity in Kansas where it truncates the Arbuckle (Fig. 19).

A more subtle example is the truncation of the Arbuckle Kindblade Formation by an intraformational (third-order) unconformity (Fig. 20). Although this unconformity shows very little truncation on a local scale, regionally it is a correlatable event and defines the top of a major third-order cycle within the Arbuckle Group.

Seismic Interpretation

Seismic data can be an important tool in the recognition of significant regional unconformities. Prior to the introduction of seismic sequence stratigraphy, geoscientists identified major unconformities on seismic lines primarily due to angular relationships between seismic events. Recognition of karst terrain is more difficult; however, good quality seismic combined with some well control and modeling can identify paleokarst, especially if the unconformity surface and related karst terrain have "ragged" profiles. New three-dimensional (3-D) seismic can further enhance recognition of unconformity surfaces by providing paleo-surface maps.

Unfortunately, in the Mid-Continent, except for southern Oklahoma, most of the unconformities are not very angular or ragged and do not show dramatic seismic response. In southern Oklahoma, structural complexity also makes it difficult to identify unconformable surfaces. Figure 21 is a very subtle example of the pre-Woodford unconformity from a line just west of the Oklahoma City field. From left to right on the seismic line (within the area colored in yellow), truncation of the Hunton and associated strata is shown by the loss of signature.

New 3-D seismic methods can also enhance recognition of exposure surfaces. Figure 22 shows an example of 3-D seismic for the pre-Woodford unconformity surface in the so-called "triangle zone" of the Arkoma basin (Valderrama et al., 1994).

Sequence Evaluation

Although a relatively new means of recognizing unconformities, modern sequence stratigraphy is one of the best methods to analyze time gaps in the geologic record. Sequence analysis methods which can be used to recognize unconformable surfaces include eustatic sea-level curves and various cycle stacking techniques, such as Fischer plots.

Figure 23 shows the projected rise and fall of sea-level curves in the Mid-Continent during the Paleozoic. The accuracy and resolution for sea-level curves in the Paleozoic are poor, and this diagram shows only the major unconformities. Recent refinements of parasequences or cyclothems (transgressive cycles) of the Middle and Upper Pennsylvanian are important developments and will help improve the Paleozoic eustatic sea-level curve (Heckel, 1986).

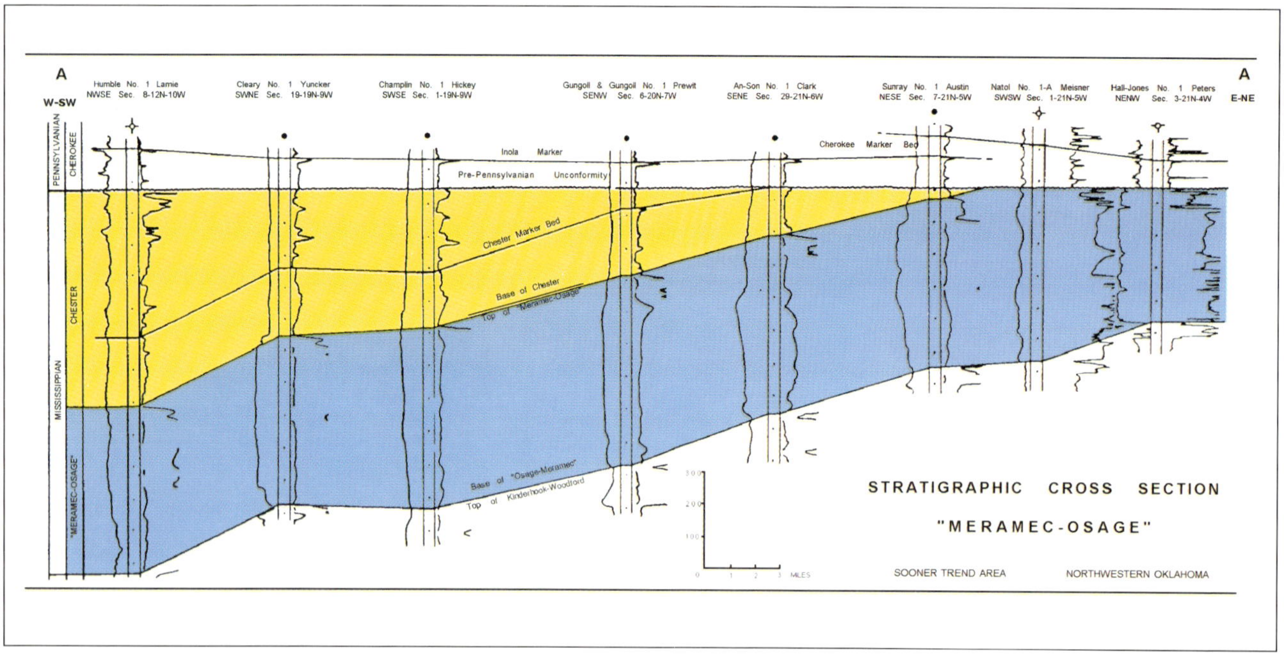

FIGURE 18 — Stratigraphic cross section of the Mississippian across the Sooner Trend (Harris, 1975).

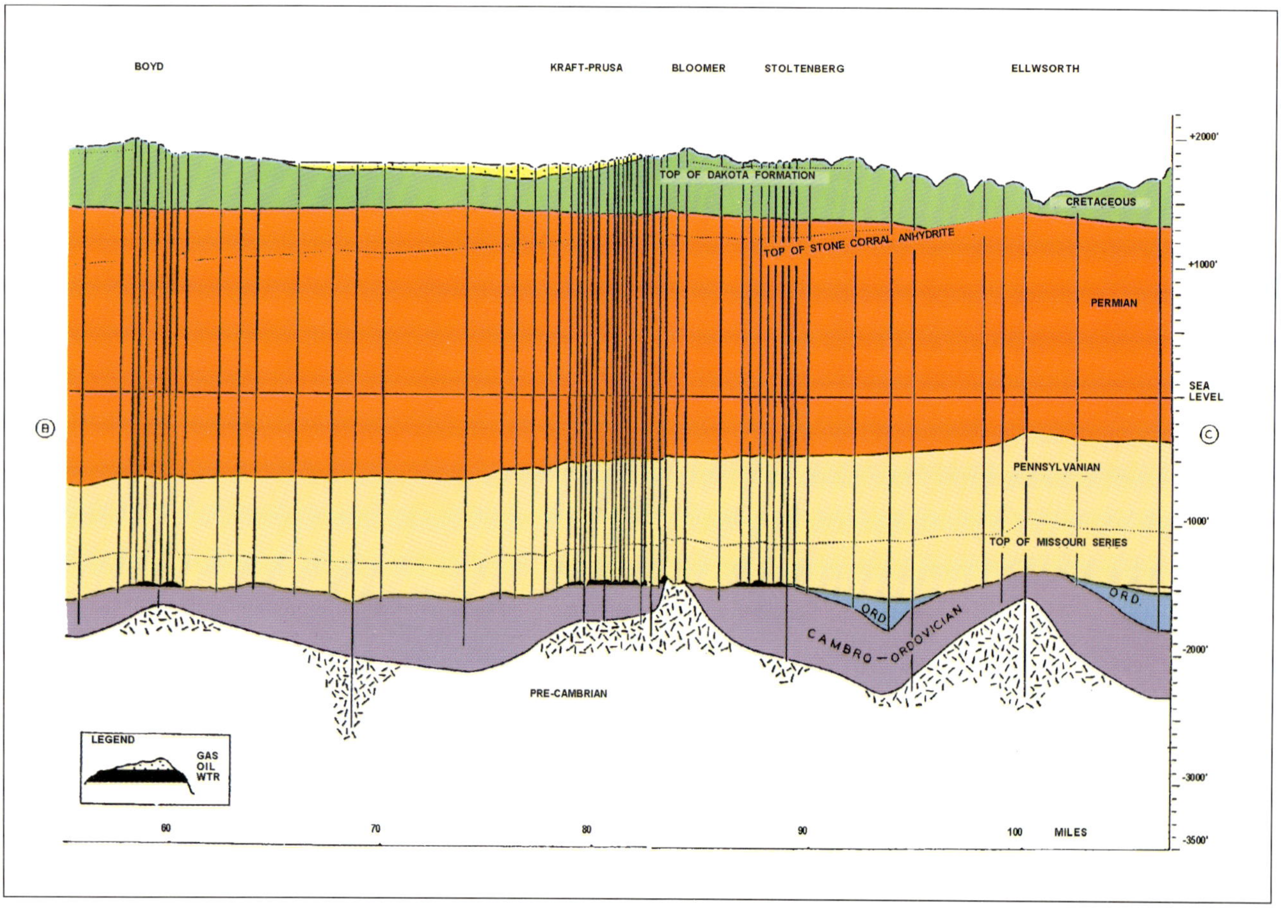

FIGURE 19 — Cross section across Boyd and Kraft-Pusa fields of Barton County, Stoltenberg field of Ellsworth County, and Ellsworth anticline in Kansas (Walters, 1958).

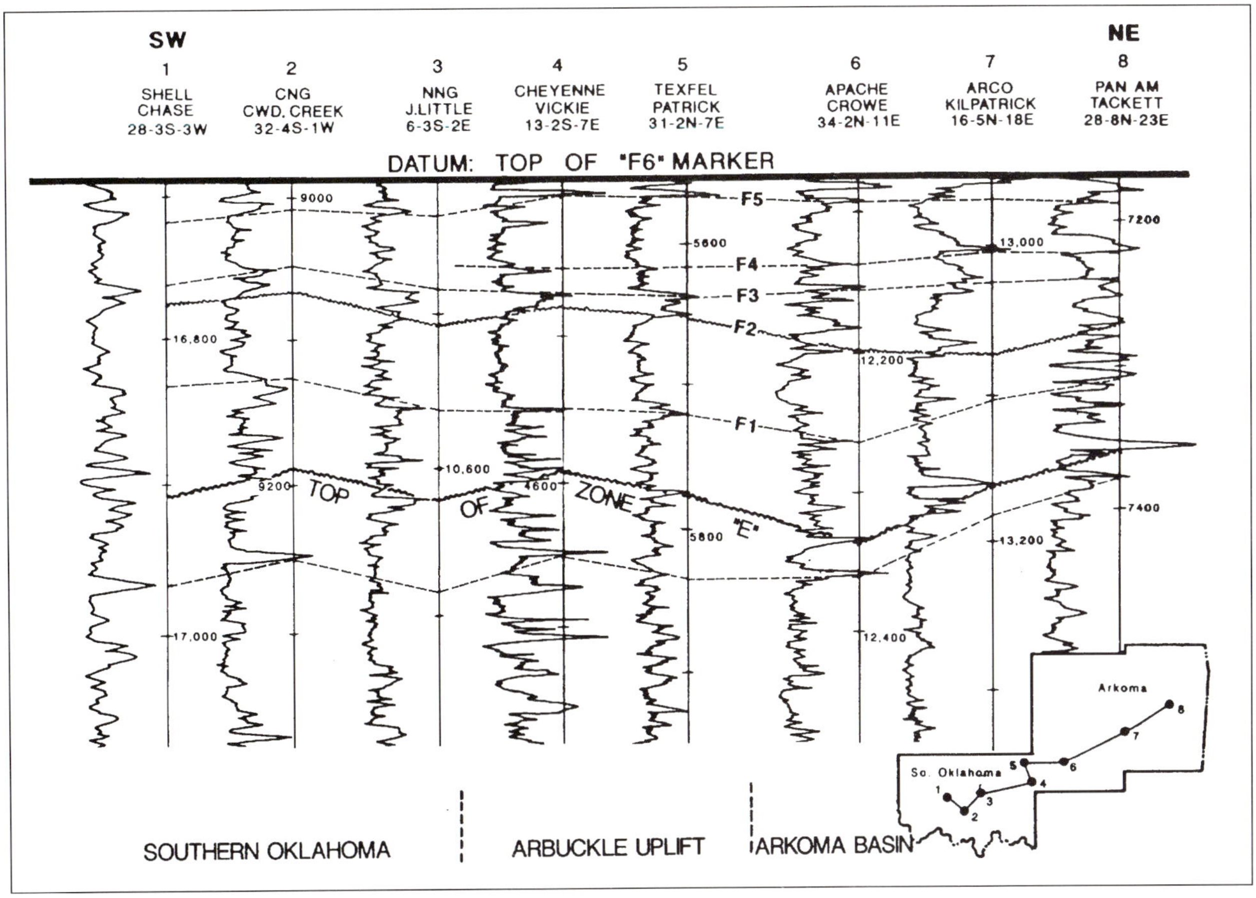

FIGURE 20 — Regional southwest to northeast stratigraphic cross section from southern Oklahoma to the Arkoma basin showing the regional unconformity at the top of the Kindblade Formation of the Arbuckle Group.

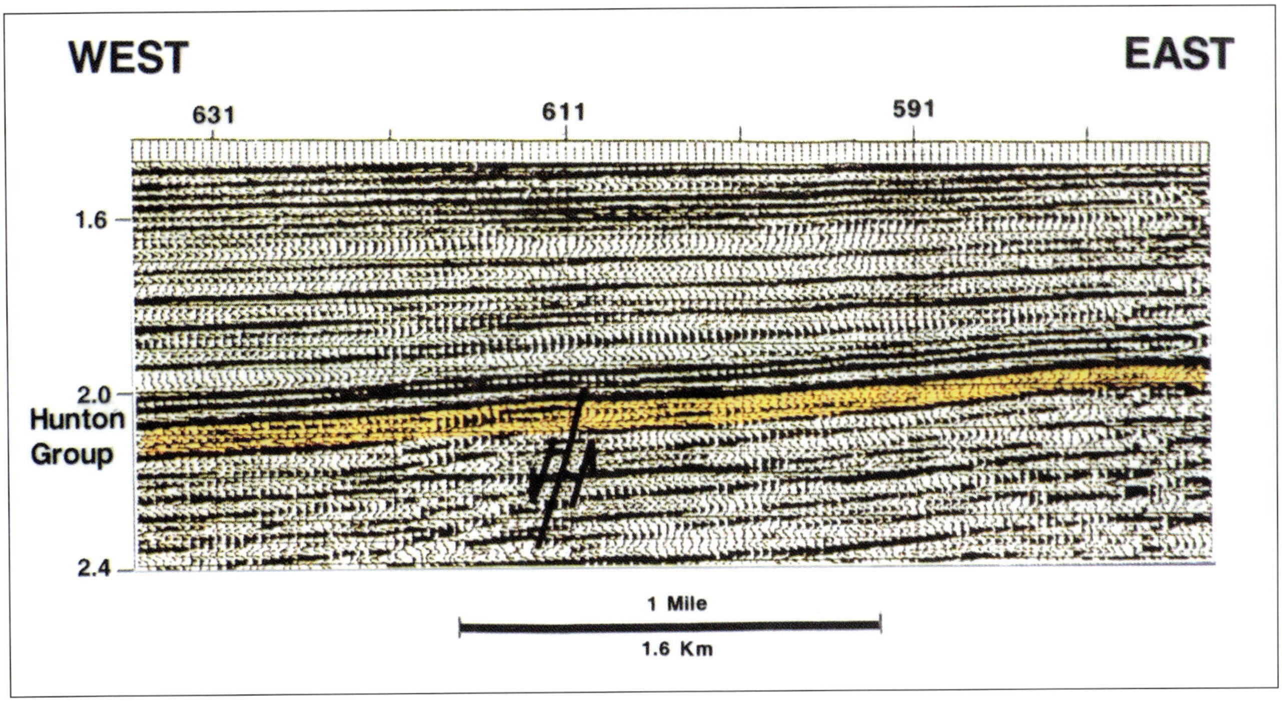

FIGURE 21 — East-west seismic line (No. 218 ext) across Hunton field in Canadian County, Oklahoma. Pre-Woodford unconformity is at the top of the yellow (Morgan et al., 1982).

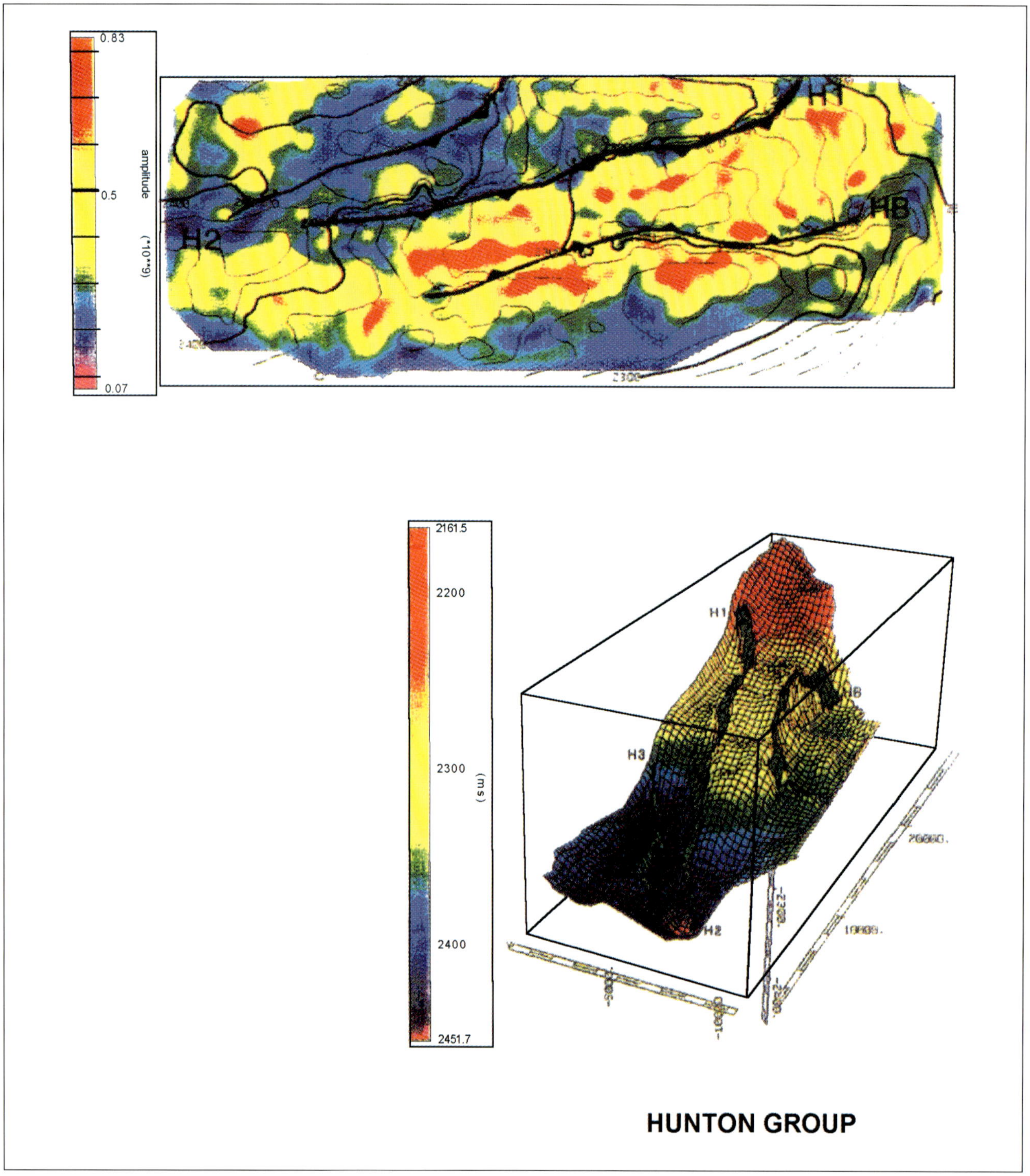

FIGURE 22 — Example of 3-D seismic in the Arkoma basin: (a) amplitude and time-contour maps, and (b) 3-D perspective of Hunton surface (Valderrama et al., 1994).

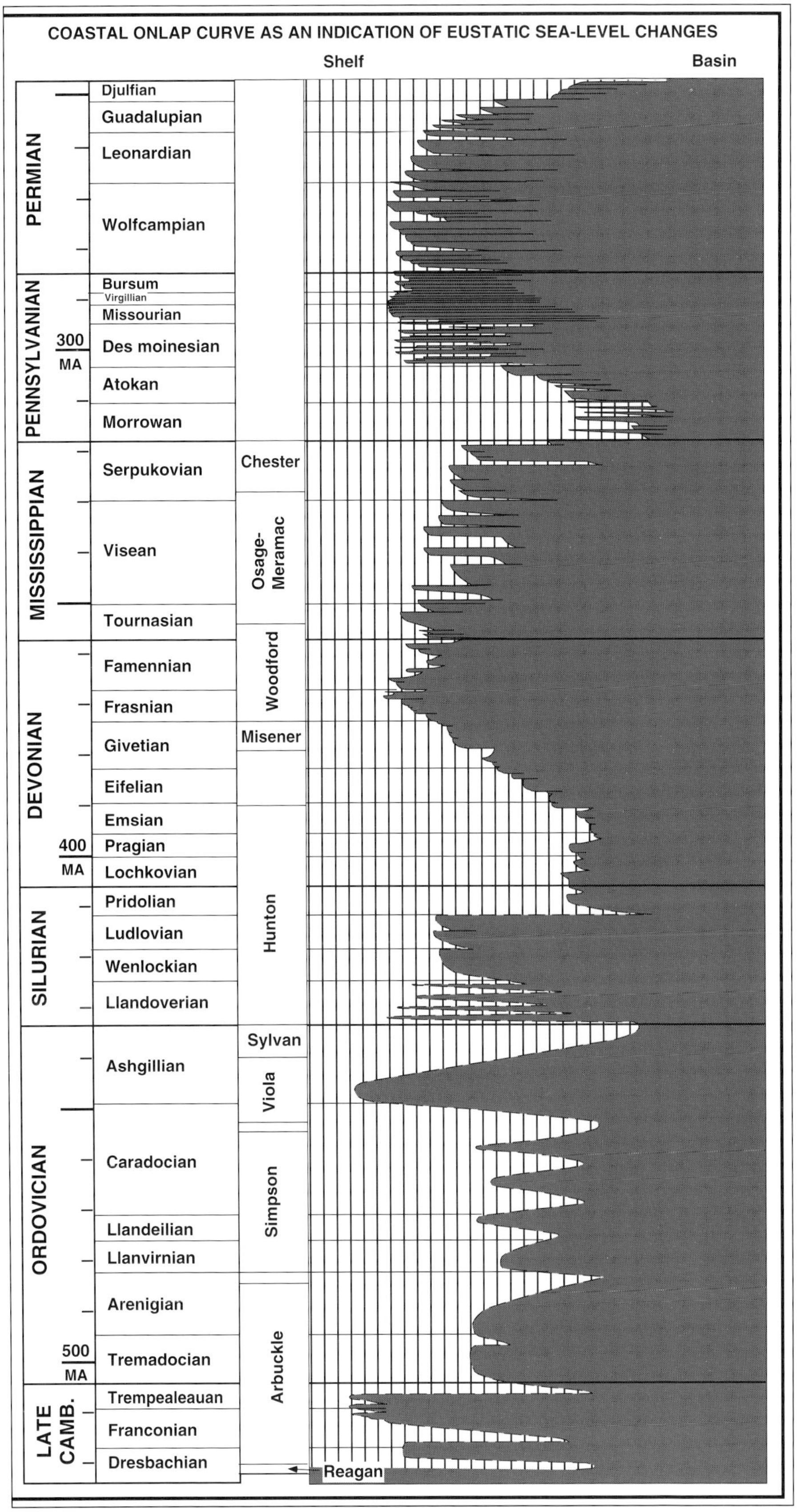

FIGURE 23 — Stratigraphic column showing Sloss sequences and eustatic sea-level changes for the Paleozoic of the Mid-Continent (Shelton and Gerken, 1995 [this volume]).

In the Permian of the Hugoton embayment, similar parasequences or cyclothems of transgressive and regressive units developed in the Wolfcampian (Caldwell, 1991). The unconformities associated with the regressive units of the Chase Group (paleoslides and tidal flat deposits) resulted in the dolomitization of the underlying carbonates by hypersaline brines. In the regressive phase, unconformities may develop that can be identified by soil profiles and desiccation features.

Recognition of cycle stacking patterns has been advanced recently by Read and Goldhammer (1988) by the use of a graphic technique suggested by Fischer (1964). In the Arbuckle Group of Oklahoma, this technique has helped to identify the location of intraformation unconformities (Fig. 24). Cycle thickness is plotted against time with resulting cycles showing overall regression or highstand and thin cycles showing overall transgression or

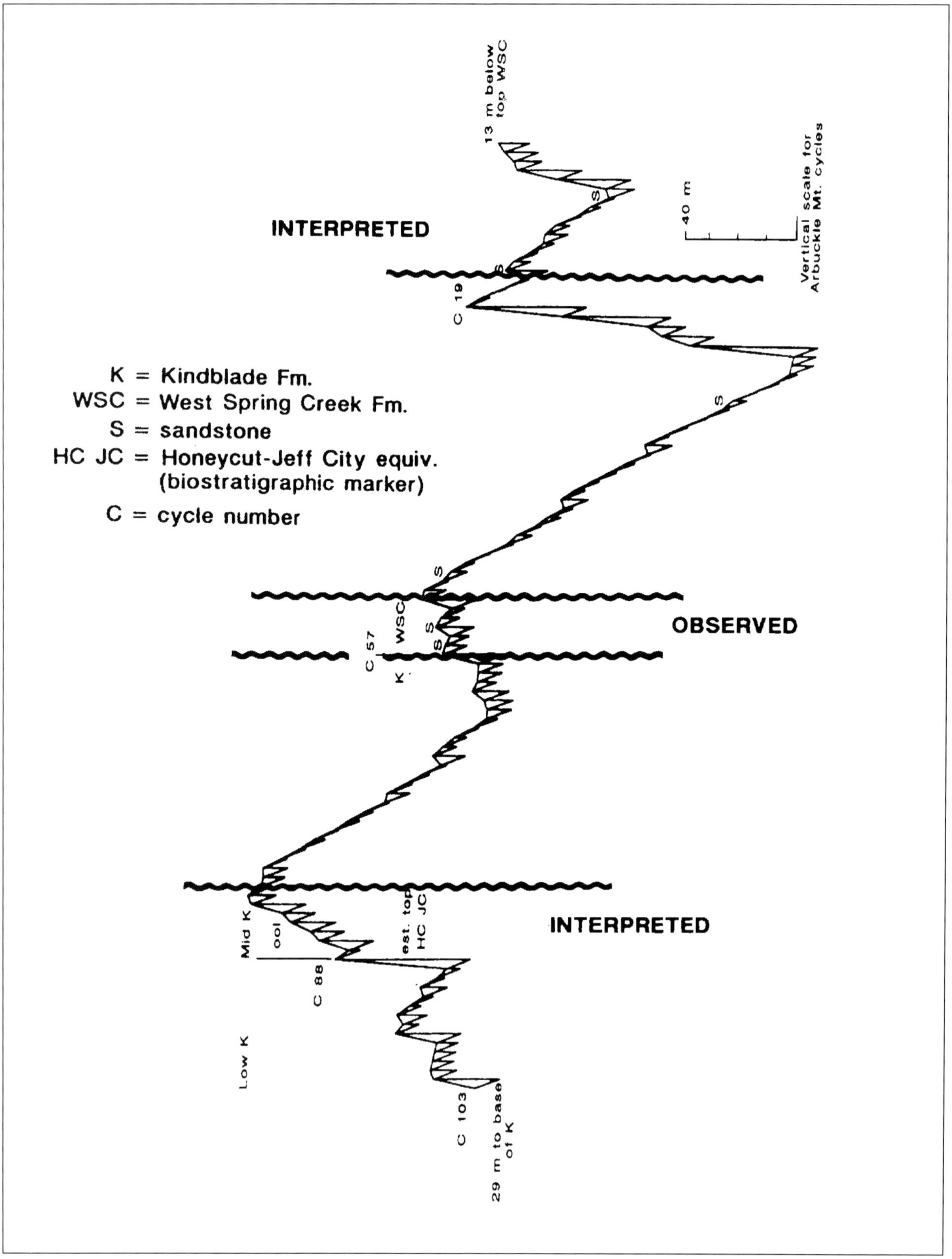

FIGURE 24 — Fischer plot developed from outcrop descriptions of the Arbuckle Group along I-35 in the Arbuckle uplift (the plot is shown with cycle thickness as the abscissa).

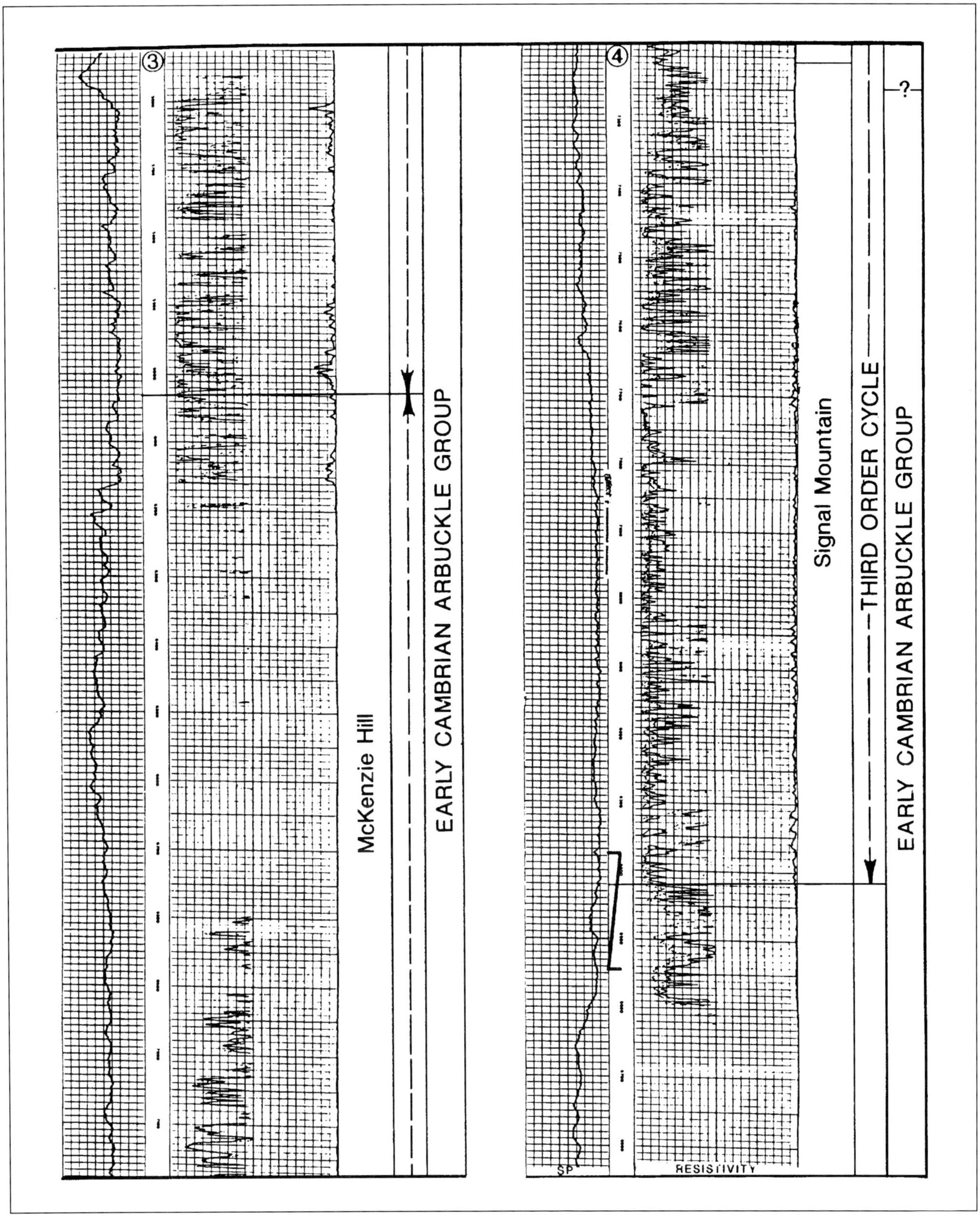

FIGURE 25 — Wireline log from the ARCO Healdton Deep Unit-Mullen No. 1 in the Healdton field (Section 3 T4S-R3W). Thin intertidally dominated cycles are shown by 10-ft resistivity spikes at the base of the Signal Mountain Formation, while thick subtidally dominated cycles are shown by off-scale resistivity at the top of the McKenzie Hill Formation.

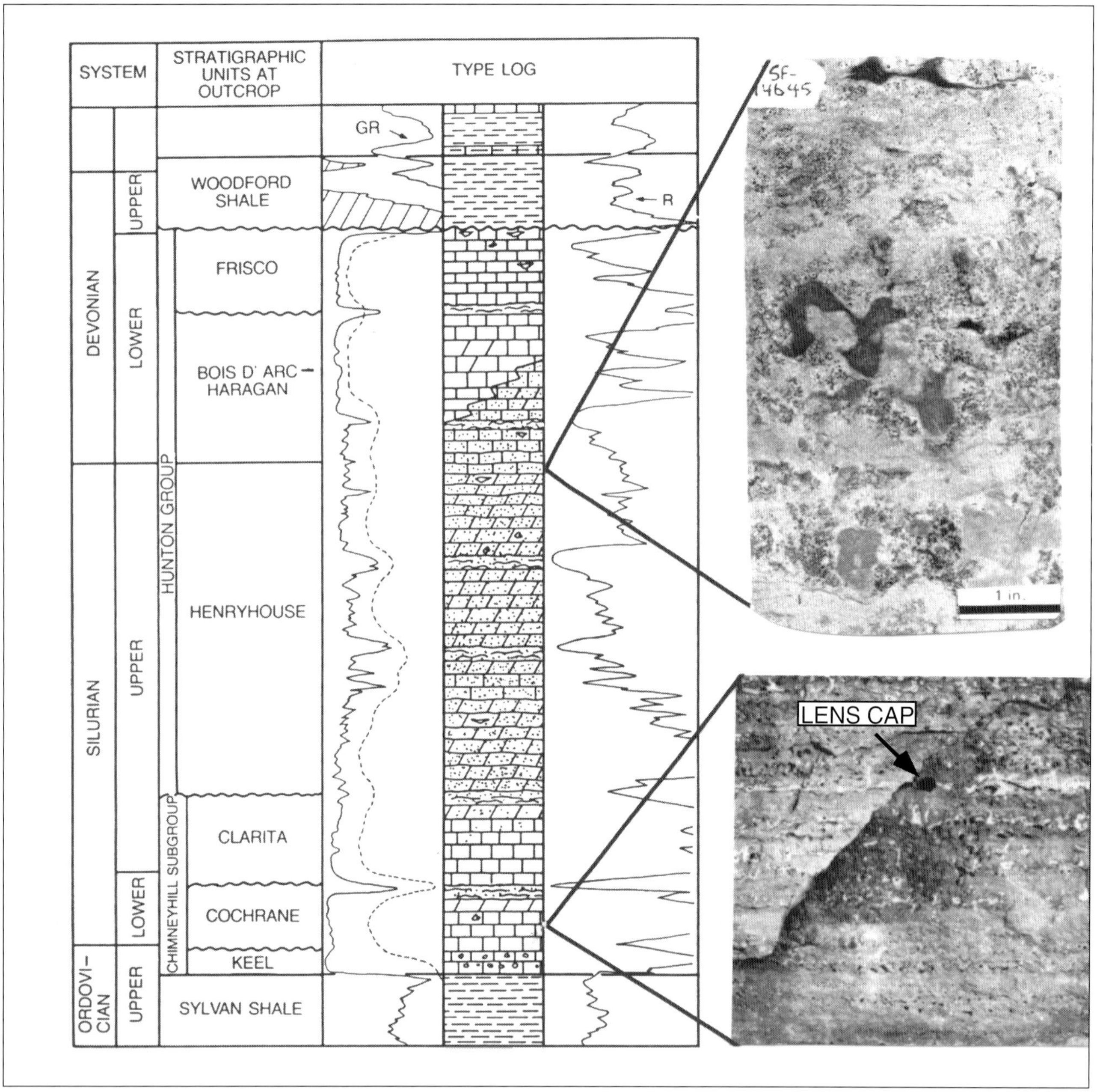

FIGURE 26 — Type log showing stratigraphic subdivisions of the Hunton Group and photographs of Hunton paleokarst: (a) core from the Henryhouse Formation in the SunRay-DX Frans No. 1 well, Section 3 T5N-R16W, North Custer field, and (b) outcrop from the Cochrane (Brassfield) Formation along the Buffalo River near the town of Gilbert in north-central Arkansas.

lowstand. Exposure surfaces can be predicted from the position of maximum flooding surfaces.

Fischer plots are typically generated from outcrop data; however, in some cases where cycles or cycle patterns can be recognized on wireline logs, similar evaluations are possible. This is especially true of the Arbuckle in southern Oklahoma where peritidal cycles can be recognized from gamma ray and resistivity patterns caused by the amount of windblown clastic, silt, and clay in the upper intertidal and supratidal facies. Figure 25 is a log from the ARCO-Healdton Deep Unit-Mullen No. 1 in the Healdton field (Section 3 T4S-R3W), Carter County, Oklahoma, which shows thin (roughly 10 ft) inter-

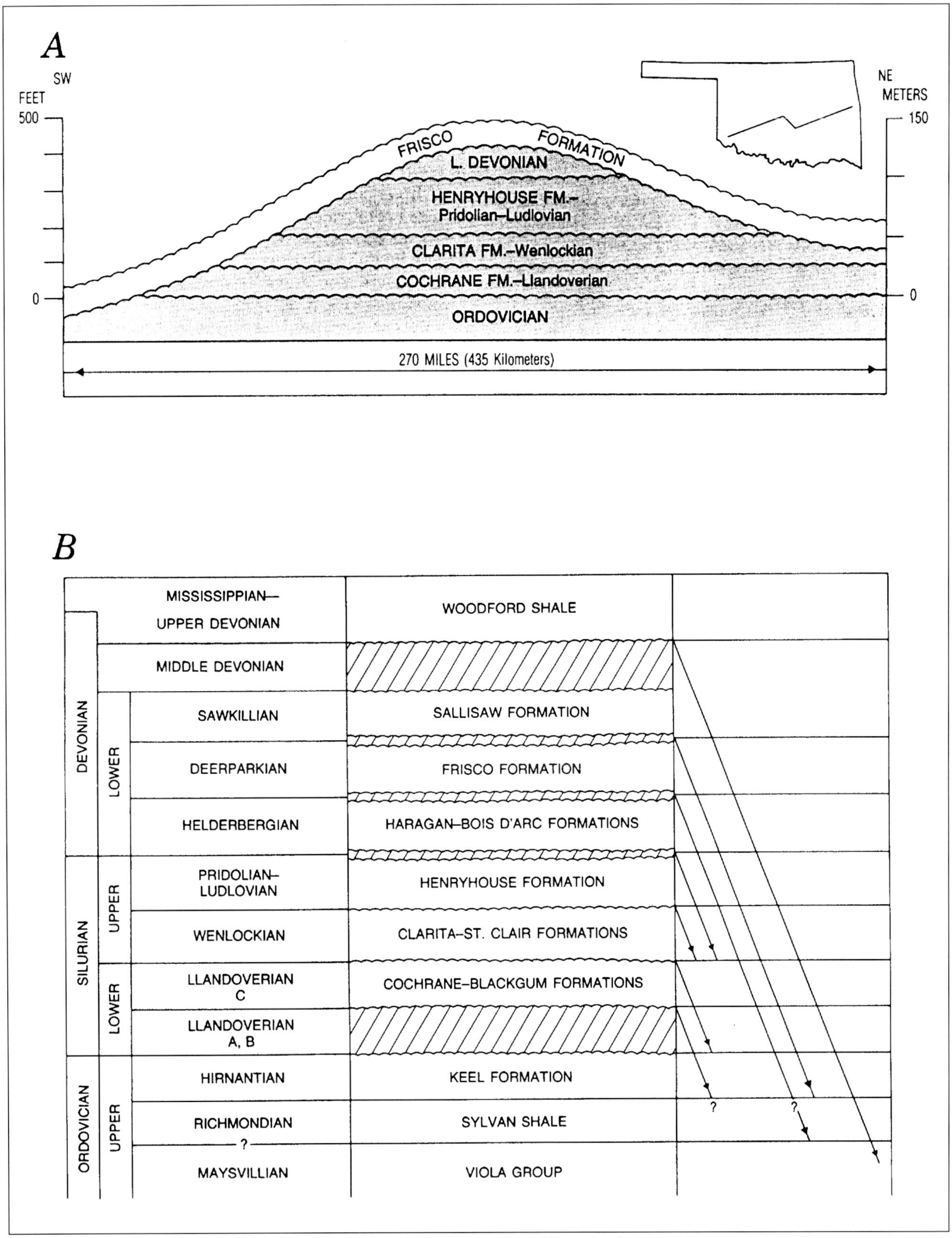

FIGURE 27 — Diagrammatic representations based on brachiopod and conodont data within the Hunton Group: (a) stratigraphic section showing the unconformable relationship between the Frisco Formation and the underlying Devonian and Silurian strata, and (b) chart showing the stratigraphic division recognized in Oklahoma and adjacent areas with arrow indicating the known point of maximum truncation.

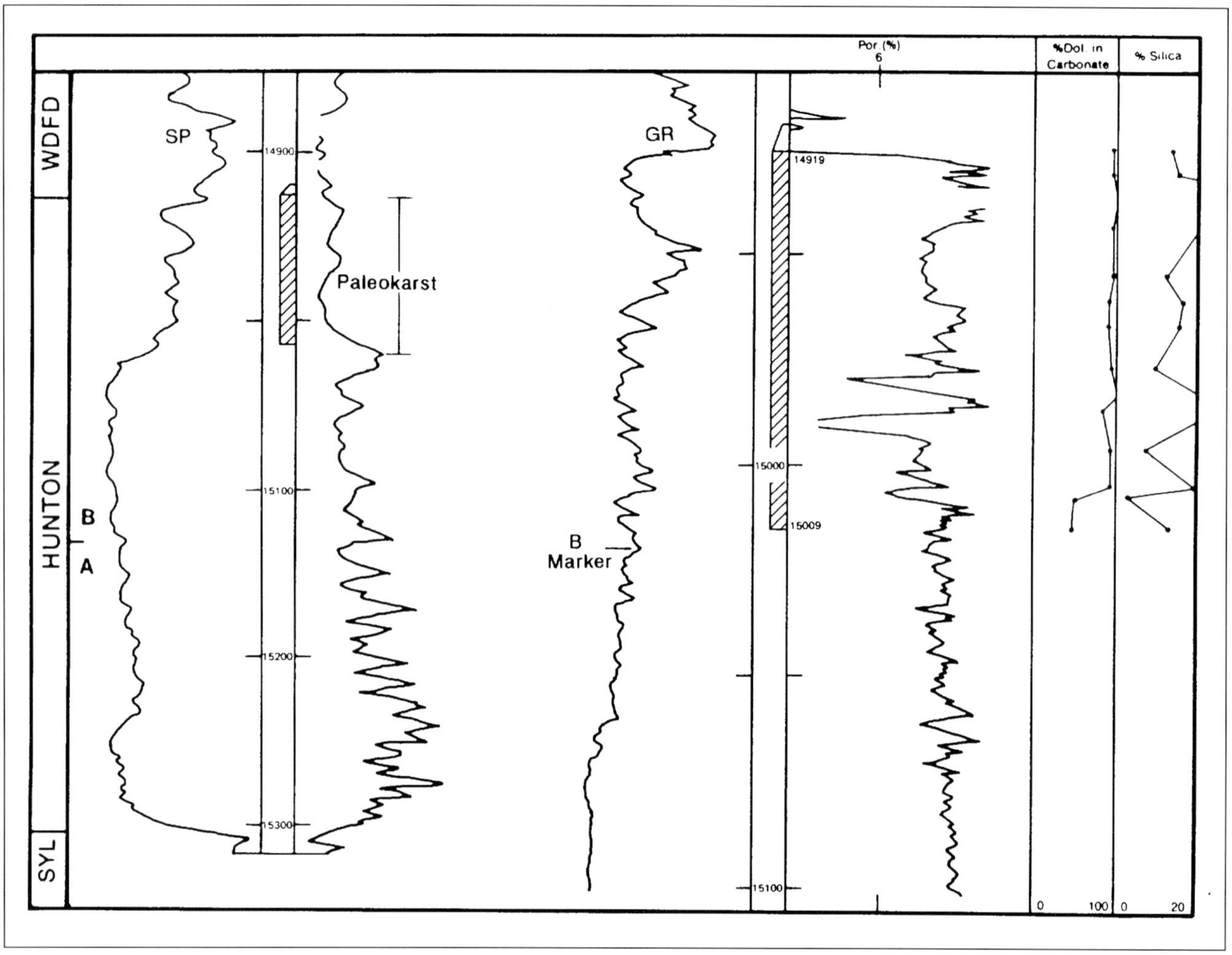

FIGURE 28 — Hunton log from the Mobil Rowland No. 1 well in Section 16 T15N-R16W, Custer City field, Custer County, Oklahoma, showing effects of paleokarst on log response. Note reaction of SP and gamma ray logs to clastic sediments in brecciated carbonate.

tidally dominated cycles at the base of the Signal Mountain Formation and thick (greater than 30 ft) subtidally dominated clues at the top of the McKenzie Hill Formation of the Arbuckle Group. It is postulated that the Signal Mountain to McKenzie Hill section represents one third-order depositional sequence.

Basin Modeling

In tectonically active areas, basin modeling can be an effective tool in the evaluation of sequence stratigraphy and stratigraphic discontinuities. Using all available geologic, geochemical, and geophysical data for certain sequences, forward models can be built that simulate depositional environment, eustatic sea-level changes, tectonic activity, and resulting platform to basin geometry (Saller and Yaremko, 1994). Once again, the problem with this method is that the accuracy of the prediction may not be reliable enough to identify relatively short exposure periods. Examples of basin modeling can be found in the work of Bice (1989), Kendall et al. (1989), and others.

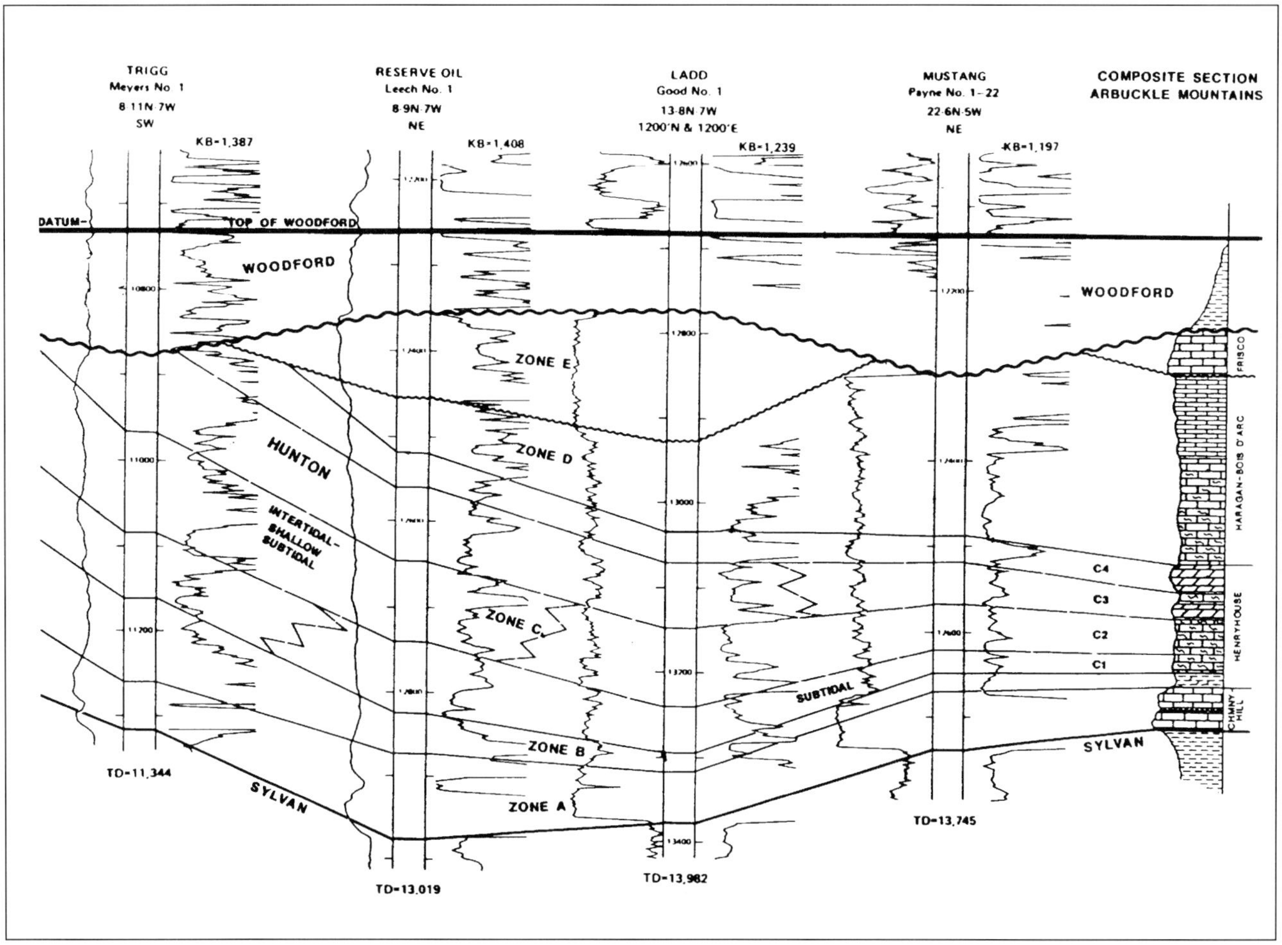

FIGURE 29 — North-south stratigraphic cross section along the Nemaha ridge showing unconformity within the Hunton Group. Note the truncation of pre-Frisco Hunton carbonates by the pre-Frisco unconformity.

EXPLORATION CONSIDERATIONS

There are many good examples of unconformity relationships in the Mid-Continent, and only a few have been discussed in this study. In most cases, recognition of unconformity and related diagenesis is based on integration of two or more of the methods outlined above. The following is an example of the integrated approach to identify and interpret unconformity and potential reservoir development.

Interpretations of outcrop and core data suggest that the Hunton Group of Oklahoma is a ramp-type carbonate composed of several third-order sequences (Fritz and Medlock, 1993). Mature and juvenile paleokarsts are observed in the outcrop and core (Fig. 26).

Paleontological data based on brachiopod and conodont ranges show that the Hunton was truncated by pre-Frisco/pre-Woodford unconformities, suggesting a major truncation episode along the pre-Frisco exposure surface (Amsden and Barrick, 1988: Fig. 27a). The paleontology also shows significant time gaps in the Chimneyhill Subgroup and overlying formations (Fig. 27b).

Wireline logs in the Hunton Group also show mature and juvenile paleokarsts (Fig. 28). The mature paleokarst is recognized by an increase in radioactivity on the gamma ray log due to sedimentation and collapse, and the juvenile paleokarst is recognized by an increase in dissolution porosity. Regional cross sections show the presence of a regional (third-order equivalent) unconformity

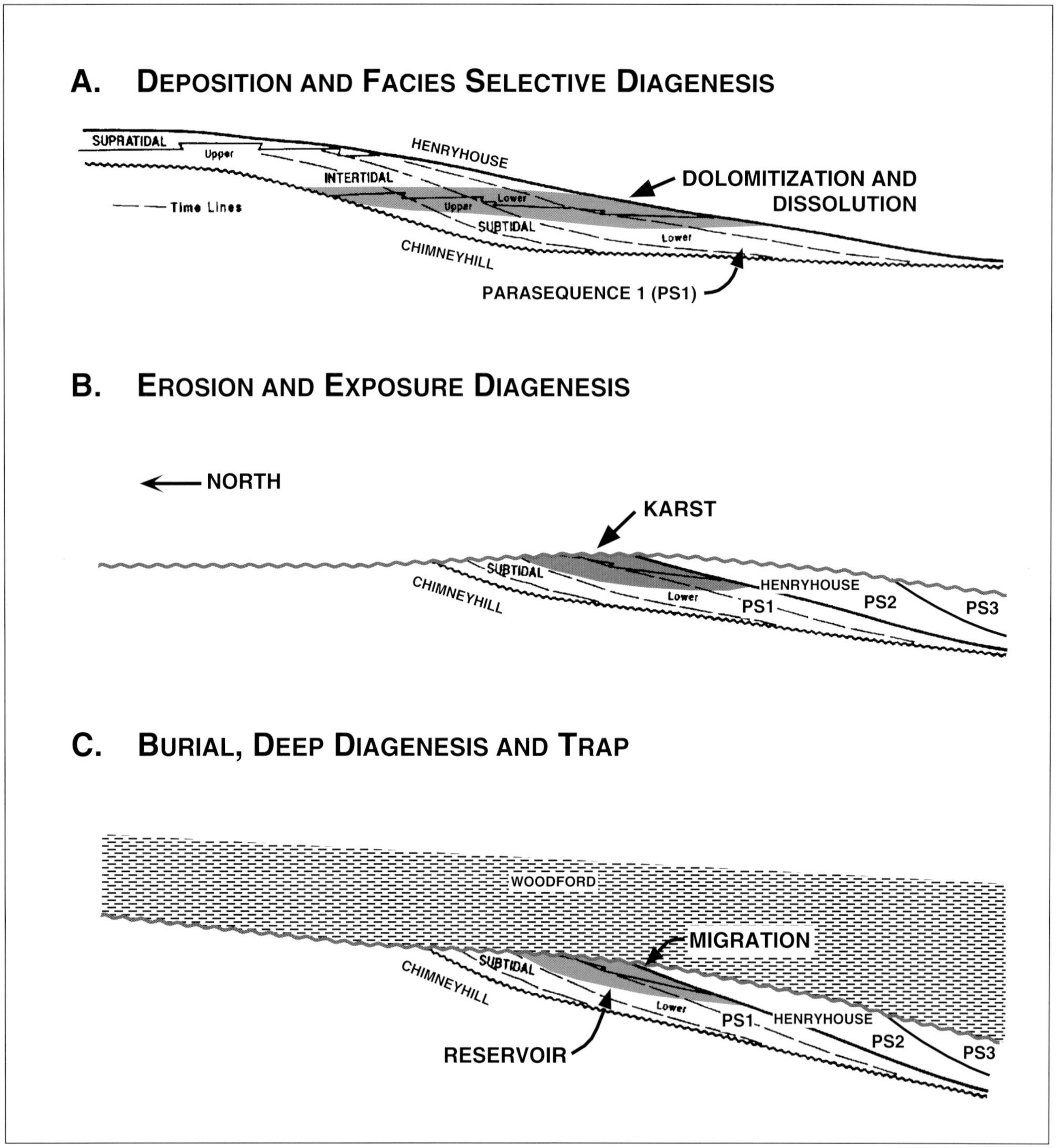

FIGURE 30 — Reservoir model for the Hunton Group: (a) depositional model for the Henryhouse through Bois-d'Arc formation, (b) karst model for pre-Woodford unconformity, and (c) typical trapping mechanism for Hunton against the Woodford Shale.

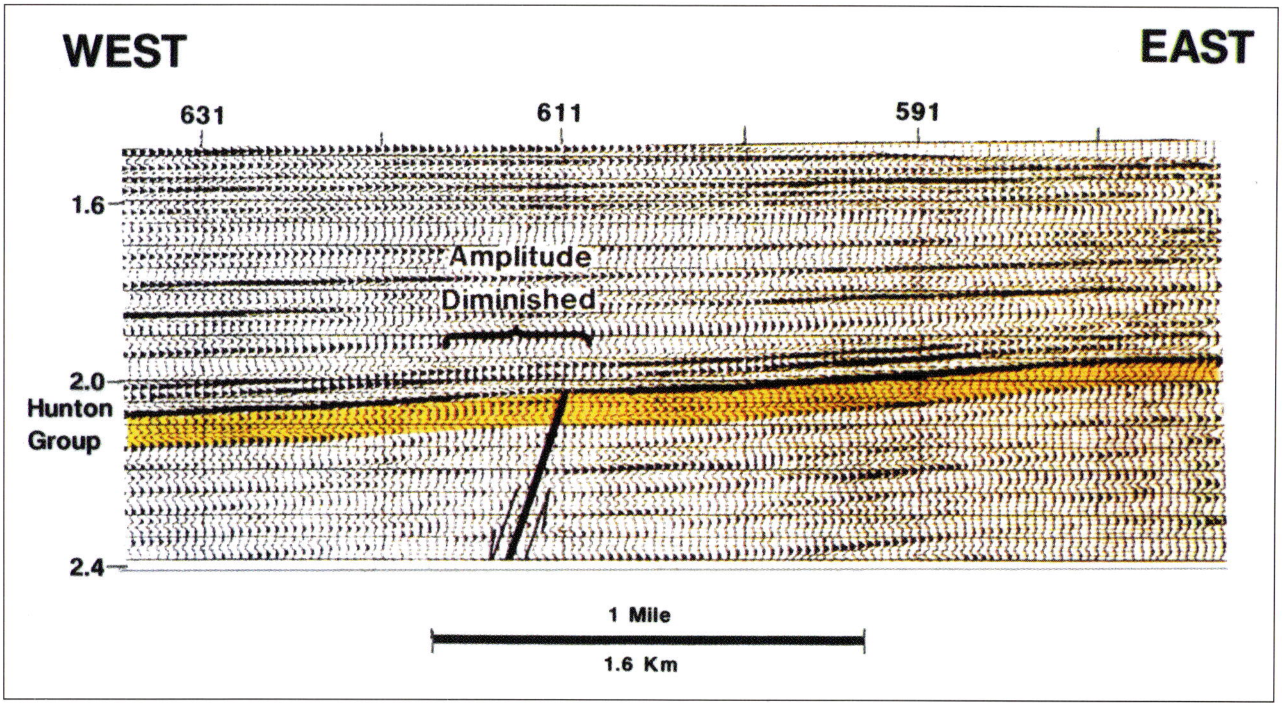

FIGURE 31 — Relative amplitude (RAM) processed version of seismic line (No. 218 ext) across Hunton field in Canadian County, Oklahoma (see Fig. 29 for geological model). Pre-Woodford unconformity is at the top of the area highlighted in yellow. Frisco is shown by the decrease in amplitude caused by lithology and porosity retarded velocity contrast. Frisco porosity is caused in part by dissolution associated with karst (Morgan et al., 1982).

between the Frisco and underlying units as well as regional and subregional discontinuities within the Hunton Group (Fig. 29). The importance of the pre-Frisco unconformity is that most of the Hunton was truncated at this surface and not by pre-Woodford erosion. This observation is supported by paleontology and correlations that show the Frisco overlies even the oldest Hunton strata. Most of Hunton karstification is interpreted to have occurred during the hiatus associated with the pre-Woodford unconformity which is estimated to represent a time gap of 10 m.y., while the associated time gap for the pre-Frisco is estimated to be only 1 m.y.

Integration of core, outcrop, wireline, and cross-section data revealed the reservoir geometry of the Hunton carbonate ramp and provided the basis for a sequence model (Fig. 30). Reservoir characterization showed that the porosity is mostly facies selective (upper subtidal/lower interitidal). This allowed the development of a comprehensive reservoir model related to diagenetic processes of the unconformities, which can be used to identify trapping mechanisms in the Hunton Group.

The Hunton Group can be mapped based on the reservoir model using either conventional subsurface mapping techniques or seismic interpretation. A good example of the latter is exploration of the Frisco Formation using relative amplitude (RAM) processing which shows the location of Frisco reservoirs sandwiched between the pre-Woodford (above) and pre-Frisco (below) unconformities. The Frisco is identified by a lower reflection coefficient between the porous Frisco Limestone and overlying Woodford Shale (Morgan et al., 1982: Fig. 31). This process can be further enhanced by the use of 3-D seismic data which can potentially detail the paleogeography of the unconformity surface.

SUMMARY

Carbonate rocks are particularly susceptible to the processes associated with stratigraphic discontinuities. Unconformities are best defined, classified, and identified according to the stratigraphic sequences to which they are related. Carbonate sequences and their sequence boundaries must be defined and described based on intrabasin origin rather than extrabasin.

Recognition of unconformities and related carbonate alterations requires an understanding of the processes associated with subaerial (or submarine) exposure such as karstification. Recognition also is accomplished by the use of certain geological methods, e.g., outcrop and core analysis, biostratigraphy, well log analyses and correlation, sequence interpretation, seismic interpretation, and basin modeling. It must be recognized that each method progresses from direct rock evaluation, to rock measurement, to some form of modeling — each requiring more interpretation and each progressively more susceptible to error.

Carbonate sequences in the Mid-Continent probably contain large undiscovered hydrocarbon reserves, as evidenced by the relatively recent discoveries of the Wilburton and Ringwood fields. These carbonates have had a unique diagenetic evolution, having been subjected to multiple episodes of exposure which have been porosity constructive and destructive. Recognition of unconformities is extremely important in the evaluation of Mid-Continent carbonate reservoirs, especially stratigraphic traps. Identification is enhanced by insightful use of the integrated geological methods described in this study.

QUANTITATIVE BIOSTRATIGRAPHY
A KEY TO SEQUENCE STRATIGRAPHY

R. W. Scott
Precision Stratigraphy Associates
Tulsa, OK 74135-5515

ABSTRACT

Sequence stratigraphy integrates diverse geophysical and geological data sets, and it has revitalized both geoscience disciplines. Biostratigraphic data have become important to the geological interpretation of seismic record sections. Quantitative analyses of fossil abundances, diversities, and ranges provide important data for the definition and constraint of sequence stratigraphic hypotheses. Within carbonate platforms, sequence boundaries may separate similar facies and become difficult to recognize. Within mixed siliciclastic and carbonate facies successions, the subtle changes in biotic associations define the stacking pattern of bed sets and aid the recognition of sequence boundaries. Quantitative biotic analyses of vertical facies successions constrain the depth limits of the beds. The timing of sea-level changes and their correlation within and between basins are essential for distinguishing the relative effects of basinal tectonics and eustasy. This distinction enlightens the prediction of potential reservoirs within a basin.

INTRODUCTION

The concept of depositional cycles is familiar to anyone raised on Mid-Continent geology. Pennsylvanian cyclothems are the basic stratigraphic unit in the division of the section and in the prediction of lithologies. It seemed natural that Sloss (1963) applied the concept to the long-term, continental-scale cycles of North America. Mitchum et al. (1977) adapted the concept for sedimentary packages in seismic record sections. Now the elemental stratigraphic unit in hydrocarbon exploration is the *sequence*, which is defined as a long-term depositional cycle bounded by unconformities across the shelf and shelf break and by correlative conformable contacts in the basin. These depositional sequences are based on either seismic or well log data.

The duration of such cycles is 0.5 to 3 m.y., depending on which scheme is used. However, the precise definition of the sequence durations remains to be statistically treated. The duration, like the thickness, is a relative matter that depends on the practicality of subdividing the section into

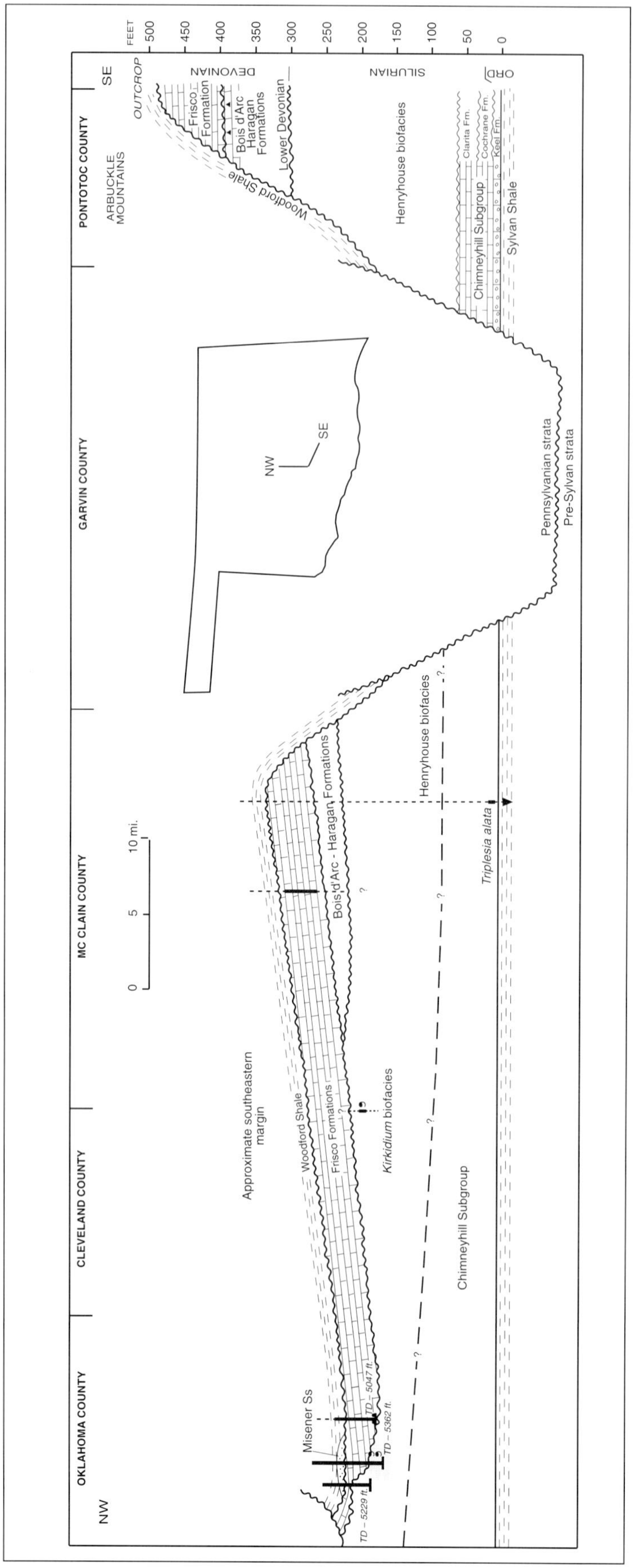

FIGURE 1 — Lithostratigraphic cross section of the Hunton Group in central Oklahoma; the datum is the base of the Hunton. The three unconformities within the Chimneyhill Subgroup can be recognized in the outcrop but not in the subsurface. The *Kirkidium* biofacies within the Henryhouse Formation marks the southeastern shelf margin. The overlying unconformities at the base of the Bois d'Arc-Haragan, Frisco, and Woodford shales are regional erosional sequence boundaries. This cross section is from Amsden (1975).

sequences based on seismic, well log, or rock data. The cyclothems of the Mid-Continent Pennsylvanian section are depositional sequences as thin as 2 m which can be resolved only by shallow, high-resolution reflection seismology with frequencies up to 500 Hz (Knapp and French, 1991). However, in deeper parts of a basin where beds thinner than 10 m are difficult to resolve, the functional sequence is much thicker. The consistent delineation of depositional or seismic sequences depends on the successful integration of biostratigraphic, geochemical, and petrographic data that define the age and paleoenvironments of the interval.

Quantitative stratigraphy is the application of statistical and other numerical techniques to the correlation of biostratigraphic and lithologic data, and to the measurement of rates and magnitudes of geological processes (Gradstein et al., 1985). When applied to biostratigraphic data, precise age determinations can be made of fossiliferous depositional sequences. Only by the accurate dating of sequences can they be related to eustatic or other depositional processes. A number of examples in the Mid-Continent illustrate the integration of paleontological data in the successful analysis of depositional cycles/sequences. The first example does not employ quantitative techniques, but does illustrate the role of paleontologic data in the recognition and dating of sequences.

SEQUENCES IN SILURO-DEVONIAN CARBONATES

This example illustrates the fundamental role of paleontological data in the recognition and dating of regional unconformities. The study was done prior to the development of the sequence stratigraphic paradigm, and shows how high-quality stratigraphic data can be reinterpreted in the light of new models.

During the Silurian and Early Devonian, an east-west carbonate platform bordered a basin in southern Oklahoma. These carbonate units have long been mapped as the Hunton Group, which contains important hydrocarbon reserves in the Mid-Continent. These platform units overlie the Upper Ordovician Sylvan Shale and underlie the Devonian-Mississippian Woodford Shale, both representing basinal settings (Amsden, 1967, 1975). The Hunton can be divided into at least seven stratigraphic units based on their lithology and fossil assemblages (Fig. 1) (Amsden, 1975). Some units are difficult to separate on well logs and in lithologic samples because many of these seven lithological units consist of similar carbonate facies. However, the section is divided by at least five important unconformities in the Arbuckle Mountains of southeastern Oklahoma. The fossil assemblages are the key to mapping these units consistently (Amsden, 1975) and to their recognition as sequences.

Oil and gas are produced from the upper sequences mapped as the Lower Devonian Frisco and the underlying Silurian Bois d'Arc formations. The Frisco Formation is a porous bioclastic limestone lithologically similar to the Bois d'Arc. In some places it overlies Silurian or Ordovician carbonate strata, but cannot be separated from them by wireline log charactersitics. However, the Frisco contains the distinctive and widespread Deerparkian fauna of brachiopods, snails, and corals.

The Silurian carbonates produce hydrocarbons from several bioclastic units that are separated by regional unconformities (Fig. 1). The pinchout of these units is difficult to map because the lithologic units cannot be recognized on logs. However, the bioclastic unit at the top of the Henryhouse Formation contains the very distinctive brachiopod *Kirkidium* biofacies, so it can be distinguished from the overlying Devonian and underlying Silurian bioclastic carbonates. Although Amsden recognized these fossils in cores, the concepts apply equally well to microfossils in well cuttings from these and rocks of other ages.

PALEOENVIRONMENTS IN PENNSYLVANIAN CYCLES

Depositional sequences in continental margin settings consist of facies sets deposited during the rise and fall of a major sea-level cycle, *depositional systems tracts* (Posamentier and Vail, 1988). In cratonic seas, systems tracts are thin and aerially extensive, subparallel beds (Watney et al., 1989). For example, black core shale units that represent the deepwater stage in the Pennsylvanian cyclothems can be traced for hundreds of kilometers (Heckel and Baesemann, 1975). These cyclothems are thin, vertical successions of interbedded siliciclastic and carbonate rocks repeated in a somewhat regular order (Heckel, 1977). Heckel's Kansas cyclothem model consists of 10 successive lithologies that relate to different water depths and phases on a theoretical sea-level curve (Fig. 2). These lithologies are grouped into four facies, each of which can be considered to be the depositional systems tracts of a cyclothem.

The boundary between superposed cyclothems is generally placed at a contact having evidence of subaerial weathering and valley incision. However, the features displayed by these contacts change from place to place. This contact commonly is placed within a facies set called the outside shale, where paleosols are developed. These intervals generally are widespread markers in Kansas and adjacent areas. The overlying rocks have coals and nearshore brackish to marine biota, indicating the

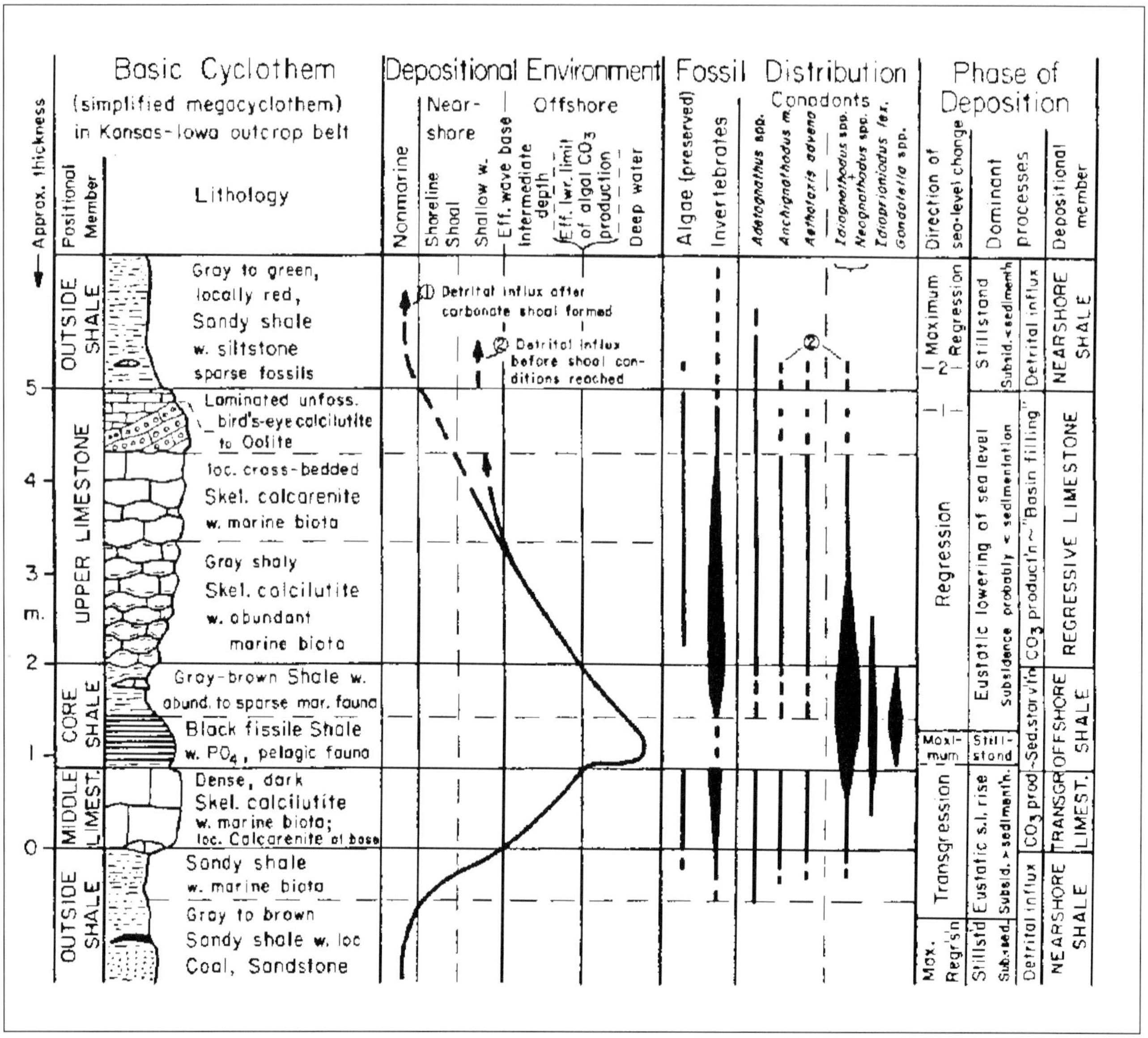

FIGURE 2 — Model of Pennsylvanian cyclothem in Kansas from Heckel (1984). One transgressive-regressive cycle ranges across five paleoenvironments having different bathymetries based on conodont assemblages.

resumption of relative sea-level rise. The lower parts of the outside shale below the paleosol commonly have well-developed deltaic deposits of the falling or regressive phase.

The next facies set is the middle or transgressive shale. This deposit is typically widespread and 1 to 15 m thick. It consists of marine wackestone deposited nearshore below the wave base.

The core shale is very thin, up to 1 m thick, and very widespread. The most distinctive lithology is black, organic-rich, phosphatic shale, but other clay rock facies are also developed. The ammonoids and conodonts of these shales may be used as time markers throughout a basin (Heckel and Baesemann, 1975; Boardman and Heckel, 1989). These and other workers concluded that the black shales were deposited in deep, stratified water columns within the epeiric seas. However, some believe that these shales were rapidly deposited in shallow waters (Maples, 1986).

The upper regressive limestone generally is the thickest part of the cyclothem and contains important hydrocarbon reservoirs (Watney et al., 1989). This unit is from 3 to 30 m thick and consists of high-energy grainstones and algal buildups. The facies and diagenetic features indicate upward shoaling.

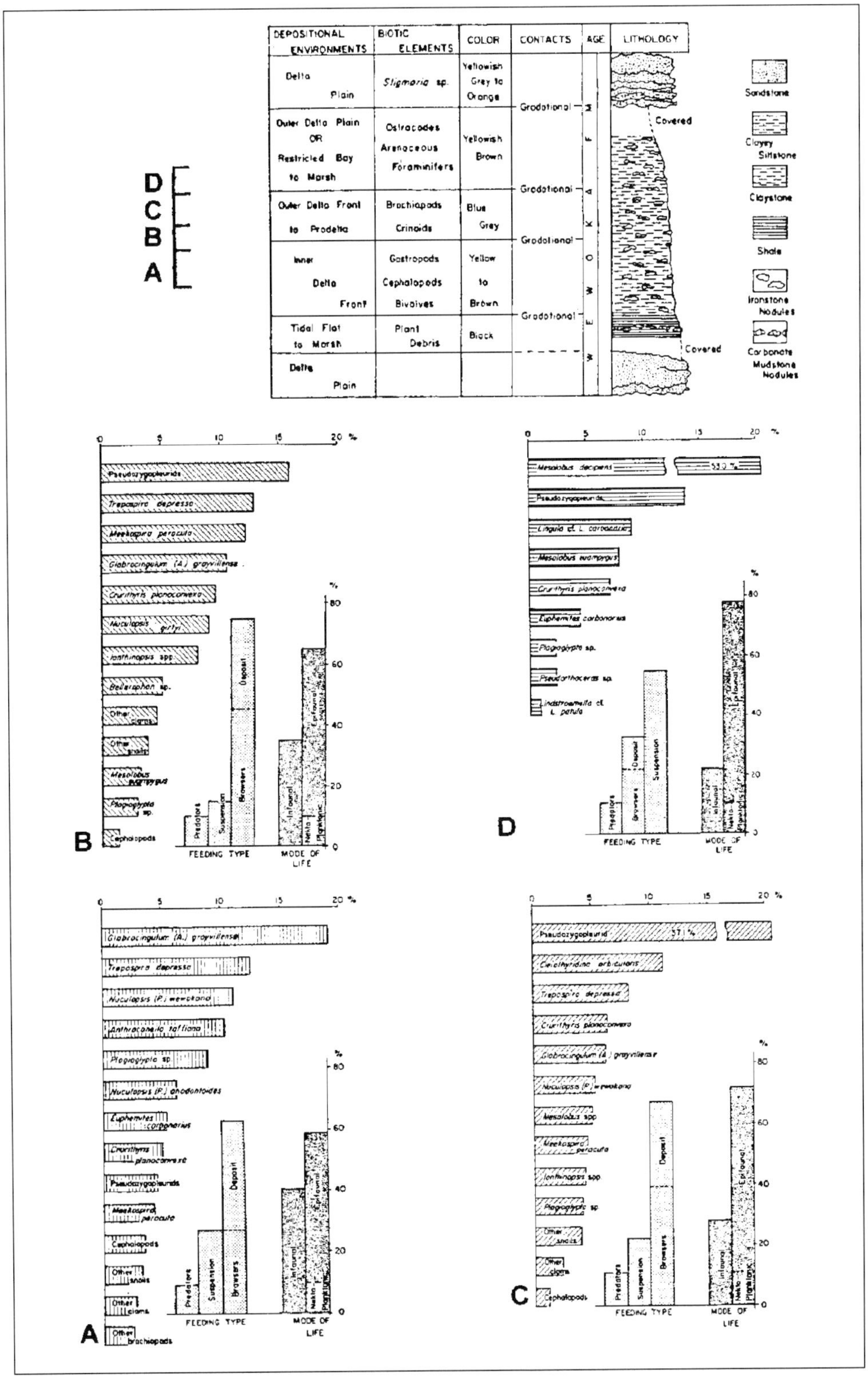

FIGURE 3 — Diagram of cyclothem dominated by siliciclastics in the Wewoka Formation in central Oklahoma from West (1972). The composition of four paleocommunities shows the relative abundances of the major taxa and the percentages of species in several feeding types and modes of life. (A) *Glabrocingulum*, (B) transitional assemblage, (C) *Cleiothyridina*, and (D) *Mesolobus*. The stratigraphic range and facies distribution of each assemblage are indicated by the brackets.

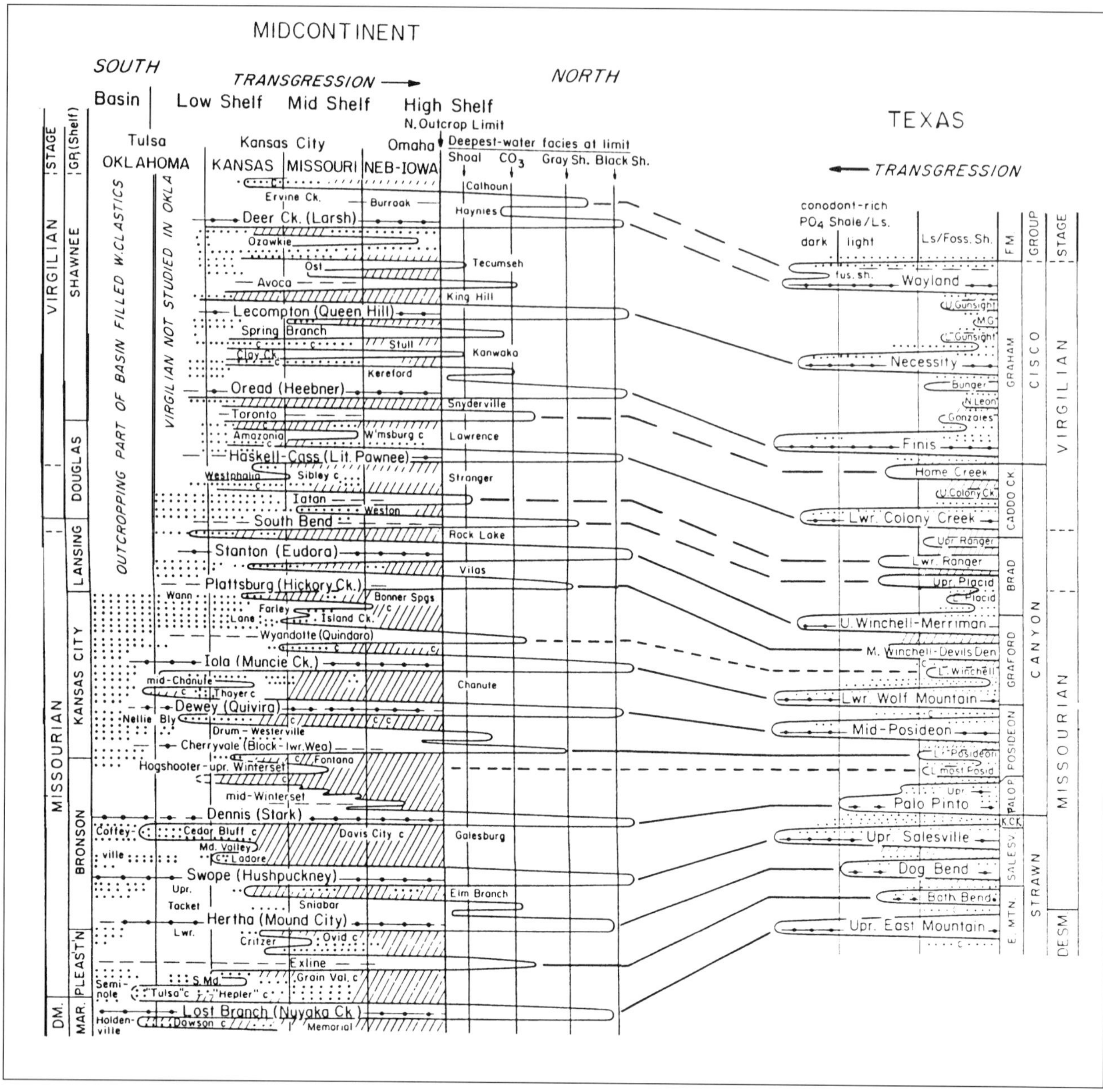

FIGURE 4 — Eustatic sea-level curve for the upper part of the Pennsylvanian sequences in the north-central Texas outcrop correlated with the curve for the section in the Mid-Continent (Boardman and Heckel, 1989). Hachured areas represent hiatuses of lowstands; stippled intervals represent dominantly sandstone units; *c* is coal; solid lines with dots represent black shale deposited during maximum flooding; and dashed lines are other maximum transgressive facies.

Three sequence stratigraphic components have been identified within the cyclothem model (Watney et al., 1989). The flooding unit (TST) is the middle limestone and consists of nearshore sandstone, siltstone, and carbonates up to 15 m thick (Fig. 2). The biota indicate upward deepening. The basal contact with the outside shale is sharp and generally indicates an abrupt facies shift from subaerial exposure or coalbeds to drowning. The condensed section (CS) is the core shale of organic-rich muds. The biota are mainly pelagic and nektic organisms. Very slow sedimentation rates are indicated by the concentration of phosphate and trace elements such as uranium, by abundant pelagic fossils, and by horizontally oriented clay minerals, suggesting sedimentation by dilute suspension (Watney et al., 1989). The third, highstand systems tract (HST) is the upper limestone which is an upward-shoaling stratal unit. It comprises

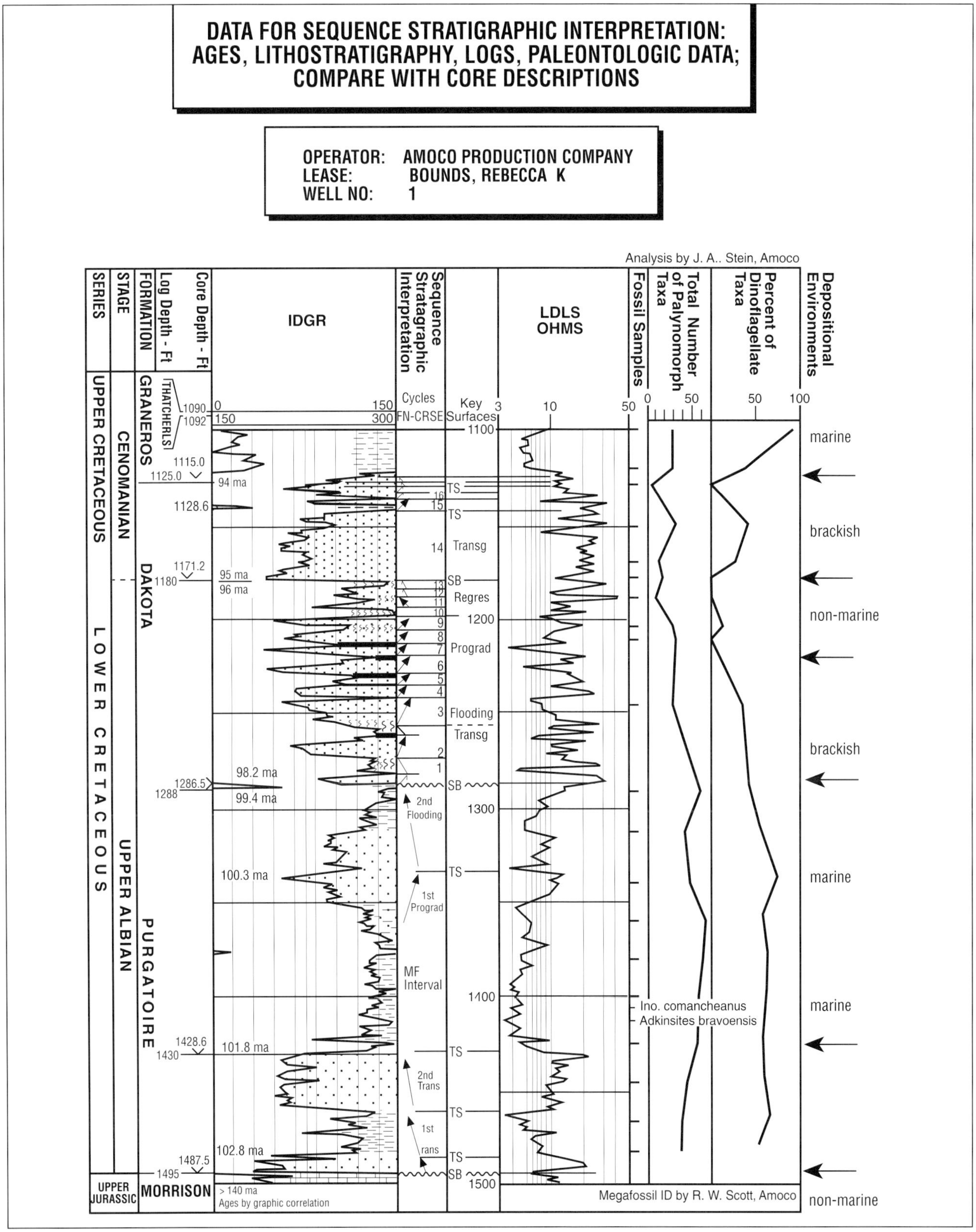

FIGURE 5 — Stratigraphy of the Purgatoire and Dakota formations in the Amoco No. 1. R. K. Bounds core, Greeley County, Kansas (Scott et al., 1994). SB = sequence boundary; TS = transgressive surface; MF = maximum flooding interval. Mega-annums are derived by graphic correlation of a composite standard calibrated to the Palmer (1983) time scale.

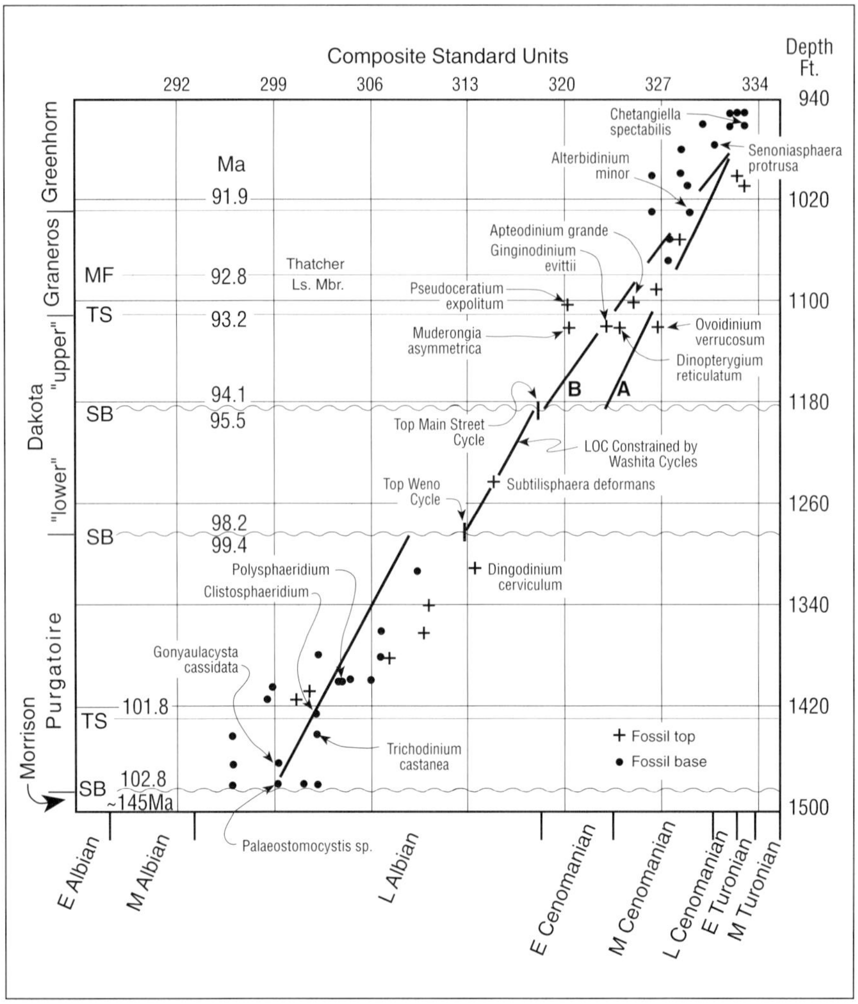

FIGURE 6 — Graphic correlation of mid-Cretaceous units in the Bounds core. Correlation of the "upper" Dakota sequence is defined by two hypotheses (A and B), which differ by the use of different fossil tops. Solution A correlates the upper Dakota with the Woodbine sequence in Texas and raises several tops. Solution B correlates the upper Dakota with the Grayson-Buda formations and raises fewer tops. The numbers on the x axis are arbitrary scaling units derived from the original composite standard, which have been converted to mega-annums (from Scott et al., 1994).

the thickest part of these cyclothems and consists of open marine carbonates and siliciclastic deltaic deposits.

Quantitative analyses of Pennsylvanian biotic assemblages are diagnostic in sequence stratigraphic interpretations of the cyclothems. Boardman et al. (1984a) recognized eight major assemblages or communities. They considered these communities to be consistent associations of organisms occurring in similar environments. Definition of the communities is based on counting specimens of each species in bulk samples and surficial collections. Some of these communities are restricted to the core shale phase of the cyclothem, and others occur in both the transgressive and regressive phases, perhaps because similar environments were present. A quantitative paleoecologic study identified the systems tracts of a Desmoinesian transgressive-regressive, delta-dominated cyclothem, where the vertical succession of clay-rich lithofacies changes little and the black core shale is absent (Fig. 3). These communities can be defined by the relative abundances of dominant and diverse taxa to show the maximum flooding interval (Fig. 3) (West, 1972). The most open marine community is *Cleiothyridina* (C in Fig. 3) which occurs in the middle part of the middle claystone representing the outer delta front to prodelta environment. This community has the highest diversity of shelled organisms among the four communities, and its trophic structure is the most complex with epifaunal suspension and infaunal deposit feeding organisms. These properties suggest that this community was far offshore in less turbid waters and on a stable substrate.

Within the highstand or regressive limestone of the Virgilian Lecompton Formation in Kansas, ostracode assemblages change as shoaling progresses, although the facies do not change (Brondos and Kaesler, 1976). Maximum diversity and equitability of the ostracodes in five measured sections were greater in the lower and more normal marine part of the highstand limestone, and decreased up section indicating more turbid or higher energy conditions. Quantitative techniques of Q-mode and R-mode cluster analyses segregated the ostracodes into assemblages that reflect subtle environmental changes. The application of these quantitative paleoecological techniques has proved useful to define critical environmental conditions, and their utility in sequence stratigraphic models will be powerful. To deal with large amounts of well data, interactive computer data bases and specialized application programs are necessary for timely results.

Finally, Boardman and Heckel (1989) correlated the glacial-eustatic, sea-level curves of sections in Kansas with those in north-central Texas by means of detailed biostratigraphy (Fig. 4). Particularly, ammonoids, conodonts, fusulinids, and corals were critical in demonstrating which maximum flooding core shales in Kansas correlate with marine facies in Texas. Paleosols defined the sequence boundaries; however, their chronostratigraphic value for correlation across the craton was derived from the correlation of the intervening marine units.

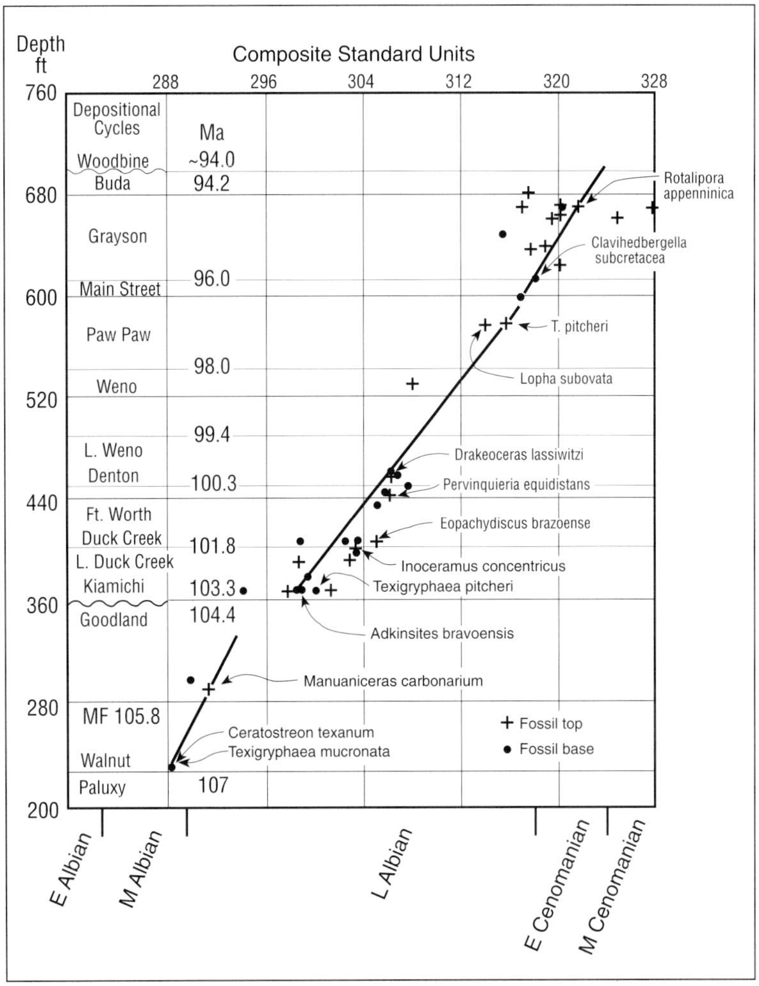

FIGURE 7 — Graphic correlation of Fredericksburg and Washita groups in the Fort Worth, Texas, area and ages of shale-limestone depositional cycles. The numbers on the x axis are arbitrary scaling units derived from the original composite standard, which have been converted to mega-annums (from Scott et al., 1994).

GRAPHIC CORRELATION OF MIDDLE CRETACEOUS CYCLES

Depositional cycles in the Middle Cretaceous siliciclastic Purgatoire/Kiowa and Dakota formations are readily defined by well log features in a continuously cored well in western Kansas — the Amoco No. 1 Bounds (Fig. 5) (Scott et al., 1994). These Albian-Cenomanian deposits in the Western Interior seaway contain significant hydrocarbon reservoirs in the Denver basin (Dolson et al., 1991). During the middle Late Albian, this seaway was open to the Tethyan sea in the Gulf Coast; it closed during the latest Albian and earliest Cenomanian. Correlation of the cycles and sequences in the Western Interior with those in the Gulf Coast is useful to determine the timing of reservoir development as well as to define the history of depositional events. Measuring the duration of depositional cycles within the Denver basin provides a basis to estimate the depositional rates. These correlations and measurements can be precisely done by graphic correlation, which demonstrates the presence of at least three scales of cycles: 2 to 3 m.y., 1 to 1.5 m.y., and 150 to 170 k.y. (Scott et al., 1994).

Graphic correlation is a quantitative technique to determine the time equivalency between a section and a standard chronostratigraphic data base (Miller, 1977). The ranges of fossils in the Bounds core were plotted on the y axis of a two-axes graph (Fig. 6). The x axis consists of the ranges of the

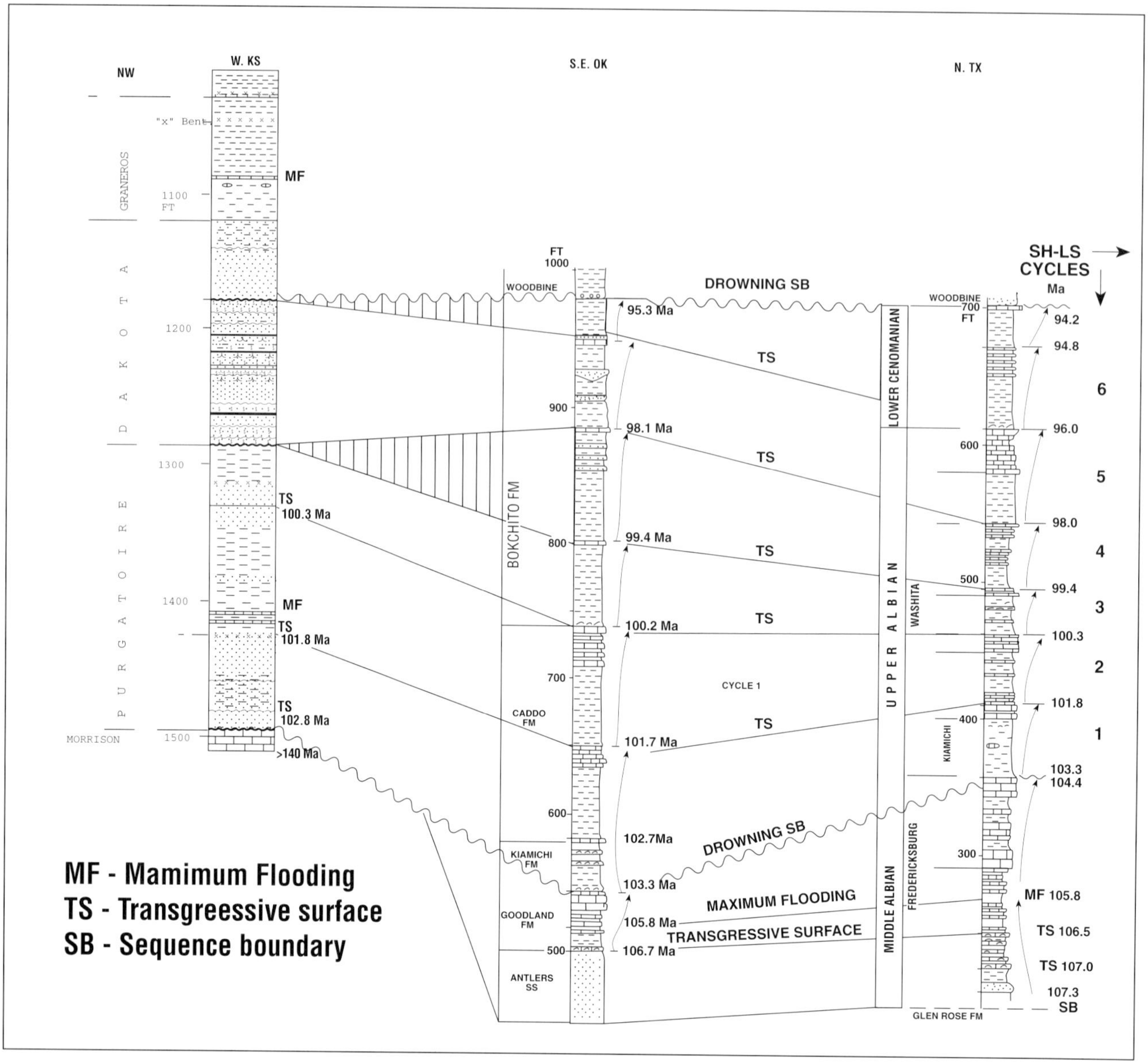

FIGURE 8 — Correlation of sequence and cycle boundaries in the Fort Worth section on the Gulf Coast with sections in southeastern Oklahoma and western Kansas using a datum of the basal Woodbine Formation and intra-Dakota unconformity (Scott et al., 1994). Arrows show depositional cycles. Abbreviations for stratigraphic surfaces are explained in Fig. 5. (Figures 5 to 8 are printed with permission of The Rocky Mountain Association of Geologists.)

same fossils determined in over 60 sections from many different basins worldwide, including many of the stadial stratotype sections. This data base of ranges evolved from the computerized graphing of these numerous sections in an attempt to identify the maximum range of each species; this is the composite standard data base. The time equivalency between the y-axis section and the composite standard is determined by the line of correlation (LOC). This line is placed by the stratigrapher to account for the most data. The placement of the LOC provides a precise interpretation that is subject, like other hypotheses, to revision as new data become available.

Two regional unconformities in the cored basal Cretaceous interval define the base of two transgressive-regressive cycles (Fig.6): the Kiowa-Skull Creek T5 cycle and the Dakota-Greenhorn T6 cycle (Kauffman, 1984). The basal Cretaceous unconformity at a 1495-ft log depth separates the

Upper Jurassic Morrison Formation from the Purgatoire Formation; this represents a hiatus of approximately 37 m.y. The unconformity between the Purgatoire T5 cycle and the Dakota Formation T6 cycle at a log depth of 1288 ft represents a hiatus of about 1.2 m.y.; this is the interval where the "J" reservoir sand is developed in Colorado (Dolson et al., 1991). The intra-Dakota unconformity at the log depth of 1180 ft is the horizon of the "D" reservoir sand in the Denver basin.

The Purgatoire Formation here represents a single depositional sequence 207 ft (63 m) thick and 3.4-m.y. duration. It is divided into five subcycles by sharp transgressive contacts between sandstone and shale or upward-coarsening and upward-fining sandstones (Fig. 5). The lower three subcycles are part of the transgressive systems tracts (TSTs) spanning 1 m.y. The maximum flooding occurred about 101.4 Ma and is represented by an interval of dark gray marine shale with carbonate nodules. The HST consists of two upward-shoaling shale-sandstone subcycles separated by a distinct gamma log kick at 1332 ft dated by graphic correlation to be 100.3 Ma. The uppermost cycle is incomplete and has been erosionally truncated.

The lower Dakota depositional sequence consists of nine upward-coarsening mud-sandstone subcycles and four upward-fining sandstone-mudstone subcycles (Fig. 5) over a time span of 2.2 m.y. However, the top of the section is bounded by a regional unconformity that, in places, has been incised and filled by the "D" sand. These small-scale subcycles represent shoaling cycles on a Dakota delta plain punctuated by numerous periods of subaerial exposure (Scott et al., 1994).

The depositional cycles of the coeval section in the Gulf Coast are illustrated by a composite section of the Upper Albian-Lower Cenomanian Washita Group in the Fort Worth, Texas, area (Scott et al., 1978). The Washita represents a long-term transgression and deepening that began about 103 Ma and ended about 94 Ma (Fig. 7). The basal contact with the underlying Fredericksburg Group, which represents a carbonate platform, is a drowned disconformity (Scott et al., 1978). The uppermost contact is an erosional unconformity at the base of the Woodbine Formation of Middle Cenomanian age. The Washita is a 9-m.y.-long sequence comprised of at least six shorter term subcycles about 1 to 2 m.y. These subcycles consist of lower dark-gray shale and upper medium-bedded limestone. This succession represents deepening and development of a low-oxygen shelf basin followed by clearing and shoaling to a carbonate shelf. These six subcycles are dated by the graphic correlation of mollusks, foraminifers, and nannofossils (Fig. 7). In this reference section, the subcycles are defined by lithostratigraphic criteria and dated by a graphic correlation.

The regional correlations of several cycle scales are based on the ages derived by graphic correlation (Fig. 8). A third section of the Washita Group in southeastern Oklahoma is added to test the graphic correlation solution of the Fort Worth section. It is also a composited section based on mapping and many years of paleontological sampling. The six subcycles in the Fort Worth section are readily identified in Oklahoma and are dated as the same age. The correlation with the Western Interior section in the Bounds core shows that the three major sequence-bounding unconformities are equivalent to the base of the Washita cycles. The transgression from the Gulf Coast into the Western Interior here is dated as spanning about 1 m.y. The basal transgressive subcycle in the Purgatoire Formation is coeval with the Kiamichi Formation in Texas. The remaining part of the Purgatoire TST is equivalent with the Duck Creek Formation. The maximum flooding and early highstand in the Purgatoire were coeval with the upper Duck Creek and Fort Worth formations. The subcycle in the upper part of the Purgatoire is equivalent with the subcycle represented by the Denton and lower part of the Weno formations. The correlation of the transgressive contacts in the Purgatoire with transgressive contacts in the Washita was unexpected, and is based on graphic correlation of different fossil groups. This result gives confidence to the hypothesis that these units represent interbasinal processes, perhaps eustatic sea-level changes.

The hiatus at the basal Dakota Formation unconformity is represented by cycle 4, the upper part of the Weno Formation in Texas (Fig. 8). This is one of the sandiest intervals in the north Texas area (Scott et al., 1978). However, while the sea level dropped in the Western Interior and valleys were cut and filled, sandy nearshore deposits accumulated in the Gulf Coast. The lower sequence in the Dakota Formation correlates with the cycle of the Pawpaw and Mainstreet formations in Texas. The unconformity within the Dakota represents the time interval of the Grayson and Buda formations in Texas. The basal Woodbine Formation is coeval with the incised valley fill sands of the "D" reservoir and the transgressive facies in the upper Dakota in the Western Interior.

CONCLUSIONS

The quantitative biostratigraphic technique of graphic correlation is a powerful tool to measure the age of the depositional sequences and key sequence contacts. It enables the testing of the time equivalency of cycles both within and between basins. It has the capability of testing and improving the charts of global depositional events.

Quantitative paleoecological techniques pioneered 20 years ago have the potential of accurately defining depositional conditions of systems tracts and of critical marker beds in sequence stratigraphy. To use these techniques, reliable paleontological, geochemical, and sedimentological data must be collected from reference wells and other important sections. The management and timely applications of these data also require an easy-to-operate computer database system. These are now available to every exploration/exploitation group.

DEPTH-RELATED MICROFAUNAL BIOFACIES MODEL FOR LATE CARBONIFEROUS AND EARLY PERMIAN CYCLOTHEMIC SEDIMENTARY SEQUENCES IN MID-CONTINENT NORTH AMERICA

Darwin R. Boardman II
Department of Geology
Oklahoma State University
Stillwater, OK 74075

Merlynd K. Nestell
Department of Geology
University of Texas at Arlington
Arlington, TX 76019-0049

Larry W. Knox
Department of Earth Sciences
Tennessee Tech
Cookeville, TN 38505

ABSTRACT

An onshore-offshore model for the distribution of microfossils (foraminifers, ostracodes, conodonts, and radiolarians) is presented for Late Carboniferous-Lower Permian cyclothemic sedimentary sequences of the North American Mid-Continent. Nearshore marginal marine strata are characterized by the low diversity of microfossil assemblages dominated by *Geisina* ostracodes, no conodonts, and rare *Thurammina* foraminifers. Nearshore normal marine strata contain the *Cavellina* ostracode biofacies, *Adetognathus* conodont biofacies, and encrusting foraminifers along with *Ammodiscus* cf. *A. semiconstrictus* var. *regularis*. Intermediate depth (open mid-shelf) strata contain the *Amphissites* ostracode biofacies, *Idiognathodus-Streptognathodus* conodont biofacies, *Tetrataxis*, *Globivalvulina*, large paleotextularids (*Climacammina*, *Deckerella*), endothyranellids (*Endothyra*, *Endothyranella*), *Ammobaculites*, and fusulinaceans. Deep-shelf marine strata contain a high-diversity pelagic microfossil assemblage but a reduced diversity benthic microfossil assemblage containing the *Mammoides* and *Healdia* ostracode biofacies, *Gondolella-Idioprioniodus* conodont biofacies, *Ammodiscus-Reophax* foraminiferal biofacies, and abundant radiolarians.

INTRODUCTION

Sequence stratigraphic methodology has been successfully applied to Late Carboniferous-Early Permian cyclothemic strata by a number of workers including Watney et al. (1989), Weibel (1991), French and Watney (1993), Youle et al. (1994), and Mazzullo et al. (this book). Implicit in developing a sequence stratigraphic framework for these cyclical deposits is the identification of significant events including transgressive surfaces, ravinement surfaces, flooding surfaces, maximum flooding surfaces (marine condensed sections), and omission surfaces. Although some of these surfaces have definite lithologic and petrographic expressions, such as transgressive lags, many of them are more readily identified by detailed paleoecologic analysis. The main purpose of this type of paleoecologic analysis is to identify vertical changes in relative water depth within the sequence, which helps to pinpoint the various surfaces where physical characteristics are obscure or ambiguous. These surfaces are used to identify sequence boundaries, parasequence boundaries, parasequence set boundaries, as well as systems tract boundaries. One of the most significant event horizons is the maximum flooding surface that commonly coincides with a marine condensed interval. The marine condensed section is significant because it represents the boundary between onlap and offlap (transgressive and highstand systems tract boundary) and is the only horizon that links the basin to the shelf. Although most outcropping Late Carboniferous (Desmoinesian-middle Virgilian) cyclothems contain a dark gray to black, and often fissile, phosphatic shale that represents the marine condensed section, many late Virgilian-Artinskian cyclothems contain carbonate marine condensed sections (commonly less than 50 cm thick) that are identified on the basis of microfaunal and detailed petrographic analysis. Furthermore, the few dark gray to black phosphatic shales that occur in late Virgilian-Artinskian cyclothems are more localized and grade updip into highly fossiliferous, glauconitic, and phosphatic wackestones that represent the proximal (nearer to shore) facies of the marine condensed section. Marine condensed sections commonly contain the maximum diversity of conodont, ammonoid, and radiolarian assemblages. These taxa are among the most useful biostratigraphical tools to correlate cyclothemic sedimentary sequences intrabasinally as well as interbasinally.

In addition to the delineation of the sequence by identification of significant surfaces, paleoecological analysis can help unravel the apparent rates of sea-level change as reflected in changes in lithofacies and biofacies. Sharp contacts between depth-controlled lithofacies and corresponding biofacies often reveal facies dislocation caused by rapid rises or falls of sea level with missing intermediate facies predicted by Walther's law. Examples include black, fissile, radiolarian-bearing phosphatic shales interpreted as deep shelf, sharply overlying a coal interpreted as coastal swamp deposit. Another common example is where a molluscan grainstone, interpreted as representing high-energy shallow marine carbonate, sharply overlies a black, fissile shale that likely represents an offshore marine condensed section. The former case is interpreted as a rapid deepening, whereas the latter is interpreted as a rapid forced regression as discussed by Posamentier et al. (1992a). In addition to delineating the internal structure of the sequence, paleoecologic analysis is also used to determine maximum water depths attained during successive sedimentary sequences. These can be used in conjunction with observed vertical stacking patterns of successive sequences to generate local sea-level fluctuation curves to correlate those generated from different basins (Boardman and Heckel, 1989; Heckel and Weibel, 1991; Connolly and Stanton, 1992; Boardman and Nestell, 1993). By using biostratigraphically calibrated sea-level curves (showing major significant evolutionary and extinction

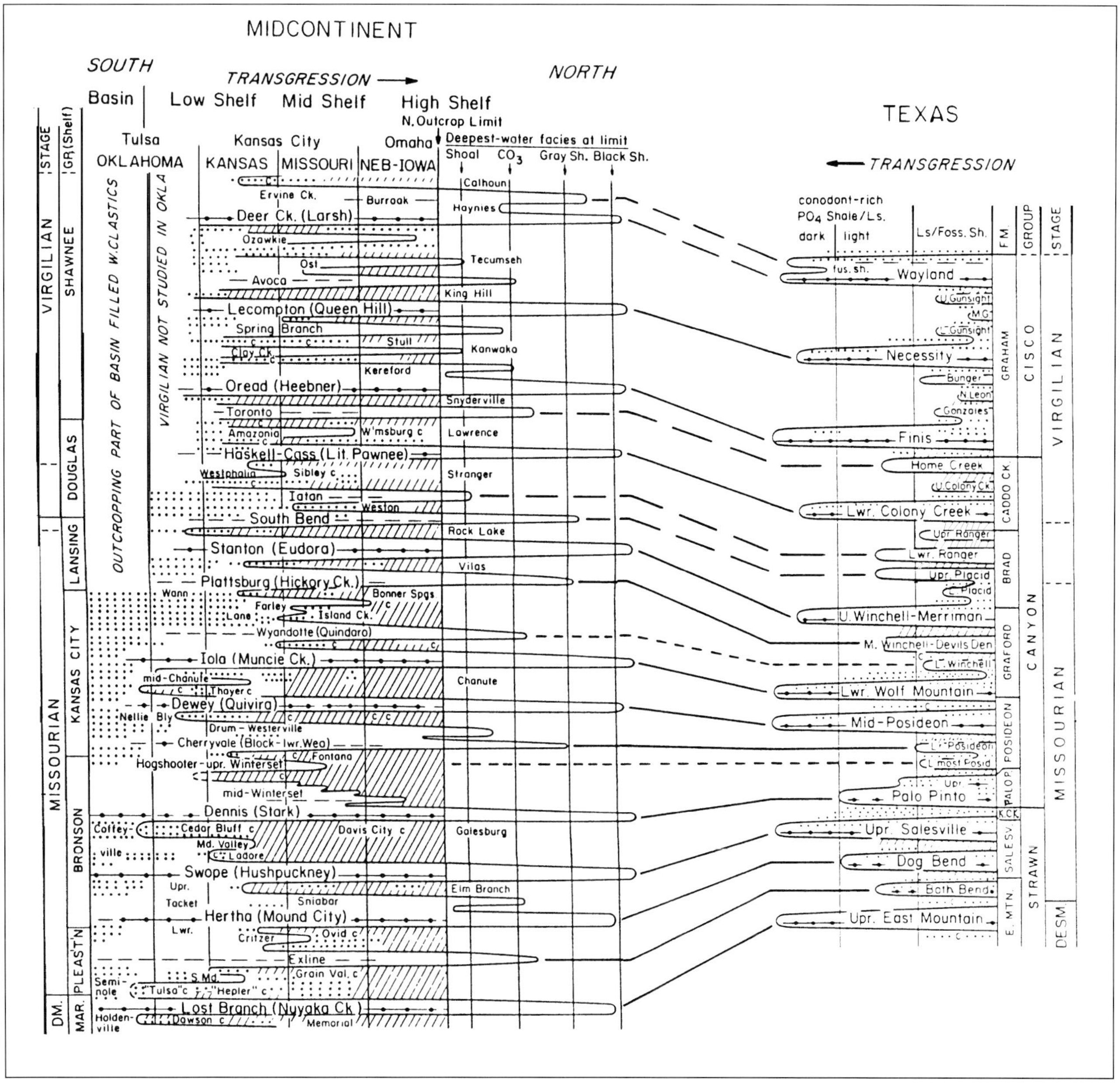

FIGURE 1 — Eustatic sea-level fluctuation curve for Upper Pennsylvanian (Missourian-middle Virgilian) strata of north-central Texas outcrop and biostratigraphic correlation with curve for North American Mid-Continent outcrop (modified from Boardman and Heckel, 1989).

events), the successive patterns of sea-level events can be used for high-resolution interbasinal and perhaps intercontinental correlation (Figs. 1 and 2). This methodology was employed by Boardman and Heckel (1989) to demonstrate cycle-by-cycle correlation between late Desmoinesian through middle Virgilian strata of the eastern shelf of the Midland basin and the northern eastern shelf of the Anadarko basin (Fig. 1).

Whereas onshore-offshore gradient models for macrofossil assemblages have been proposed for Late Carboniferous-Early Permian cyclothemic sedimentary sequences (Elias, 1937; Calver, 1968; Boardman et al., 1984a), no comprehensive model has been applied to microfossil assemblages. Biofacies models for individual groups have been proposed, but with no attempt to integrate all microfaunal components.

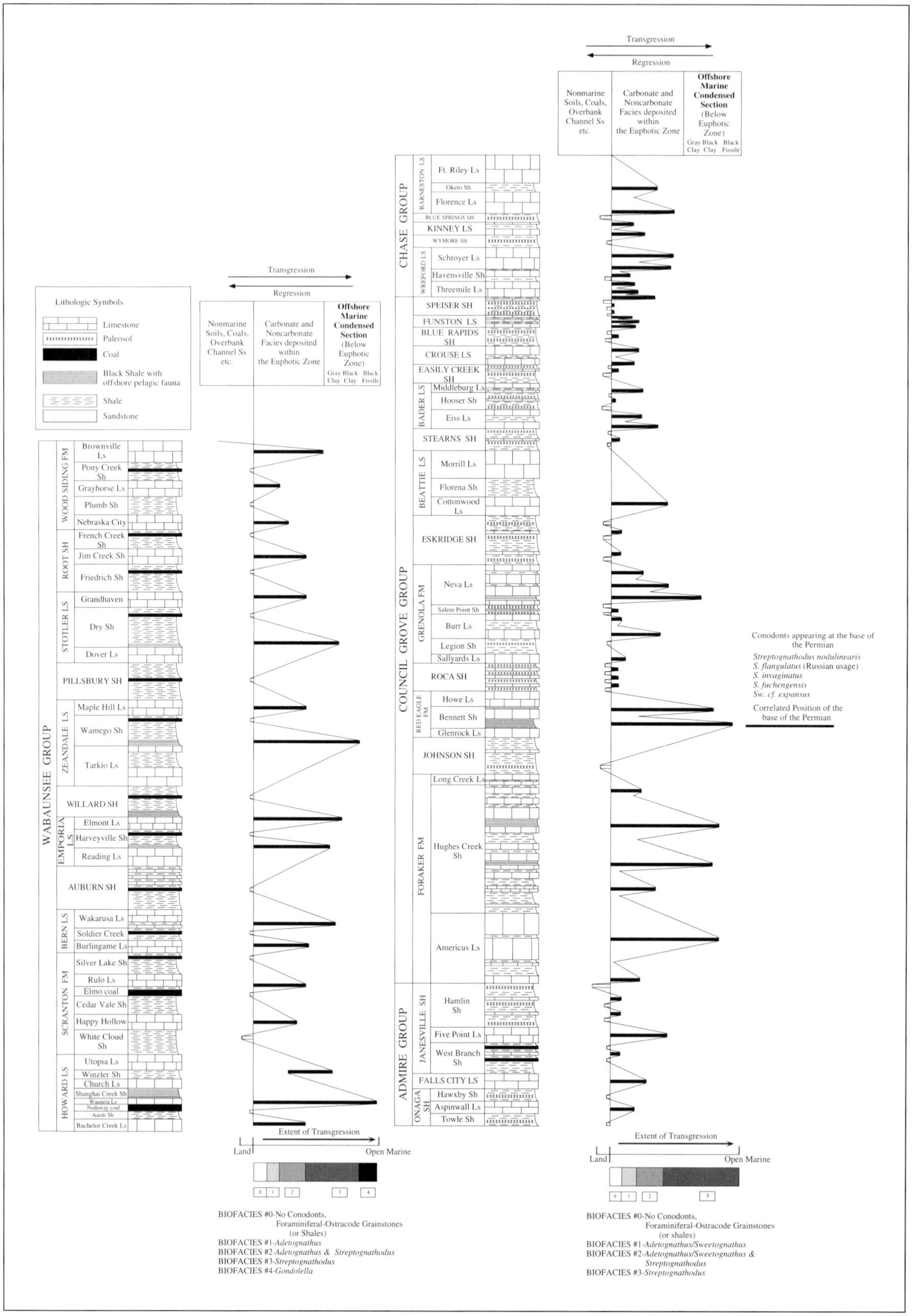

FIGURE 2 — Eustatic sea-level fluctuation curve for uppermost Pennsylvanian through lowermost Permian (upper Virgilian-lower Artinskian) strata of the North American Mid-Continent (modified from Boardman and Nestell, 1993).

The primary purpose of this study is to present preliminary findings on the distribution of microfossils (conodonts, foraminifers, ostracodes, and radiolarians) in relationship to the onshore-offshore gradient as independently determined by lithofacies and megafaunal analysis. The model presented should be regarded as preliminary because many additional data need to be analyzed so that three-dimensional (3-D) motifs showing lateral facies as well as vertical distributional data can be created for each individual cyclothemic sedimentary sequence.

Conodont Paleoecology (Plates 1 and 2)

Conodont paleoecologic models (Table 1) for Upper Carboniferous and Lower Permian strata have been proposed by Merrill (1973), von Bitter (1972), Heckel and Baesemann (1975), Merrill and von Bitter (1976, 1989), Ritter (1987), Swade (1985), and Heckel (1994). The most widely accepted model for the distribution of conodonts in Upper Carboniferous cyclothemic strata is that proposed by Heckel and Baesemann (1975) based on adaptation of the depth stratification model of Seddon and Sweet (1971). This model assumes a pelagic life mode for conodonts with different taxa living in different water masses that are vertically differentiated. It also predicts that nearshore assemblages are of low diversity in contrast to deep shelf assemblages that exhibit high diversity. The depth stratification model provides the best overall explanation for the observed stratigraphic distribution of conodonts within cyclothems. The well-documented trends of vertical distribution of conodonts documented by von Bitter (1972) and Baesemann (1973) for Virgilian and Missourian cyclothems, respectively, are most consistent with the depth stratification model. The succession of conodont genera from nearshore to offshore is as follows: *Adetognathus, Hindeodus, Aethotaxis, Idiognathodus* and *Streptognathodus, Idioprioniodus*, and *Gondolella.* The most offshore bioassemblage contains all genera of conodonts but is dominated by *Idiognathodus, Idioprioniodus,* and *Gondolella.* The most offshore bioassemblage does contain representatives from the shallow water masses (*Adetognathus, Hindeodus,* and *Aethotaxis*, but in greatly reduced numbers (<1%). Merrill and von Bitter (1976) recognized four biofacies (*Streptognathodus* I, *Streptognathodus* II, *Streptognathodus* III, and *Idiognathodus*) instead of a single *Idiognathodus* Biofacies. Baesemann (1973) and Heckel and Baesemann (1975) did not recognize *Streptognathodus* as being a valid genus and placed all representatives of *Streptognathodus* into *Idiognathodus.* Barrick and Boardman (1989) pointed out that *Streptognathodus* is a valid genus that evolved from *Idiognathodus* in early Missourian time (via the ancestral *Streptognathodus* species *S. cancellosus*). The following conodont biofacies are recognized by the current authors: *Adetognathus* Biofacies, *Streptognathodus-Idiognathodus* Biofacies, and *Gondolella-Idioprioniodus* Biofacies. Boundaries between these biofacies within a cyclothemic sequence are often gra-

von Bitter (1972)	Merrill (1973)	Merrill & von Bitter (1976)	Heckel & Baesemann (1975)	This Study
Lonchodina	Appalachian	*Aethotaxis*	Not Recognized	Not Recognized
Neoprioniodus conjunctus	Mid-Continent	*Idioprioniodus*	*Idioprioniodus*	*Idioprioniodus*
Not Recognized	Not Recognized	*Neognathodus*	Not Recognized	Not Recognized
Gondolella	*Gondolella*	*Gondolella*	*Gondolella*	*Gondolella*
Cavusgnathus	*Cavusgnathus*	*Cavusgnathus*	*Adetognathus*	*Adetognathus*
Streptognathodus	*Streptognathodus*	*Streptognathodus I*	*Idiognathodus*	*Streptognathodus I*
Streptognathodus gracilis	Not recognized	*Streptognathodus II*	*Idiognathodus*	*Streptognathodus II*
Not recognized	Not recognized	*Streptognathodus III*	*Idiognathodus*	*Streptognathodus III*

TABLE 1 — Conodont biofacies.

PLATE 1

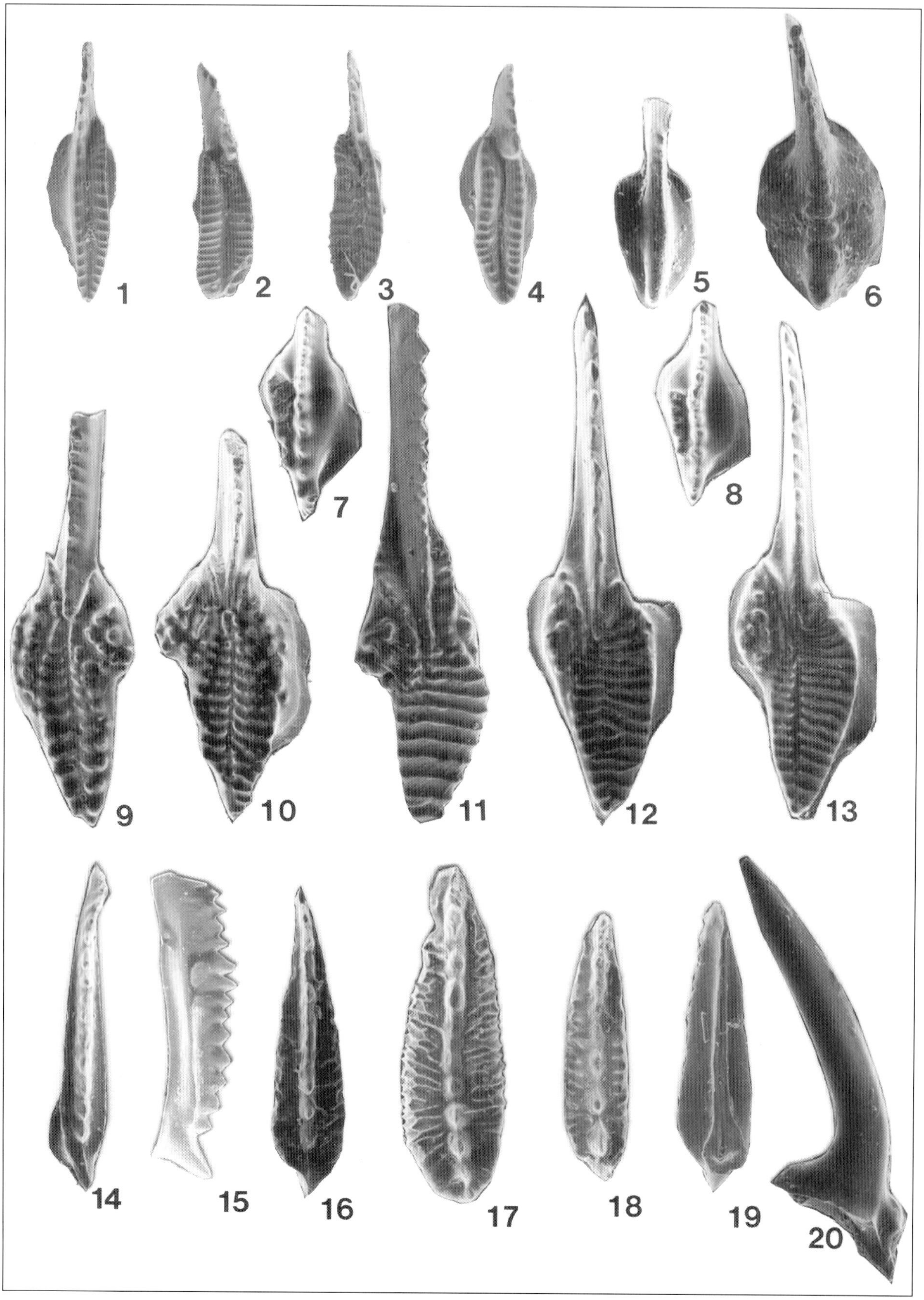

PLATE 1

CONODONT GENERA *ADETOGNATHUS, SWEETOGNATHUS, NEOGNATHODUS, IDIOGNATHODUS, IDIOPRIONIODUS, GONDOLELLA*

FIGURE DESCRIPTION

Figures 1 to 13 oral view of PA element; Figs. 14, 16, 17, and 18 oral views of PA element; Fig. 15 lateral view of PA element; Fig. 19 aboral view of PA element; Fig. 20 lateral view of PA element. All views X70.

Figures 1 to 6	Represent the nearshore *Adetognathus-Sweetognathus* Biofacies.
Figures 7 to 8	*Neognathodus Biofacies* (not recognized in this study). *Neognathodus* is most abundant in the *Streptognathodus* III Subfacies, the *Idiognathodus* Subfacies, or the *Idioprioniodus-Gondolella* Biofacies.
Figures 9 to 10	Represent the *Streptognathodus* III Subfacies.
Figures 11 to 13	Represent the *Idiognathodus* Subfacies.
Figures 14 to 20	Represent the offshore *Idioprioniodus-Gondolella* Biofacies.

Figure 1 — *Adetognathus* sp. A, SUI 87802, from the lower Towle Shale (6 ft above top of Brownville Limestone), Onaga Shale, Admire Group, upper Virgilian, of Oklahoma (Loc. 1).

Figure 2 — *Adetognathus* sp. B, SUI 87803, from the base of the Brownville Limestone Member, Woodsiding Formation, Wabaunsee Group, upper Virgilian of Kansas (Loc. 2).

Figure 3 — *Adetognathus* sp. C, SUI 87804, from the lower Americus Limestone Member, Foraker Limestone, Council Grove Group, upper Virgilian of Kansas (Loc. 3).

Figure 4 — *Adetognathus* sp. D, SUI 87805, from the upper ledge Aspinwall Limestone Member, Onaga Shale, Admire Group, upper Virgilian of Kansas (Loc. 4).

Figure 5 — *Sweetognathus expansus*, USNM 484185, from the Neva Limestone Member, Grenola Formation, Lower Permian of Kansas (Loc. 5).

Figure 6 — *Sweetognathus inornatus*, SUI 87806, from the Threemile Limestone Member, Wreford Formation of Kansas (Loc. 6).

Figure 7 — *Neognathodus dilatus*, SUI 87807, from the upper black shale bed, East Mountain Shale, Strawn Group, late Desmoinesian of north-central Texas (Loc. 7).

Figure 8 — *Neognathodus dilatus*, SUI 87808, from the upper black shale bed, East Mountain Shale, Strawn Group, late Desmoinesian of north-central Texas (Loc. 7).

Figure 9 — ?*Idiognathodus nodocarinatus*, SUI 87809, (= *Streptognathodus nodocarinatus* Jones, this species possesses a *Streptognathodus*-like groove but has no carina), from the upper black shale bed, East Mountain Shale, Strawn Group, late Desmoinesian of north-central Texas (Loc. 7).

Figure 10 — ?*Idiognathodus nodocarinatus*, SUI 87810, (= *Streptognathodus nodocarinatus* Jones, this species possesses a *Streptognathodus*-like groove but has no carina), from the upper black shale bed, East Mountain Shale, Strawn Group, late Desmoinesian of north-central Texas (Loc. 7).

Figure 11 — *Idiognathodus magnificus*, SUI 87811, from the PP3 Member of the Posideon Shale, Canyon Group, middle Missourian of north-central Texas (Loc. 8).

Figure 12 — *Idiognathodus* sp. A, SUI 87812 (without *Streptognathodus*-like groove), from the Finis Shale Member, Graham Formation, Cisco Group, middle Virgilian, north-central Texas (Loc. 9).

Figure 13 — *Idiognathodus* sp. A, SUI 87813 (with *Streptognathodus*-like groove), from the Finis Shale Member, Graham Formation, Cisco Group, middle Virgilian, north-central Texas (Loc. 9).

Figure 14 — *Gondolella aff. G. elegantula*, SUI 87814, from the Finis Shale Member, Graham Formation, Cisco Group, middle Virgilian, north-central Texas (Loc. 9).

Figure 15 — *Gondolella aff. G. elegantula*, SUI 87815, from the Finis Shale Member, Graham Formation, Cisco Group, middle Virgilian, north-central Texas (Loc. 9).

Figure 16 — *Gondolella* sp., SUI 87816, from the Wolf Mountain Shale Member, Graford Formation, Canyon Group, middle Missourian of north-central Texas (Loc. 10).

Figure 17 — *Gondolella cf. G. costata*, SUI 87817, from the PP3 Member of the Posideon Shale, Canyon Group, middle Missourian of north-central Texas (Loc. 8).

Figure 18 — *Gondolella magna*, SUI 87818, from the upper black shale bed, East Mountain Shale, Strawn Group, late Desmoinesian of north-central Texas (Loc. 7).

Figure 19 — *Gondolella magna*, SUI 87819, from the upper black shale bed, East Mountain Shale, Strawn Group, late Desmoinesian of north-central Texas (Loc. 7).

Figure 20 — *Idioprioniodus conjunctus*, SUI 87820, from the PP3 Member of the Posideon Shale, Canyon Group, middle Missourian of north-central Texas (Loc. 8).

dational except where extreme rates of sea-level rise or sea-level fall resulted in facies dislocation.

The *Adetognathus* Biofacies is characterized by the nominate taxon *Adetognathus*, which typically dominates the biofacies, along with representatives of *Hindeodus* and *Aethotaxis*. The *Adetognathus* Biofacies is lithology independent as it is found in carbonate as well as siliciclastic rocks. However, *Hindeodus* and *Aethotaxis* appear more prevalent in carbonate lithologies than in terrigenous clastic facies. *Adetognathus* is missing in the Mid-Continent due to extinction slightly above the base of the Permian (highest occurrence in the Red Eagle Limestone). Above this level, the *Adetognathus* conodont biofacies is replaced by the *Sweetognathus* Biofacies in the nearshore regime (Ritter, 1987).

The *Idiognathodus-Streptognathodus* Biofacies is characterized by the nominate taxa which typically dominate the biofacies. This biofacies is also lithology independent and is found in carbonate and siliciclastic rocks. Subordinate numbers of *Hindeodus* and *Aethotaxis* are present, but are more common in carbonate lithologies. The subdivisions within this biofacies presented by Merrill and von Bitter (1976) are also recognized in this study.

(a) *Streptognathodus* I Subfacies — This subfacies was established by Merrill and von Bitter (1976) to include narrow, node-less *Streptognathodus* such as *S. elegantulus.* Based on the distributional data of the current authors, narrow slightly noded taxa are included within this subfacies. *Streptognathodus gracilis, S. pawhuskensis* (= *S. alekseevi* of Russian workers), *S. brownvillensis, S. simplex, S. elongatus, S. longissimus,* and *S. constrictus* in addition to *S. elegantulus* are recognized as typical representatives of the biofacies. Merrill and von Bitter (1976) suggested that this subfacies represented the majority of environments dominated by the *Streptognathodus-Idiognathodus* Biofacies. Currently, this subfacies is viewed to represent the shallow end of the *Streptognathodus-Idiognathodus* Biofacies. This conclusion is based on observational data on the spatial distribution of conodonts within individual cyclothems. Commonly, cyclothems that have robust heavily noded representatives of the *Streptognathodus-Idiognathodus* Biofacies grade shoreward and over paleo-topographic highs into biofacies with narrow, nodeless, or reduced noded forms before grading into the *Adetognathus* Biofacies. The *Streptognathodus* I Subfacies becomes well established in the middle Missourian and extends up into Sakmarian strata of the Mid-Continent.

(b) *Streptognathodus* II Subfacies — This subfacies was established by Merrill and von Bitter (1976) to include noded species of *Streptognathodus* including *S. gracilis* and *S. excelsus.* The current authors also include robust unnoded species of *Streptognathodus* in this subfacies. Merrill and von Bitter (1976) concluded that this subfacies was represented in restricted shallow-water environments. As previously mentioned, *S. gracilis* is included in the *Streptognathodus* I Subfacies. The *Streptognathodus* II Subfacies includes *S. cancellosus, S. confragus, S. corrugatus, S. firmis, S. zethus, S. wabaunsensis, S. flangulatus, S. fuchengensis, S. nodulinearis, S. barskovi,* and *S. farmeri.* This subfacies is considered to represent more open marine offshore environments. The *Streptognathodus* II Subfacies becomes well established in the middle Missourian and extends up into lower Artinskian strata of the Mid-Continent.

(c) *Streptognathodus* III Subfacies — This subfacies was established by Merrill and von Bitter (1976) to include deeply grooved late Desmoinesian streptognathodid-like forms (e.g., *Streptognathodus aff. excelsus*-Moscow basin, *S.* sp. 5 of Swade [1985], *S.* sp. 6 Swade [1985], and *S. nodocarinatus* Jones). This subfacies is typically found in dark gray to black phosphatic shales in the Mid-Continent. This subfacies is unnecessary in the Mid-Continent because these species of *Streptognathodus* occur with *Idioprioniodus* and *Gondolella.* However, it is possible that in regions where *Gondolella* and *Idioprioniodus* are not present that this subfacies would be significant and would represent the maximum offshore fauna. This interpretation is in stark contrast to the Merrill and von Bitter conclusion that this subfacies represents a nearshore restricted biofacies. This subfacies is present only in late Desmoinesian strata or equivalent (upper Moscovian) strata.

(d) *Idiognathodus* Subfacies — This subfacies was been recognized by numerous workers. The current concept of the composition of this subfacies is similar to that of Merrill and von Bitter (1976). Merrill and von Bitter (1976) demonstrated that through the Desmoinesian, *Idiognathodus* dominated most environments within the *Streptognathodus-Idiognathodus* Biofacies. However, they believed that during the Missourian *Idiognathodus* was more common in restricted shallow-water environments. In addition to *Idiognathodus*, Merrill and von Bitter (1976) included *Streptognathodus eccentricus* and *Streptognathodus simulator* within the *Idiognathodus* Subfacies. Barrick and Boardman (1989) agreed that *S. simulator* and *S. eccentricus* were more idiognathodid-like and placed them into *Idiognathodus*. In the Mid-Continent, this subfacies is found with *Idioprioniodus* and *Gondolella*, whereas *Gondolella* is rarely found in association in the Appalachian basin. Due to this difference, the Missourian and Virgilian *Idiognathodus* Subfacies actually is in large part equivalent to the *Gondolella-Idioprioniodus* Subfacies in the Mid-Continent but has no other counterpart in the Appalachian basin. It is for this reason that the current authors also recognize this subfacies. They agree with Merrill and von Bitter (1976) that Missourian and Virgilian idiognathodids are typically associated with stressed marine environments, but they view this subfacies as representing a low-oxygen, offshore marine setting instead of a shallow, marginal marine, low-pH environment.

The *Gondolella-Idioprioniodus* Biofacies is characterized by the nominate taxa along with abundant representatives of the *Streptognathodus-Idiognathodus* Biofacies and locally with minor representation (<1%) of the *Adetognathus* Biofacies. This biofacies is abundant in dark gray to black phosphatic shales of north-central Texas, the Mid-Continent, and Illinois basin, but is rare in the Appalachian basin. As correctly pointed out by Merrill and von Bitter (1989), this biofacies is not restricted to black phosphatic shales. The current authors have recovered this biofacies from highly fossiliferous phosphatic and glauconitic wackestones and packstone carbonates as well as glauconitic sandstones. Additionally, *Gondolella* and *Idioprioniodus* have been observed in carbonates interpreted to represent shelf edge limestones in the Marathon uplift. Although the *Gondolella-Idioprioniodus* Biofacies is lithology independent, this biofacies is viewed as representing offshore outer shelf water depths instead of the nearshore restricted marine model proposed by Merrill and von Bitter (1976). Merrill and von Bitter considered these taxa to be associated with low-pH water conditions, whereas the authors of this study view these taxa to have inhabited lower oxygen and cooler water regimes. Two subfacies are recognized within the *Gondolella-Idioprioniodus* Biofacies: *Idioprioniodus* Subfacies and *Gondolella* Subfacies.

(a) *Idioprioniodus* Subfacies — This subfacies containing *Idioprioniodus* without *Gondolella* was recognized by Heckel and Baesemann (1975) and Merrill and von Bitter (1976). This subfacies is abundant in dark gray to black phosphatic shales, but is also present in highly fossiliferous glauconitic and phosphatic wackestones. This subfacies is viewed as representing a slightly shallower end of the biofacies as evidenced by co-occurrence with fusulinaceans and red phylloid algae in some carbonates. However, some black, fissile phosphatic shales contain only *Idioprioniodus* without *Gondolella* (Little Osage, Anna, and Lake Neosho as well as coeval strata in north-central Texas). The absence of *Gondolella* during that part of the late Desmoinesian in the Midland, Mid-Continent, and Illinois basins is anomalous, and may represent local extinction with later re-migration of the genus.

(b) *Gondolella* Subfacies — This subfacies recognized by numerous workers contains both *Gondolella* and *Idioprioniodus*. This biofacies is thought to represent the most offshore conodont biofacies in Upper Carboniferous cyclothemic strata. The highest stratigraphic occurrence of the *Gondolella* subfacies in the Mid-Continent occurs in the Shanghai Creek Shale Member of the Howard Limestone (basal Wabaunsee Group-late Virgilian = Aarde of Heckel, 1986). This level also corresponds to the last black, fissile, phosphatic shale in the Mid-Continent. The association of radiolari-

PLATE 2

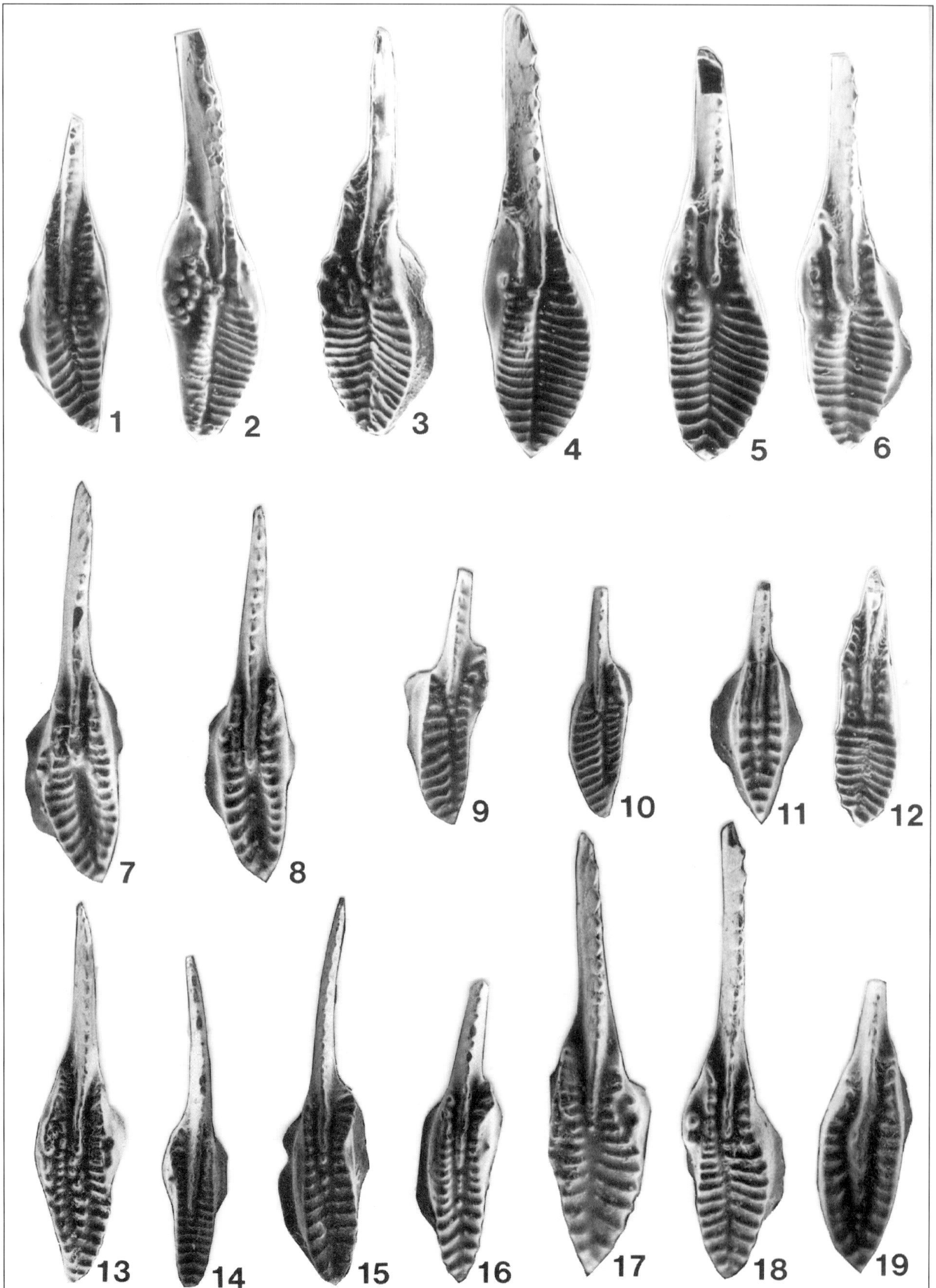

PLATE 2
SPECIES OF THE CONODONT GENUS *STREPTOGNATHODUS*
FIGURE DESCRIPTION

All figures oral views of PA element

Figures 1-7, 8, 12-14, 17, 19	Represent the *Streptognathodus* II Subfacies.
Figures 11, 15-16	Represent the *Streptognathodus* I Subfacies.
Figures 9-10, 18	Represent the *Idiognathodus* Subfacies.

Figure 1 — *Streptognathodus aff. S. flangulatus*, USNM 484089, from the upper part of the Bennett Shale Member, Red Eagle Limestone, Council Grove Group, basal Permian of Kansas (Loc. 11) X70.

Figure 2 — *Streptognathodus aff. S. flangulatus*, USNM 484094, from the upper part of the Bennett Shale Member, Red Eagle Limestone, Council Grove Group, basal Permian of Kansas (Loc. 11) X70.

Figure 3 — *Streptognathodus aff. S. flangulatus*, USNM 484081, from the upper part of the Bennett Shale Member, Red Eagle Limestone, Council Grove Group, basal Permian of Kansas (Loc. 11) X70.

Figure 4 — *Streptognathodus fuchengensis*, USNM 484085, from the upper part of the Bennett Shale Member, Red Eagle Limestone, Council Grove Group, basal Permian of Kansas (Loc. 11) X70.

Figure 5 — *Streptognathodus fuchengensis*, USNM 484084, from the upper part of the Bennett Shale Member, Red Eagle Limestone, Council Grove Group, basal Permian of Kansas (Loc. 11) X70.

Figure 6 — *Streptognathodus nodulinearis*, USNM 484087, from the upper part of the Bennett Shale Member, Red Eagle Limestone, Council Grove Group, basal Permian of Kansas (Loc. 11) X70.

Figure 7 — *Streptognathodus* cf. *S. pawhuskensis* (= *S. alekseevi* of Russian workers), SUI 56362, from the Colony Creek Shale Member, Caddo Creek Formation, Canyon Group, basal Virgilian of north-central Texas (Loc. 12). From Barrick and Boardman (1989, Fig. 22, Plate 3) X50.

Figure 8 — *Streptognathodus zethus*, SUI 87826, from the Colony Creek Shale Member, Caddo Creek Formation, Canyon Group, basal Virgilian of north-central Texas (Loc. 12). From Barrick and Boardman (1989, Fig. 23, Plate 3) X50.

Figure 9 — *Idiognathodus simulator* (= *Streptognathodus simulator* of most workers), SUI 56349, from the Finis Shale Member, Graham Formation, Cisco Group, middle Virgilian of north-central Texas (Loc. 13). From Barrick and Boardman (1989, Fig. 9, Plate 3) X60.

Figure 10 — *Idiognathodus simulator* (= *Streptognathodus simulator* of most workers), SUI 56348, from the Finis Shale Member, Graham Formation, Cisco Group, middle Virgilian of north-central Texas (Loc. 14). From Barrick and Boardman (1989, Fig. 8, Plate 3) X60.

Figure 11 — *Streptognathodus ruzhencevi*, SUI 56353, from the Necessity Shale Member, Graham Formation, Cisco Group, middle Virgilian of north-central Texas (Loc. 15). From Barrick and Boardman (1989, Fig. 13, Plate 3) X60.

Figure 12 — *Streptognathodus conjunctus*, USNM 484034, from the lower Hughes Creek Shale Member, Foraker Limestone, Council Grove Group, latest Carboniferous of Kansas (Loc. 11) X70.

Figure 13 — *Streptognathodus cancellosus*, SUI 56315, from the upper Salesville Shale, Strawn Group, lower Missourian of north-central Texas (Loc. 16). From Barrick and Boardman (1989, Fig. 18, Plate 1) X45.

Figure 14 — *Streptognathodus confragus*, SUI 56302, from the basal Palo Pinto Limestone (Wynn Limestone), Canyon Group, middle Missourian of north-central Texas (Loc. 17). From Barrick and Boardman (1989, Fig. 5, Plate 1) X55.

Figure 15 — *Streptognathodus elegantulus*, BEG 19166-Holotype, from the Wolf Mountain Shale Member, Graford Formation, middle Missourian of north-central Texas (Loc. 18). From Barrick and Boardman (1989, Fig. 14, Plate 2) X45.

Figure 16 — *Streptognathodus gracilis*, SUI 56332, from the PP3 Member of the Posideon Shale, Canyon Group, middle Missourian of north-central Texas (Loc. 8). From Barrick and Boardman (1989, Fig. 12, Plate 2) X60.

Figure 17 — *Streptognathodus excelsus*, SUI 56339, from the Wolf Mountain Shale Member, Graford Formation, middle Missourian of north-central Texas (Loc. 10). From Barrick and Boardman (1989, Fig. 22, Plate 2) X55.

Figure 18 — *Idiognathodus* cf. *simulator* (= *Streptognathodus simulator* of most authors), SUI 56360, from the upper Winchell Limestone Member, Graford Formation, Canyon Group, upper Missourian of north-central Texas (Loc. 19). From Barrick and Boardman (1989, Fig. 20, Plate 3) X55.

Figure 19 — *Streptognathodus firmis*, SUI 56365, from the basal Ranger Limestone Member, Brad Formation, Canyon Group, uppermost Missourian of north-central Texas (Loc. 20). From Barrick and Boardman (1989, Fig. 25, Plate 3) X75.

PLATE 3

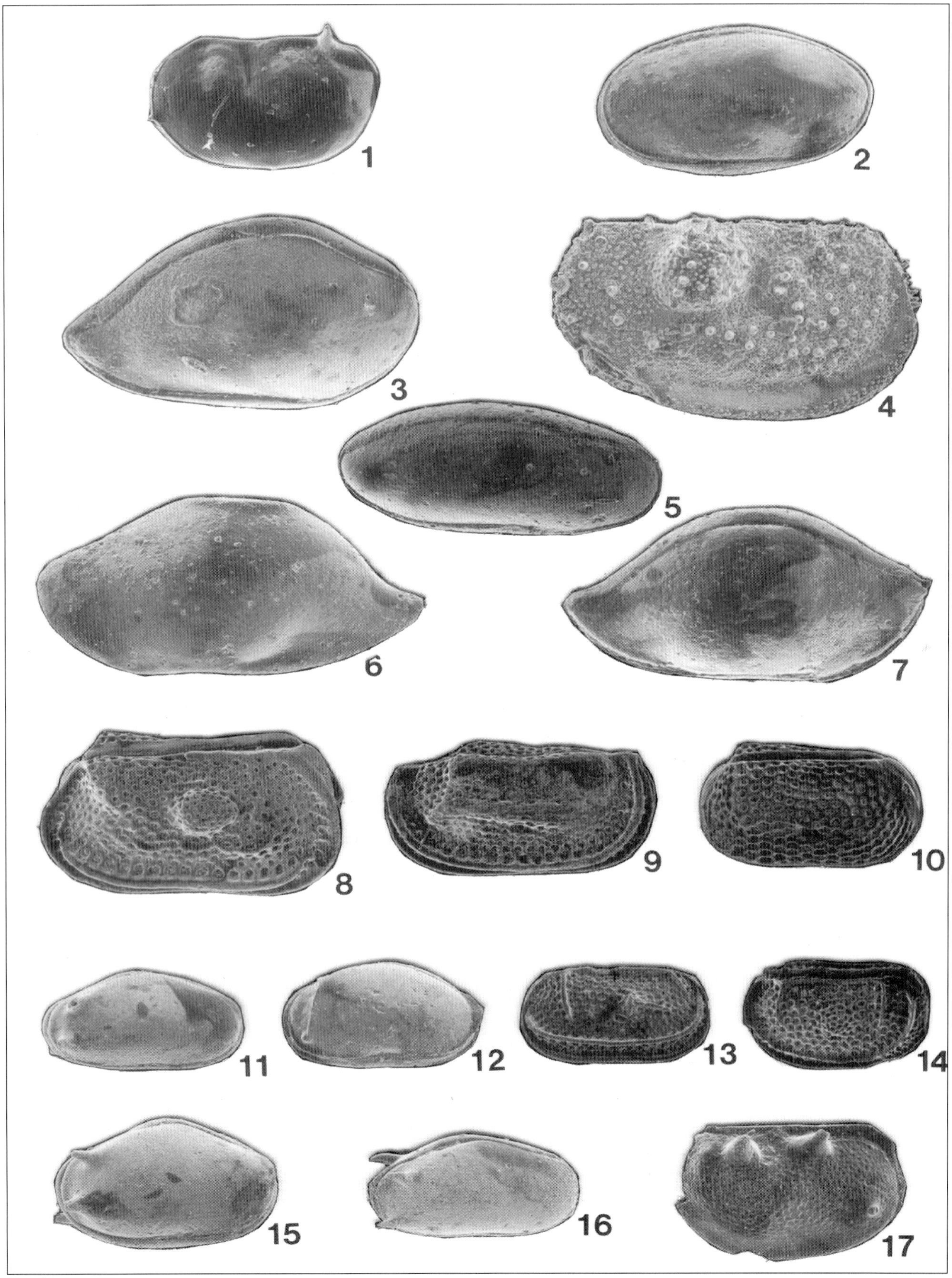

PLATE 3
OSTRACODES
FIGURE DESCRIPTION

Plate 3 — All magnifications approximately X50; Fig. 1 from the Cedar Vale Shale Member, Scranton Formation, Wabaunsee Group, late Virgilian of Nebraska (Loc. 21); Figs. 2 to 11, 13 to 14, and 16 from the Dog Bend Limestone Member, Salesville Formation, Strawn Group, lower Missourian of north-central Texas (Loc. 22); Figs. 12, 15, and 17 from the upper black shale bed, East Mountain Shale, Strawn Group, late Desmoinesian from north-central Texas (Loc. 7).

Figure 1 — *Geisina spinosa*, SUI 87821, lateral view of left valve.

Figure 2 — *Cavellina pulchella*, SUI 87822, left view of carapace.

Figure 3 — *Orthobairdia oklahomaensis*, SUI 87823, right view of carapace.

Figure 4 — *Hollinella (Hollinella) ulrichi*, SUI 87824, lateral view of right valve.

Figure 5 — *acetalata*, SUI 87825, right view of carapace.

Figure 6 — *pompilioides*, SUI 87826, lateral view of left valve.

Figure 7 — *Bairdia* sp., SUI 87827, right view of carapace.

Figure 8 — *Kegelites aff. K.* sp. B Sohn, SUI 87828, right view of carapace.

Figure 9 — *Aurikirkbya knighti*, SUI 87829, right view of carapace.

Figure 10 — *Shlessha pinguis*, SUI 87830, right view of carapace.

Figure 11 — *Healdia* sp., SUI 87831, right view of carapace.

Figure 12 — *Healdia ehlersi*, SUI 87832, right view of carapace.

Figure 13 — *Amphissites girtyi*, SUI 87833, right view of carapace.

Figure 14 — *Amphissites centronotus*, SUI 87834, right view of carapace.

Figure 15 — *Healdia elegans*, SUI 87835, right view of carapace.

Figure 16 — *Healdia oblongata*, SUI 87836, right view of carapace.

Figure 17 — *Mammoides* cf. *M. mammillata*, SUI 87837, left view of carapace.

---

an-bearing nonskeletal phosphates and *Gondolella* in Desmoinesian-middle Virgilian cyclothems is attributed by Heckel (1977, 1986, 1990, 1991a, 1994) and Boardman et al. (1984) to represent deposition in relatively deep, cooler nutrient-rich waters that resulted from a thermocline and upwelling of nutrient-rich waters from the basin onto the shelves. In the uppermost Carboniferous-lowermost Permian strata (Foraker Limestone, Red Eagle Limestone, and Neva Limestone), black moderately fissile shales are present in northern Kansas and Nebraska. These shales contain only skeletal phosphate and contain no evidence of radiolarians. The Mid-Continent Anadarko basin apparently was undergoing greatly reduced rates of subsidence along with rapid infilling by late Virgilian-Lower Permian time (Rascoe, 1962; Rascoe and Adler, 1983), thereby reducing access to much nutrient-rich cool water that would enhance phosphate precipitation and support *Gondolella* and (later) *Neogondolella*. Coeval basinal and shelf edge strata in the southern Urals and western basins of the United States commonly contain radiolarian-bearing phosphates and carbonate nodules, and a more or less continuous sequence of *Gondolella* and *Neogondolella* species. *Gondolella* disappears near the Carboniferous-Permian boundary, whereas *Neogondolella* is long ranging through the Permian.

Ostracodes (Plate 3)

Ostracodes are often present in large numbers in certain lithic members of Mid-Continent Pennsylvanian and Lower Permian cyclothems. Numerous ostracode species have been described from these rocks, mostly by authors sponsored by the petroleum industry more than 50 years ago (Sohn, 1982). Largely because of their complex morphology, these ostracodes have been studied by specialists who have devoted much of their efforts to taxonomic analysis and revision of the group.

Recently Melynk and Maddocks (1988) detailed the distribution of some 226 Permo-Carboniferous ostracode species along an onshore-offshore gradient by means of an independent coenocline derived from what they considered to be paleoecologically significant taxonomic units summarized from the literature. They concluded that *Mammoides mammillata, Healdia simplex, Healdia elegans,* and *Pseudobythocypris pediformis* inhabited nearshore environments, whereas *Amphissites centronotus, A. girtyi,* and *Aurikirkbya knighti* lived offshore. Due to the principal components of analysis methodology used in their study, the onshore versus offshore gradient was determined from inferred paleoecological analysis of other taxa (informal taxonomic units — ITUs). The distribution of conodonts and foraminifers within the section played a big role in their determination of the onshore-offshore positions. The foraminifer *Ammodiscus* was considered to be very shallow water, whereas all conodonts were placed in very shallow to deep. No discrimination of conodont genera was used to determine shallow versus deep, and their conodonts were all assumed to be shallow water. The current authors demonstrate in this study that some species of *Ammodiscus* inhabited shallow waters while others inhabited intermediate and deep waters. Likewise, the conodont *Adetognathus* is widely regarded as representing shallow water, whereas *Gondolella* is thought to represent an offshore environment (Heckel and Baesemann, 1975). The present study reveals that an onshore-offshore distribution of ostracodes can indeed be attained, but disagree with Melynk and Maddocks (1988) as to which were onshore and which were offshore. The ordination method used by Melynk and Maddocks (1988) is viewed by the current authors as robust, but they maintain that a more refined analysis of the ITUs is necessary to arrive at an acceptable onshore-offshore gradient.

This analysis reveals that five distinct ostracode biofacies can be recognized within Pennsylvanian and Lower Permian cyclothems. These biofacies are defined on easily recognized taxonomic groups, simple morphological features, and relative species diversity and dominance. From the most nearshore to offshore, the following biofacies are recognized: *Geisina spinosa* Biofacies, *Cavellina* Biofacies, *Amphissites* Biofacies, *Healdia* Biofacies, and the *Mammoides cf. mammillata* Biofacies.

The *Geisina spinosa* Biofacies is characterized by very low diversity assemblages dominated by tremendous numbers of *Geisina spinosa*. This biofacies occurs in dark gray to black nonphosphatic shales with rare pectinid, myalinid, and *Dunbarella* clams, and *Donaldina* gastropods. The distribution of this biofacies is restricted to very late regressive or very early transgressive deposits, and is interpreted to represent restricted bodies of water (probably lagoons of variable salinity regimes from brackish to hypersaline).

The *Cavellina* Biofacies is characterized by a very high dominance of species belonging to the genus *Cavellina*. Additionally, *Kegelites dattonensis, Pseudobythocypris pediformis,* and species of *Hollinella* and *Bairdia* round out the assemblage. *Cavellina* usually constitutes more than 75% of the ostracode specimens within the fauna. This biofacies is thought to reflect nearshore but fully marine conditions. Typically, the megafauna associated with the *Cavellina* is molluscan dominated. This biofacies is very common in the shallower parts of Pennsylvanian and Lower Permian cyclothems from north-central Texas and the Mid-Continent.

The *Amphissites* Biofacies is characterized by the nominate taxa *Amphissites centronotus*, and *Amphissites girtyi* along with several other reticulate ostracode species including *Kirkbya punctata, Aurikirkbya knighti, Kegelites dattonensis, Roundyella simplicissima,* and *Shleesha pinguis.* Additionally, *Healdia ehlersi, Healdia elegans, Bairdia pompilioides, Fabalicypris acetalata, Orthobairdia oklahomensis, Hollinella ulrichi, Mooreites minutus, Pseudobythocypris pediformis, Pseudoparparchites oblongus, Shivaella brazoensis, Silenites fabalis,* and *Sulcella sulcata* are also present. The *Amphissites* Biofacies is characterized by high species diversity (>16 species), and equitable species distribution. The *Amphissites* Biofacies is represented in the gray core shale facies in the carbonate-dominated northern Mid-Continent cyclothems as well as in shale partings within the lower part of the "regressive limestone" facies in that region. In the southern Mid-Continent, the *Amphissites* Biofacies is similarily well represented in gray core shale facies and shale partings in the lower "regressive limestone" facies. The *Amphissites* Biofacies displays the highest species diversity of any ostracode biofacies, and also is found with open marine assemblages of highly diverse megafossils.

The *Healdia* Biofacies is characterized by a low species diversity of ostracodes dominated by several species of *Healdia* including *Healdia ehlersi, Healdia elegans, Healdia oblongata*, and *Healdia simplex*, along with other taxa, including *Orthobairdia oklahomensis, Cavellina pulchella,* and *Pseudobythrops pediformis*. This biofacies is gradational with the *Mammoides* Biofacies with which it shares most of the same taxa. This biofacies occurs in dark gray pyritic, slightly phosphatic shales that overlie and underlie the black phosphatic shales. This biofacies is also apparently restricted to the Oklahoma and north Texas regions with higher siliclastic influx than northward.

The *Mammoides cf. M. mammillata* Biofacies is characterized by the nominate species along with *Healdia ehlersi, Healdia elegans, Healdia oblongata, Healdia simplex,Waylandella symmetrica*, and *Cavellina pulchella*. This biofacies is of low species diversity with the nominate taxon commonly composing 40% of the ostracode fauna, and is present in black, phosphatic clay shales present in Pennsylvanian cyclothems in Oklahoma and Texas but rare-to-absent in coeval strata in northern Kansas and Nebraska. These Texas and Oklahoma black shales are nonfissile compared to their highly fissile northern Mid-Continent shelf (Kansas and Nebraska) counterparts. The current authors attribute the lack of fissility to less organic carbon contrasted to the high organic-rich black fissile shales typical of the northern Mid-Continent carbonate-dominated shelf. The lowered amounts of total organic carbon seen in the Texas and Oklahoma dark shales are due to sediment dilution in those regions caused by highstand deltaic progradation. Thick highstand deltaic sequences commonly directly overlie this facies (Hentz, 1994). The authors of the current study suggest that the highstand deltaic progradation was at least a factor in partial oxygenating of the lower part of the water column which may have enabled oxygen levels to be sufficient to support this assemblage of ostracodes. A parallel pattern for megafossils was noted by Boardman et al. (1984). Typical examples of this biofacies are found in the upper East Mountain black shale bed, Strawn Group, uppermost Desmoinesian, and the coeval upper Holdenville Nuyaka Creek Black shale bed of Oklahoma. Due to the presence of the radiolarian-bearing nonskeletal phosphate, high concentration of diverse conodonts (>1000/kg), pyritized mollusks including ammonoids, and chondrites ichnofacies, this biofacies is viewed as representing an oxygen-stressed outer shelf depth environment.

Foraminifers (Plate 4)

Recently, Murray (1991) presented an extensive discussion and bibliography of the ecology of benthic foraminifers. Perusal of this monograph will quickly convince the reader of the ecological complexity of the distribution of Holocene foraminifers. Ecological parameters noted include mode of life (epifaunal, semi-infaunal, and infaunal) and related test architecture, ecophenotypical variation, relationship of test structure to substrate, and dimorphism. The collection and interpretation of data recoverable from fossil foraminiferal assemblages are just as complex, with the further addition of such factors as diagenesis and preservation. In this study, the authors present a few observations of the occurrences of several foraminifer taxa that are ubiquitous in north-central Texas, Oklahoma, and Kansas Late Pennsylvanian-Lower Permian cyclical lithofacies. Although closely spaced samples have been taken through several complete cycles, it is not the purpose here to analyze distribution of all foraminiferal taxa in these strata because such a study is beyond the scope of this work, and would require updated classification of the taxa described by Cushman and Waters (1927, 1928a, 1928b, 1930), Plummer (1930, 1945), Harlton (1927a, b), Waters (1927), and others. In several master's the-

PLATE 4

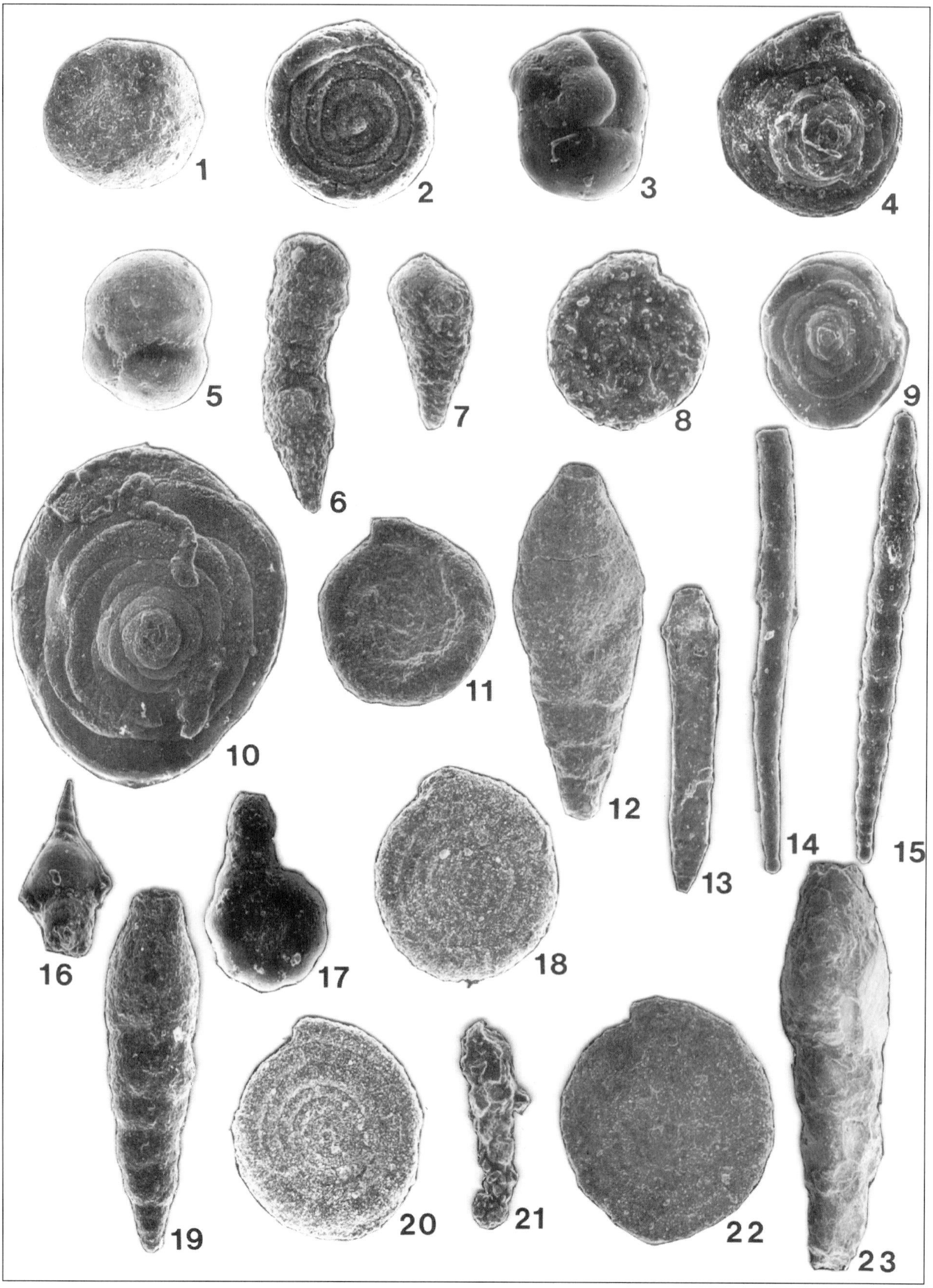

PLATE 4
FORAMINIFERS AND RADIOLARIA
FIGURE DESCRIPTION

Figure 1 from shale immediately below Salem School Limestone Member, Graham Formation, Cisco Group, middle Virgilian north-central Texas (Loc. 9), Figs. 2 to 4 from the Salem School Limestone interval, Graham Formation, Cisco Group, middle Virgilian of north-central Texas (Loc. 9), Figs. 5 to 9 from top of the Finis Shale Member, Graham Formation, Cisco Group of north-central Texas, Figs. 10 to 11 from shale above the Jacksboro Limestone Member, Graham Formation, Cisco Group, middle Virgilian of north-central Texas (Loc. 23), Figs. 12 to 18 from the Wolf Mountain Shale Member, Graford Formation, Canyon Group, middle Missourian of north-central Texas (Loc. 18), Figs. 19 to 20 from base of black Finis Shale Member, Graham Formation, Cisco Group, middle Virgilian of north-central Texas (Loc. 9), Figs. 21 to 23 from upper black shale bed East Mountain Shale, Strawn Group, late Desmoinesian from north-central Texas (Loc. 7).

Figure 1 — *Thurammina sphaeroidalis*, SUI 87838, X50.

Figure 2 — *Ammodiscus* cf. *A. semiconstrictus*, SUI 87839, X100.

Figure 3 — *Globivalvulina* cf. *G. ovata*, SUI 87840, X100.

Figure 4 — *Tetrataxis* sp., SUI 87841, X50.

Figure 5 — *Globivalvulina* sp, SUI 87842, X50.

Figure 6 — *Climacammina cylindrica*, SUI 87843, X25.

Figure 7 — *Climacammina cylindrica*, SUI 87844, X25 (juvenile).

Figure 8 — *Ammodiscus* sp., SUI 87845, X50.

Figure 9 — *Tetrataxis corona*, SUI 87846, X50.

Figure 10 — *Tetrataxis corona*, SUI 87847 X75 (with encrusting foraminifer).

Figure 11 — *Ammodiscus* cf. *A. semiconstrictus* var. *regularis*, SUI 87848, X100.

Figure 12 — *Reophax* cf. *R. fittsi*, SUI 87852, X50.

Figure 13 — *Sansabina elegantissima*, SUI 87853, X50.

Figure 14 — *Erlandia* cf. *E. minuta*, SUI 87854, X75.

Figure 15 — *Nodosinella perelegans*, SUI 87855, X75.

Figure 16 — *Albaillella* sp., SUI 87849, X100 (Radiolarian).

Figure 17 — *Endothyranella* cf. *E. powersi*, SUI 87850, X50.

Figure 18 — *Ammodiscus* cf. *A. semiconstrictus*, SUI 87851, X90.

Figure 19 — *Reophax* cf. *R. fittsi*, SUI 87856, X50.

Figure 20 — *Ammodiscus* cf. *A. semiconstrictus*, SUI 87857, X90.

Figure 21 — *Ammobaculites stenomecus*, SUI 87858, X90.

Figure 22 — *Ammodiscus* cf. *A. semiconstrictus*, SUI 87859, X90.

Figure 23 — *Reophax* cf. *R. fittsi*, SUI 87860, X50.

---

ses (Early, 1951; Wilson, 1954; Drake, 1958; Echols, 1959), data regarding the distribution of a number of microfossil taxa are presented with some environmental interpretation, but these data are not presented within the modern framework of the transgressive/regressive cyclical nature of these strata. In this study, a simplified depth zonation is presented for foraminifers consisting of three bioassemblages that can be easily recognized by distinctive lithofacies association, stratigraphic position in transgressive or regressive sequences, or by accompanying megafossils.

As is to be expected, residues containing the shallow bioassemblage are usually dominated by a few species, sometimes in high numbers as in the case of a species referred to as *Ammodiscus* cf. *A. semiconstrictus* var. *regularis*, which occurs often in the samples from *Myalina* banks. *Tetrataxis*, minute paleotextularids, and other forms such as lenticular fusulinaceans (*Ozwainella*) are sometimes present. Often great numbers of encrusting foraminifers of calcitornellid and apterrinellid types are present in residues obtained from both shale and limestone. *Thurammina sphaerodalis* was the sole taxon found in association with leaf impressions in a green mudstone at the base of a typical transgressive cycle (Finis Shale) in north-central Texas. This occurrence is thought to correspond to a marginal marine brackish water association.

Intermediate depth bioassemblages are highly diverse and contain taxa that occupy a variety of life modes. For example, *Tetrataxis* and *Globivalvulina*, are epifaunal. Other taxa present include ammodiscids, large paleotextularids (*Climacammina, Deckerella*), endothyranellids (*Endothyra, Endothyranella*), *Ammobaculites*, and fusulinaceans. Further subdivision of these bioassemblages is very difficult, as a great variety of lithofacies are present which affect the type of taxa present.

Recognition of the deep shelf bioassemblage is probably the easiest of the three categories. Detailed sampling within "core shales" (marine condensed sections) revealed several common taxa. For example, *Reophax fittsi* appears in nearly every sample in varying numbers. Often this distinctive vase-shaped agglutinated form incorporates conodont elements, other fossil debris, and quartz grains into its test. Other taxa present include *Ammodiscus semiconstrictus, Sansabaina, Hyperammina, Earlandia, Nodosinella,* and *Endothyranella.* Distinctive large specimens of *Ammodiscus semiconstrictus* are often present in large numbers. Radiolarians are present in dark gray to black shales in Oklahoma, Kansas, and north-central Texas and have been reported from phosphate nodules in black shales in Kansas (Nodine-Zeller, 1978; Nodine Zeller et al. 1985; Kidder 1985). The specimen illustrated in Plate 4, Fig. 16, represents the first documented radiolaria from late Pennsylvanian strata in north-central Texas.

COMPOSITE LITHOFACIES AND BIOFACIES MODEL

The basic model for the northern Mid-Continent cyclothem showing vertical changes in lithofacies, sequence stratigraphic terminology, inferred vertical shifts in depositional environments, and sea-level fluctuation curve is shown in Fig. 3. Although this model adequately represents a "typical Kansas

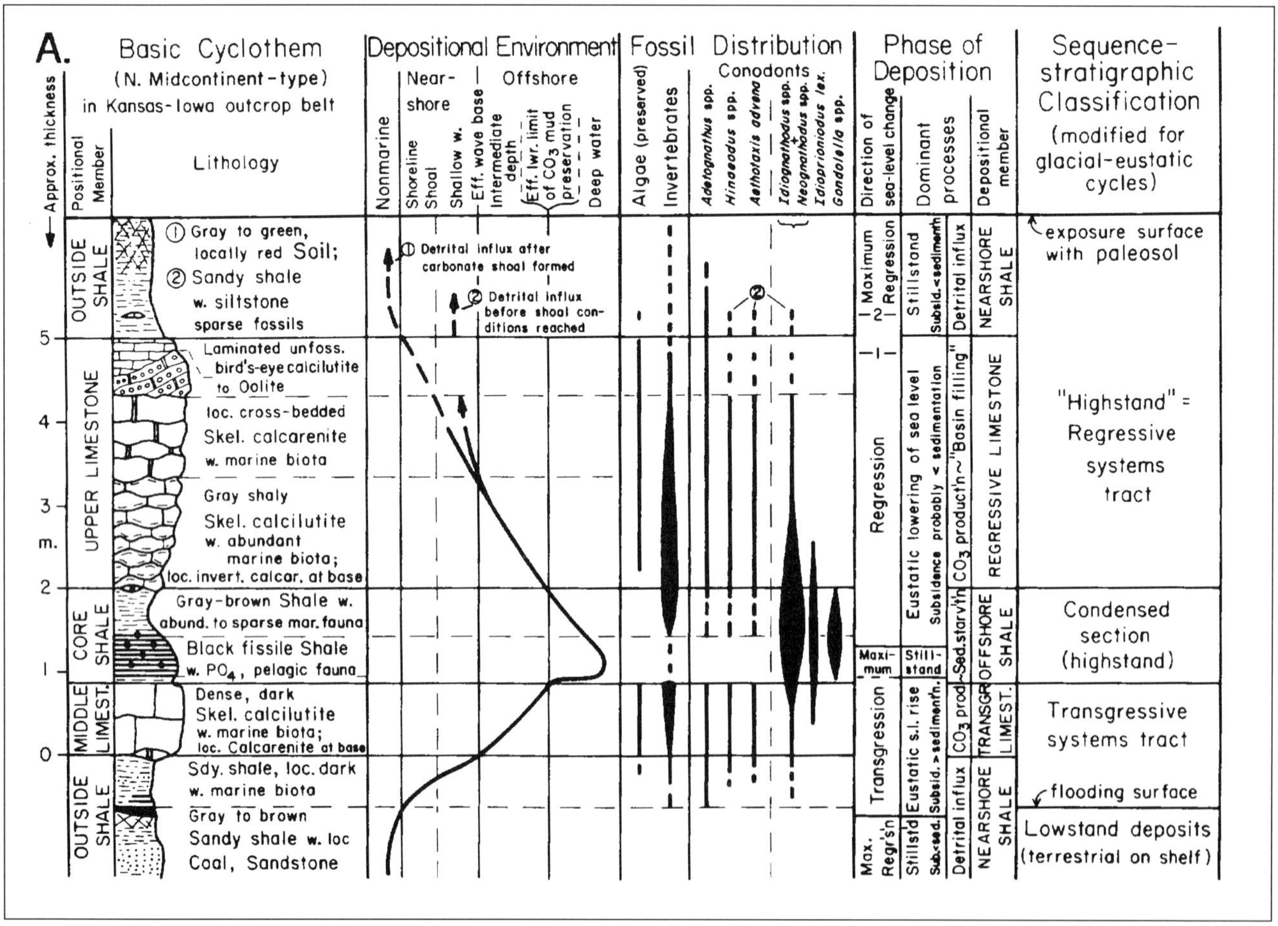

FIGURE 3 — Idealized northern Mid-Continent cyclothem on mid to low shelf representing one complete cycle of marine transgression and regression across northern shelf, showing lithofacies, biofacies, and sequence stratigraphic nomenclature, modified from Heckel (1994).

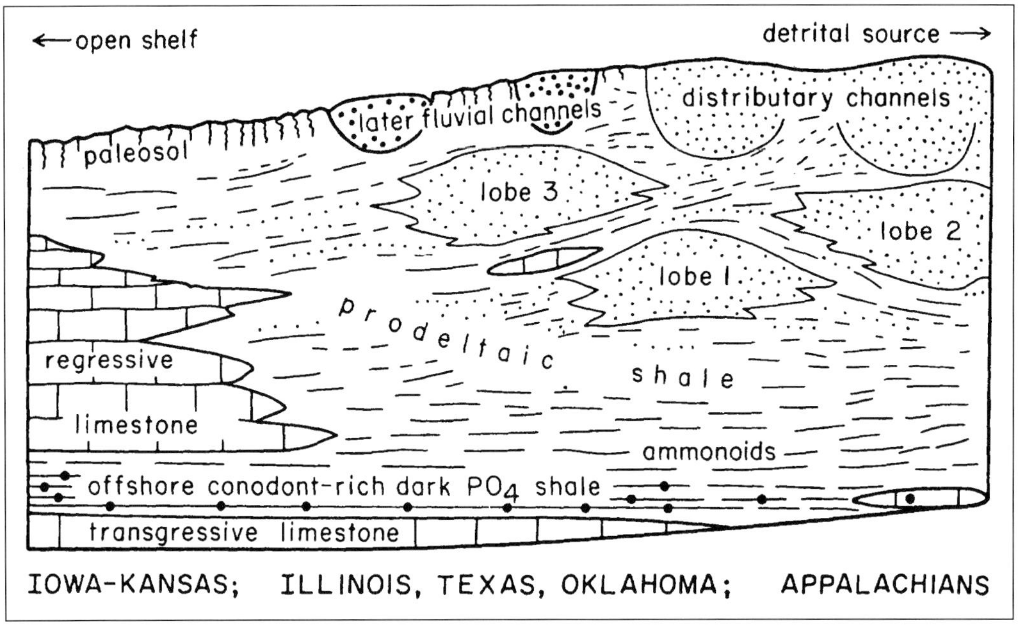

FIGURE 4 — Composite eustatic-autocyclic (deltaic) model for marine Pennsylvanian cyclothems showing the distribution of lithofacies as related to geographic distribution of siliciclastic influx from highstand and forced regressive deltas (modified from Boardman and Heckel, 1989; partly derived from Moussavi-Harami and Brenner, 1984).

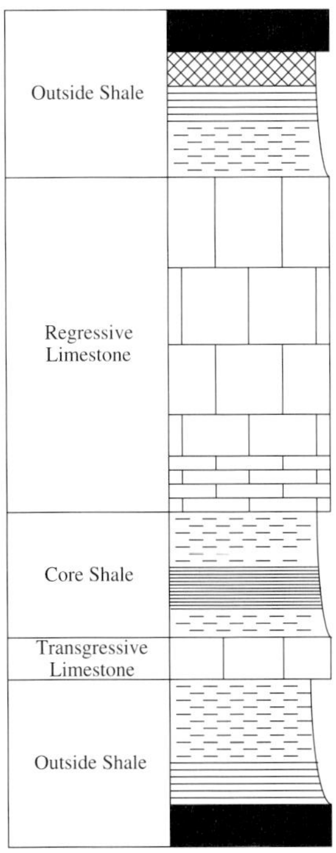

Lithology	Ostracode Biofacies	Conodont Biofacies	Foraminifer Biofacies	
Gray Blocky Mudstone	Nonmarine Biofacies	Nonmarine Biofacies	Nonmarine Biofacies	
Dark Gray to Black Shale	*Geisina* Biofacies	No Conodonts	No Foraminifers	
Gray to Green Shale	*Cavellina* Biofacies	*Adetognathus* Biofacies	Encrusting Foraminifers-*Ammodiscus* Biofacies	
Sparsely Fossiliferous Massive Bedded to Cross-bedded Locally Oolitic Packstone to Grainstone Limestone				
Fossiliferous Wackestone to Packstone Limestone	*Amphissites* Biofacies	*Streptognathodus* - *Idiognathodus* Biofacies	Open Marine Calcareous Foraminifer Biofacies; Fusulinids	
Gray Shale		*Idioprioniodus* Biofacies		
Black, Fissile, Phosphatic Shale	No Ostracodes	*Gondolella* Biofacies	*Ammodiscus-Reophax* Biofacies	Ammonoids
Gray Shale	*Amphissites* Biofacies	*Idioprioniodus* Biofacies	Open Marine Calcareous Foraminifer Biofacies	
Dense Wackestone to Packstone Limestone		*Streptognathodus* - *Idiognathodus* Biofacies	Fus.	
Gray to Green Shale	*Cavellina* Biofacies	*Adetognathus* Biofacies	Encrusting Foraminifers-*Ammodiscus* Biofacies	
Dark Gray to Black Sh	*Geisina* Biofacies	No Conodonts	No Foraminifers	
Coal	Nonmarine Biofacies	Nonmarine Biofacies	Nonmarine Biofacies	

FIGURE 5 — Idealized northern Mid-Continent Middle to Upper Pennsylvanian (Desmoinesian-lower upper Virgilian) cyclothem representing one complete cycle of marine transgression and regression across the northern shelf, showing lithofacies and microfossil (ostracodes, conodonts, foraminifers) distribution. This cyclothem is developed at a considerable geographic distance from any source of detrital clastic influx. Commonly, the nearshore and terrestrial facies show extensive lateral as well as vertical variations, and may be considerably different from shown on this diagram.

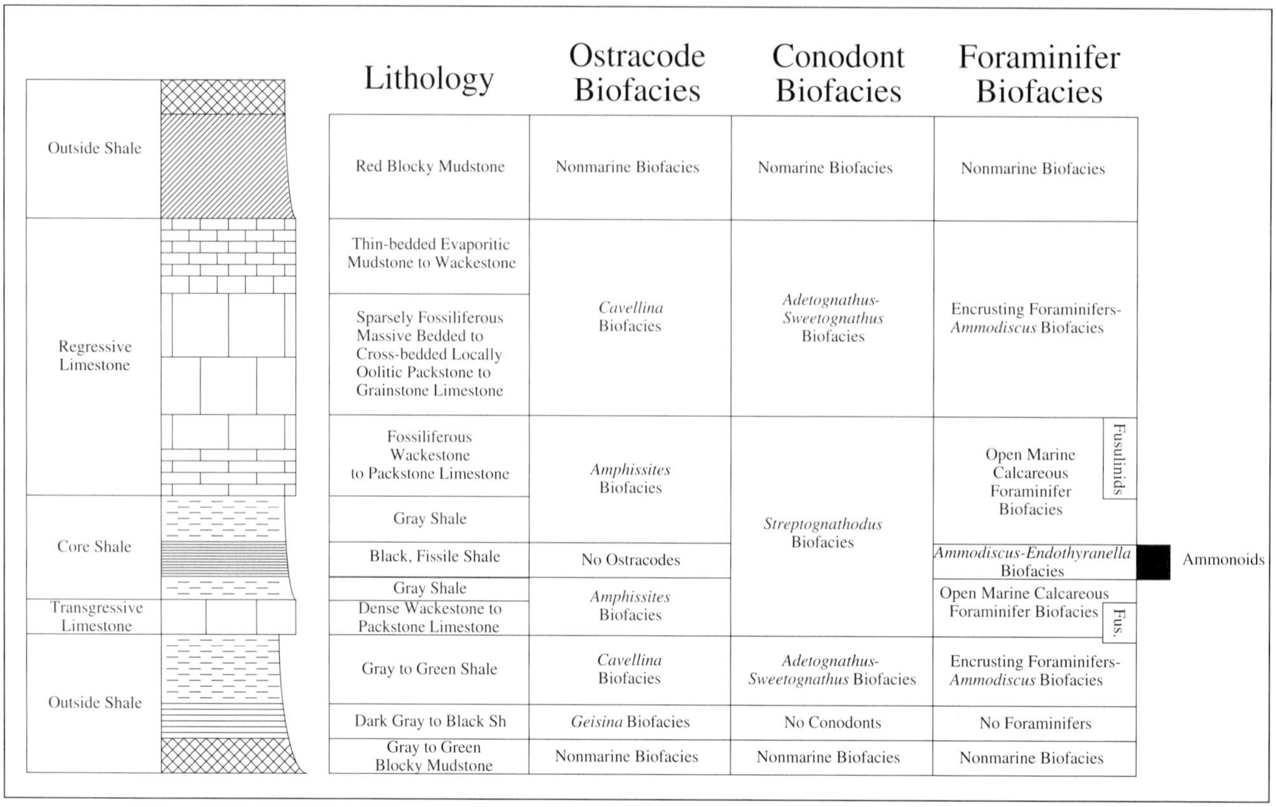

FIGURE 6 — Idealized northern Mid-Continent uppermost Pennsylvanian-Permian (upper Virgilian-lower Artinskian) cyclothem representing one complete cycle of marine transgression and regression across the northern shelf, showing lithofacies and microfossil (ostracodes, conodonts, foraminifers) distribution. This cyclothem is developed at a considerable geographic distance from any source of detrital clastic influx. Commonly, the nearshore and terrestrial facies show extensive lateral as well as vertical variations, and may be considerably different from shown on this diagram. Additionally, the most offshore facies of the black shales are developed only in the lower one-half of the Council Grove Group. Cyclothems with lesser magnitudes of inferred magnitudes of sea-level rise have fossiliferous carbonates representing the maximum water depths or maximum flooding.

cyclothem," it does not illustrate the model for cyclothems developed in a mixed carbonate-siliciclastic shelf such as occurs in North-Central Texas, Oklahoma, Illinois, and Appalachian basins. In regions where deltaic sedimentation is pervasive, the locations of siliciclastic influx during highstands and forced regression largely determine the lithic nature of the cyclothem. This combination model (Fig. 4) that uses autocyclic controls as well as an overall eustatic signature was independently proposed by Boardman et al. (1984) and Moussavi-Harami and Brenner (1984).

For the purposes of this study, different facies tracts are used to illustrate the lithic as well as biotic response to different levels of terrigenous clastic influx. Late Carboniferous and Lower Permian cyclothems developed in regions distant from siliciclastic influx (such as in the northern Mid-Continent) are summarized in Figs. 5 and 6. In regions that are greatly affected by deltaic progradation, three basic facies tracts are noted: Nondeltaic Facies Tract (Fig. 7) for regions only moderately influenced by terrigenous clastic influx, Interdeltaic Facies Tract (Fig. 8) for regions adjacent to the margins of deltaic lobes, and Deltaic Facies Tract (Fig. 9) for regions directly affected by deltaic progradation.

Cyclothems developed in the Nondeltaic Facies Tract are similar to those of the northern Mid-Continent, but differ in having a much thicker core shale facies. This thicker core shale lithofacies con-

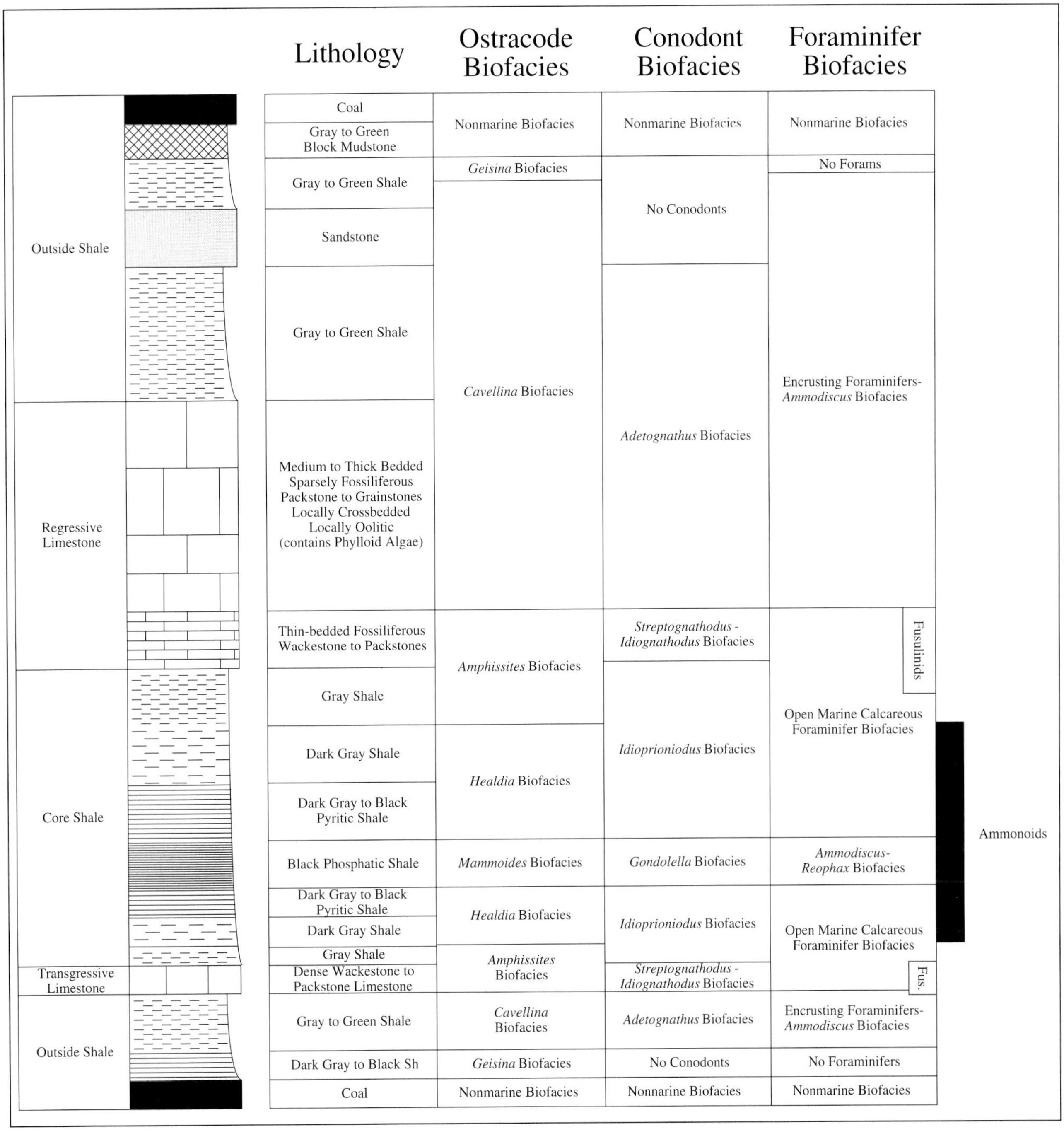

FIGURE 7 — Idealized North-Central Texas, Oklahoma, Illinois, and Appalachian basins Middle to Upper Pennsylvanian (Desmoinesian-lower upper Virgilian) cyclothem representing one complete cycle of marine transgression and regression across the northern shelf, showing lithofacies and microfossil (ostracodes, conodonts, foraminifers) distribution. This cyclothem is developed in a Nondeltaic Influenced Facies Tract which was located at a moderately far geographic distance from any source of detrital clastic influx. Commonly, the nearshore and terrestrial facies show extensive lateral as well as vertical variations, and may be considerably different from those shown on this diagram.

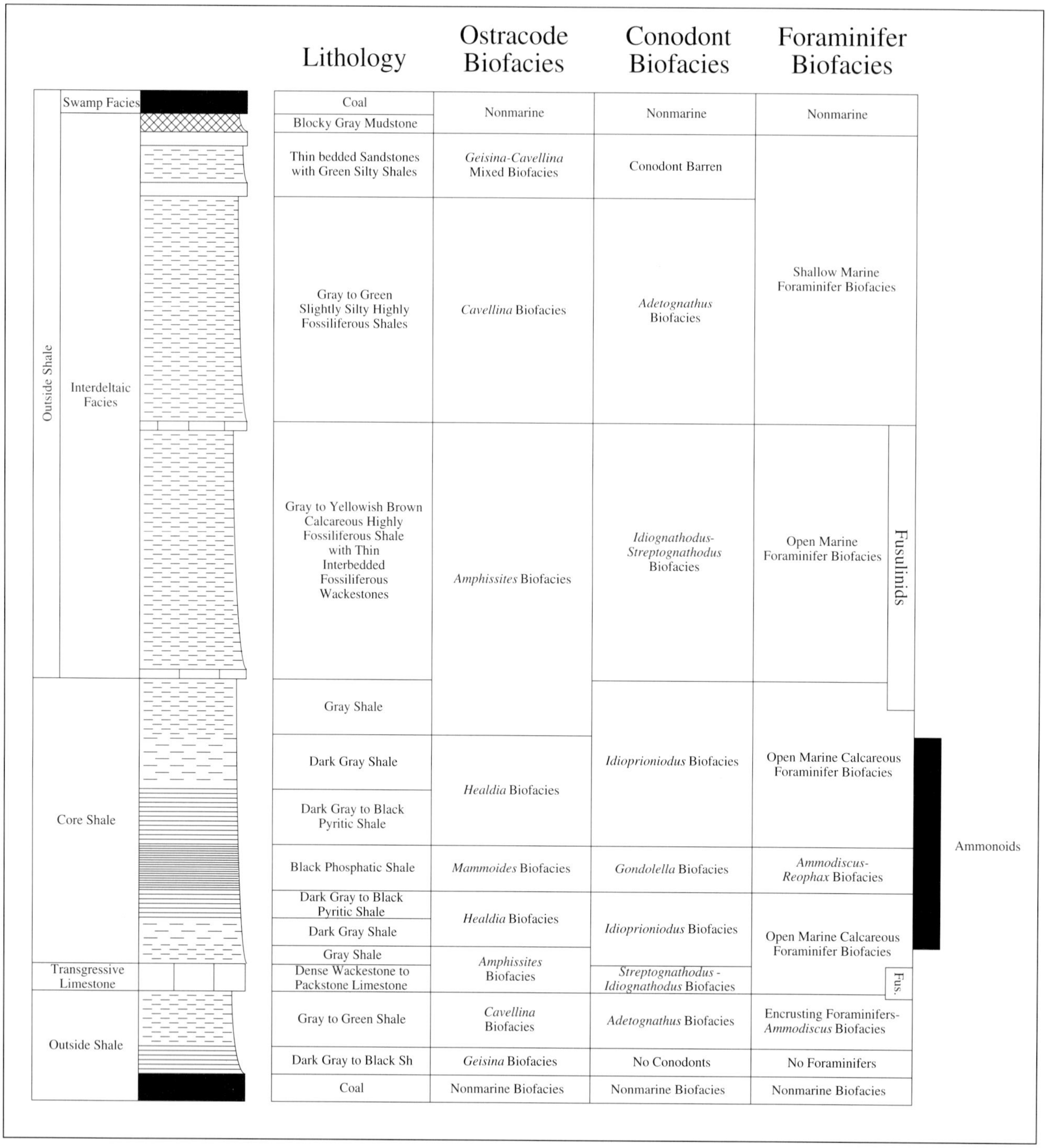

FIGURE 8 — Idealized North-Central Texas, Oklahoma, Illinois, and Appalachian basins Middle to Upper Pennsylvanian (Desmoinesian-lower upper Virgilian) cyclothem representing one complete cycle of marine transgression and regression across the northern shelf, showing lithofacies and microfossil (ostracodes, conodonts, foraminifers) distribution. This cyclothem is developed in an Interdeltaic Influenced Facies Tract which was located at a moderately close geographic distance from any source of detrital clastic influx. Commonly, the nearshore and terrestrial facies show extensive lateral as well as vertical variations, and may be considerably different from those shown on this diagram.

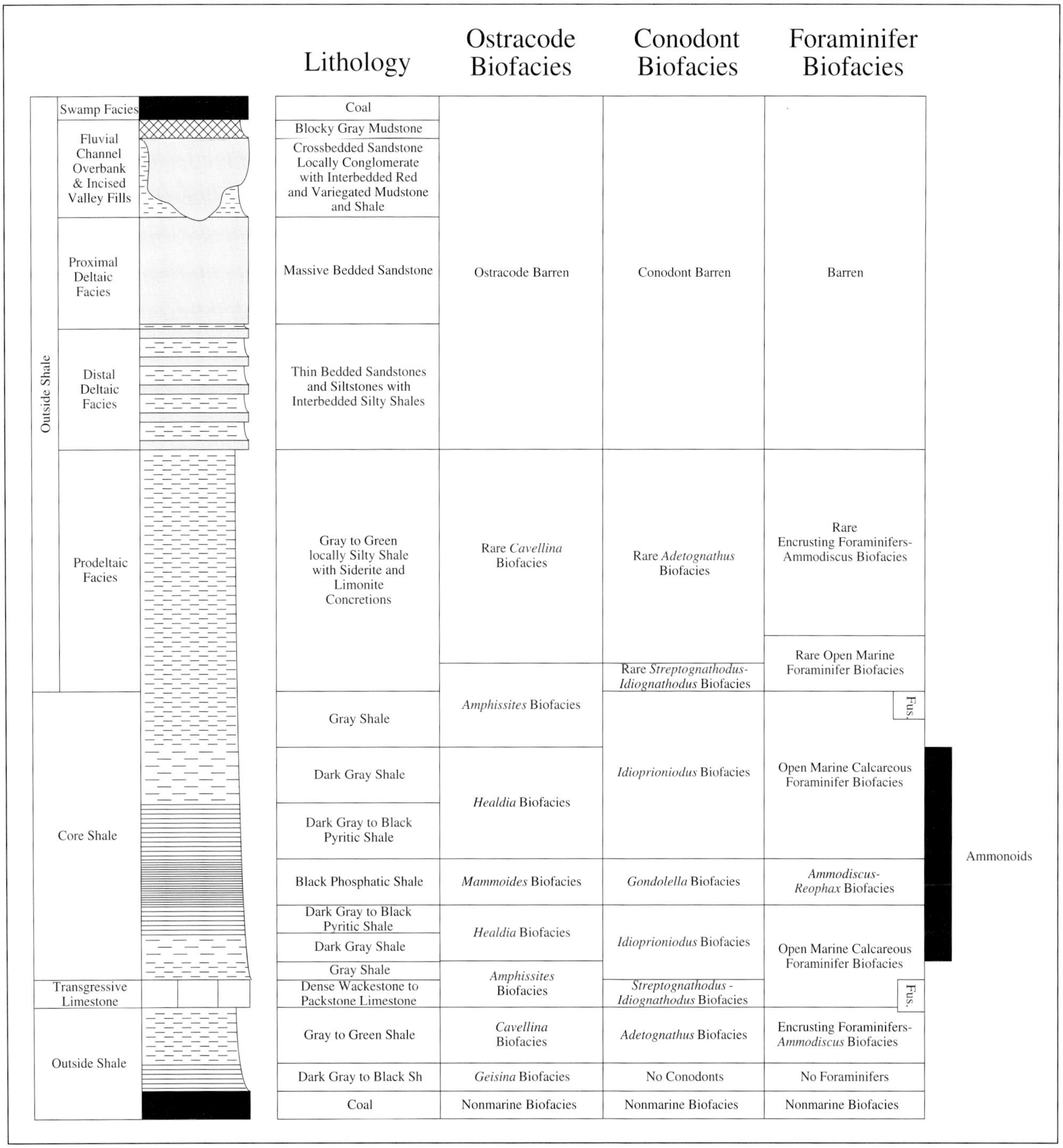

FIGURE 9 — Idealized North-Central Texas, Oklahoma, Illinois, and Appalachian basins Middle to Upper Pennsylvanian (Desmoinesian-lower upper Virgilian) cyclothem representing one complete cycle of marine transgression and regression across the northern shelf, showing lithofacies and microfossil (ostracodes, conodonts, foraminifers) distribution. This cyclothem is developed in a Deltaic Influenced Facies Tract which was located in a geographic location directly affected by detrital clastic influx from a prograding delta during early regression. Many cyclothems show the deltaic facies directly overlying the deepest marine beds of the core shale, and therefore are interpreted as highstand deltas (Hentz, 1994). Commonly, the nearshore and terrestrial facies show extensive lateral as well as vertical variations, and may be considerably different from those shown on this diagram.

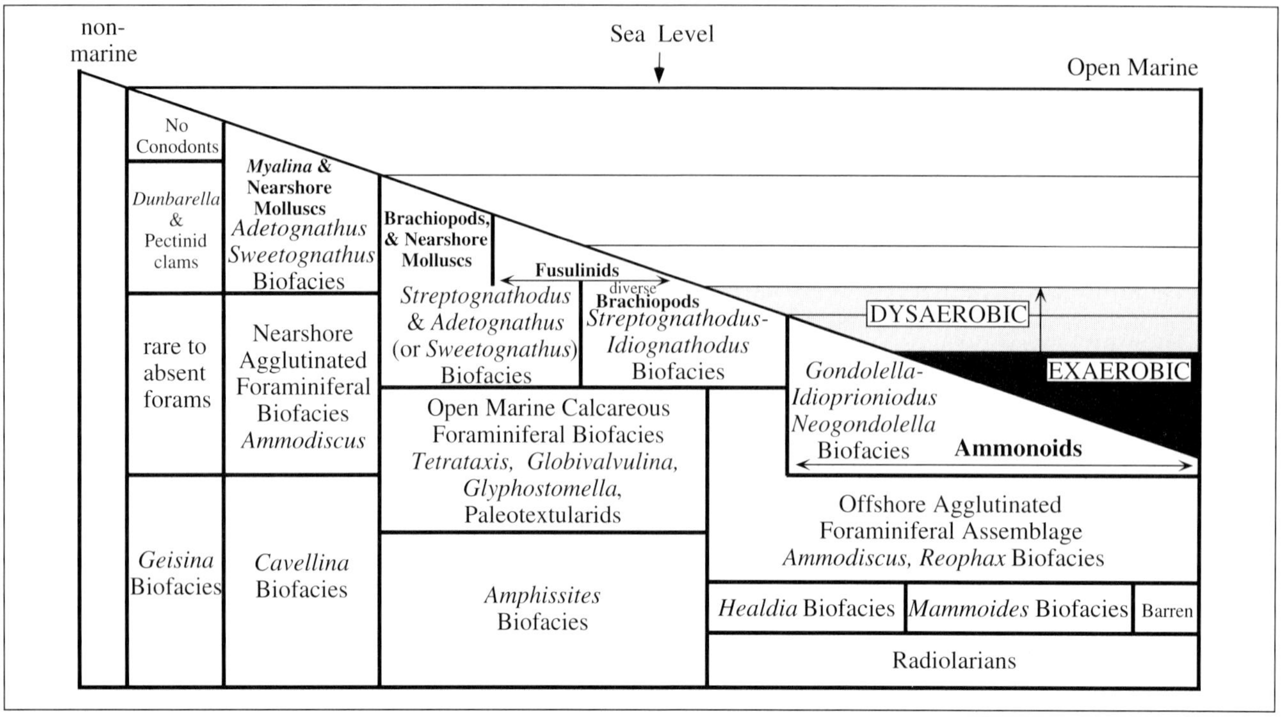

FIGURE 10 — Onshore-offshore model for Pennsylvanian depth and oxygen related biofacies associations related (modified from Boardman et al. 1984; Boardman and Nestell, 1993).

tains additional lithofacies and biofacies not developed in core shales of the northern Mid-Continent. Cyclothems developed in the Interdeltaic Facies Tract have a thick core shale lithofacies followed vertically by highly fossiliferous shales and thin interbedded limestones and sandstones higher in the sequence. Cyclothems developed in the Deltaic Facies Tract have a thick core shale lithofacies followed vertically by prodeltaic, distal, and proximal deltaic sediments.

A summary cross section illustrating onshore-offshore biotic gradients is shown in Fig. 10. This diagram shows trends compatible to those showing macrofossil distributions presented by Boardman et al. (1984).

SUMMARY AND CONCLUSIONS

1. Microfossils can be used for delineating onshore-offshore trends in Late Carboniferous and Early Permian cyclothemic sedimentary sequences.
2. Nearshore marginal marine strata are characterized by a low diversity of microfossil assemblages dominated by *Geisina* ostracodes, an absence of conodonts, and a foraminiferal assemblage with rare *Thurammina.*
3. Nearshore marine strata thought to have been deposited in a nearly normal marine salinity environment are characterized by a low diversity of microfossil assemblages dominated by the *Cavellina* ostracode biofacies, *Adetognathus* conodont biofacies, and encrusting foraminifers along with *Ammodiscus* cf. *A. semiconstrictus* var. *regularis.*
4. Mid-shelf open strata contain the greatest diversity of benthic microfossils in the cyclothem including *Amphissites* ostracode biofacies, *Idiognathodus-Streptognathodus* conodont biofacies, *Tetrataxis, Globivalvulina,* large paleotextularids (*Climacammina, Deckerella*), endothyranellids

(*Endothyra, Endothyranella*), *Ammobaculites*, and fusulinaceans.

5. Deep-shelf marine strata contain a high diversity pelagic microfossil assemblage but a reduced diversity benthic microfossil assemblage containing the *Mammoides* and *Healdia* ostracode biofacies, *Gondolella-Idioprioniodus* conodont biofacies, *Ammodiscus-Reophax* foraminiferal biofacies, and abundant radiolarians.

ACKNOWLEDGMENTS

The authors would like to express gratitude to Brad Vann, Carter Keairns, and Mike Daugherty of Oklahoma State University for sharing micropaleontological data from their master's thesis projects. This report benefited from critical review by P. H. Heckel of The University of Iowa.

LOCALITY REGISTER

1. Towle Shale Member, Onaga Shale, Admire Group, upper Virgilian; spillway over Acker Creek, c/wl, NE/4, Section 16 T29N-R7E, Osage County, Oklahoma.

2. Brownville Limestone Member, Woodsiding Formation, Wabaunsee Group, upper Virgilian; road cut on secondary unimproved road, SW/4, SE/4, Section 23 T23S-R10E, Greenwood County, Kansas.

3. Americus Limestone Member (lower ledge), Foraker Limestone, Council Grove Group, uppermost Virgilian; road cut on Kansas Highway 166, NE/4, NW/4, Section 7 T33S-R8E, Cowley County, Kansas.

4. Aspinwall Limestone Member (upper ledge), Onaga Shale, Admire Group, upper Virgilian; road cut on Kansas Highway K-38, SW/4, SW/4, Section 22 T32S-R8E, Chautauqua County, Kansas.

5. Neva Limestone Member, Grenola Formation, Council Grove Group, Asselian; stream cutbank, SE/4, SW/4, SW/4, Section 4 T31S-R8E Cowley County, Kansas. The measured section for this locality is modified from Mudge and Yochelson (1963) and equates to their measured Section 288, and is the type locality for the Grenola Formation.

6. Threemile Limestone Member, Wreford Formation, Chase Group, Sakmarian; road cut on Kansas Highway K-38, NL/ Section 30 (irregular section) T32S-R8E, Cowley County, Kansas.

7. East Mountain Shale (upper black shale bed), Strawn Group, uppermost Desmoinesian; steep hillside exposure on East Mountain UTM14SNM^{5}8341^{36}3096 in Mineral Wells, Palo Pinto County, Texas.

8. PP3 Shale Member, Posideon Formation, Canyon Group, middle Missourian; road cut on west side of TX 2123, UTM14SPM^{6}1552^{36}7302 just south of Bridgeport, Wise County, Texas.

9. Finis Shale Member, Graham Formation, Cisco Group, middle Virgilian; new spillway for Lake Jacksboro, UTM14SNM^{5}8275^{36}7825, Jack County, Texas.

10. Wolf Mountain Shale (basal black shale), Graford Formation, Canyon Group, middle Missourian; lake edge exposure on Runaway Bay on Lake Bridgeport, UTM14SPM^{6}0820^{36}7370, Wise County, Texas.

11. Bennett Shale Member, Red Eagle Limestone, Council Grove Group, Asselian; spillway for Tuttle Creek Reservoir, Section 19 T9S-R8E Pottawatomie County, Kansas.

12. Colony Creek Shale Member, Caddo Creek Formation, Canyon Group, basal Virgilian; road cut on U.S. 180, UTM14SNM^{5}4362^{36}2305, Stephens County, Texas.

13. Finis Shale Member, Graham Formation, Cisco Group, middle Virgilian; pond dam and adjacent hillside exposure, UTM14SNM^{5}2285^{36}2045, Stephens County, Texas.

14. Finis Shale Member, Graham Formation, Cisco Group, middle Virgilian; road cut on old road, UTM14SNM^{5}9467^{36}8758, Jack County, Texas.

15. Necessity Shale Member, Graham Formation, Cisco Group, middle Virgilian, hillside exposure, UTM14SNM^{5}3031^{36}5094, Young County, Texas.

16. Salesville Shale (upper black shale bed), Strawn Group, lower Missourian, road cut on Texas Highway 337, UTM14SNM^{5}7867^{36}3156, Palo Pinto County, Texas.

17. Palo Pinto Limestone (basal-Wynn Limestone Type Section), Canyon Group, middle Missourian, road cut on U.S. 180, UTM14SNM^{5}6922^{36}2632, Palo Pinto County, Texas.

18. Wolf Mountain Shale Member, Graford Formation, Canyon Group, middle Missourian; exposure near base of shale pit, UTM14SPM^{6}1582^{36}7580, Wise County, Texas.

19. Winchell Limestone Member (upper Winchell), Graford Formation, Canyon Group, upper Missourian; spillway exposure for Lake Brownwood, UTM14RML^{4}9976^{35}2234, Brown County, Texas.

20. Ranger Limestone Member (basal Ranger), Brad Formation, Canyon Group, uppermost Missourian; road cut on Texas 717, UTM14SNL^{5}2946^{35}9585, Eastland County, Texas.

21. Cedar Vale Shale Member, Scranton Formation, Wabaunsee Group, upper Virgilian; hillside exposure, SE/4, SE/4, Section 3 T2N-R12E, Pawnee County, Nebraska.

22. Dog Bend Limestone Member, Salesville Formation, Strawn Group, lower Missourian; road cut on Texas Highway 4, UTM14SNM^{5}6734^{36}2161, Palo Pinto County, Texas.

23. Finis Shale Member, Graham Formation, Cisco Group, middle Virgilian; limestone quarry (Dyer Quarry) exposure, UTM14SNM^{5}8040^{36}7790, Jack County, Texas.

INCISED VALLEYS AND ESTUARINE FACIES OF THE DOUGLAS GROUP (VIRGILIAN): IMPLICATIONS FOR SIMILAR PENNSYLVANIAN SEQUENCES IN THE U.S. MID-CONTINENT

Allen W. Archer
Department of Geology
Kansas State University
Manhattan, KS 66506

Howard R. Feldman
Kansas Geological Survey
University of Kansas
Lawrence, KS 66047

ABSTRACT

Incised valleys are an important, but little studied, component of Upper Pennsylvanian cyclothems in the Mid-Continent. During the Upper Pennsylvanian, glacio-eustatic cycles resulted in alternate exposure and flooding of vast areas of the Mid-Continent. Near equatorial paleolatitudes and extensive chemical weathering of the exposed craton resulted in mud-rich siliciclastic sediment influx to drainage networks that included incised valleys. During transgressions, incised valleys flooded to form tide-dominated, muddy estuaries. The Douglas Group (Virgilian) in eastern Kansas contains several incised valley fills (IVF) and a range of tidal facies that can be best understood through comparison with modern deposits, mostly in macrotidal estuaries. Recognition of tidal influence in facies that are devoid of macrofossils is important for interpreting sea-level history, and lack of the recognition of tidal influence commonly leads to incorrect interpretations. Transitions from fluvial to estuarine facies, which were formed during marine flooding, can be recognized by the occurrence of mud-draped bedforms. Also the presence of facies that are likely the result of macrotidal conditions may have implications for paleogeographic reconstructions, because very high tidal ranges commonly occur in funnel-shaped estuaries. The IVFs occur throughout much of the Pennsylvanian in the Mid-Continent, and are probably even more common than previously has been recognized.

Several factors in addition to sea-level history influenced formation of IVFs. Climate may have had a profound influence. During wetter periods, rivers incised more deeply and developed extensive valley systems. During the drier periods, such as the Permian, valleys apparently did not develop at all. Other controls on valley morphology include bedrock lithologies exposed during incision, and basement structure.

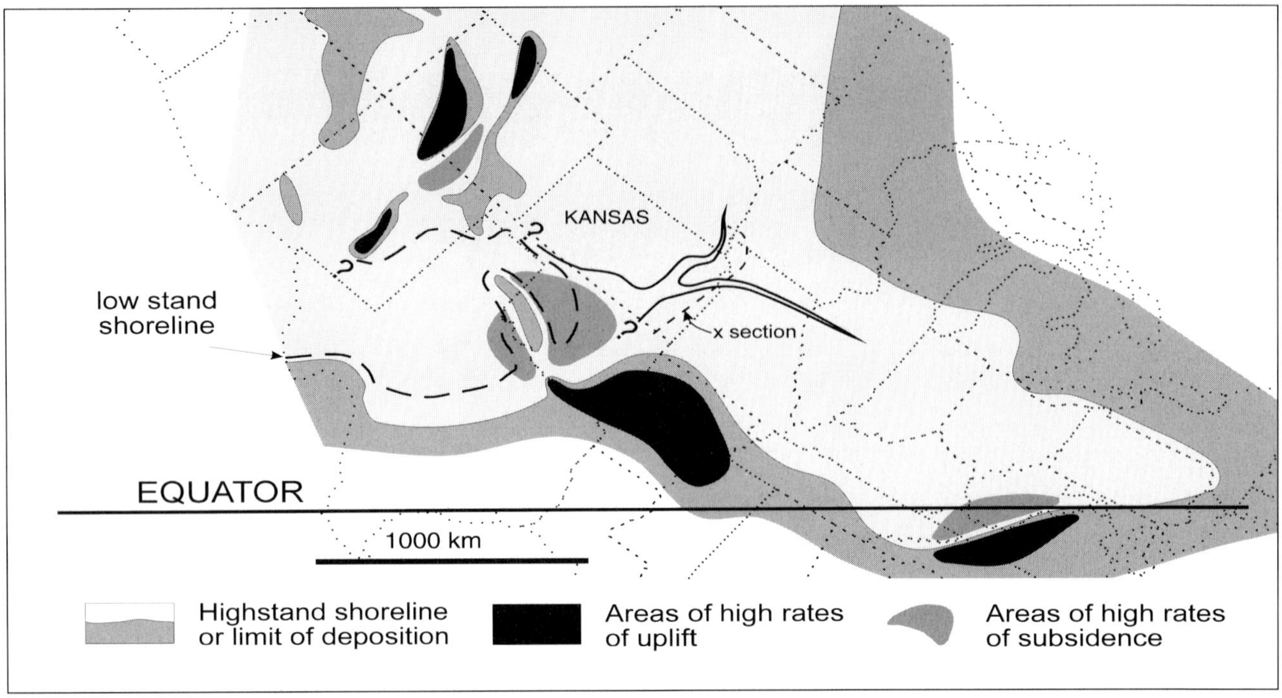

FIGURE 1 — Paleogeographic reconstruction for interval comprising Douglas Group of Kansas and lateral equivalents. Coastal position in Oklahoma based on interpretation of Stalnaker Sandstone in southern Kansas as a delta (Walton and Griffith, 1985) and interpretations of Tonkawa Sandstone as submarine-fan and slope facies (Kumar and Slatt, 1984) in the Arkoma basin. The location of the cross section of Fig. 2 is also indicated. Modified from McKee and Crosby (1975) with Douglas Group information from Archer et al. (1994b) and Feldman et al. (1995).

INTRODUCTION

During the Pennsylvanian, the U.S. Mid-Continent was located in near equatorial paleolatitudes. Vast portions of the craton were exposed when the sea level was low, allowing large fluvial systems to develop in the low-relief, tropical, humid setting (Fig. 1). Lowstand watersheds for larger river systems potentially extended from Kansas or Oklahoma east to the Appalachians, delivering large volumes of sediment to trunk valleys, and ultimately to the coast. The Upper Pennsylvanian Douglas Group in northeastern Kansas contains several examples of IVFs including both small valleys that likely had small drainage basins, and larger valleys that may have been the downstream portions of large drainage systems (Figs. 1 and 2). These IVFs are dominated by muddy facies, which may reflect the intense chemical weathering of the continental interior just as modern tropical coastlines are dominated by mud-rich sediment that is derived from the craton (Hayes, 1967) (Fig. 3). During rising sea level the valleys were flooded, creating large muddy estuaries. Critical to interpreting sea-level history in these siliciclastic settings is the ability to distinguish fluvial versus estuarine versus marine facies. Many of the facies are superficially similar and, in the absence of shelly benthos, may be difficult to delineate. Of particular importance is the recognition of tidal influence. A range of types of tidal bedding present within the Douglas Group IVFs are very similar to modern estuarine facies found only in macrotidal environments. Modern macrotidal coastlines commonly exhibit funnel-shaped estuaries that result from flooding of river valleys (Hayes, 1975) (Fig. 4), because these are common sites of tidal amplification (Defant, 1961); thus, recognition of certain tidal facies may have implications for paleogeographic reconstructions.

The common occurrence of a variety of tidally influenced facies within the Douglas Group indi-

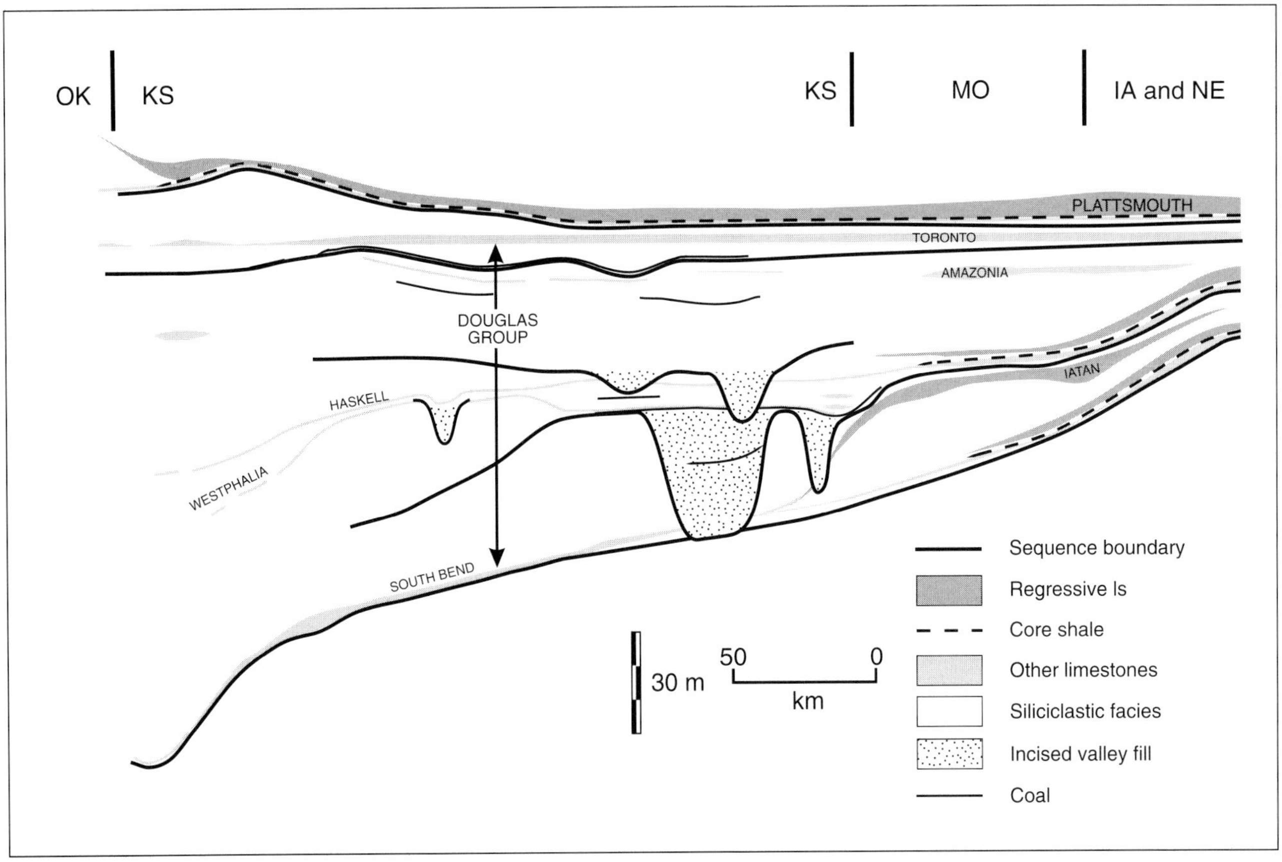

FIGURE 2 *(above)* — Outcrop cross section of the Douglas Group and immediately overlying and underlying units from northeastern Oklahoma to southeastern Nebraska. Only coals have been exaggerated in thickness. Coals resting on sequence boundaries have been translated slightly upward. Sequence boundaries have not been correlated into areas where their placement is uncertain. See Fig. 1 for cross section location. Data for this cross section were compiled from D. S. Ball (1985), S. M. Ball (1964, 1971), Bowsher and Jewett (1943), Feldman et al. (1995), Gibling et al. (1993), Goebel et al. (1989), Griffith (1981), Heckel (1977), Heckel et al. (1979), Howe (1986), Joeckel (1994), and Spriggs (1989).

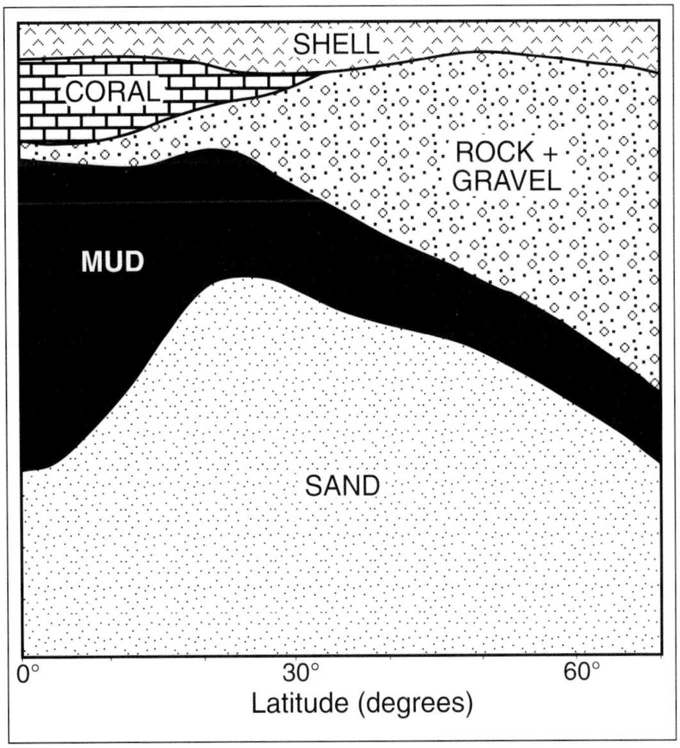

FIGURE 3 — Relative proportions of various types of coastal and inner continental shelf sediments as related to latitude. Note the high proportion of mud at low latitudes (adapted from Hayes, 1967).

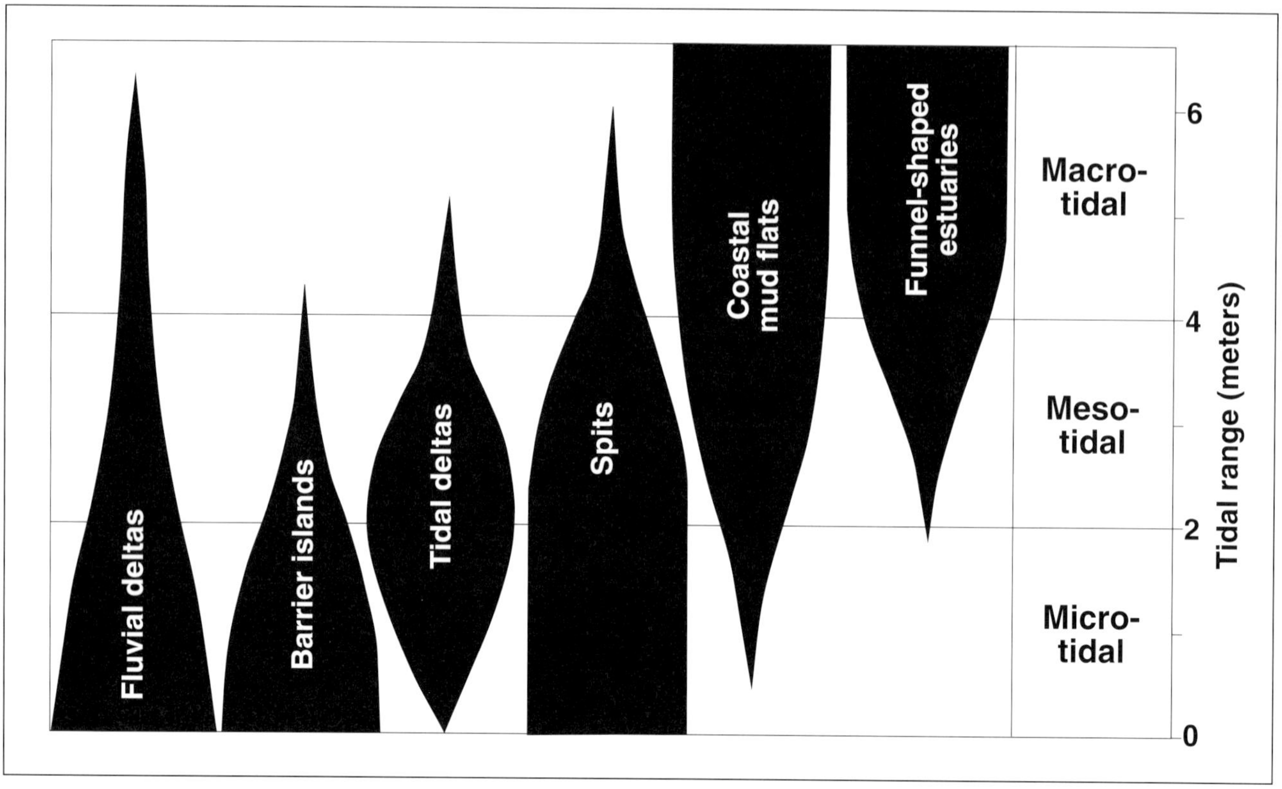

FIGURE 4 — Type and relative frequency of a variety of coastal depositional environments as related to tidal range. Note the decreasing frequency of deltas and increasing frequency of estuaries with elevated tidal range (adapted from Hayes, 1975).

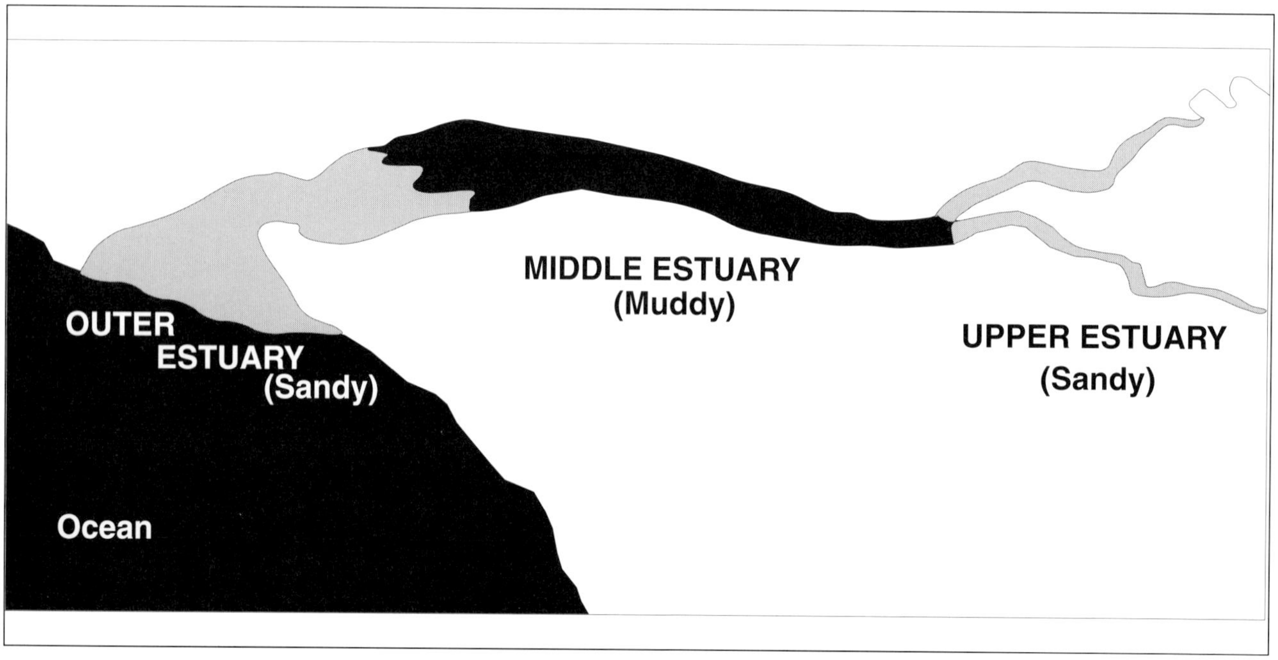

FIGURE 5 — Generalized estuarine depositional model showing tripartite arrangement of major facies. The inner part of system contains channelized fluvial sands. The middle flaring estuary contains a variety of rippled sandstones and heterolithic facies including tidal rhythmites. The outer coastal zone contains sheet-like marine sandstones. Adapted from Allen (1991) and Rahmani (1989).

cates that tides were a significant factor in controlling sediment transport and deposition. The valley-fill depositional models developed for the Douglas Group can be directly compared to vertical and lateral sequences developed within modern tidal estuarine sequences, thus providing modern facies analogues for interpreting sea-level history from facies patterns.

The goal of this study is to illustrate how IVFs are recognized in Pennsylvanian cyclothems using examples from the Upper Pennsylvanian (upper Missourian to lower Virgilian) of Kansas. Recognition of IVFs is critical because these features are commonly the most obvious indications of a lowering of relative sea level, and development of type 1 sequence boundaries (*sensu* Van Wagoner et al., 1988, 1990). Key to interpreting IVFs is recognizing estuarine facies, and a variety of criteria are presented that should be of general use to help distinguish tidal estuarine facies from related fluvial and marine facies. Lastly, some additional Pennsylvanian valley-fill sequences that are known to, or which potentially contain, tidal estuarine facies are briefly reviewed.

ESTUARINE FACIES MODELS

Estuarine facies have the greatest potential for development and preservation during transgressive periods (Dalrymple et al., 1992). Conversely, during falling relative sea level, coastal areas become emergent, and rivers incise valleys by erosion. During subsequent base-level rise, incised valleys are flooded and transformed into estuarine systems as marine environments step landward. The flooded valley estuaries serve as sediment sinks and commonly fill rapidly with sediment derived from fluvial and marine sources. Similar to deltas, estuarine systems exhibit a spectrum of facies-stacking patterns controlled by the relative contribution of fluvial, wave, and tidal energies (Dalrymple et al., 1992) as well as the position relative to the maximum transgressive and regressive shorelines (Dalrymple et al., 1994; Zaitlin et al., 1994).

During the Pennsylvanian, glacio-eustasy was the likely driving mechanism that caused alternate flooding and exposure of large portions of the Mid-Continent (for review, see Heckel, 1994). Although carbonate rocks are abundant within parts of the Pennsylvanian and have received much attention, siliciclastic facies are the volumetrically dominant lithology throughout much of the exposed Pennsylvanian section. Deltaic systems may characterize the progradational, or regressive, upper half of the siliciclastic component of depositional cycles; however, it is likely that estuarine systems characterize many of the transgressive siliciclastic intervals. Thus, the recognition of estuarine depositional systems is of fundamental importance for interpreting sea-level history. In addition, tidal facies are now widely recognized in Pennsylvanian interior basins (Appalachian basin [Greb and Archer, 1995], Illinois basin [Kvale et al., 1989], and Forest City basin [Lanier et al., 1993]). The common occurrence of tidally influenced facies in these basins strongly refutes traditional notions of nontidal interior seaways, such as those advocated by Shaw (1964).

The complex fresh- and saltwater circulation within estuaries can create a zone of concentrated suspended sediment, or turbidity maximum. Within tidal estuaries, the turbidity maximum can be maintained by strong tidal currents (Allen et al., 1980). Modern analogues include the Bay of Fundy in Canada, Bay of Mont Saint Michel, Gironde Estuary in France, and Severn River Estuary in England (Dalrymple et al., 1992, for additional modern analogues). Estuarine circulation and the presence of a turbidity maximum can result in a tripartite facies distribution that is characterized by inner fluvio-estuarine sands, middle estuarine muds, and outer estuarine marine sands (Fig. 5). Coarse, river-derived sediment is deposited in the narrow inner parts of estuaries where cross-sectional areas are small and fluvial energies are dominant. As the channels widen in the middle estuary, fluvial processes merge with estuarine processes. Dechannelization results in rapid deposition of coarser bedload

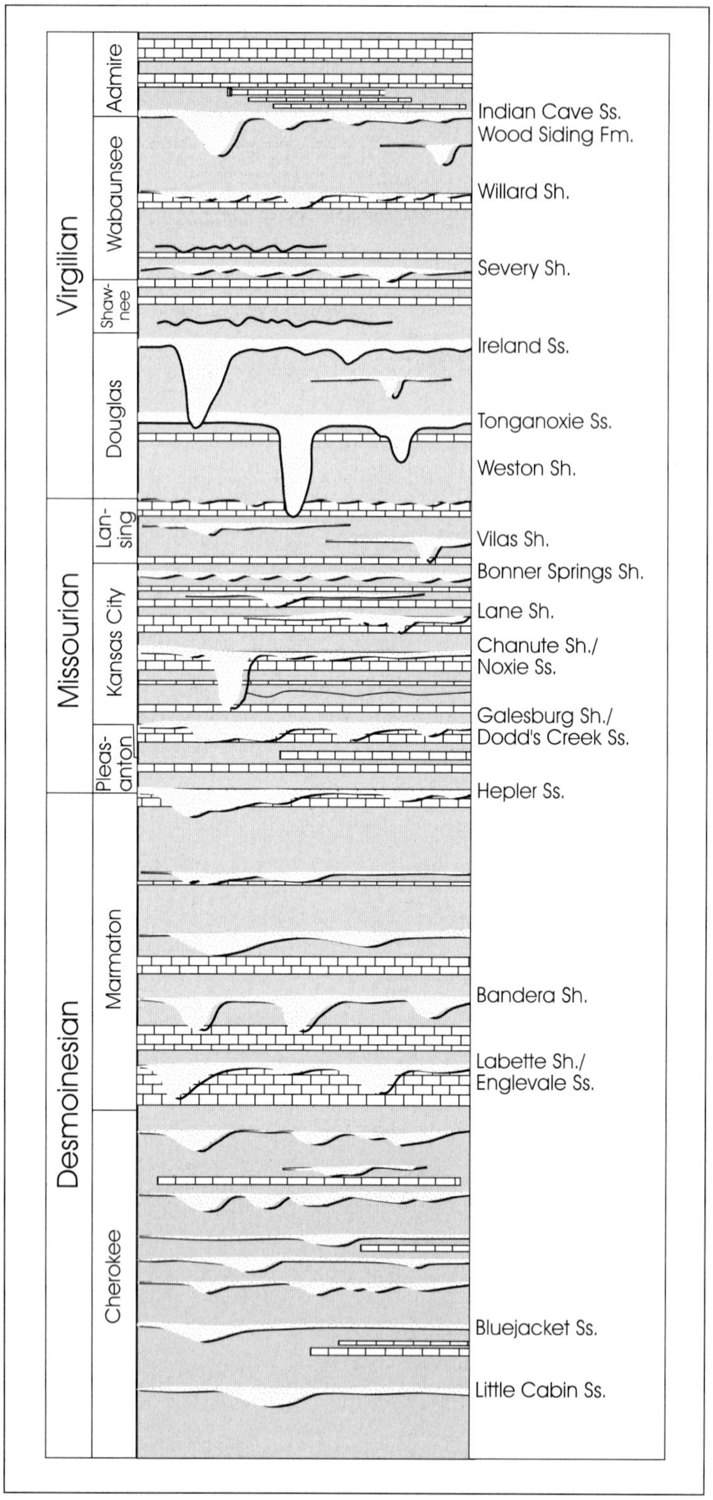

FIGURE 6 — Generalized stratigraphic section of Middle and Upper Pennsylvanian rocks of eastern Kansas. Note the common occurrence of major sandstones near disconformities; many of these sandstones are apparently related to major incised valley systems. Modified from Ross and Ross, 1988, p. 232; updated for revision of Pennsylvanian/Permian boundary (Baars et al., 1994b).

sediments. Tidal forces may cause flow reduction, and current-velocity stillstands induced by tidal influx can result in large-scale mud draping of bedforms. In the middle estuary, bottom currents are relatively weak, turbidity is high, and sediment is generally muddy. Much of the suspended mud is deposited during high- and low-tide stillstands. As tidal velocities decrease, the suspended sediment is deposited as thin, normally graded layers. Muddy tidal flats can be well developed in the middle estuary and commonly consist of tidal bedding (flaser- to lenticular-type bedding). At the estuary mouth, marine processes such as waves tend to keep the finer sediment in suspension and a variety of relatively clean, mud-free sands can be deposited.

The tripartite facies distribution has been discussed for several estuarine systems (Dalrymple et al., 1991, 1992; Allen, 1991; Dorjes and Howard, 1975; Nichols et al., 1991) and provides a model useful for recognition of ancient estuarine systems. In combination with recent developments in sequence stratigraphy, some successions of sandy fluvial to muddy facies have been reinterpreted as estuarine incised valley deposits (Bowen et al., 1993). Previously, some facies (now recognized as estuarine) were misinterpreted as deltaic or prodeltaic. Similarly, marine sands at the mouth of an estuary have been misinterpreted as offshore bars or barrier islands. Much of the older literature that has invoked such traditional fluvio-deltaic interpretations needs to be re-evaluated in terms of estuarine models. A summary of criteria useful for such evaluations is presented in subsequent sections.

GEOLOGIC SETTING OF THE PENNSYLVANIAN OF THE MID-CONTINENT

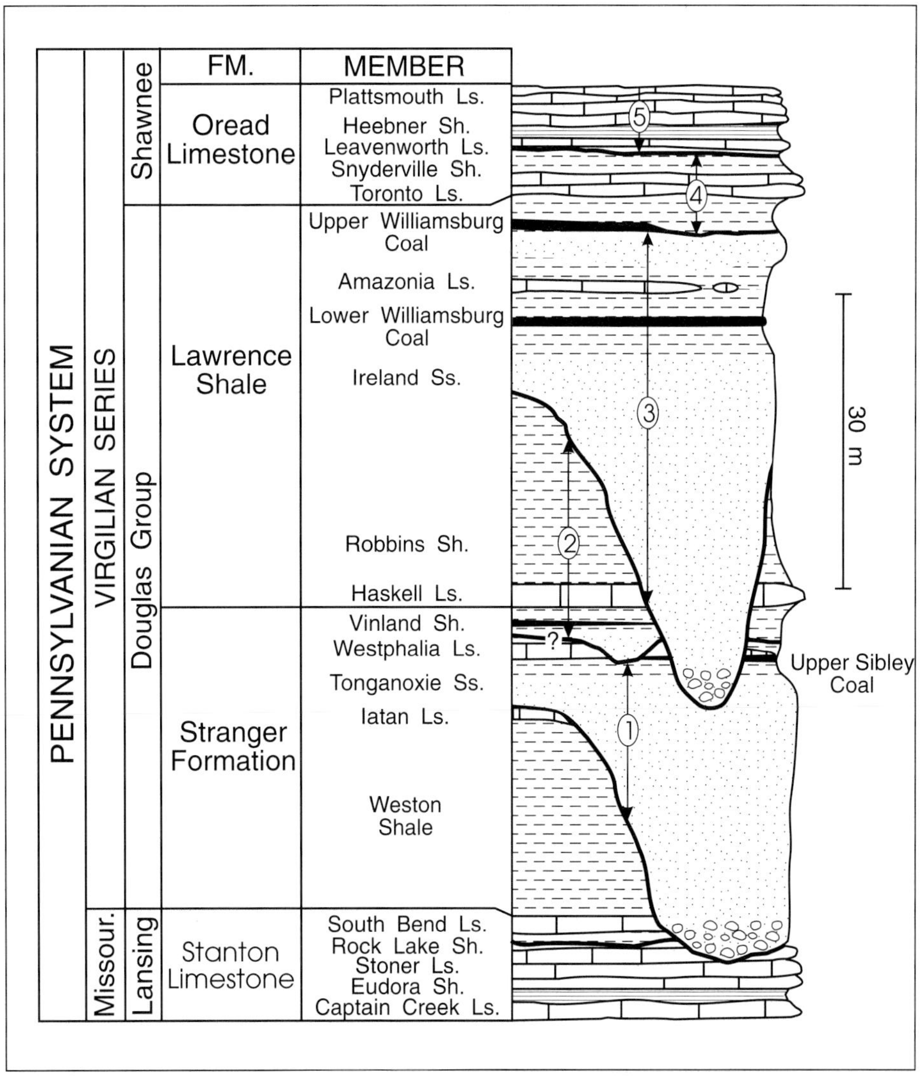

FIGURE 7 — Generalized stratigraphy of the Douglas Group showing the deeply incised Tonganoxie and Ireland sandstones. Lowstand surfaces of erosion (thick lines) delineate the sequence boundaries, which are named for the most widespread marine unit. (1) Westphalia sequence; (2) Haskell sequence; (3) Amazonia sequence; (4) Toronto sequence; and (5) Heebner sequence. Top of Heebner sequence is not known.

The Middle and Upper Pennsylvanian of Kansas consist of alternations of siliciclastic and carbonate strata (Fig. 6). This repetition led to the cyclothem concept (Wanless and Weller, 1932) as originally applied to lithologic cycles in the Illinois basin and subsequently applied to the Mid-Continent (Moore, 1935). The megacyclothem concept of Moore (1935) was later reinterpreted as the now widely accepted Kansas cyclothem model of Heckel (1977, 1994). Within a Kansas cyclothem, the apparently most marine parts (which included thickest limestones and black shales) were defined as the center or core of the cycles, and the most nonmarine, dominantly siliciclastic and coal-bearing portions were defined as the outside parts of the cycles (Heckel, 1977).

The biostratigraphic correlation and establishment of an extremely detailed, formal lithostratigraphy have long been fundamental components of Mid-Continent geologic research, particularly in Kansas. A result of this approach has been a research focus on marine facies including organic-rich core shales that commonly exhibit a low degree of lateral variability and are relatively easy to correlate in areas typically having limited data (Heckel, 1977, 1994; Watney, 1993). Core shales commonly contain radioactive, organic-rich units and are particularly easy to correlate in the subsurface using gamma ray logs. Conversely, outside shales have been less extensively studied. The outside shales are highly variable over short distances and key surfaces, such as unconformities, are not easily recognizable on wireline well logs. A model equating each core-shale-based, Kansas-type cyclothem with a single unconformity-bounded sequence (Heckel, 1994) has not been adequately documented.

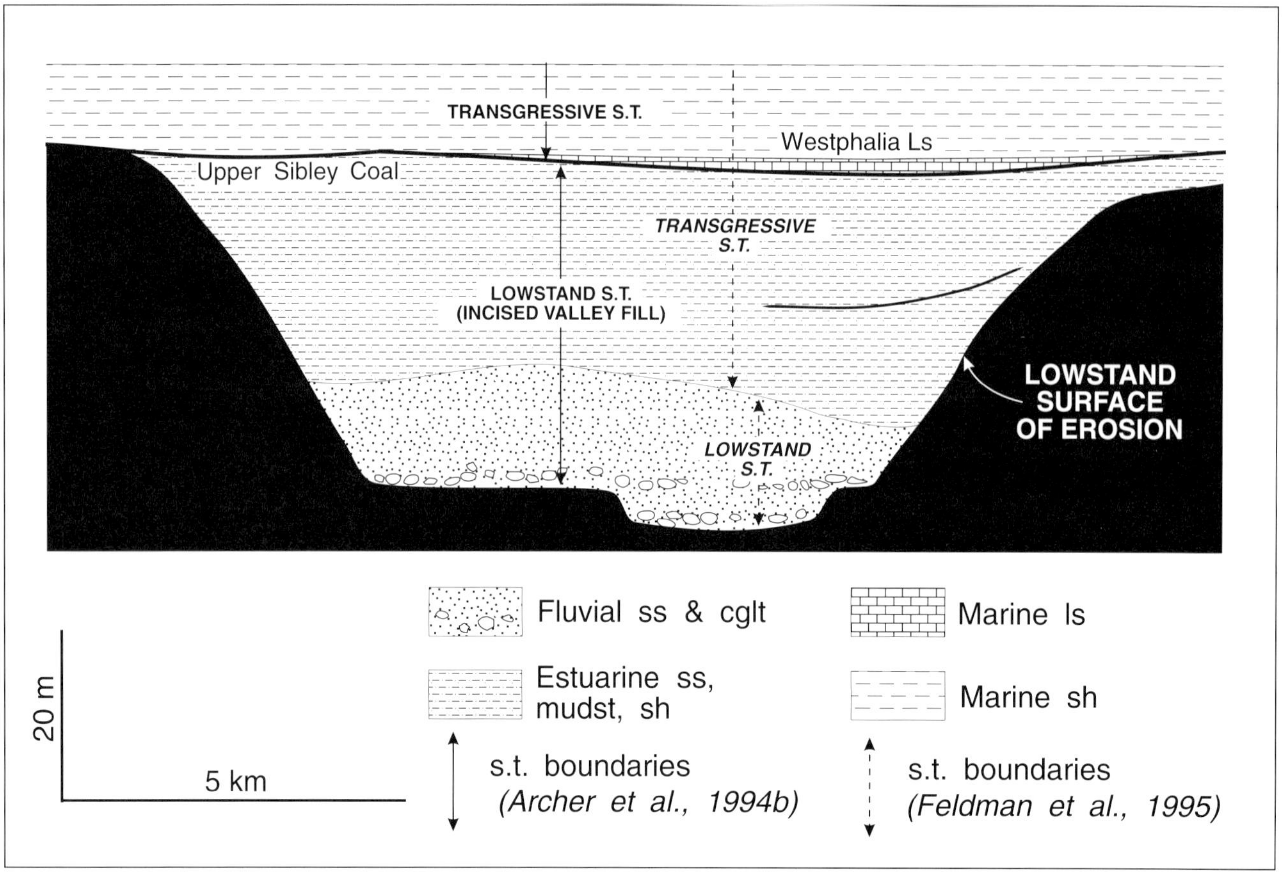

FIGURE 8 — Generalized cross section of the Tonganoxie paleovalley in northeastern Kansas. Two sets of criteria are used to delineate systems tract boundaries depending on the placement of the lower part of the transgressive systems tract (TST). Adapted from Archer et al. (1994b) and Feldman et al. (1995).

The lower Virgilian Douglas Group in eastern Kansas contains a wide range of estuarine facies within at least four unconformity-bounded sequences (Fig. 2). The estuarine facies are similar to those present in siliciclastic rocks throughout the Mid-Continent (Archer, 1994; Lanier et al., 1993; Archer et al., 1994a, b). The Douglas Group provides a unique opportunity for the sedimentological and stratigraphic analysis of well-developed outside shales. The Douglas Group is relatively well exposed (for Kansas) and the outcrop belt intersects the trend of major incised valleys and also contains well-developed marine markers that are critical for delineating deep incision.

The current study considers four unconformity-bounded sequences within the uppermost Lansing Group, Douglas Group, and lower Oread Group, which are named for the most prominent marine lithostratigraphic unit within each: the (1) Westphalia, (2) Haskell, (3) Amazonia, and (4) Toronto sequences (Fig. 7). Previous reference has been made to sequences named for valley-fill sandstones, such as the Tonganoxie sequence for the Westphalia sequence, and the Ireland sequence for the Amazonia sequence (Archer et al., 1994b). However, incised sandstones occur in only a geographically limited part of the depositional sequences of the Kansas Pennsylvanian section. To establish a more widely applicable genetic stratigraphy, each sequence is named here for its most widespread marine marker (Fig. 7). These sequence names are informal and are being proposed in order that the sequences may be correlated far beyond the limits of recognizable incisions. Within the study interval, the Westphalia sequence (previously the Tonganoxie sequence) is the best understood because of its incision into well-defined and laterally persistent markers of the Iatan Limestone and members of the

Stanton Limestone (upper Lansing Group) and because of the fortuitous intersection of the outcrop belt with a major IVF (Feldman et al., 1995). In addition, the Stanton Limestone contains a widely correlated organic-rich unit (Eudora Shale) that provides a useful lower marker for subsurface work involving the overlying Douglas Group.

The lowest sequence considered in this study is the Westphalia sequence of the lower Douglas Group. The base of the Westphalia sequence occurs at a prominent paleosol near the top of the Iatan (Goebel et al., 1989) and extends to the incision at the base of the Tonganoxie Sandstone (Figs. 2 and 7). The basal Tonganoxie Sandstone contains conglomerates, consisting of limestone, shale, and siltstone clasts, which were derived from the underlying units. The Tonganoxie Sandstone generally exhibits an upward-fining trend within the portion confined within the paleovalley (Lins, 1950; Feldman et al., 1995) (Figs. 8 and 9). Analysis of the subsurface distribution of the Tonganoxie throughout eastern Kansas indicates a change from incised valley fills in the north, to a more sheet-like phase in the south (Fig. 9). This southernmost extent of the Tonganoxie, which has been informally termed the Stalnaker Sandstone, is at least locally deltaic (Walton and Griffith, 1985).

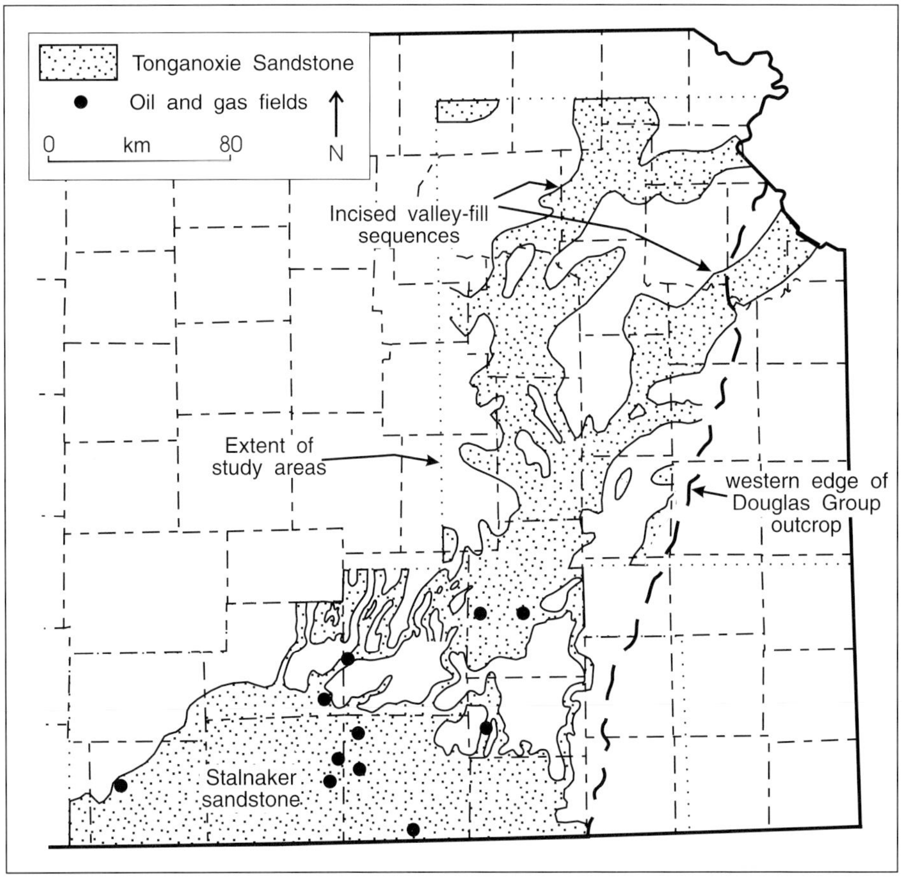

FIGURE 9 — Map of eastern Kansas showing subsurface extent of Tonganoxie Sandstone (adapted from Archer et al., 1994b, original data from Lins, 1950; Winchell, 1957; Sanders, 1959; and Griffith, 1981). Incised valley-fill sequences characterize the unit in northeastern Kansas, whereas a sheet-like deltaic phase (Stalnaker Sandstone) occurs in southeastern Kansas.

A series of thin coals occur within the upper part of the Tonganoxie valley fill and become generally thicker and more widespread upward (Bowsher and Jewett, 1943). The thickest coal is the Upper Sibley Coal, which is immediately overlain by marine facies including the Westphalia Limestone (Ball, 1964, 1971). The Westphalia extends beyond the southern limits of the Upper Sibley Coal, becoming a fusulinid-rich packstone to the south.

A sequence boundary occurs within the lower Vinland Shale, but has not been correlated beyond small IVFs that occur in east-central Kansas (Bowsher and Jewett, 1943) (Figs. 2 and 10). These paleovalleys are incised as deep as 20 m, and are filled with sandstone, heterolithic facies, and coal. The overlying Vinland Shale is generally a marine unit, which is overlain by the widespread, thin Haskell Limestone. The Haskell is the best marker within the Douglas Group and has been correlated from central and western Oklahoma through Kansas and into Missouri and southern Nebraska (Ball, 1985;

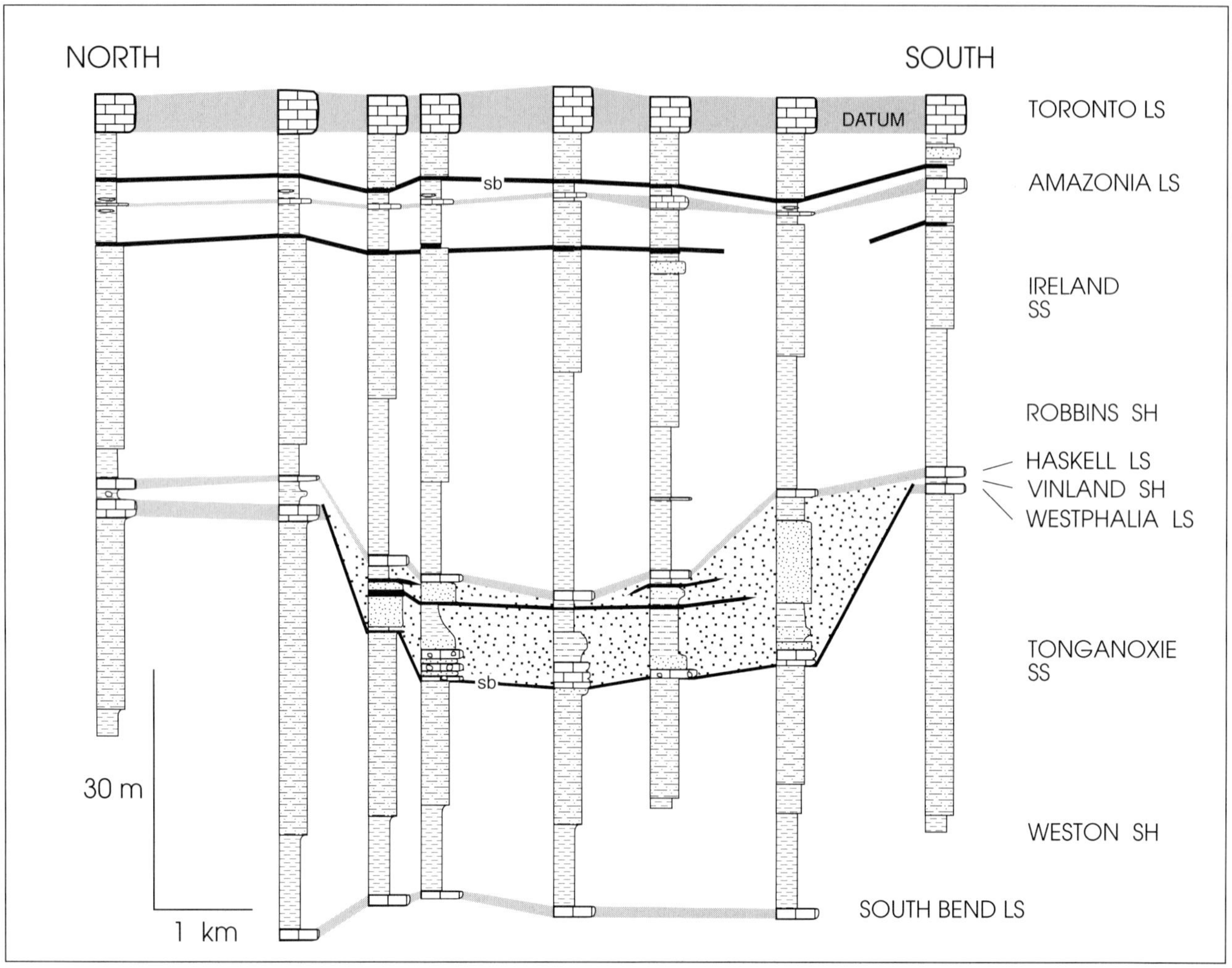

FIGURE 10 — Cross section of Vinland Shale incised valley from Coffee County, Kansas, based on cores described in Kansas City Power and Light Company (1975). Limestones at the base of the IVF are fossiliferous grainstones to limestone clast conglomerates. Fossils were probably derived from the Westphalia Limestone. The position of the basal Westphalia sequence boundary outside the IVF is not known. The position of the basal Tonganoxie sequence boundary is also not known in this area. The sequence boundary at the top of the Amazonia sequence is shown near the top of the Douglas Group. This sequence boundary is coincident with a coal in this area. For symbols used in stratigraphic columns see Fig. 8, and for symbols used in background correlations see Fig. 2. Sequence boundaries are labeled sb.

Kumar and Slatt, 1984). The overlying Robbins Shale is a fossiliferous, marine shale and includes the most marine facies developed within the Douglas Group.

The Ireland Sandstone Member of the Lawrence Formation erosionally incises the Stranger Formation (Bowsher and Jewett, 1943). The Ireland valley fill is generally similar to the Tonganoxie and includes basal limestone-clast conglomerates. Within the upper Lawrence Formation, recognized members include the Amazonia Limestone and the Williamsburg Coal. A well-developed paleosol also occurs in this interval (Joeckel, 1994), which separates the Amazonia sequence from the overlying Toronto sequence. The base of the Toronto Limestone Member of the Oread Limestone (Shawnee Group) marks the lithostratigraphic top of the Douglas Group. The top of the Toronto sequence is at a prominent, laterally extensive paleosol in the Snyderville Shale (Joeckel, 1994) (Figs. 2 and 7).

FACIES INTERPRETATIONS FOR THE DOUGLAS GROUP

Valley-fill facies within the Douglas Group are best studied in the Ireland Sandstone of the Amazonia sequence (Rutan, 1980), and within the Tonganoxie Sandstone of the Westphalia sequence (Lins, 1950; Spriggs, 1989; Feldman et al., 1995). The following facies descriptions and interpretations are based on these studies starting with the fluvial facies at the base of the IVFs and moving upward to more marine-influenced facies.

Fluvial Facies

Conglomeratic Sandstones — The basal conglomeratic sandstone of the Tonganoxie paleovalley ranges from 3 to 18.3 m thick and is confined to the trunk paleovalley and largest tributary paleovalleys. The conglomerates are clast supported, poorly sorted, form discrete beds approximately 2 m thick, and consist of clasts derived from underlying limestones, shales, and siltstones. No quartz pebbles have been noted. Poorly developed crossbed sets ranging from 10 to 30 cm thick indicate unidirectional paleoflow generally to the southwest. Thinner conglomeratic lags are present at the bases of some sandstones that occur above the basal conglomerates. The conglomeratic sandstone facies is similar to deposits occurring in modern braided rivers that are subject to strongly episodic, seasonal flow variations (Feldman et al., 1995). The limited thickness of the crossbedding suggests water depths in the fluvial channels of only a few meters.

Crossbedded Sandstones — Crossbedded sandstones lacking clay drapes are present in the incised valleys of the Westphalia, Haskell, and Amazonia sequences. Trough crossbedded, generally fine-grained sandstones are the most common component of the valley-fill sandstones. These sandstones form units, composed of 80-cm-thick sets, that reach thicknesses of 7 m. Within the Tonganoxie Sandstone, erosionally based stories a few meters thick composed primarily of trough crossbedded sandstone suggest that the strata were laid down in a network of shallow, braided channels floored by three-dimensional (3-D) dunes. Plane beds indicate periodic upper flow regime. Planar crossbedding is also common and forms units approximately 3.5 m thick that consist of cross-sets up to 1.2 m thick. Paleoflow for trough and planar crossbedded sandstones is generally down valley. The sandstones suggest a braided fluvial origin (Feldman et al., 1995).

Estuarine Facies

Within tidally influenced estuaries, current-flow velocities undergo significant variation in velocity and flow direction during individual tidal cycles (Nio and Yang, 1991). Stillstands occur at highest or lowest tides and can result in mud deposition. A muddy layer, or drape, can form directly on foresets during such stillstands. Similarly, sand ripples formed during periods of higher flow velocities can be mud-draped during stillstands. Flow variations result in a range of facies, termed heterolithic, in which there are abundant muddy drapes overlying coarser (generally sandy or silty) sediment. Drapes on large-scale foreset crossbedding result in inclined heterolithic strata (Smith, 1988). On a smaller scale, discontinuous drapes on rippled sands are termed flaser bedding, continuous drapes on rippled sands are termed wavy bedding, and discontinuous sandy lenses (starved ripples) or sandy streaks in muds are termed lenticular bedding (Reineck and Wunderlich, 1968). These facies have traditionally been widely misinterpreted within the Mid-Continent. Specifically, many examples of flaser to lenticular bedding have been incorrectly described as distal floodplain and even prodeltaic facies. Such misinterpretations probably reflect a common lack of awareness of tidal sedimentology as well as a reluctance, based on misguided theoretical models (Shaw, 1964), to invoke tidal processes in epicontinental seas. In addition, although the lack of benthic marine fossils in these facies has been used as evidence of freshwater deposition, it may in fact be due to the environmental stresses related to the high depositional rate, high turbidity, and fluctuating salinity (Feldman et al., 1993a).

Inclined Heterolithic Bedding — Within valley-fill crossbedded fluvial sandstones there exist gradations from planar crossbed foresets that lack muddy drapes, to foresets that exhibit well-formed and common clay drapes. The former were deposited in fluvial environments that were not tidally influenced. The occurrence of numerous mud-draped foresets suggests tidally induced variations in flow velocities during deposition. However, these mud drapes do not indicate coastal proximity, because such variations can occur at significant distances inland. For example, within the St. Lawrence River and associated estuarine system, macrotidal conditions extend from about 400 to 650 km inland from the mouth of the seaway (Archer et al., 1994b). Within this zone of the St. Lawrence, tidal flats and associated facies occur within a completely freshwater setting.

Thus, mud-draped foresets are a good indicator of the development of estuarine conditions within the fluvial to estuarine transition. The large-scale planar foresets indicate significant channelized flow, whereas the muddy drapes indicate that there was sufficient incoming tidal flow to temporarily reduce down-channel fluvial flow. The relative position within the estuarine environment of mud-draped facies varies depending on the relative tidal and fluvial flux. In addition to shifts related to tidal cycles (neap-spring periods), there can be seasonal variations in tidal heights (Archer et al., 1991) and fluvial flux. The fluvio-estuarine transition shifts seaward during wetter seasons and landward during drier seasons. Such effects have been described from Pennsylvanian estuarine facies (Archer et al., 1994a; Kvale et al., 1994).

Inclined heterolithic bedding is readily identifiable in cores (Reinson, 1989). In some cases, goethite-cemented foreset boundaries appear to have replaced original muddy drapes (Kvale and Archer, 1991). Reactivation surfaces, especially if mud draped, are also a good indicator of the presence of tidal controls (de Mowbray and Visser, 1984), but these are not always particularly easy to identify in cores.

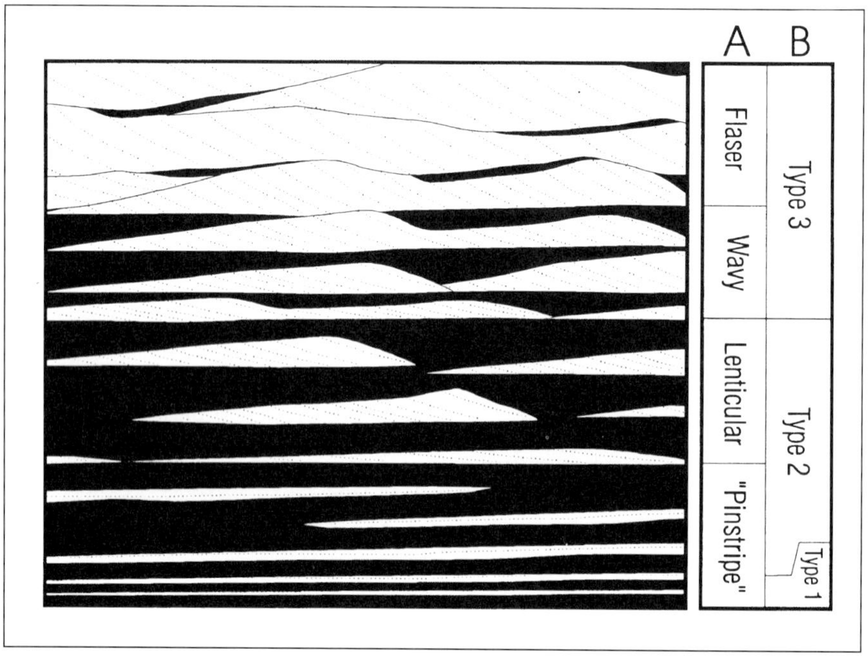

FIGURE 11 — Comparison of (A) flaser, wavy, and lenticular bedding (Reineck and Wunderlich, 1968) and "pinstripe" bedding (Klein, 1977), and (B) analogous types of heterolithic rhythmites described from roof strata of Pennsylvanian coals of the Illinois basin (Kvale and Archer, 1990). These types of fine-scale bedding are characteristic of, but not necessarily restricted to, tidal environments where sedimentation alternates between bed- and suspended-load deposition.

Silty Rhythmites — Planar laminated and bedded siltstones with clay drapes occur within the upper parts of valley-fill sequences. There may be systematic variations in lamina and bed thicknesses that can be directly related to neap-spring tidal periods (Kvale et al., 1989). Bedding surfaces preserve a wide range of surface features including flat-topped ripples, raindrop imprints, insect trackways, and tetrapod trackways, all indicating an intertidal setting. This is a relatively rare facies, but has been described from Morrowan and Desmoinesian valley-fill successions from Indiana and Illinois (Kvale et al., 1989; Kuecher et al., 1990; Archer et al., 1995) and from the Douglas Group in Kansas (Lanier et al., 1993; Feldman et al., 1995).

Silty rhythmites where observed to date are not bioturbated, but can exhibit a diverse suite of arthropod trackways, insect markings, fish-fin drag marks, and tetrapod (amphibian and reptile) trackways (Archer and Maples, 1984; Lanier et al., 1993). Well-preserved, articulated, terrestrial plant fossils are also common and rare insect fossils also occur (Kvale and Archer, 1991). The siltstone laminae tend to be normally graded with a very thin, upper muddy drape, and resemble annual varves. This varve-like appearance, abundance of plant fossils, and lack of marine invertebrates lead to misinterpretation of silty rhythmites as indicative of lacustrine settings (Archer and Maples, 1984; Maples and Archer, 1987). However, the pervasive occurrence of tidal periods inferred from lamina-thickness variations makes a strong case for tidal controls during sedimentation (Kvale et al., 1989).

Excellent modern analogues for silty rhythmites occur in the upper reaches of macrotidal estuaries in the Bay of Fundy, Canada (Dalrymple et al., 1991) and Bay of Mont Saint Michel, France (Tessier, 1993). These areas are macrotidal and have tidal ranges up to 15 m. On fluvio-tidal point bars in these areas, mm- to cm-thick layers of silt are deposited during single high tides. Since the best modern analogues for silty rhythmites occur in these macrotidal settings, it is tempting to propose that similar conditions existed during the formation of the Pennsylvanian rhythmites. Although this cannot be absolutely confirmed, it is likely that mesotidal or higher tidal ranges are necessary for the deposition of silty rhythmites.

Heterolithic Rhythmites — These rhythmites comprise common forms of tidal bedding, which include flaser, wavy, lenticular, and pinstripe bedding (Fig. 11). Heterolithic rhythmites can exhibit lamina-thickness cycles similar to silty rhythmites, and can be directly related to tidal periodicities (Kvale and Archer, 1990). Heterolithic rhythmites commonly occur directly overlying coals and can contain abundant upright trees (Kvale and Archer, 1990; Archer et al., 1994a). Significant thicknesses of heterolithic rhythmites compose the upper parts of valley-fill sequences in the Douglas Group (Feldman et al., 1995). Small-scale (2 to 10 m wide) tidal channels occur within the rhythmites and commonly contain abundant climbing ripple lamination and have either ebb- or flood-dominated paleoflow (Archer et al., 1994b).

Heterolithic rhythmites apparently form near a middle estuarine position and are related to deposition within the turbidity maximum zone. Very similar rhythmites are being deposited within the Salmon River Estuary in the Bay of Fundy (Dalrymple et al., 1991), and also occur within the estuarine-like deltaic distributaries of the Rajang River Delta of Malaysia (J. Staub, pers. comm., 1994).

The Pennsylvanian rhythmites exhibit gradations from being nonbioturbated to nearly completely bioturbated by sediment-feeding organisms. The degree of bioturbation apparently relates to rates of deposition and to the paleosalinity. Highest rates of vertical accretion probably occurred in mid-estuarine positions where rates of deposition and changes in salinity are very high. Such conditions precluded extensive infaunal habitation and bioturbation. Conversely, in the seaward portions of the estuaries, rates of sedimentation were probably lower, and salinity variation was less pronounced. Thus, these rhythmites were more extensively bioturbated. Tidal range also directly affects sedimentation rates and salinity fluctuations. Therefore, less bioturbated and more cyclic rhythmites may indicate higher tidal ranges. Within otherwise similar Holocene valley-fill sequences, microtidal estuaries are generally more extensively bioturbated (Nichols et al., 1991) than macrotidal estuaries (Dalrymple et al., 1991). If meso- to macrotidal conditions existed during valley-fill sedimentation during the Pennsylvanian, then much of the heterolithic facies would not have been extensively bioturbated. This lack of bioturbation may in part help to explain why such facies have commonly been misinterpreted in the Mid-Continent as floodplain, lacustrine, or other related terrestrial facies.

Outer Estuarine Sandstones

In the outer parts of estuaries, marine processes such as waves will significantly winnow the sediments and result in the deposition of relatively clean sand. In the Bay of Fundy, for example, there are extensive sand flats and bars in the outer parts of the estuary (Dalrymple et al., 1991, 1992). Muddy drapes are formed during high-tide stillstands, but wave activity and high flow velocities during flood and ebb tides tend to rework and erode these drapes. Thus, the preservational potential of mud drapes is low in the outer, marine zones.

Within the Douglas Group, sheet-like sandstones are best developed overlying the valley fills. Within the Lawrence Shale (upper Douglas Group), sheet-like sandstones exhibit a variety of features suggestive of sand-flat and bar deposition (Archer et al., 1994b). These include flat-topped ripples and ladder-back ripples, which are formed during falling water phases. Ripple fans with a variety of orientations are also common and are related to the development of complex, 3-D bedforms. Parallel lamination is also common and is apparently related to upper flow regime conditions. These parallel laminated sandstones are similar to the braided sand flats that form in macrotidal estuaries, such as the Bay of Fundy (Dalrymple et al., 1991). Extensively bioturbated sheet-like sandstones also occur directly below the Haskell Limestone. Such sandstones probably indicate lower shoreface conditions (Feldman et al., 1995).

Marine Facies

Because of the biostratigraphic and paleoecological focus of much work in the Mid-Continent, there are many well-established criteria for the delineation of the degree of marine influence. Aspects of the tremendous variability of fossil occurrences within the Pennsylvanian cyclothems are summarized in Moore (1964). Within the Douglas Group, bivalves are commonly one of the stratigraphically lowest, or earliest, indicators of marine influence in the valley-fill sequences. *Carbonicola*-type clams and associated trace fossils occur within the middle to upper parts of valley-fill sandstones, but apparently represent only brackish water conditions (Archer et al., 1994a). Myalinid bivalves form transgressive lags within the Westphalia Limestone (Archer et al., 1994b) and also occur as a widespread marker, ranging from northeastern Kansas south to Oklahoma, within the upper Vinland Shale (Moore, 1935; Ball, 1985). Fusulinids are abundant with the Westphalia and Haskell limestones indicating open-marine conditions.

SEQUENCE STRATIGRAPHY

Types of Unconformities

The most prominent unconformities within the Douglas Group are type 1 sequence boundaries (Van Wagoner et al., 1990), which were formed during sea-level lowstand and resultant subaerial exposure. The lower sequence boundaries at the base of the South Bend, Westphalia, Haskell, and Amazonia sequences are deeply incised (compared to other Mid-Continent Pennsylvanian IVFs), and basal valley fills contain conglomerates composed of clasts eroded from older strata exposed in valley walls. The lower sequence boundary of the Toronto sequence in the uppermost Douglas Group is not associated with deep incisement, but is related to features such as laterally persistent paleosols (Joeckel, 1994).

Other prominent, but less regionally extensive, erosional surfaces are associated with transgressive surfaces. The basal Westphalia Limestone has been interpreted (Archer et al., 1994b) as having formed as a ravinement or transgressive surface of erosion. The base of this unit exhibits localized channeling and shell lags of myalinid bivalves. There are various facies within the Westphalia (Ball, 1971); however, in much of the southern half of the Douglas Group outcrop belt in Kansas, the dominant lithofacies is that of a fusulinid packstone with an erosional base, which led Ball (1971) to interpret the facies as a storm deposit. More work needs to be done on this unit; however, it appears that the formation of the transgressive surface of erosion may be associated with the stratigraphically lowest indication of marine flooding.

Estimation of Base-Level Changes

Magnitude of the lowering of base level associated with incision and subsequent flooding of the Tonganoxie paleovalley cannot be directly measured, but can be estimated by measuring the depth of incision, thickness of the valley fill, and estimations of depositional depth. A minimum estimate for the fall in relative sea level is the depth of incision, which is 40 m for the Tonganoxie valley. The valley is completely filled with fluvial through marine (Westphalia Limestone) so there had to be at least a 40-m rise in relative sea level. The thickness of marine strata overlying the valley fill is not known because the sequence boundary at the base of the Haskell sequence is not known in the area of the Tonganoxie paleovalley.

The 40-m flux of water depth can be increased based on depth estimates of facies and sediment decompaction. The maximum preincision water depth was probably associated with deposition of the Iatan Limestone. Depth estimates for such carbonates vary, but modern analogues suggest 10 m or less (Heckel and Cocke, 1969). Relative sea-level fall from Iatan Limestone to valley incision would then be 50 m. Decompacting the valley fill would increase its thickness by 30% or so based on the amount of shale in this section, giving an estimate of 65 m of base-level rise.

Fall of base level as indicated by valley incision must have resulted in exposure of large interfluvial areas. Although surfaces may be difficult to correlate beyond incision, this evidence of a falling base level justifies the search for such surfaces and correlations of these surfaces among widely spaced outcrops.

Correlating Sequences Beyond Incised Valleys

Correlating sequence boundaries beyond the incised valleys is not easy because the deepest incised valleys are not all associated with well-developed paleosols in adjacent interfluves. It is assumed that laterally extensive marine units (such as core shales) are correctly correlated, and that sequence boundaries do not cross these units. For example, if a single paleosol occurs in most outcrops between two marine strata (such as the paleosol in the Snyderville Shale between the Toronto and Leavenworth limestones), then that horizon of exposure can be reasonably correlated.

The sequence boundary for the Westphalia sequence occurs at the base of the Tonganoxie paleovalley in northeastern Kansas (Feldman et al., 1995). At the sequence boundary in the paleovalley, fluvial to estuarine facies rest unconformably on older marine strata. The sequence boundary is correlated from the base of the Tonganoxie incised valley to a paleosol between the Iatan and Haskell limestones in Nebraska, Iowa, and Missouri (Goebel et al., 1989). A paleosol is apparently not well developed at the boundary in Kansas (Figs. 2 and 10). Localized paleosols are present on valley walls, and rarely on the valley floor. Correlations of wireline well logs indicate that a coal occurs at the boundary in some interfluvial areas (Feldman et al., 1995). No direct observations of interfluvial areas adjacent to the incised valley have yet been made.

The Tonganoxie paleovalley extends to southeastern Kansas but does not extend to the stratigraphically equivalent Stalnaker Sandstone (Fig. 9). The basal Westphalia sequence boundary apparently correlates with the base of the Stalnaker Sandstone (Watney, 1993). The northern portion of the Stalnaker Sandstone is incised in a series of narrow (1 to 2 km) valleys that feed into the main body of the Stalnaker Sandstone, a broad sandstone sheet (Winchell, 1957). Most marine limestone beds (except the Haskell Limestone) do not continue south into the Anadarko basin, where large quantities of immature siliciclastic sediment were being shed from tectonic highlands (Ford, 1975; Kumar and Slatt, 1984).

The incised valleys that feed into the Stalnaker do not appear to be organized into a tributary and trunk valley system, which is very different from the organization of river incision in northeastern Kansas (Feldman et al., 1995). The abrupt southern end of the Tonganoxie paleovalley may reflect the position of a nick point at which river incision was initiated (Schumm, 1993; Wescott, 1993). After the sea level dropped below the nick point, incision would have proceeded upstream by headward erosion. Downstream of the nick point, the valley may have broken up into many smaller valleys; if sea-

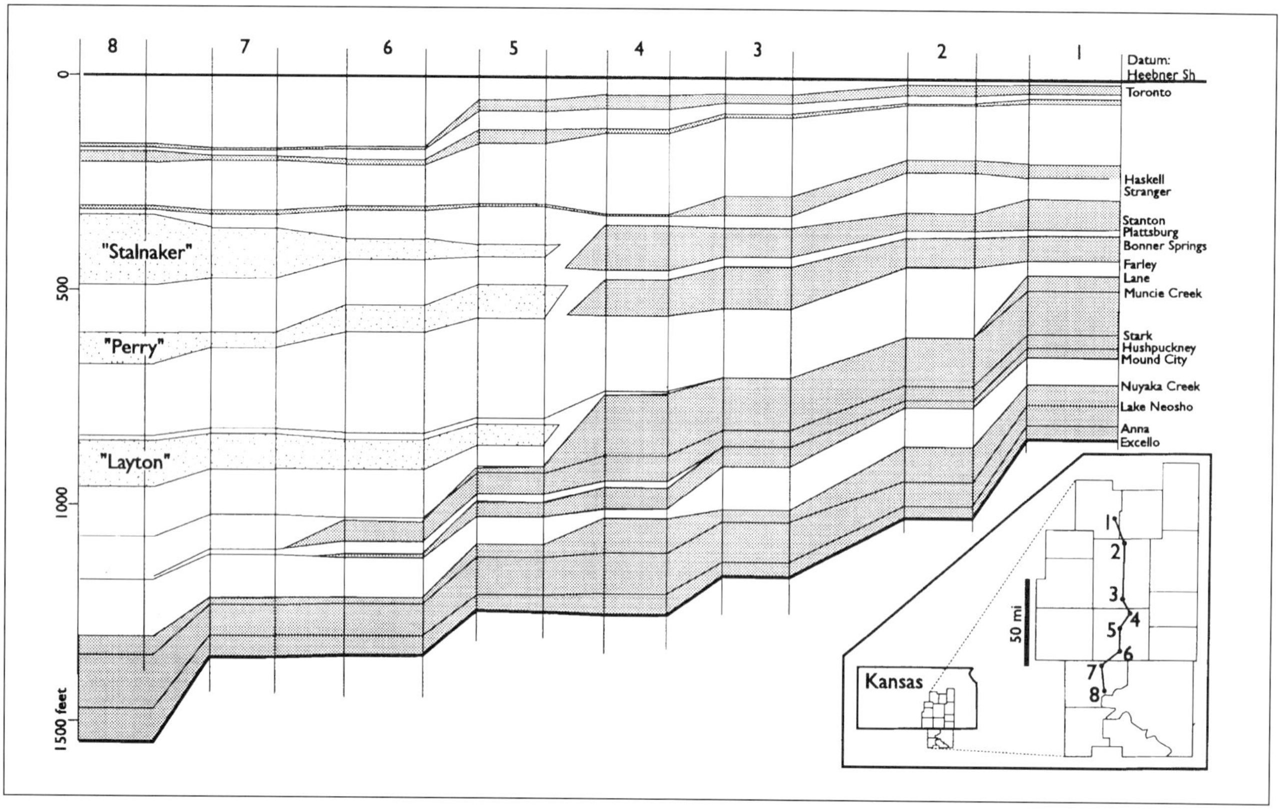

FIGURE 12 — North-to-south cross section of interval encompassing Douglas Group of south-central Kansas to northern Oklahoma. Stalnaker Sandstone is probably stratigraphically equivalent to the Tonganoxie Sandstone. Modified from Watney (1993).

level fall was sufficiently rapid, no organized stream system may have reached the coast (Van Wagoner, 1995), thus the downstream limit of valley incision is not necessarily at the coastline. A nick point in this case may have formed at a tectonic hinge that bounded a stable shelf to the north, and more rapidly subsiding basin to the south. This boundary is nearly coincident with the edge of several carbonate banks that also responded to the hinge (Watney, 1993) (Fig. 12).

The basal Haskell sequence boundary has not yet been correlated beyond a series of incised valleys in Coffee County. There is no obvious boundary in the Vinland Shale in northeastern Kansas; however, the sequence boundary must merge with the basal Tonganoxie sequence boundary in Missouri, Nebraska, and Iowa, where the interval between the Iatan and Haskell limestones thins.

The basal Amazonia sequence boundary also has not been correlated beyond incised valleys in northeastern Kansas. Within these incised valleys, fluvial sandstones rest unconformably on older marine strata (Bowsher and Jewett, 1943), yet no potentially correlative paleosols are known.

The sequence boundaries at the bases of the Toronto and Heebner sequences are both based on widespread and deeply weathered paleosols (Joeckel, 1994; Feldman and Archer 1995). In both boundaries, underlying marine strata were profoundly weathered by vadose zone processes (Joeckel, 1994).

The best developed paleosols tend to be associated with sequence boundaries that lack incised valleys, such as the sequence boundaries at the bases of the Toronto and Heebner sequences. The lack of paleosols on interfluves of well-developed incised valleys may be a result of preservation. Ravinement during transgression could have removed the loose, weathered sediment, leaving a sharp erosional surface at the sequence boundary. Likewise the contact at the base of the incised valleys was formed mostly by fluvial erosion. The valley walls may have developed soils that were buried

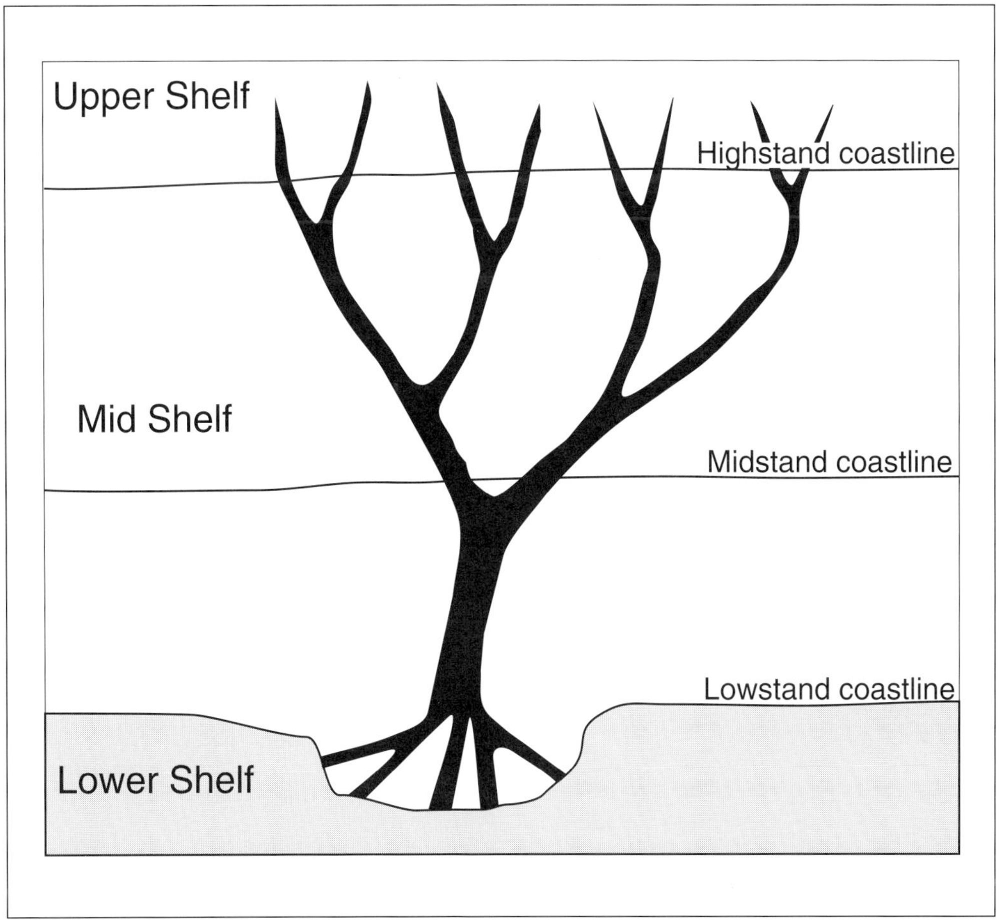

FIGURE 13 — Diagrammatic representation of lowstand fluvial and associated deltaic system. The lowstand coastline would have an associated large-scale deltaic system. During transgression of the shelf, a variety of flooded fluvial valleys (estuaries) would be produced. Mid-shelf flooding would result in relatively larger estuarine systems, whereas flooding of the upper shelf would result in relatively smaller estuarine systems.

by lower energy, fine-grained estuarine deposits that are associated with preservation of delicate sedimentary structures and fossils (Feldman et al., 1993a; Lanier et al., 1993). The greatest potential for preservation of paleosols associated with these incised valleys is on gently dipping valley walls, not in valley floors, or interfluves.

During transgression, there would be a progression of marine flooding of larger scale (higher order) to smaller scale (lower order) incised valleys. During a continued transgression, a series of successively smaller, estuarine systems would be formed and filled as the fluvial valleys were intersected by the backstepping coastline. Valley-fill successions formed during higher stand conditions would thus be significantly smaller than those developed during lower stand conditions. Transgression of a drainage network would result in the development of smaller, but more numerous estuarine and associated drainage systems as lower order tributaries were intersected by the coastline (Fig. 13). Thus only at basinward positions would there occur valley-fill sequences of Douglas Group size.

The highest stand coastlines are not well known for any of the studied sequences. Recent correlations indicate that highstand deposits, if preserved, would be found in other basins. The Haskell Limestone of the Haskell sequence has been correlated to marine strata in the Illinois basin, and the Heebner Shale of the Heebner sequence has been correlated with marine strata in Pennsylvania (Heckel, 1994). Highstand coastlines for these sequences are far shelfward of Kansas.

Controls on Valley Incision and Filling

The primary control on valley incision and filling is falling and rising sea level. Valleys are generally filled with upward increasing marine facies. Sequence stratigraphic systems tracts are based on parasequence stacking patterns; however, within the Douglas Group IVFs, parasequences are not well developed. In the absence of easily delineated parasequences, the boundary between the lowstand systems tract (LST) and transgressive systems tract (TST) may be variously identified based on how the first major flooding surface or transgressive surface is recognized (Van Wagoner et al., 1990). The top of the LST was previously placed at the top of the fluvial sandstone (Feldman et al., 1995), and at the erosional surface at the base of the Westphalia Limestone (Archer et al. 1994b) (Fig. 8).

Clearly, in the absence of recognizable parasequences, placement of the systems tract boundaries remains problematic and should probably be avoided. Given that the total eustatic flux was of very high magnitude, the valley-fill successions would have backfilled at various times as the progressively more basinal to more shelfal areas were flooded. Thus, the shift from fluvial to estuarine conditions may be time transgressive, and not coincide with any systems tract boundaries, but may be more directly related to position of the incised valley relative to the maximum and minimum shorelines.

Climate during lowstand may also have influenced the development of IVFs. Douglas Group IVFs contain rooted coals that indicated high water tables. Fluvial sandstones derived from the craton also suggest well-developed river systems, and wet climates. Sequence boundaries at widespread carbonate-rich paleosols, such as vertisols, typically lack incised valleys, at least in the areas where the vertisols are well developed. Examples include the paleosols at the bases of the Toronto and Heebner sequences (Fig. 7). These vertisols contain pedogenic carbonate, and formed in low relief coastal plains, with seasonal rainfall (Joeckel, 1994). The climate of these paleosols was drier than climates associated with the coal-rich deeply incised valleys of the Westphalia and Amazonia sequences. Reduced rainfall may have resulted in few incisions, or perhaps major incisions are present, but have not yet been recognized.

There was a pronounced drying trend that occurs throughout the Late Pennsylvanian and up through the Permian. This trend is interpreted from the general decrease in coals and increase in evaporites during this time (Archer and West, 1993). All other factors being equal, the progressive decrease in yearly precipitation could have resulted in reduced fluvial flux and a reduced ability to erode paleovalleys. Thus, in east-central Kansas no Permian-age paleovalleys have yet been recognized. However, a continuing problem is that the relationship among soil development, soil preservation, depth of incision, climate, and rate and magnitude of sea-level change has not been resolved.

The cross-sectional geometries, or cross-valley profiles, of paleovalley systems are also strongly controlled by the nature of the pre-erosional sedimentary sequence. In general, incision was apparently more restricted by underlying carbonate units than by siliciclastic-rich units. In subsurface analyses, incision and erosional truncation of regional carbonate markers provide strong evidence for paleovalley development. A carbonate-bedrock constrained paleovalley is also likely to be deeply incised and relatively narrow; thus the lateral stratigraphic variability is more likely to be observed. Conversely, a wide, shallower incised paleovalley within a completely siliciclastic sequence is much more difficult to recognize. Even if the erosion is detected, however, it may not be possible to determine if it is related to incision during sea-level lowstand or autogenic processes, such as a channel developed within a deltaic complex.

Basement structure may also have influenced the development of drainage networks. Many of the world's largest fluvial systems are strongly correlated with structural controls (Potter, 1978). Inspection of modern drainage networks in the Midwest reveals a pronounced tendency for streams to parallel basement structures. Similar structural controls may have operated during the Pennsylvanian. Mapping of the Tonganoxie paleovalley in northeastern Kansas revealed a trunk valley oriented northeast-southwest, and tributaries oriented roughly normal to it (Feldman et al., 1995). Both the trunk valley and tributaries are parallel to basement structures in the area (Baars, 1992).

EXAMPLES OF OTHER PENNSYLVANIAN VALLEY-FILL SEQUENCES

Morrowan Paleovalleys

Paleovalleys are easiest to delineate when they are developed on older carbonate-rich strata. They are also best developed if the fall in base level was extreme and protracted. These types of circumstances led to the development of numerous Morrowan paleovalleys in the Central Appalachian basin (Greb, 1993) and Illinois basin (Archer et al., 1994a; Kvale and Barnhill, 1994; Greb et al., 1992; Bristol and Howard, 1974). The development of these paleovalleys, which overlie a major unconformity, has long been understood even before the development of modern sequence stratigraphic concepts. In the Morrow of the Hugoton embayment, a stacked series of paleovalleys occur above the Mississippian-Pennsylvanian unconformity (Sonnenberg, 1985; Sonnenberg et al., 1990; Wheeler et al., 1990). These incised valleys were filled with fluvial, estuarine, and marine facies during rising sea level.

Desmoinesian Paleovalleys

In Kansas and Oklahoma, portions of the Cherokee Group strata appear to offer significant potential for the recognition of incised valley systems. The shoestring sandstones of southern Kansas

(Bass, 1936), commonly termed the Bartlesville or Bluejacket sandstone, exhibit geometries similar to valley-fill sequences. Bass (1936) presented what was, at that time, a very thoughtful application of modern analogues, but concluded that the shoestring sandstones were coastal, barrier islands. Subsequently, the shoestring sandstones have been widely interpreted as channel deposits within deltaic complexes (Brenner, 1989). The vertical sequences commonly developed within these sandstones, however, are much more typical of a fluvio-estuarine valley-fill system (Zaitlin et al., 1994). Basal facies within the shoestring sandstones commonly contain cherty conglomerates and coarse-grained sandstones. The actual sand bodies commonly have been described as being punctuated by shales and finely laminated, fine-grained siliciclastic facies. All these criteria match the typical sequences that have been described from valley-fill sequences of the Douglas Group.

In the Illinois basin, major fluvial systems and paleovalleys are related to major economic coals. Tidal estuarine facies are developed in the upper parts of the valley-fill sequences and form the roof strata over areas of low-sulfur coal. Tidal rhythmites have been documented overlying the Seelyville Coal (Archer et al., 1995), Colchester Coal (Kuecher et al., 1990), and Springfield and Herrin coals (Archer et al., 1994a). In Kansas and Oklahoma, there are lateral correlatives of most of these coals, although they are not mined as extensively as in the Illinois basin. The Labette Shale interval, for example, is laterally correlative of the interval that includes the Herrin Coal of Illinois. Field reconnaissance indicates that estuarine facies occur in channel- or perhaps valley-fill facies of the Labette Shale and Chanute Shale (Pennsylvanian) of Kansas and Oklahoma. A prominent black-shale marker, the Anna Shale, overlies the Herrin Coal in the Illinois basin and marks the top of the Labette Shale in the Mid-Continent. Within the Illinois basin, a clastic wedge (Wanless, 1964) occurs above the Herrin Coal and the Anna Shale. This wedge has been interpreted as an estuarine valley-fill sequence (Archer and Kvale, 1993) and appears to correlate with the lower part of the Labette, which contains the incised Englevale Sandstone Member. The Englevale has also been called the Warrenburg and, in the subsurface, the Peru sandstone in Kansas and northeastern Oklahoma (Moore et al., 1951). Preliminary field work in northeastern Oklahoma indicates that the Englevale/Peru contains tidal rhythmite facies, which suggest a potential for estuarine and valley-fill facies.

Missourian Paleovalleys

Much of this series in Kansas is dominated by carbonate strata. It is within this part of the section that the typical Kansas-type cyclothems (Heckel, 1986) are well developed. However, the Chanute Shale occurs near the middle of this series and contains a valley-confined sandstone in the lower part, which has been referred to as the Noxie Sandstone, and a sandstone in the upper part termed the Cottage Grove Sandstone (Schulte, 1959). These sandstones have also been referred to as the Layton Sandstone in the subsurface. Ongoing field work in southern Kansas indicates that tidal rhythmites are common within various intervals of the Chanute Shale. The Garnett paleovalley occurs within the upper Missourian Stanton Limestone in central eastern Kansas. The incision is approximately 9.8 m deep and 300 m wide (Feldman et al., 1993b). The valley fill is composed of a lower limestone conglomerate of uncertain origin, and an upper laminated mudstone deposited in shallow to intertidal estuarine conditions (Feldman et al., 1993b; Reisz et al., 1982; Woodruff, 1984). The Garnett IVF is well known for the well-preserved reptiles that have been recovered from laminated estuarine mudstones (Reisz et al., 1982).

Virgilian Paleovalleys

There are also several sandstone-filled incisions within the upper Virgilian Series in Kansas in addition to those described above. Incised valley-fill sandstones occur in the Wabaunsee Group and are apparently related to paleovalley-fill sequences developed within the Pillsbury Shale, Root Shale, and Wood Siding Formation. This part of the stratigraphic succession has not been particularly well studied from a sedimentological perspective and deserves more detailed attention. These sandstones were previously considered to be Lower Permian prior to the upward revision of the Pennsylvanian/Permian boundary by Baars et al. (1994b).

The Hamilton paleovalley has also been well studied because of its diverse and well-preserved fossil assemblage of plants, fish, and tetrapods. The sequence boundary is within the Topeka Limestone of the upper Shawnee Group (see Fig. 6 for position of Shawnee Group). The valley fill is similar to the fill of the Garnett incised valley; there is a lower limestone conglomerate, and an upper estuarine mudstone and laminated limestone (Cunningham et al., 1994).

CONCLUSIONS

The Douglas Group in northeastern Kansas provides useful models for Pennsylvanian IVF sequences. Based on the vertical stacking patterns developed within the Douglas Group IVFs, particularly those of the Westphalia and Amazonia sequences, criteria can be established for recognition of the transition from fluvial to estuarine conditions. The occurrence of significant mud draping of sandstone foresets and occurrence of heterolithic facies suggest tidal estuarine influences. The common occurrence of tidal facies within the Douglas Group, as well as in other apparent valley-fill sequences, indicates the importance of tidally mitigated sedimentation in the Mid-Continent.

The cyclic nature of the Pennsylvanian sequences in Kansas has long been noted; however, only recently have the depositional settings of thicker siliciclastic units (such as the Douglas Group) been well understood. A variety of sandstone facies occur, particularly within the Tonganoxie Sandstone. These include braided fluvial, deltaic, tidal estuarine, and marine facies. Application of sequence stratigraphic models allows a greater understanding of stratigraphic relationships. In particular, sequence stratigraphy provides a more useful framework for interpretations than do the more traditional facies models; however, systems tracts are not easily delineated and may not apply to these settings. There is probably not a one-to-one correspondence of Kansas-type cyclothems and unconformity-bounded sequences. Because of low paleoslopes and high magnitude of eustatic flux, the sequences are spread out over considerable lateral distances. Thus, multistate regional analyses will be needed to adequately address the entire sequence stratigraphic history.

The depth and morphology of Pennsylvanian incised valleys were influenced primarily by sea-level fluctuations. During lowstands, vast land areas were exposed and eroded by rivers forming incised valleys. These were filled with fluvial through estuarine facies during transgressions. Primary control on valley incision and filling is sea-level history. Preliminary results suggest that an important secondary factor may be climate. Douglas Group IVFs formed during relatively wetter periods. During the driest times, valleys may not have formed, or may be so rare that they have not been encountered.

The IVFs and associated estuarine facies are common throughout the Pennsylvanian of Kansas, parts of northern Oklahoma, as well as other Mid-Continent basins such as the Illinois basin. In the current analyses of the Douglas Group IVFs, the degree of incision is relatively easy to determine in subsurface analyses, because of under- and overlying marine-carbonate markers. In parts of the section that lack well-defined, laterally persistent markers, it may be difficult to adequately delineate incision. However, the recognition of specific facies, facies-stacking patterns, and other features should be a useful guide to the determination of potential for IVF-related sequences.

ACKNOWLEDGMENTS

Research related to this study was performed in conjunction with several projects. The general understanding of estuarine facies and delineation of incised valleys related to fossil-rich facies, such as the Garnett and Hamilton paleovalleys, was funded by the National Science Foundation (Grant EAR-9018079). Ongoing coring of the Douglas Group is being supported by the National Science Foundation (Grant EAR-9405123). Earlier work on the Lawrence Shale (Douglas Group) was supported by the Kansas Department of Transportation (Grant K-TRAN: KSU-91-6). Various aspects of the field work were supported by Kansas State University and the Kansas Geological Survey. Ongoing collaborative work has greatly assisted our current understanding of the depositional systems. In particular, we have greatly benefited from interactions with Martin Gibling and William Lanier on Douglas Group sedimentology, Lynn Watney on stratigraphic relationships, Bernadette Tessier on modern analogues, and Erik Kvale on similar facies in the Illinois basin. This manuscript was greatly improved by comments of Evan Franseen, Tim Carr, Lynn Watney, S.J. Mazzullo, and Rod Tillman.

CYCLE HIERARCHY AND GENETIC STRATIGRAPHY OF MIDDLE AND UPPER PENNSYLVANIAN STRATA IN THE UPPER MID-CONTINENT

W.L. Watney,[1,2] J.A. French,[3] J.H. Doveton,[1] J.C. Youle,[4] and W.J. Guy[1]

ABSTRACT

A hierarchy of sedimentary cycles has been recognized in Middle and Upper Pennsylvanian strata in the northern Mid-Continent, U.S.A., and a genetic and sequence stratigraphic classification has been used to characterize them. The basic building block in this succession is the cyclothem-scale depositional sequence, which is bounded by subaerial unconformities on the shelf. Each of these sequences includes several lithogenetic units: a basal flooding unit, a condensed section, a late highstand carbonate or siliciclastic unit, and a paleosol. The thin condensed sections are generally radioactive, conodont, and ammonoid-rich marine black shales that can be readily correlated over large distances in the surface and subsurface using gamma ray logs and conodont and ammonoid biostratigraphy. The subaerial exposure surfaces and paleosols that mark sequence boundaries are difficult to recognize with well log data alone, but the radioactive condensed sections, which in many cases closely overlie the sequence boundaries, can be used to correlate the sequences. Because of these limitations, genetic stratigraphic units, bounded by these correlatable condensed sections, have been used in this study as proxies for depositional sequences. These genetic stratigraphic units are practical for regional subsurface mapping, and can be linked to their corresponding depositional sequences in areas where cores are available. The boundaries of a genetic stratigraphic unit and the associated depositional sequence virtually coincide on the shelf, but diverge in shelf margin and basinal settings, due to significant sedimentation during the early stages of relative sea-level rise.

Sets of these genetic stratigraphic units that exhibit uniform stacking patterns (termed genetic

[1]Kansas Geological Survey, The University of Kansas, Lawrence, KS 66047
[2]Energy Research Center, The University of Kansas, Lawrence, KS 66047
[3]Consultant, Littleton, CO 80127
[4]Consultant, Buena Vista, CO 81211

sets) are developed along the shelf margin of the Anadarko and Arkoma basins. Each genetic set is approximately 330 ft (100 m) thick and contains from five to seven primary genetic stratigraphic units. Internally, each set is characterized at the base by an abrupt backstepping of up to 50 mi (80 km) relative to the top of the underlying set, followed by more gradual "lateral accretion," as progressively younger units step basinward. The backstepping marking the bases of individual genetic sets is apparently associated with major readjustments of the shelf, which clearly exerted a significant control on the development of sequences and genetic units. Recognition and characterization of these sets of genetic stratigraphic units in the interior regions of the craton present challenges due to lack of distinctive stacking patterns. Depth profiles of the thorium-to-uranium (Th:U) ratios derived from spectral gamma logs are used in these interior regions to extend this longer term cyclicity originally established along the shelf margins.

A two-dimensional (2-D) computer model was used as a means to investigate the authors' understanding of the processes (including eustasy, subsidence, sedimentation rates, and shelf configuration) that may have controlled the observed stratal hierarchy. Similarities in the patterns of observed and modeled third- and fourth-order cyclicity suggest a high degree of deterministic cause and effect between these processes and the observed stratigraphy, suggesting that stratal patterns in areas away from known control may be predictable via modeling. Increasingly advanced simulation models may be useful in synthesizing observations and interpretations with a goal of identifying areas prone to subtle stratigraphic traps.

INTRODUCTION

Cyclic Pennsylvanian sandstones and carbonates produce oil and gas from large areas in and adjacent to the Anadarko and Arkoma basins in the Mid-Continent United States. An estimated 9 billion barrels of oil will be produced from these reservoirs (Rascoe and Adler, 1983), and over 29.6 trillion cubic feet of natural gas has been produced from Pennsylvanian reservoirs in the Kansas and Oklahoma portion of the Mid-Continent. These petroleum-bearing, stacked, generally thin and discontinuous reservoirs include valley-fill and deltaic sandstones as well as carbonate grainstones and phylloid- algal buildups.

To date, the regional stratigraphic variations of these reservoirs have been described only at the group level. Higher resolution studies that include a more precise temporal definition of regional stratigraphic relationships are needed to locate remaining subtle stratigraphic oil and gas accumulations, and to provide frameworks for reservoir models to better predict reservoir heterogeneities. Furthermore, this improved data base is needed to apply quantitative deterministic geologic models to aid in predicting stratigraphic and spatial locations of petroleum reservoirs in this region.

An approach is presented to systematically and quantitatively characterize temporal and spatial changes in Pennsylvanian sedimentary cycles to help find new petroleum production and to assist in establishing more precise reservoir analogues. Resulting quantitative experiments, e.g., stratigraphic simulations that examine the dynamic interactions of inferred geologic processes, can be used to test and refine a geologic model as knowledge improves about the system.

GOALS AND OBJECTIVES

Goals of this study are to (1) more precisely and accurately characterize and interpret temporal stratigraphic units within the late Middle and Upper Pennsylvanian cyclothemic succession, (2) provide another method to improve this interpretation of a relative sea-level curve for use in reconstructing and inferring stratal patterns in new areas and predicting reservoir development, and (3) relate the variations in high-frequency Pennsylvanian cycles to a cycle hierarchy.

METHODOLOGY

An integrated approach is used in this study to characterize the multiscale nature of the stratigraphic cycles encountered in the Pennsylvanian, including (1) studies of oil and gas fields and near surface and surface analogues using cores and wireline logs; (2) regional mapping and cross-section analysis of depositional sequences and their proxies-genetic units and facies from shelf-to-basin; (3) time series analysis of wireline log data, specifically Th:U ratios to define cycle hierarchy; and (4) stratigraphic simulation modeling to integrate derived process parameters.

This study considers several scales of cycles and uses a variety of information to characterize the subaerial unconformity-bounded depositional sequences (Vail et al., 1977b) and genetic stratigraphic units (GSUs) bounded by condensed sections (Galloway, 1989a, b). Information includes (1) measured sections of surface exposures and cores to establish the genetic units that make up the depositional sequences, (2) hand-held gamma ray scintillometer measurements that record the natural gamma radiation of surface sections, and (3) core descriptions correlated to wireline logs, which include natural gamma ray, neutron, density, resistivity, and natural gamma ray spectra used to derive potassium (K), thorium (Th), and uranium (U) concentrations.

Surface exposures and near surface study areas provide a framework within which to effectively apply sequence stratigraphic methods to these rocks. In these areas it is possible to correlate lithofacies and genetic units to wireline logs. Optimum sites can be targeted to discern detailed three-dimensional (3-D) stratal geometries and stacking patterns in places where there were relatively steep paleoslopes. Such slopes effectively compressed depositional geometries to a much smaller area than is typical in the Mid-Continent, allowing for the recognition of depositional sequence components such that they can be used to reconstruct relative sea-level history (Watney et al., 1989; Youle et al., 1994). Regional mapping is used to integrate the detailed studies and to view the depositional sequences in a regional 3-D perspective to assess sediment accommodation space, trends, and patterns.

Regional 3-D mapping (primarily in the subsurface) that extends local, detailed sequence stratigraphic studies is used to assess trends in sediment accommodation space, with a goal of recognizing and predicting facies patterns. The lack of ability to definitively correlate subaerial unconformities in these thin sequences using typical well log suites and conventional seismic methods prevents the subsurface mapping of sequences over large distances. Thus, a more reliable unit is needed that can be a proxy for the depositional sequence, and which can constrain the stratigraphic position of the sequence so that it can be re-established in local situations where core control is adequate. Such a proxy is the GSU introduced by Galloway (1989a, b), which is bounded by a lithologic unit that represents "maximum transgressive flooding" (Galloway, 1989a, p. 133). In this study such units are the dark radioactive shales representing condensed sections. These condensed sections are commonly distinguishable and correlatable using well logs, and many can be uniquely identified using ammonoid and conodont biostratigraphy (Boardman and Heckel, 1989).

The condensed sections closely overlie the subaerial unconformities (i.e., the sequence boundaries) in shelf cycles, which are the subjects of most of the mapping and analysis conducted in this study. However, condensed sections diverge from sequence boundaries where there is significant depositional topography and/or variations in sediment supply. In such situations, the boundaries of GSUs begin to deviate from sequence boundaries.

Attempts are being made to glean more sequence stratigraphic information from well logs, in particular the distinction of lithogenetic units that comprise sequences. For example, a neural network program (KIPLING developed by J. Doveton and G. Bohling) and Markov chain analysis were used together to classify a modern well log suite of porosity, resistivity, and spectral gamma ray tools into lithogenetic units comprising a succession of Middle and Late Pennsylvanian depositional sequences (cyclothems). Lithogenetic units included flooding units, condensed sections, highstand and late highstand (regressive) siliciclastics and carbonates, and paleosols (Watney et al., 1994).

The detailed procedure for this stratigraphic analysis used in this study is described below.

(1) Vertical-Sequence Analysis

(a) Describe strata and interpret depositional environment; estimate relative water depth and define evidence for shallowing or deepening trends; identify potential marker beds (thin distinctive units to aid in correlation).

(b) Describe surfaces: assess nature of bedding planes (frequency and nature; distinguish diagenetic from depositional); associate surfaces with facies dislocation; rank facies dislocation according to water-depth change; establish evidence of subaerial exposure or prolonged nondeposition, e.g., hardground developed in subaqueous marine environment.

(c) Draw profiles of sections providing interpretation of genetic units and water depth; correlate to well log signature; recognize the following genetic units that consist of (from base to top)

- *flooding or transgressive units* — usually associated with base of depositional sequence; usually thin limestone or coal on shelf areas in Pennsylvanian depositional sequences; distinguish initial flooding unit (possibly within thicker clastic succession, below transgressive (flooding) carbonate unit
- *condensed sections* — these may be associated with abundant organic matter (e.g., the black shales), or may be manifested as hiatal surfaces without significant clastics (e.g., hardgrounds)
- *late highstand units* —these are most commonly an overall upward-shallowing carbonate or clastic succession (or combination of the two), and are typically the thickest and most complex component of a given sequence (in many cases there may be multiple smaller scale units embedded in a single regressive unit)
- *paleosols and subaerial exposure surfaces* — although not strictly genetic units, these important features in most cases represent sequence boundaries
- any repetition of the above.

(2) Correlation Between Localities

(a) Establish correlations of marker beds (radioactive, black shales) and surfaces (paleosols), using lithostratigraphic, paleontologic, geophysical, or geochemical data (preferably through continuous or detailed systematic sampling); recognize any evidence that can help to define internal geometries (e.g, toplap, downlap, onlap).

(b) Identify genetic stratigraphic units (bounded by condensed sections) and depositional sequences (bounded by subaerially exposed unconformity surfaces); correlate major genetic units and bounding surfaces associated with a depositional sequence or GSU using all information available; compare characteristics and patterns of genetic units in successive stratal packages.

(c) If possible, extend control to the third dimension regionally to evaluate allogenic and autogenic causal mechanisms.

The temporal distinction of sedimentary units provides the foundation to

(a) define detailed paleogeography

(b) address rates, duration, and magnitude of events responsible for sedimentation

(c) possibly improve the ability to predict facies for economic development.

A procedure for classifying and naming depositional sequences and GSUs is not clearly defined in the literature. The following approach used in this study is a preliminary methodology that is continually tested. This approach may be generally useful to the study of thin depositional sequences in the subsurface of the Mid-Continent.

A depositional sequence is by definition bounded by subaerial unconformities (and their correlative conformities) (Mitchum et al., 1977), without regard to the lithostratigraphic components that occur between the unconformities. In this study, *primary* Pennsylvanian depositional sequences in the northern Mid-Continent are considered as those that contain a well-developed condensed section (generally a radioactive marine black shale) that is regionally correlatable using physical and biostratigraphic means providing constraints as to the location of the bounding subaerial unconformities (Fig. 1). In terms of naming such sequences, it is proposed that the name be based on the condensed section that is derived from the age of the stratigraphic interval containing the sequence, e.g., Missourian. Thus, the name given to each of these sequences will not be confused or limited by names given to lithostratigraphic units. Also, the classification and naming would be derived from and associated with a locality in which the properties of the sequences can be easily recognized and well studied. Ideally, the locality is a representative continuous core or outcrop in lower shelf or shelf margin setting where the depositional sequences are well developed. As higher shelf positions are encountered, sequences may lap out (recognized by the loss of one of the condensed sections) and the subaerial unconformities will become amalgamated. In this case, the remaining sequence is referred to as the combined name of the originally defined units, e.g., M1 and M2 become M1/2 (Fig. 1).

Sequences located in relatively downdip positions may have developed in response to a more complex sea-level history than those on the shelf. In some cases, additional subaerial unconformities representing higher frequency sequences embedded within the primary sequences may be developed in downdip locations (Fig. 1). These higher frequency sequences are currently below the resolution of regional correlation methods, and their names have only local significance unless otherwise noted. These higher frequency sequences are identified by lower case letters following the primary sequence designation beginning at the base of the primary sequence and working upward (e.g., M1a, M1b, M1c) (Fig. 1).

Far downdip, in areas of limited sediment supply and little or no subaerial exposure, amalgamated sequences lacking subaerial unconformities occur. These are distinguished by addition of the codes for the individual sequences that are thought to be amalgamated (e.g., M1 to M3, etc).

The GSUs, bounded by the condensed sections, are designated by the name of the lower bounding condensed section, usually a member or bed in the formal stratigraphic nomenclature, e.g., the Mound City Shale. Informal names can be used if formal names are not available, although care should be given to stratigraphic names currently in use. Widespread condensed sections are common and may have different names in different areas. Use of local names is encouraged, with synonyms provided as needed and useful.

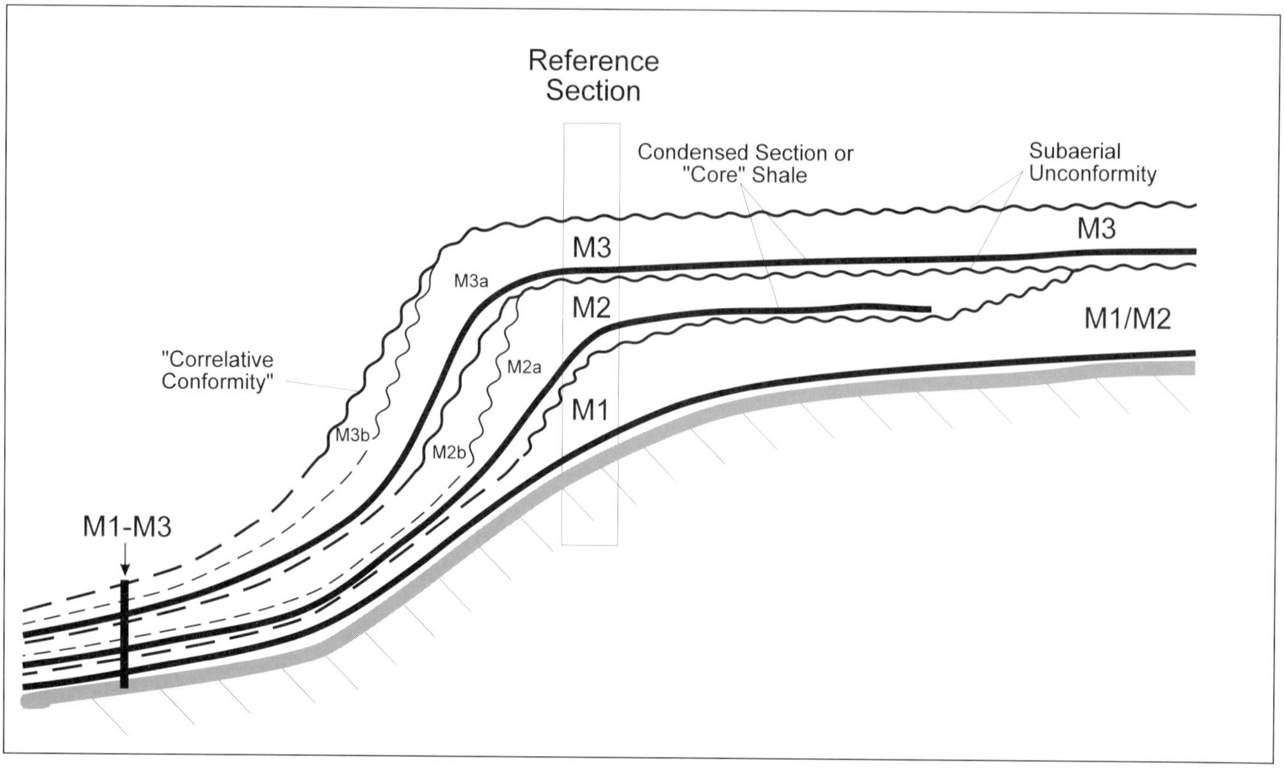

FIGURE 1 — Conceptual cross section from shelf-to-basin showing depositional sequence classification and nomenclature introduced in this study. A reference section is chosen along lower shelf with optimal stratigraphic resolution of the sequences and bounding surfaces. Adjustments are made to names of sequences both landward as subaerial unconformities amalgamate and basinward as unconformities diverge and extend to their correlative conformities. GSUs are defined to include intervals between successive condensed sections. This informal working nomenclature is presented in the context of this study and may continue to be useful as it is refined and adapted to results of continued investigations.

STRATIGRAPHY

The units examined in this study include the Late Desmoinesian-to-Early Virgilian Marmaton, Pleasanton, Kansas City, Lansing, and Douglas groups (Middle and Upper Pennsylvanian)(Fig. 2). In the northern Mid-Continent, these strata are characterized by 10-ft (3-m) to over 150-ft (about 50-m) thick cyclic successions that contain alternating marine carbonates and nonmarine to shallow marine siliciclastics. These units are the classic Mid-Continent cyclothems first described in the Illinois basin by Udden (1912), and subsequently considered in great detail by numerous workers, including Moore (1930, 1931), Weller (1930, 1931), Wanless (1931), Wanless and Weller (1932), Wanless and Shepard (1936), Merriam (1964), and Heckel (1977, 1980, 1984, 1985, 1994).

Mid-Continent shelf cyclothems generally consist of a lower, relatively thin but extremely widespread transgressive unit overlain by a thicker regressive unit (making up the well-known "Kansas cyclothem" of Heckel, 1977) (Figs. 2 and 3). Well-developed paleosols and other surfaces that exhibit diagenetic and textural features indicative of subaerial weathering are common at or near the boundaries of many of these cyclothems (Watney and Ebanks, 1978; Watney, 1980; Prather, 1981; Schutter and Heckel, 1985; Goebel et al., 1989). Numerous mechanisms have been offered to explain the origin of these cyclothems, with the most widely accepted model invoking glacial eustatic fluctuations (e.g., Wanless and Shepard, 1936; Heckel, 1977).

Figures 3 and 4 show the components of a typical Kansas cyclothem and the boundaries of both the depositional sequence (subaerial unconformities) and the GSU (condensed sections, equivalent

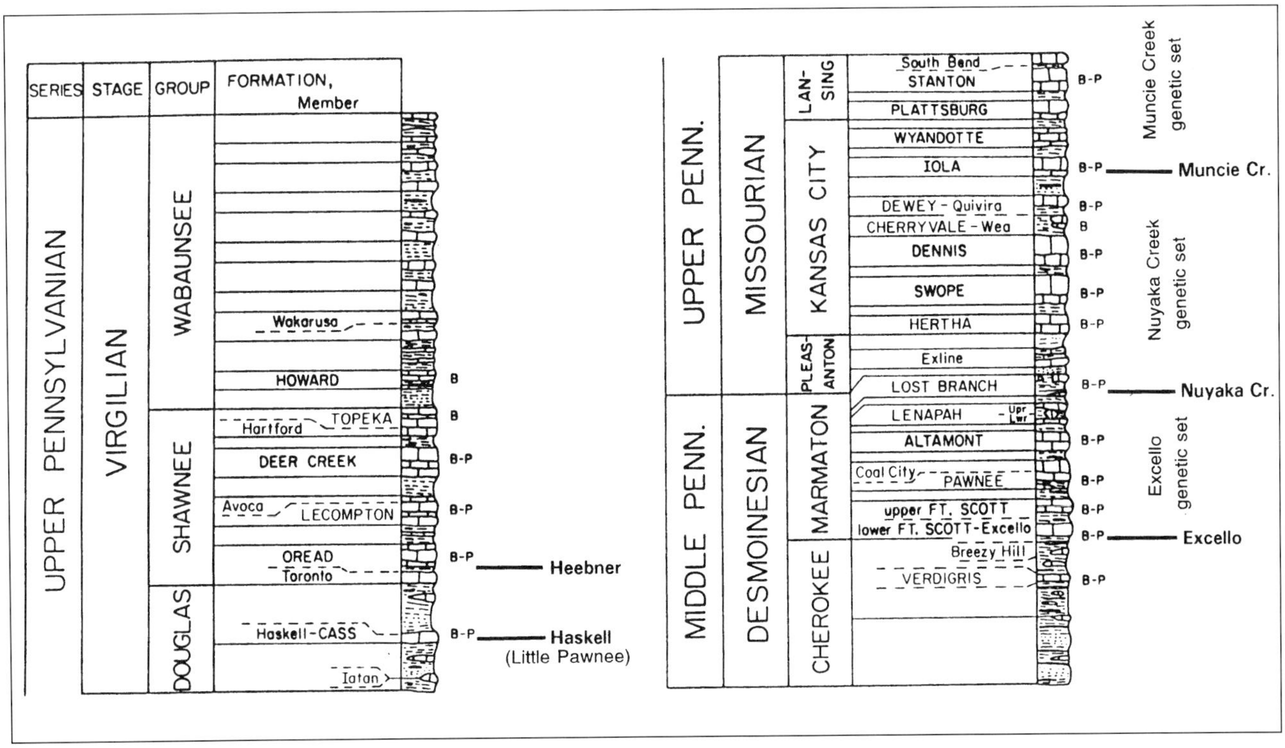

FIGURE 2 — Stratigraphic section of Middle and Upper Pennsylvanian in Kansas (adapted from Heckel, 1990). Each formation corresponds roughly to a single cyclothem. Positions of widespread phosphatic black shales are delimited by "B-P," black shales by "B." In addition to formal stratigraphic nomenclature, a new subdivision of third-order genetic sets is identified as described in this study. These third-order genetic sets are distinguished by name — Excello, Nuyaka Creek, and Muncie Creek. Major bounding stratal units of these genetic sets include the Excello, Nuyaka Creek, Muncie Creek, and Little Pawnee shales, the latter associated with the Haskell Limestone. The Heebner and Haskell (Little Pawnee) are both noted because the upper boundary of the Muncie Creek third-order genetic set begins at the Haskell horizon but is transitional with the Heebner Shale.

to the core shales in the cyclothem model). The intervals and thicknesses of these stratal units are similar in this shelf setting, which is the focus of GSU mapping (shown later). This cyclothem-scale depositional sequence and associated GSU (Fig. 4) are considered to be the thinnest regionally correlatable units.

Pennsylvanian cyclic successions are common in shelfal deposits in many basins, but accurate interbasinal correlations were not possible because cyclothem durations were too short for the existing biostratigraphic resolution. Recently, an integration of ammonoid, conodont, fusulinid foraminifera, and coral data has resulted in the correlation of 13 major Virgilian and Missourian cyclothems between the eastern shelf of the Midland basin in Texas and the northern Mid-Continent (Boardman and Heckel, 1989). These correlations suggest that the individual marine inundations believed responsible for fourth-order cyclothem deposition were synchronous in both areas. Foraminiferal, ammonoid, and conodont zones have also been used by Ross and Ross (1987) to globally extend correlations of similar Pennsylvanian marine inundations. Boardman and Heckel (1989) and Ross and Ross (1987) attribute the sea-level fluctuation to Late Paleozoic continental glaciation. Other aspects of climate change have recently been invoked to explain lithologic variations and to some degree the overall stratigraphic succession (Cecil, 1990; Miller et al., 1992).

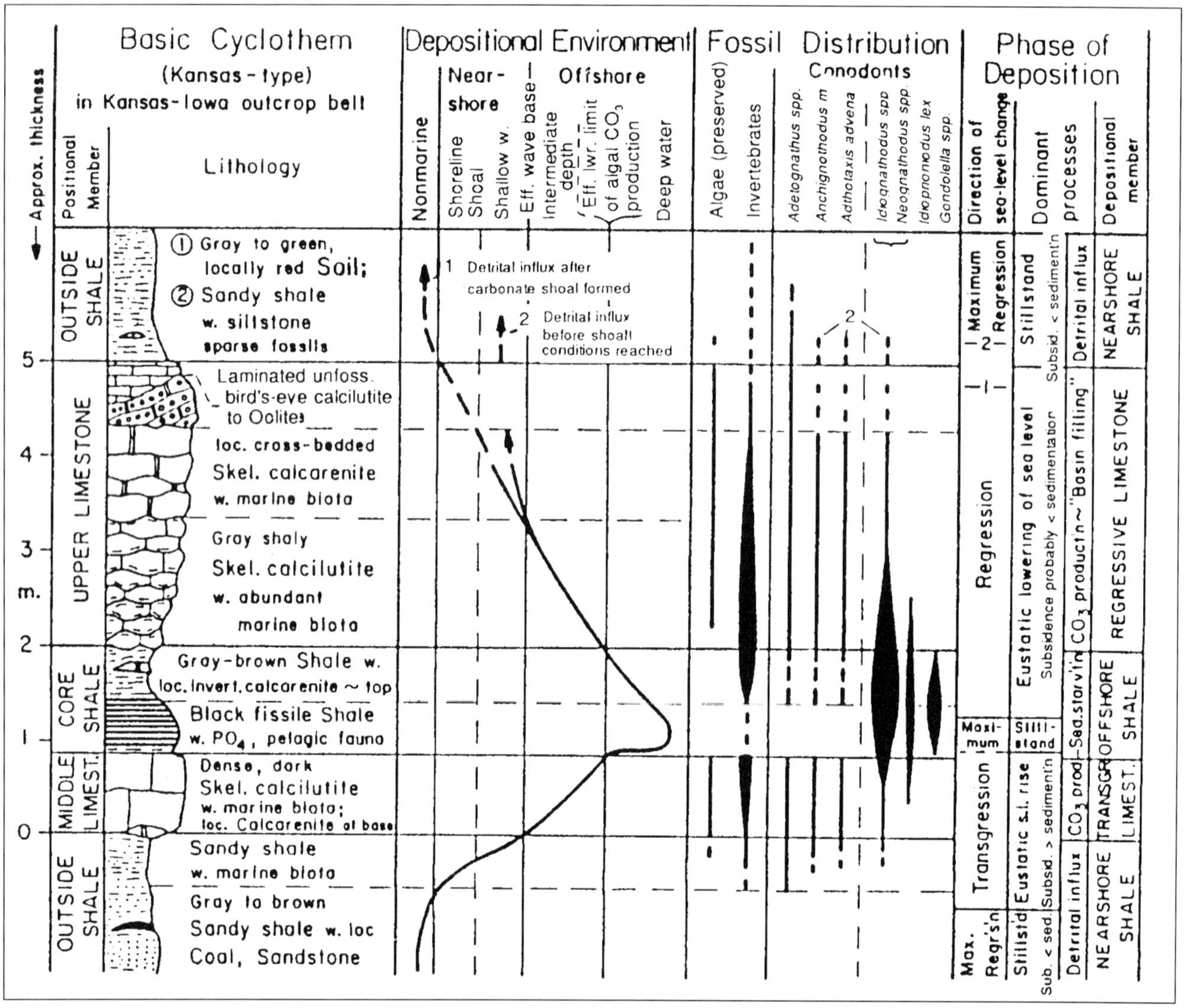

FIGURE 3 — Generalized Kansas cyclothem and its stratal components common to carbonate-dominated strata developed in the Missourian Upper Pennsylvanian of the northern Mid-Continent shelf (from Heckel, 1977).

Cyclothem Concept

Several commonly recognized, lithologically distinct units comprising many Middle and Upper Pennsylvanian cyclothems are described below, along with their interpretation and significance. These units include the middle limestone, core shale, upper limestone, and outside shale. The following discussion of these units is focused on providing insights into the rates, durations, and magnitudes of the processes that controlled their deposition. An attempt will also be made to indicate the position of these lithologic units in regard to sequence stratigraphic terminology.

Middle Limestone

Using the nomenclature of Heckel (1977), the lowermost cyclothem unit is the middle limestone (Fig. 3), which is a widespread deposit of variable carbonate type that is typically a few feet thick or less. Most preserved beds consist of subtidal marine wackestones and peloidal packstones that contain a diverse assemblage of normal marine biota. Middle limestones are normally less than 3 ft (1 m) thick, although locally they can be relatively thick (50 ft [15 m]). In high shelf positions, the middle lime-

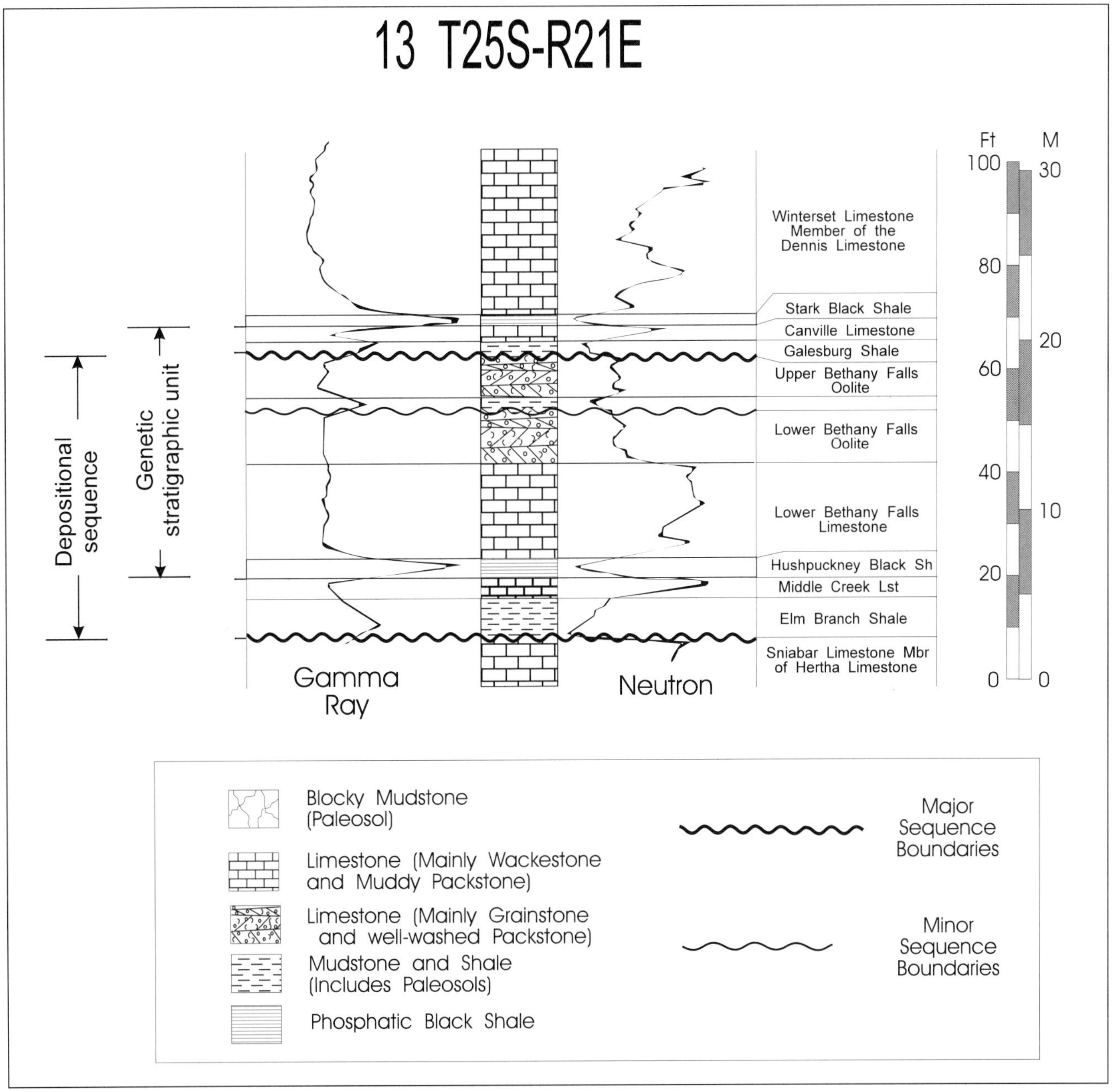

FIGURE 4 — Wireline log and graphic core description of the Swope Limestone, K2 depositional sequence, and Hushpuckney GSU from a well in western Bourbon County, southeastern Kansas.

stone may be very thin or absent altogether. The unit typically abruptly overlies a thin, marginal marine siliciclastic unit, which is classified with the outside shale (see below). In some cases the middle limestone sits directly atop the subaerial exposure surface or paleosol that forms the boundary of the underlying sequence. The middle limestone is usually sharply overlain by the core shale.

Middle limestones are interpreted to have been deposited during a rapid transition from nonmarine or marginal marine conditions to open marine environments. Where abundant siliciclastic sediment was available during the initial stages of transgression, many tens of meters may lie between the initial flooding surface at the base of the cycle and the transgressive carbonate. In places, these deposits are oil and gas reservoirs. An example is the thick Galesburg Shale (lower Layton Sandstone) deposited on the southern lower shelf (proximal to Arkoma basin) in southeastern Kansas.

Here nearly 100 ft (30 m) of estuarine and shallow marine clastics was deposited, then capped by the transgressive Canville Limestone, which is a classic middle limestone (p. 9, p. 202, Watney et al., 1989). The initial flooding unit in this clastic package is a widespread coal (Cedar Bluff) that overlies a paleosol. This surface is the sequence boundary, diverging significantly from the base of the middle limestone. In more northerly (shelfward) settings, the Galesburg Shale is only a meter or less of mainly paleosol, sharply overlain by the thin transgressive Canville Limestone. Here the initial flooding surface and base of the transgressive carbonate are essentially the same, i.e., the sequence boundary coincides with the base of the middle limestone. It appears that on the lower shelf, the early stages of relative sea-level rise may have been slow, allowing significant volumes of siliciclastic sediment to reach the area. In contrast, upper shelf inundation was apparently quite rapid, effectively trapping sediment before it could reach mid to high shelf positions.

Core Shale

The core shale overlies the middle (transgressive) limestone in the typical Kansas cyclothem (Fig. 3). Core shales are typically thin, about 1 to 3 ft (0.3 to 0.9 m) and, like the middle limestones, are aerially extensive. There is strong evidence (see below) that core shales are condensed sections in sequence stratigraphic terminology that can also be used as boundaries of GSUs.

Some core shales are black and organic rich (>4% carbon), and commonly contain phosphate nodules. The black core shales are readily recognizable in surface exposures and are frequently excellent subsurface markers due to their high natural gamma radioactivity. For example, the Hushpuckney and Stark shales in the lower Kansas City Group can be traced throughout the outcrop belt from Iowa and southward deep into clastic cycles in the Arkoma basin south of Tulsa, Oklahoma. They also extend over 400 mi (640 km) to the west into western Kansas and eastern Colorado, where they can be readily identified through their strong radioactive response on gamma ray logs (Watney et al., 1989, Figs. 25 and 26, p. 32-33). Nevertheless, the black shales grade rather abruptly to gray fossiliferous shales (or in some cycles are no longer distinguishable) along the upper shelf in northwestern Kansas and southwestern Nebraska, and locally over inferred paleohighs in central Kansas, e.g., the Central Kansas uplift (Watney et al., 1989, Figs. 25 and 26).

Wanless (1964) used these black shales, which are also common to a number of Middle Pennsylvanian cyclothems, to physically correlate clastic-dominated cyclothems in Illinois with equivalent, more marine-dominated carbonate successions in the western Mid-Continent. Although biostratigraphy is now verifying some of these correlations, Wanless emphasized that a particular cyclothem could be distinguished through correlation of the core shale regardless of the variation in the remaining succession of sediments, a critical relationship important in bridging from cyclothems to sequences. Heckel (1977) reemphasized the significance of the regional correlatability demonstrated by Wanless and stressed that this lateral continuity, coupled with the faunal composition and presence of nonskeletal phosphate, made the black shale the deepest water deposit of the cyclothem.

The origin of black core shales has been vigorously debated with regard to whether the core shale was deposited in deep or shallow water. The apparent controversies in interpreted depth, rate of sediment accumulation, and duration of black shale deposition are addressed in Coveney et al. (1991). A knowledge of the water depth represented by core shales is critical to understanding the changes in relative sea level that occurred during cyclothem deposition. An accurate interpretation of relative sea-level change and depositional setting of the core shale is also important with regard to quantitative modeling. Workers such as Zangerl and Richardson (1963), Maples (1986), and Coveney and Martin (1983) have provided paleontologic and inorganic and organic geochemical evidence that supports a shallow-water origin and rapid sediment accumulation over a relatively short term for certain Middle Pennsylvanian black shales in the Illinois and Appalachian basins. These shales correlate with shales in the western Mid-Continent, but have distinctive differences in these particular shelf settings. It appears the productivity and preservation of organic matter are important variables in these shales and may not be as closely tied to depth as previously thought. A number of process models have been proposed to explain black shale deposition, namely upwelling and quasi-estuarine circulation (Heckel, 1977), upwelling through Ekman transport (Parrish, 1982), halocline formation (the Black Sea model with freshwater influx, Demaison and Moore, 1980, or basinal brine upwelling, Hite, 1978), and the presence of a thermocline (Rossignol-Strick, 1982; Heckel, 1991b).

In the western craton along the outcrop belt, the indigenous fauna in these black shales is composed primarily of conodonts, ammonoids (from the southern exposures), and dissociated fish debris, suggesting very slow sediment accumulation rates. Minor elements such as uranium and various other metals are also abundant in the black shale along with phosphate. Nonskeletal phosphate (apatite) is common as nodules and laminae in black shales. Abundant and diverse radiolarians, nautiloids, and fish debris have been found in the nodules in black shales in eastern Kansas (Kidder, 1985). The source of the phosphate could be planktonic organisms, fecal material, or perhaps solution and suspension from river water. Heckel (1977, 1994) proposed that the water column in which black shale accumulated contained a thermocline leading to quasi-estuarine circulation and upwelling, which accounts for the conspicuous phosphate. Water depths would necessarily be deep to accommodate formation of a long-term, stable thermocline.

Boardman et al. (1984) provided a paleoecologic model that related faunal communities to dissolved oxygen, which in turn is related to a depth-defined bottom oxycline associated with vertical stratification resulting from the thermocline. Sufficient depths and stratification can result in anoxic bottomwaters where organic matter typically accumulates. Accumulation rates would be slow and duration of black shale deposition would be long in a deep, stratified water column.

Much of the evidence from core shales of the Pennsylvanian in the western Mid-Continent suggests that they accumulated very slowly in deeper (quiet, below photic zone), low oxygen, marine water (Heckel, 1977; Boardman et al., 1984; Coveney et al., 1991). Biota in the Upper Pennsylvanian black shales in the western Mid-Continent is limited to nektonic and nektobenthic organisms, abundant pelagic organisms, and rare benthic forms. Large amounts of conodonts and fish debris are common, but these shales generally lack benthic marine invertebrates. The microstratigraphy is complex as indicated by interlayered bioturbated zones, thin carbonate layers, and marked variations in minor elemental composition. Uranium concentrations in black shales range from <20 ppm to more than 250 ppm, accounting for most of the high natural gamma radiation emitted by black shales. Greatly reduced sedimentation rates are suggested by (1) commonly abundant phosphate, (2) high concentrations of normally sparse pelagic fossils, (3) horizontal orientation of clay minerals composing the shale (suggesting dilute suspension sedimentation; James, 1970), and (4) elevated concentrations of minor elements such as uranium.

These attributes suggest that the core shales in the study area are predominantly condensed sections — thin deposits representing reduced sedimentation. Furthermore, based on this evidence, the core shale incorporates the maximum flooding surface. Also, it follows that excess sediment accommodation space existed during the span of time that the core shale was deposited, because water depths during deposition of the core shale almost certainly exceeded the total thickness of the cyclothem. Accommodation could be added through any subsidence that occurred during accumulation of the core shale. How much excess depth occurred is subject to the interpretation of the depth of the photic zone, processes that led to stagnation and accumulation of organic matter, etc. Shelf areas that are more distant from the basins with thinner strata are commonly characterized by the changes of the black shale facies to green or gray shale characterized by oxic and dysoxic benthic faunal assemblages (Boardman et al., 1984). These facies changes suggest shallower settings during accumulation of the core shale. The bottom probably remained more oxygenated on the higher shelf due to greater circulation and mixing of the water column. At other shelf locations, core shale is not present, but instead only carbonate facies are deposited for the entire marine cycle, e.g., the Cherryvale Formation. The lack of a definitive core shale suggests that water depths were shallow enough or water was clear enough to where the depth of the photic zone was not exceeded during maximum marine inundation. These variations are end members of the core shales. A particular core shale will vary within these end members. Systematic changes are commonly observed in a single core shale along a shelf-to-basin transect and strongly indicate elevation changes on the shelf. Boardman and Heckel have also used the extent and facies of core shales to help establish a relative sea-level curve for the Mid-Continent.

Hatch and Leventhal (1992) described detailed internal stratigraphy of the "black" shale facies of the Stark Shale (Missourian) from Waubaunsee County, Kansas. Four lithologic/geochemical cycles were found in the continuous core of the shale. They attributed these stratal variations to changes in redox conditions from strongly anoxic to thin (brief) episodes of dysoxic conditions. Minor cycles were attributed to climate or sea-level fluctuations. Further sampling of the Stark Shale by Hatch in southern Kansas in proximity to the Arkoma basin shows continuous anoxic shale in the center of the core shale perhaps suggesting deeper shelf (Hatch, pers. comm., 1994), consistent with stratigraphic and facies indicators.

A good modern overview of black shales is provided by Arthur and Sageman (1994).

Upper Limestone

In upper Middle and Upper Pennsylvanian strata throughout the northern Mid-Continent shelf, the marine core shale is typically overlain by a generally upward-shallowing upper limestone ranging in thickness from 5 to over 65 ft (2 to 20 m). A laterally extensive subaerial exposure surface or noncalcareous paleosol may be developed at or near the top of the regressive carbonate unit in locations where clastic influx was reduced. This upward-shallowing limestone is often the thickest unit of the succession on the shelf in the Middle and Upper Pennsylvanian. Many of the better oil and gas reservoirs occur in grainstones present in the upper part of this limestone.

The overall upward-shallowing (regressive) trend in these upper limestone units is typically interrupted by one or more thin, transgressive-regressive (T-R) cycles. The bases of these minor cycles are typically characterized by evidence of abrupt deepening. These minor cycles may be capped by thin paleosols or certain diagenetic evidence suggesting subaerial exposure (French and Watney, 1993). Subaerial exposure on subtidal carbonates and units showing abrupt shallowing both suggest that these regressive events are forced by external processes rather than by simple aggradation. The asymmetric nature of slow regression versus rapid transgression suggested by the stratigraphy across most of the northern Mid-Continent shelf is probably partly responsible for the generally thicker sections of regressive carbonate and siliciclastics. Also, accumulation rates for carbonate sediment are in general relatively high provided time is available in the euphotic zone for carbonate sediment to accumulate (Enos, 1991).

Outside Shale

The outside shale, the uppermost unit of the Kansas-type Missourian cyclothem, exhibits considerable variability (Fig. 3). It overlies the upper limestone in the cyclothem but contains both regressive and transgressive deposits. The subaerial unconformity and a sequence boundary can be found anywhere within this unit or at either of its boundaries. Thus, the outside shale can be part of two different depositional sequences. Outside shales are locally thin or missing in parts of western Kansas due to greater distance from an apparent lack of clastic influx. When a significant source of clastics was available, thicknesses of the clastic units may be as much as 300 ft (91 m) in eastern Kansas and can dominate both regressive and locally transgressive sedimentation (Brenner, 1989). Thicker packages of shale, sandstone, and siltstone represent deltaic clastic influx and shallow shelf sedimentation. These clastic deposits often contain invertebrate-rich horizons and thin limestones. The platform deltaic to fluvial deposits form broad aprons of mainly shale extending for tens or hundreds of miles. Estuarine and fluvial deposits may also be deposited in valleys incised into previously deposited marine sediments.

Deltaic progradation and accumulation of relatively thick outside shales occurred during times when adequate sediment accommodation space was available. Deltaic influx in northeastern Kansas occurred late in the marine cycles while the shelf was still submerged. For example, clastics in the upper Kansas City and Lansing groups reached the east-central Kansas shelf prior to extensive subaerial exposure and paleosol development on the upper limestones (Watney et al., 1989). In these situations, clastic sedimentation ended when the shelf was again exposed, except for sediment that was locally transported across the shelf through incised valleys (Lins, 1950; Archer et al., this volume, 1994b; Feldman et al., 1995).

Paleosol and Subaerial Exposure Surfaces

Widespread subaerial exposure is clearly evident in most outside shales and on the tops of many of the upper limestones (Watney and Ebanks, 1978; Schutter and Heckel, 1985; Goebel et al., 1989). The following diagnostic features are commonly present in Mid-Continent Pennsylvanian paleosols (Joeckel, 1991, 1994):

(1) rhizoliths (rootlets)

(2) ped surfaces in the blocky mudstones (from Retallack, 1988)

(3) color mottling or isolated horizons of color and textural variation in the mudstone due to differential oxidation and hydrolization of iron, and redistribution and formation of clay minerals (illuviation)

(4) micritic carbonate nodules or casement around rhizoliths, or carbonate crusts (calcrete) (Prather, 1985).

Paleosols rich in calcium carbonate are referred to as *caliche*, defined as a strataform to irregular deposit, formed primarily of calcium carbonate, with concretional, pisolitic, banded, or massive structure that is formed in the soil or subsoil of arid and semiarid regions (Gonzales-Bonorino and Terruggi, 1952).

In contrast to caliche, the following are some of the identifying criteria for paleosols developed under moist, subaerial conditions, provided sufficient time has elapsed for soils to develop:

(1) terra rossa (red residuum derived from carbonate dissolution)

(2) paleokarst and solution piping, commonly plugged with clay of paleosol or wall-rock debris

(3) well-developed clayey soil profile exhibiting illuviation.

In many cases, the outside shale (at least the lower portion) consists of only a paleosol capping the upper carbonate unit. Moreover, these paleosols can be traced across the northern carbonate shelf from Iowa to southeastern Kansas; they also are present westward in cores throughout the subsurface of western Kansas and southwestern Nebraska. Spatial changes in a single subaerial exposure surface have also been recognized on the shelf (Watney, 1980). Northern areas in northwestern Kansas and southwestern Nebraska are interpreted to be higher shelf locations that generally exhibit evidence of more intense subaerial weathering and more intense early meteoric diagenesis than do units deposited in lower shelf settings (Watney and Ebanks, 1978; Watney, 1980; Heckel, 1983). Missourian sequences in northwestern Kansas are commonly capped by chalky caliches found in red paleosols or as laminated calcrete at the top or near the top and filling in the carbonates (Watney, 1980). Their abundance in western Kansas indicates drier conditions than in southeastern Kansas, which fits with paleoclimatic reconstructions (Heckel, 1980). While varying from area to area, these paleosols appear to be shelf-wide developments.

Some paleosols exhibit relationships of both wet and dry climates. Dubois (1985) described karsting as prevalent in topographically lower areas at the top of the Dennis sequence in southwestern Nebraska. Solution piping extends through the carbonate (20 ft [6 m]). Proximity to the water table and greater residence time of water in the low areas are likely reasons for karsting. Topographically higher locations contain caliche paleosols indicative of locally drier conditions (Dubois, 1985).

Paleosols can exhibit apparent subaerial exposure on internal surfaces (penetrative calcretes) containing rhizoliths and associated diagenetic features (Rossinsky et al., 1992). Penetrative calcretes mark internal surfaces associated with topographically higher subaerial exposure surfaces.

Investigations along the outcrop and in shallow coreholes of southeastern Kansas indicate that subaerial surfaces can be traced to conformable surfaces on the lower shelf that show no evidence of exposure (Watney et al., 1989). When paleosols are present, they provide a critical, temporal break in the sedimentary record. However, problems remain in the distinction and correlation of these surfaces regionally, particularly in thick outside shale sections where paleosols are often not as well developed or preserved.

Variations in Cyclothems

The lithofacies successions of these cyclothems have been generalized in ideal conceptual models since they were first described by Weller (1930) and Moore (1930). Yet, as a given cyclothem is examined in more detail, the more restricted the applicability of the generalized model of the Kansas cyclothem (i.e., the uniqueness of each cyclothem becomes apparent). An example of this complexity is the distinct shelf-to-basin pattern in the variation of individual cyclothems comprising the upper Atokan to lower Desmoinesian interval deposited along a ramped shelf in west-central Kansas (Youle, 1992; Youle et al., 1994) (Fig. 5). The inferred steepness of the shelf provides the basis to associate these stratal and lithologic changes to progressive variations in sediment accommodation and relative sea-level history at each shelf location. Other subsurface studies in the Mid-Continent have recognized similar stratal geometries and have made similar inferences about sea-level history (Rich, 1951; Van Siclen, 1958 — cycle; Busch, 1974 — genetic increments of strata; Rascoe, 1978 — formats; and others).

The "Kansas cyclothem," which is the idealized, carbonate-dominated lithologic succession of Heckel (1977) (Fig. 3), typifies the cyclic successions along only a narrow portion of this ramp in the steeper shelf example of Youle et al. (1994) (Fig. 5). While other shelf locations have the same correlatable marine cycle, the respective lithologic successions constituting the cycle are very different. The nature of the unconformity-bounded lithologic succession deposited at a given location was apparently dependent on a specific set of conditions including shelf elevation, siliciclastic sediment supply, and residence time in the photic zone (for carbonate production). Subsidence will also be shown to be another important variable.

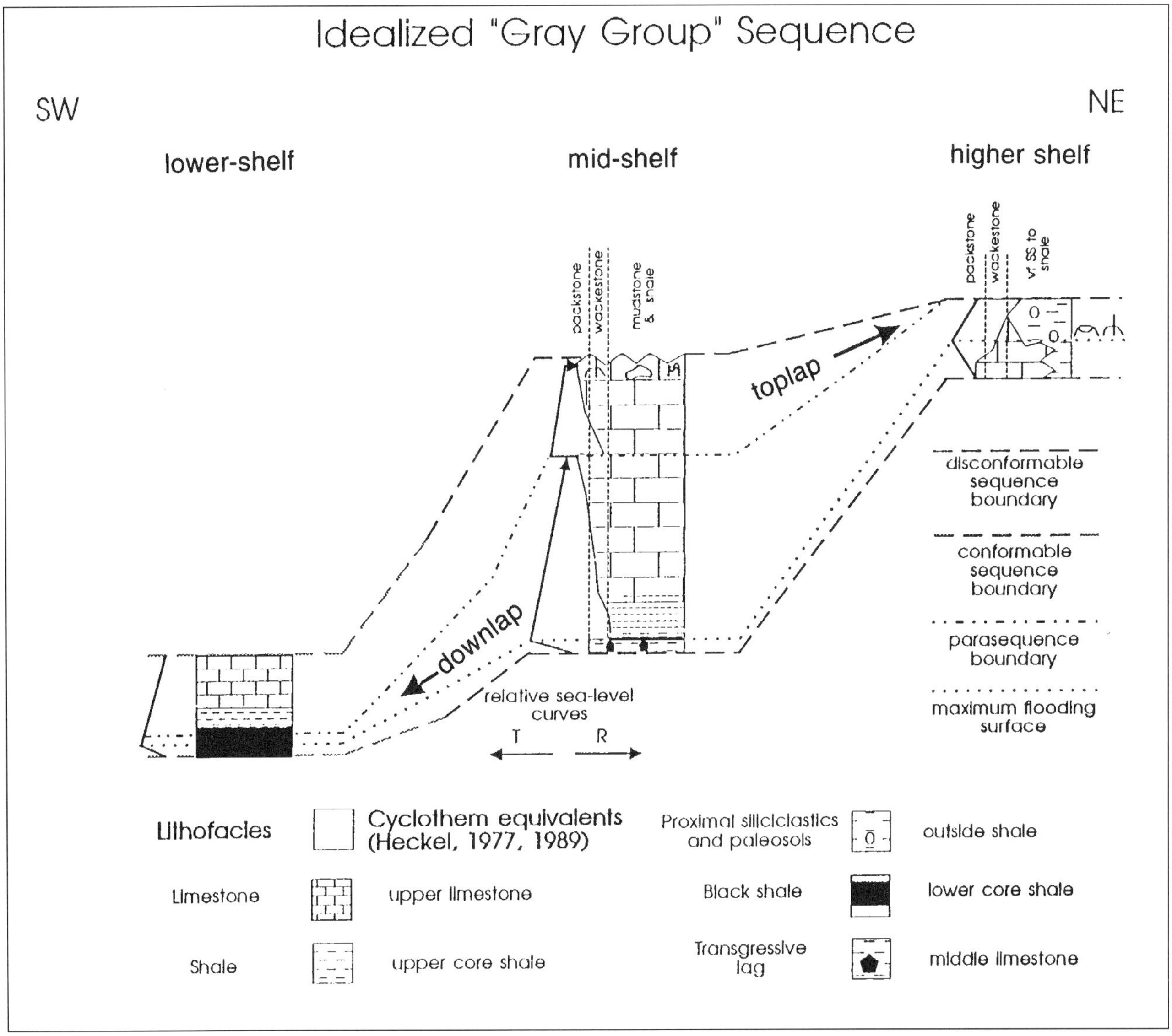

FIGURE 5—Generalized distribution of lithofacies succession and genetic components in an upper Atokan "Gray Group" depositional sequence developed along the eastern shelf of the Hugoton embayment, western Kansas (from Youle et al., 1994).

Cycles on the higher shelf in this Atokan example are characterized by amalgamated paleosols and incised valley-fill deposits, apparently the result of limited (short duration and low amplitude) marine incursions onto the high shelf. It is probable that considerable amounts of time were spent under subaerially exposed conditions. Equivalent cycles on the lower shelf contain only thin limestones and relatively thick dark gray and black shales. In contrast to the upper shelf, sedimentation on the lower shelf appears to have been limited to lowered sea level when water depths were less than the depth of the photic zone. During other times, the lower shelf was drowned and sediment starved. While cycles on the upper shelf are bounded by unconformities, the successions on the lower shelf are separated by essentially amalgamated and currently indistinguishable conformable surfaces in a condensed section. These lower shelf cycles apparently reflect extended times of sediment-starved conditions (Fig. 5).

The bounding subaerial exposure surfaces provide the means for temporal distinction of individual marine inundations. However, multiple subaerial exposure surfaces are often developed within a cyclothem and provide additional opportunities to resolve stratal packages (Fig. 4). The age of the primary bounding surface of a cyclothem probably changes with elevation and distance onto the higher shelf (see example of landward increase in intensity of subaerial diagenesis along a bounding surface in Watney, 1980). Minor unconformity-bounded cycles in the regressive unit of a cyclothem suggest a staggered fall in relative sea level.

The expression and impact of forced regression depend on the location on the shelf. Higher areas of the shelf may be locations where marine sedimentation was brief and sporadic, and where each marine event is separated by long hiatuses of subaerial exposure. As minor cycles developed, subaerial surfaces would merge and amalgamate on the higher shelf and split in a basinward direction (Figs. 1 and 5). Downdip, subaerial exposure would pass laterally to a conformable surface identifiable as the basal surface of an upward-shallowing stratigraphic unit. At some lower elevation of the shelf, minor cycles would lead to alternating condensed sections and regressive units deposited as sea level fluctuated in and out of the photic and oxygenated zones. On the lower shelf that was not exposed and in relatively deep water below the photic zone, most of the duration of the cycle is represented in the slowly deposited condensed section. The age and timing of these minor cycles using new correlation techniques will assist in improving the understanding of relative sea-level history.

Correlatability of Cyclothems

Individual cyclothems are widespread on broad reaches of the shelf, with physical correlations of basic lithologic units described above often extending over tens of thousands of km^2 (Heckel, 1977, 1980; Watney, 1985; Watney et al. 1989). Correlations by Boardman and Heckel (1989) between the Mid-Continent and Texas, and by Heckel and Weibel (1991) between the Mid-Continent and Illinois suggested that virtually synchronous (within biostratigrapic limits) marine inundations in these areas were responsible for the deposition of individual cyclothems, although the areas are in separate basins. While the marine inundations are generally correlatable, the internal components comprising lithologic units may cross time lines and present a challenge to the understanding of temporally distinct packages, i.e., to the application of depositional sequence analysis and distinction of lithostratigraphy versus genetic stratigraphy.

Neogene Analogues to Cyclothems

The cyclothem-scale changes in relative sea level have generally been attributed to eustasy resulting from Late Paleozoic glaciation (Wanless and Shepard, 1936; Wheeler and Murray, 1957; Crowell, 1978; Crowell and Frakes, 1975; Heckel, 1977, 1980, 1986, 1990, 1994; Crowley and Baum, 1991). Marine Pleistocene strata deposited in response to glacio-eustasy are probably good analogues for these Paleozoic marine cycles because temporal and spatial and cause and effects relationships can be better delimited. The information and insights into processes affecting sedimentation are invaluable to the development of quantitative models (Enos, 1991). Wide ranging and lithologically contrasting strata consisting of flooding units, condensed sections, regressive strata, and paleosols resulting from rapid inundation and forced regression are also a hallmark of cyclic marine Pleistocene deposits (Evans, 1979; Piper and Aksu, 1992; Abbott and Carter, 1994). Pleistocene glacial advances and retreats produced high-frequency sea-level changes with magnitudes on the order of 330 to 500 ft (100 to 150 m), periodicity around 100,000 years, and rates of sea-level change around 33 ft/k.y. (10 m/k.y.) (Donovan and Jones, 1979). The view of the current authors of Pleistocene sea-level history continues to get more complex through refinements provided by the latest studies of deep-sea sediments, ice cores, and glacial deposits. Abrupt climate change and unstable continental ice have led to inferred rapid sea-level fluctuations at a very high-frequency, subcyclothem scale (Broecker and Denton, 1989; Hughes, 1992) with magnitudes of meters and rates over 40 m/k.y. The later rates of sea-level rise were sufficient to drown healthy reefs of *Acropora palmata* (Blanchon and Shaw, 1995).

Results from studies of Holocene coastal sedimentation provide important analogues to the Pennsylvanian flooding units. The Holocene work suggests that at a certain threshold rate of deepening, effective carbonate sedimentation can significantly decrease (Neumann and Macintyre, 1985). Sedimentation does not keep up with rapidly increasing water depth, probably due, in part, to loss of opportunities for sediment progradation and aggradation, e.g., inundation of deltas and estuaries or rates of deepening exceeding effective carbonate accumulation rates (Heckel, 1984, 1994).

In their studies of Holocene carbonate settings, Hallock and Schlager (1986) concluded that reefs and most carbonate-sediment-producing organisms flourish in nutrient-deficient systems. However, these organisms undergo significant reduction in sediment production during increasing nutrient supply, which often occurs during submergence of a carbonate shelf. Flooding is thought to produce a biotic crisis due to nutrient excess. While increased nitrate and phosphate would stimulate growth of some organisms, other more deleterious effects apparently lead to net decreased sediment production, e.g., reduced water transparency limiting the depth range for some corals and calcareous algae, and consequently reduced sediment production. Hallock and Schlager (1986) identified criteria for recognizing changes in nutrient levels in the sedimentary record: nondeposition (hiatal) surfaces, bioerosion such as clam borings, and reduced redox potential encouraging preservation of organic matter. The carbonate surface is commonly covered by fine-grained siliciclastics if conditions of nutrient excess persist. The net result is reduced sedimentation during flooding, contributing what is referred to as lag time. This is precisely what is observed in the Pennsylvanian strata.

The sensitivity of a carbonate platform to flooding is a function of local conditions. For example, the euphotic zone, below which reefs composed of zooxathellate corals will not survive, ranges from 561 ft (170 m) in the Gulf of Aqaba to 59 ft (18 m) around Barbados (Hallock and Schlager, 1986: Fig. 17). The shallower the euphotic zone, the more susceptible the area is to diminished $CaCO_3$ accumulation during flooding events. Thus, in spite of the fact that short-term reef growth has been estimated at rates in excess of 33 ft/k.y. (10 m/k.y.), environmental conditions must remain nearly constant while accommodation space must be increased in order for these rates to be maintained. Neumann and Macintyre (1985) and Adey et al. (1977) further described the widespread rapid drowning associated with the early Holocene eustatic rise due to glacial melting. Neumann and Macintyre (1985) suggested that the reefs were shot in the back by their lagoons. Some reefs essentially give up during the rapid rise (26 ft/k.y. [~8 m/k.y.]) in sea level. Stratigraphic evidence noted by these workers for drowning includes bored hardgrounds, ferromanganese-oxide accumulations, phosphates, and glauconite separating neritic from overlying deeper water deposits. The nutrient influx and increased organic productivity led to reduced water clarity. This could effectively lead to a shallowing of the euphotic zone. Thus the euphotic zone, i.e., zone of carbonate production, could be dependent on the rate of sea-level rise. Oscillatory sea-level fluctuations in even shallow water could lead to retardation or termination of carbonate production.

The Pennsylvanian Mid-Continent seaway was a tropical inland sea with limited connection to the open ocean (Heckel, 1977, 1980). Conditions probably favored a high nutrient supply during marine flooding due to freshwater runoff contributing terrestrial organic matter and inorganic compounds to the marine realm. The tendency for water stratification due to an equitable climate promoting a stable water column supports conditions of lowered oxygen and preservation of organic matter. For these reasons, the effective depth of the euphotic zone may have been shallow in the interior Pennsylvanian seaway and perhaps quite variable, especially in the areas of detrital influx. The duration and rates of carbonate sedimentation would have been severely affected. Moreover, flooding units may be linked to the overlying condensed sections through the nutrient excess potentially created by the flooding process itself.

DEPOSITIONAL SEQUENCE INTERPRETATION

The overall conceptual differences between a cyclothem and a depositional sequence are relatively small, but are focused on different aspects of the rock, e.g., both are unconformity-bounded packages with components generally interpreted to represent a response to a rise and fall of relative sea level. Depositional sequences by definition are unconformity bounded. Sequences also are scale independent as defined below; thus, a lithologically defined cyclothem may contain multiple sequences or sequences may comprise multiple cyclothems. Other causal mechanisms have been advanced as important in their formation, e.g., climate change and sediment supply for cyclothems (Cecil, 1990) and for depositional sequences (Schlager, 1992; Perlmutter and Matthews, 1993). The concepts and methods used in sequence stratigraphy can provide new perspectives and insights to refine and possibly quantify interpretations. Sequence stratigraphic analysis can provide crucial information about the genesis of stratigraphic units through analysis of stacking geometries (Haq et al., 1987). Vail (1987) pointed out the following conclusions:

- Relatively deep water leads to preservation of depositional topography. Sedimentary onlap onto an irregular depositional surface can be used to characterize submarine topography.
- Stratal patterns can be used to quantify minimum paleowater depths by analyzing height of prograding clinoforms.
- Extent of inundations (onlap) onto a shelf and the lower basinward extent of subaerial exposure can be used to quantify the relative sea-level change or rise associated with inundation. Some inundations have been correlated on a global basis from which a eustatic sea-level curve can be inferred.
- Apparent truncation and downlap indicate sediment starvation. Truncation occurs along topset beds beneath a type 1 sequence boundary, and downlap is associated with basinward terminations of sedimentary wedges.

The depositional sequence is the fundamental unit of sequence stratigraphy, and is defined as a genetically related succession of strata bounded by unconformity surfaces and their correlative conformities (Mitchum et al., 1977). The cyclothem fits the definition of a depositional sequence, with the sequence boundaries being the laterally continuous subaerial exposure surfaces. However, the former is a lithostratigraphic unit and the sequence is dependent on the bounding surface and is independent of the lithostratigraphy.

Van Wagoner (1985) defined a parasequence as "a relatively conformable succession of genetically related beds bounded by marine-flooding surfaces and their correlative surfaces." Parasequences stack into sets that delineate systems tracts (transgressive, highstand, lowstand) within a depositional sequence. Parasequences are interpreted to occur within the cyclothem-scale sequences, but are usually very difficult to correlate at sufficient distances on the shelf to recognize stratal geometries and systems tracts within the individual sequence.

Although most sequence stratigraphers agree that the depositional sequence is the "fundamental unit of sequence stratigraphy" (Vail et al., 1991), they disagree on its scaler properties. Vail et al.

(1991) referred to the depositional sequence as a third-order cycle that ranges from 0.5 to 3 m.y. in duration; however, Mitchum and Van Wagoner (1991) and others (e.g., Posamentier et al., 1992) believed the sequence is scale independent. A nonscaler sequence stratigraphic terminology is used in this study. Since unconformities cannot be readily correlated directly on well logs, proxies for depositional sequences called GSUs (Galloway, 1989a, b) are used to regionally map associated depositional sequences. The GSUs are regionally correlative, based on their bounding condensed sections (as described earlier). These condensed sections constrain the presumably laterally extensive subaerial unconformities of the depositional sequence, approximating the delineation of temporally distinct units for purposes of high-resolution paleogeographic reconstructions and establishing detailed facies relationships. Cyclothem-scale sequences and their genetic-stratigraphic proxies currently represent the highest resolvable and regionally mappable Pennsylvanian units that can be reliably defined with current methods and technologies, and are named primary depositional sequences and primary GSUs, respectively. These cyclothem units are analogous to the "major cycles" of Heckel (1994).

The sequence model identified by Youle et al. (1994) is a good example of how a cyclothem and sequence (*sensu stricto*) are associated. The Youle et al. model is very similar to the Exxon sequence "slug" model (Fig. 3). The "Kansas cyclothem" (Heckel, 1977) is equivalent to the lithofacies succession that developed at mid-shelf position of the depositional sequence. The core shale is restricted to the mid and lower shelf, and represents the condensed section and position of the maximum flooding surface. Incised valleys and amalgamated paleosols on the upper shelf contain the last vestiges of a marine cycle consisting of a transgressive-regressive estuarine sandstone with the maximum flooding surface represented by a marine calcareous horizon. The regressive carbonate consists of parasequences when correlated in detail with well logs over the extent of the shelf (75 mi [120 km]) (Youle et al., 1994). The parasequences toplap landward and downlap toward the basin. The geometries suggest forced regression under staggered falling sea level. The unconformities bounding the sequence on the mid and upper shelf give way to conformable sequences on the lower shelf. The physically based model of Youle et al. (1994) is thought to apply to sequences of the later Pennsylvanian with modification although variations are described, e.g., Archer and Feldman (this volume). Actual internal packaging may be much more complicated, and analogous to the Pleistocene deposits. As lower depositional slopes decrease, the "slug" model gets flattened and the lateral accretion rapidly becomes a three-dimensional problem with deposition filling in local irregular topography. Another modification is that internal unconformity-bounded units (minor cycles) also punctuate this sequence as described earlier.

Sequence sets are groups of cyclothem-scale sequences that stack into similar progradational, aggradational, or retrogradational motifs (Youle et al., 1994). These sets are analogous to systems tracts within a single depositional sequence (Mitchum and Van Wagoner, 1991) except that each cyclothem-scale sequence is subaerially exposed. The sequence sets are recognized and characterized later in this study.

A cycle hierarchy is defined following that of Goldhammer et al. (1991). The unconformity-bounded sequence sets and sets of GSUs are herein referred to as third-order cycles (of 1- to 10-m.y. duration), primary sequences and GSUs are equivalent to fourth-order cycles (0.1- to 1-m.y. duration), and unconformity-bounded minor sequences are synonymous with fifth-order cycles (0.01 to 0.1 m.y.). Unconformity-bounded fourth- and fifth-order cycles possess all of the components of Vail's third-order depositional sequences based on estimated durations of each cycle ranging from 0.25 to 0.5 m.y. (Heckel, 1986; Klein, 1990a).

Third-order, unconformity-bounded sequence sets and sets of GSUs (genetic sets) contain bundles of fourth-order cycles (primary depositional sequences and GSUs). In some cases where strati-

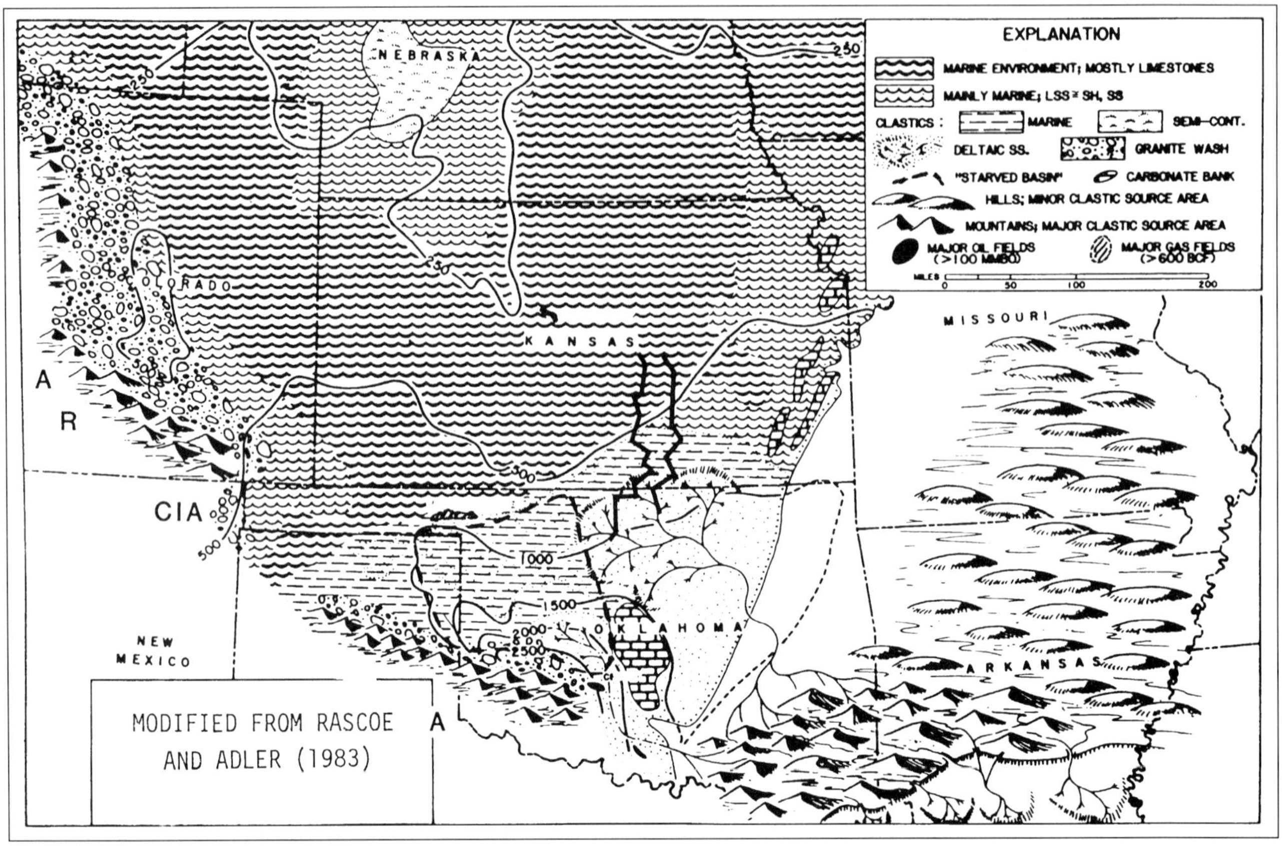

FIGURE 6 — Lithofacies and paleogeography of Missourian in Mid-Continent showing average conditions during the period. Dashed line in Anadarko basin indicates area of predominantly deepwater conditions based on work of Kumar and Slatt (1984). Map modified from Rascoe and Adler (1983).

graphic resolution is high, fifth-order unconformity-bounded minor cycles and parasequence-type cycles have also been recognized in these strata.

All the cyclicity described here cannot be solely attributed to eustasy because of the limited area of study. These cycles are related to changes in *relative sea* level that resulted from dynamic interactions among tectonism, eustasy, and sediment supply to produce the inferred sedimentation and accommodation history. Climate is indirectly linked to sediment supply which can affect accommodation and basic stratigraphic relationships. Until the rates, durations, magnitudes, and interactions of these processes are understood and differentiated, their effects must be considered *together*, perhaps one dominating during various times or interacting with the other. Conditions are dynamic and complicated even where one factor such as glacial eustasy predominates, if the Pleistocene is any indication.

TECTONIC SETTING IN THE MID-CONTINENT DURING THE PENNSYLVANIAN

Tectonic activity in the adjoining foreland basins was important in molding the configuration and elevation of the shelf. The Anadarko basin, which was responsible for subsidence along the western Kansas shelf, is a hybrid foreland basin that owes at least some of its subsidence to overthrusting of crustal blocks now exposed in the Wichita Mountains in western Oklahoma. Uplift of the Amarillo-

Cycle Order	Duration (million yrs., Ma)	Amplitude (m)	Rise/Fall rates (m/ka)	Sequence terminology	This study
1st order	>100		<.01		
2nd order	10-100	50-100	0.01-0.03	Supersequence	
3rd order	1-10	50-100	0.01-0.1	Sequence	Sequence Set Genetic Set
4th order	0.1-1	1-150	0.4-5	Sequence, Cycle	Sequence, Genetic Stratigraphic Unit (GSU)
5th order	0.01-0.1	1-150	0.6-7	Parasequence,Cycle	Minor Cycle

TABLE 1 — Temporal cycle orders (modified after Goldhammer et al., 1993).

Wichita-Arbuckle mountains began in the Early Pennsylvanian (Atokan) coincident with the onset of significant subsidence and definition of the Anadarko basin (Brewer et al., 1983). Some 5 to 6 mi (8 to 9 km) of northward thrusting in the Wichitas is indicated by deep-reflection seismic profiling (Brewer et al., 1983). Thrusting is attributed to the plate collision along the Ouachitas, and perhaps is ultimately responsible for subsidence of the Arkoma basin. Uplift along the mountain front is recorded as major episodes of conglomerate progradation into the southern margin of the Anadarko basin (Ham and Wilson, 1967). These episodes appear to have each lasted several million years and led to considerable subsidence in the basin and adjoining shelves. A paleogeographic reconstruction during the Missourian is depicted in Fig. 6 (from Rascoe and Adler, 1983).

Maximum subsidence in the western Anadarko basin immediately south of the western Kansas shelf is recognized by sediment-starved conditions (Galloway et al., 1977; Kumar and Slatt, 1984; Rascoe and Adler, 1983). Eastern limits of the basin near the Ouachita Mountains received reciprocally deposited clastic sediments similar to the Arkoma basin. Estimated relief across the shelf margin and the Anadarko basin during the Late Pennsylvanian was 1100 ft (335 m) (Kumar and Slatt, 1984).

The areal variation of average subsidence rates on the northern shelf during the Missourian, for example, conforms to basin development in the southern Mid-Continent (Kluth, 1986). The average subsidence rates calculated from backstripping calculations vary considerably, ranging from more than 1 ft/k.y. (0.3 m/k.y.) in the sediment-starved, drowned basin to less than 0.17 ft/k.y. (0.05 m/k.y.). on the northern shelf (Figs. 7 and 8). The average subsidence probably includes episodic thrust-induced subsidence, characterized by pulses of rapid downwarp followed by longer periods of gradually reduced subsidence (Beaumont, 1981). The precise duration of these episodes is not well known.

Differential subsidence on the Kansas shelf led to large-scale polygonal patterns of contrasting thickness patterns apparently responding to varying amounts of reactivation of structural and compositional heterogeneities in the Precambrian basement (Watney et al., 1993). The orientation and sense of reactivation were apparently related to changing vectors of laterally directed stresses associated with deformation in the adjoining foreland basins.

Although maximum rates of subsidence began to diminish during Upper Pennsylvanian time, areas of higher subsidence rates progressively extended northward toward the interior of the craton throughout the Virgilian (Kluth, 1986). This broadened subsidence is typical of foreland basin deformation (Beaumont, 1981). Tectonic activity in the Ouachita Mountains also diminished during the Virgilian while subsidence in the Anadarko basin neared its maximum, still fostering sediment-starved conditions in the Oklahoma and Texas panhandles (Dickinson and Yarborough, 1979).

Subsidence created significant accommodation space for sediments and produced a yet to be resolved signal in the sedimentary record. Average basinal subsidence rates of 1 ft/k.y. (0.3 m/k.y.) could result in creation of 344 ft (105 m) of space for sediment during a single fourth-order sequence (350 k.y.) while upper shelf rates (0.16 ft/k.y. [0.05 m/k.y.]) could result in 82 ft (25 m) of new accommodation space. The complexities of this varying subsidence history must be understood and accounted for in any enlightened attempts to quantitatively model the strata and account for observed stratigraphy.

CORRELATION AND MAPPING OF SEQUENCES AND GSUS

Portions of the Upper Pennsylvanian interval were mapped from the outcrop westward some 400 mi (644 km) using over 4000 wells (mostly gamma ray/neutron/density). Maps prepared include isopachs of GSUs and key stratigraphic intervals, thicknesses of porous carbonate, and structure.

Shelf-to-Basin Cross Section of a Portion of the Upper Pennsylvanian Kansas City Group

The Upper Pennsylvanian Pleasanton Group and the Hertha, Swope, and a portion of the Dennis formations are depicted in a 124-mi (200-km) long stratigraphic well log cross section (Fig. 9) that extends from near Kansas City, Missouri, to near Coffeyville, in extreme southeastern Kansas (Fig. 10). Equivalent sequences based on correlation of logs to numerous cores and outcrops are identified as M1 through M4 (Missourian age sequences). The GSUs would be bounded by and named after the condensed sections: Nuyaka Creek, Mound City, Hushpuckney, and Stark (all black phosphatic shales). The cross section illustrates the marked stratigraphic and lithologic changes within four primary sequences and associated GSUs along a mid shelf-to-lower shelf (north to south) transect that extends southward toward the margin of the Arkoma basin. Each sequence and GSU underwent substantial lateral changes in sedimentation due to apparent changes in factors (e.g., shelf elevation, proximity of siliciclastic sources, and sediment accommodation space).

The datum of the stratigraphic section is the Nuyaka Creek Shale, a regionally correlatable high gamma ray black shale (a condensed section). The datum coincides with a major change in the stratal framework. Correlations below the Nuyaka Creek Shale indicate general thickening of cycles toward the basin. In contrast, the overlying Pleasanton sequence abruptly thins southward, setting up inferred significant local topography (230+ ft [70+ m]) on which subsequent cycles were deposited. The black shales, highlighted on the cross section, serve as excellent subsurface markers owing to elevated natural gamma radiation that typically exceeds 160 API units.

A suite of well logs and corresponding core from the Amoco No. A-4 Cox in Haskell County in southwestern Kansas (Fig. 11) shows the similarity in stratigraphy to equivalent sections (Figs. 11 and 12) from eastern Kansas, over 280 mi (450 km) away. Cores and outcrop locations provide both definition of depositional sequences and GSUs.

Deposition during the M1, M2, and lower portions of the M3 sequences and analogous Nuyaka Creek, Mound City, and Hushpuckney GSUs was characterized by sediment-starved conditions in the northern Arkoma basin as shown on the southern part of the cross section in Fig. 9. Each sequence

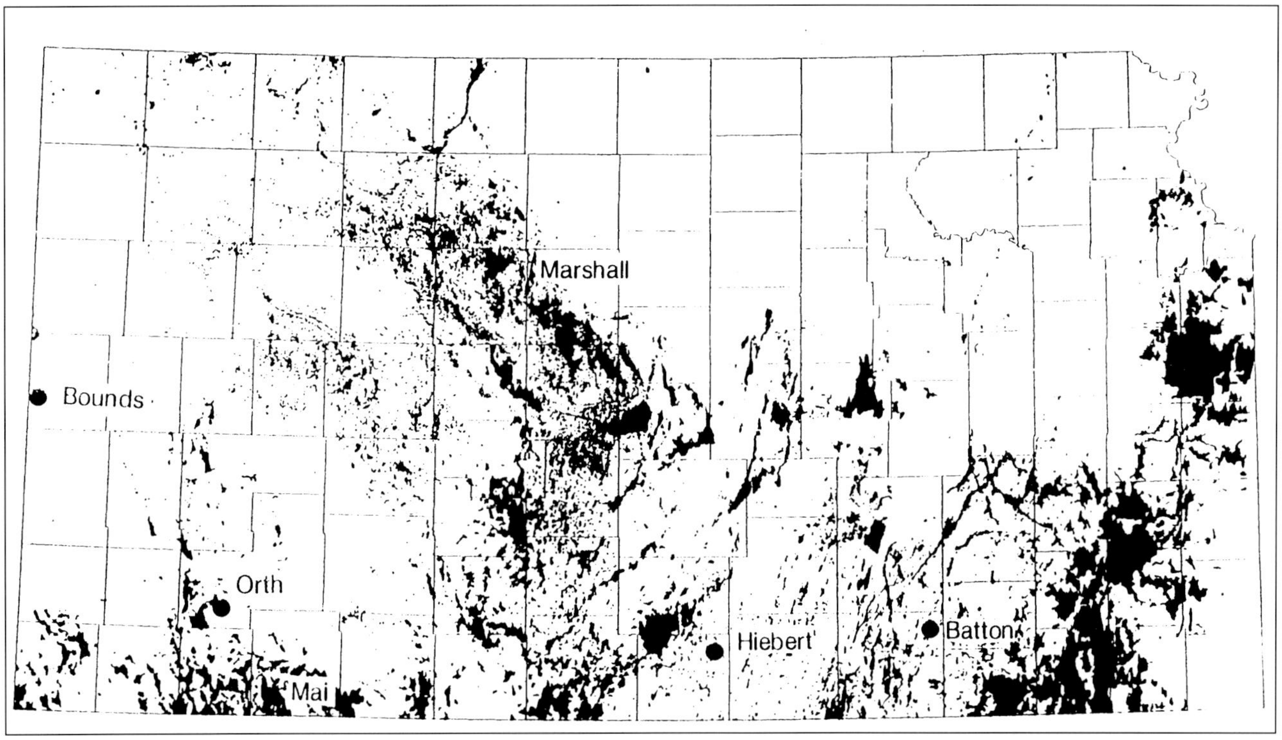

FIGURE 7 — Oilfield map of Kansas showing wells used in backstripping (decompaction) subsidence analysis.

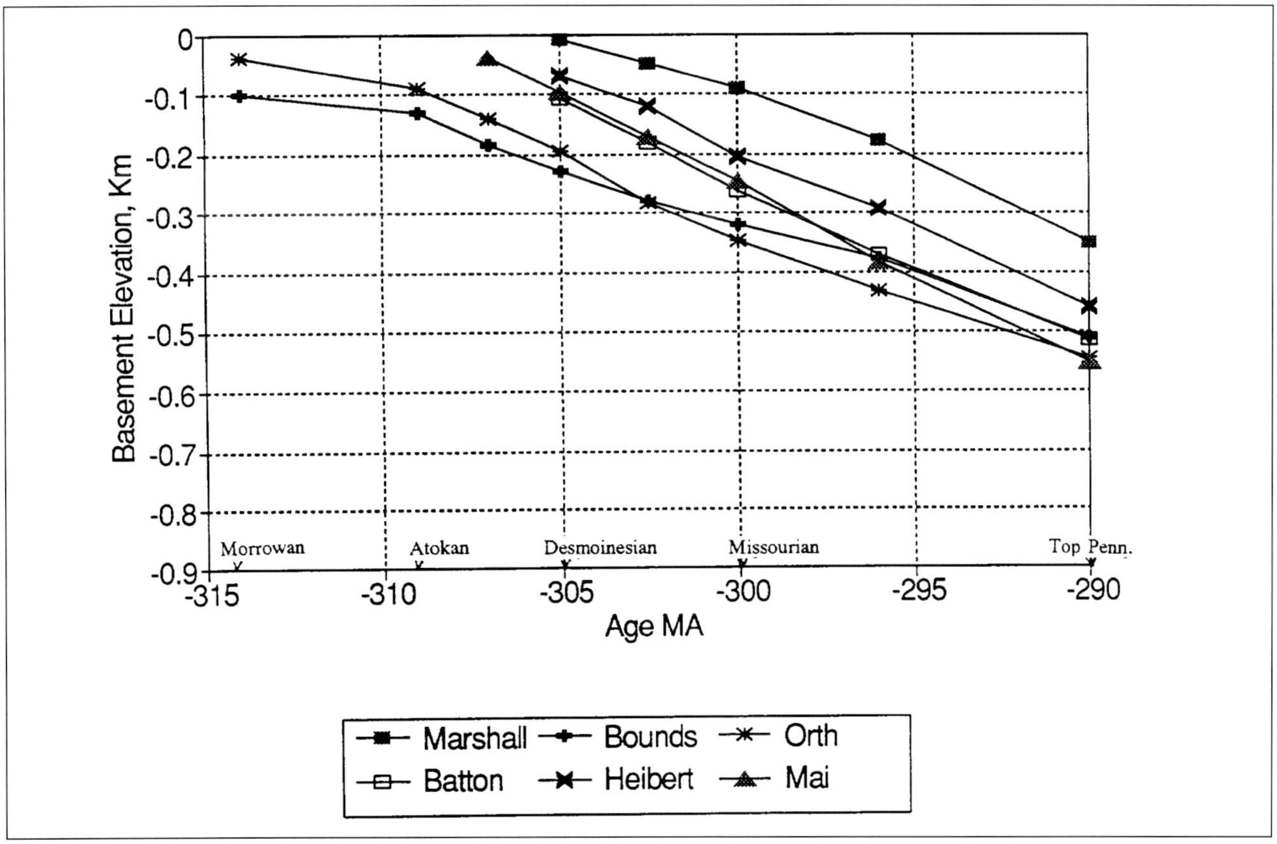

FIGURE 8—Burial history plot (depth versus time) of Pennsylvanian strata from six wells in Kansas shown on map in Fig. 7.

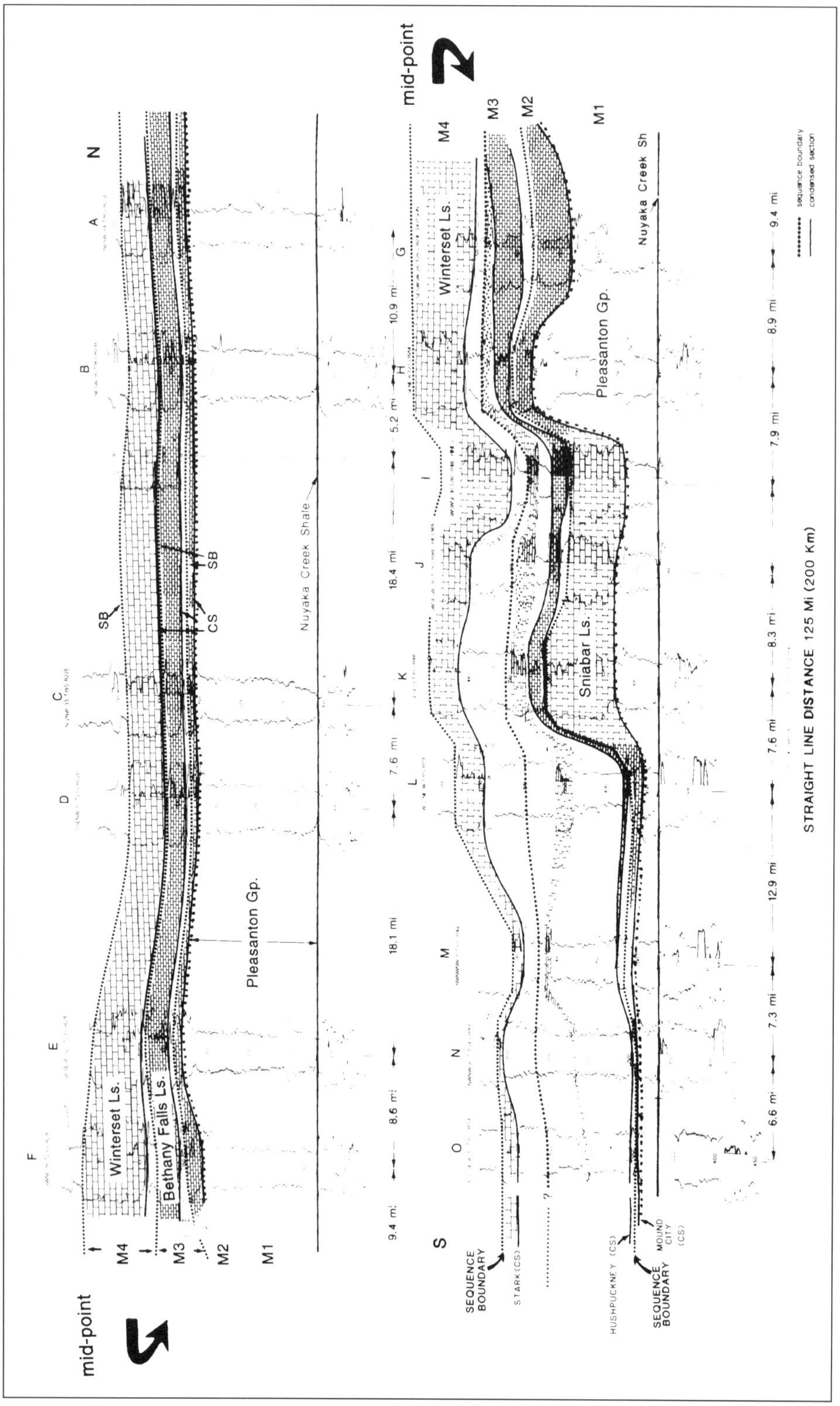

FIGURE 9 — Regional south-to-north stratigraphic cross section extending from 50 mi (81 km) south of Kansas City to the Oklahoma line. Datum is Nuyaka Creek Shale (base of Kansas City third-order sequence set). Central portion of the southern half of the cross section is shelf-to-basin transition containing successive basinward, lateral accretion of fourth-order depositional sequences.

in the basinal setting contains a thin (2 m), organic-rich marine shale with no significant associated carbonate units. The carbonates in the M2 and M3 apparently downlap and pinch out into the basin, resulting in the amalgamation of the Nuyaka Creek, Mound City, and Hushpuckney black shales there. The sediment-starved succession of these basinal shales is referred to locally as the Tacket Formation, which extends southward into Oklahoma.

Some of the bounding subaerial exposure surfaces that delineate each sequence on the northern shelf change downdip to conformable surfaces before reaching the sediment-starved lower shelf. Cores and surface exposures (Heckel, 1991a) indicate that paleosols immediately adjacent above and below the Nuyaka Creek Shale persist across the line of section while exposure surfaces in the overlying M1 and M2 sequences do not reach the lower shelf as the associated clastic wedge and limestones downlap into the more basinal setting. This suggests that the sea level fell to expose a large portion of the decidedly higher upper shelf, but not the lower reaches. In contrast, the sea level on both sides of the cycle that includes the Nuyaka Creek fell low enough to expose the lower shelf, suggesting a fundamental change in relative sea level between Nuyaka Creek and later intervals.

The upper part of the M3 sequence thickens significantly on the lower shelf to a nearly 300-ft (90-m) thick section of shallow marine and estuarine clastics with an additional marine limestone (the Mound Valley) (Watney et al., 1989). These clastics thin landward onto inferred higher elevations of the shelf to a single, thin, widespread, well-developed paleosol (the Galesburg Shale) that occurs on top of the main regressive carbonate (Bethany Falls). The Mound Valley Limestone has been traced up onto the edge of the shelf where it laps out (French and Watney, 1993), apparently as accommodation was lost at higher elevations of the shelf.

Farther south beyond the sediment-starved area of the M1, M2, and lower M3 sequences, these sequences thicken substantially into the clastic-dominated Skiatook Group of Oklahoma. These clastics were likely sourced from the Ouachita Mountains, and during lowstand conditions the clastics episodically onlapped the carbonate shelf in southern Kansas (Bennison [1985], p. 22 *in* Watney et al., 1989). These clastic-dominated sequences are comparable in thickness to their correlative carbonate-dominated sequences on the northern shelf and also contain equivalent, similarly developed radioactive marine shales. The axis of the depositional basin during this particular time period is probably the sediment-starved area between Tulsa and the Kansas-Oklahoma line, lying north of the Ouachita clastic wedge and south of the carbonate shelf (Bennison [1985], p. 22 and Heckel, p. 164 *in* Watney et al., 1989).

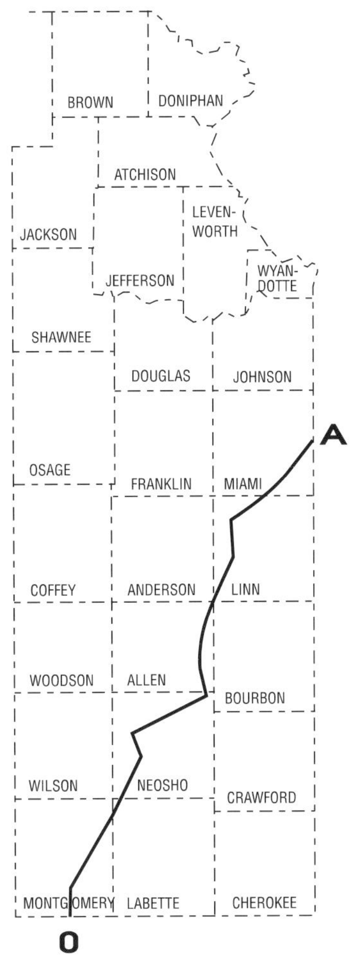

FIGURE 10 — Index map for cross section shown in Fig. 9.

Shelf-to-Basin Pennsylvanian Cross Section — Central Kansas to North-Central Oklahoma

Another regional north-south stratigraphic cross section of Middle and Upper Pennsylvanian strata illustrates changes associated with a shelf-to-basin transition along an approximately 125-mi (200-km) long section from the northern shelf in central Kansas into the northern edge of the Arkoma basin in northern Oklahoma (Fig. 13). The section is some 60 mi (97 km) west of the section in Fig. 9. The section closely follows that prepared by Lukert (1949).

The stratigraphic interval shown in the cross section in Fig. 13 extends from the base of the Pennsylvanian upward to the Heebner Shale (the cross-section datum). The interval thickness doubles from 900 ft (274 m) in the north to nearly 1800 ft (548 m) in the south. Correlations were made using analogous stratal relationships developed from the study of near surface and surface analogues farther east in southeastern Kansas. Maximum flooding surfaces in Fig. 13 delimit primary GSUs and sets of GSUs. Shaded intervals denote the carbonate-dominated sets of GSUs. The genetic units

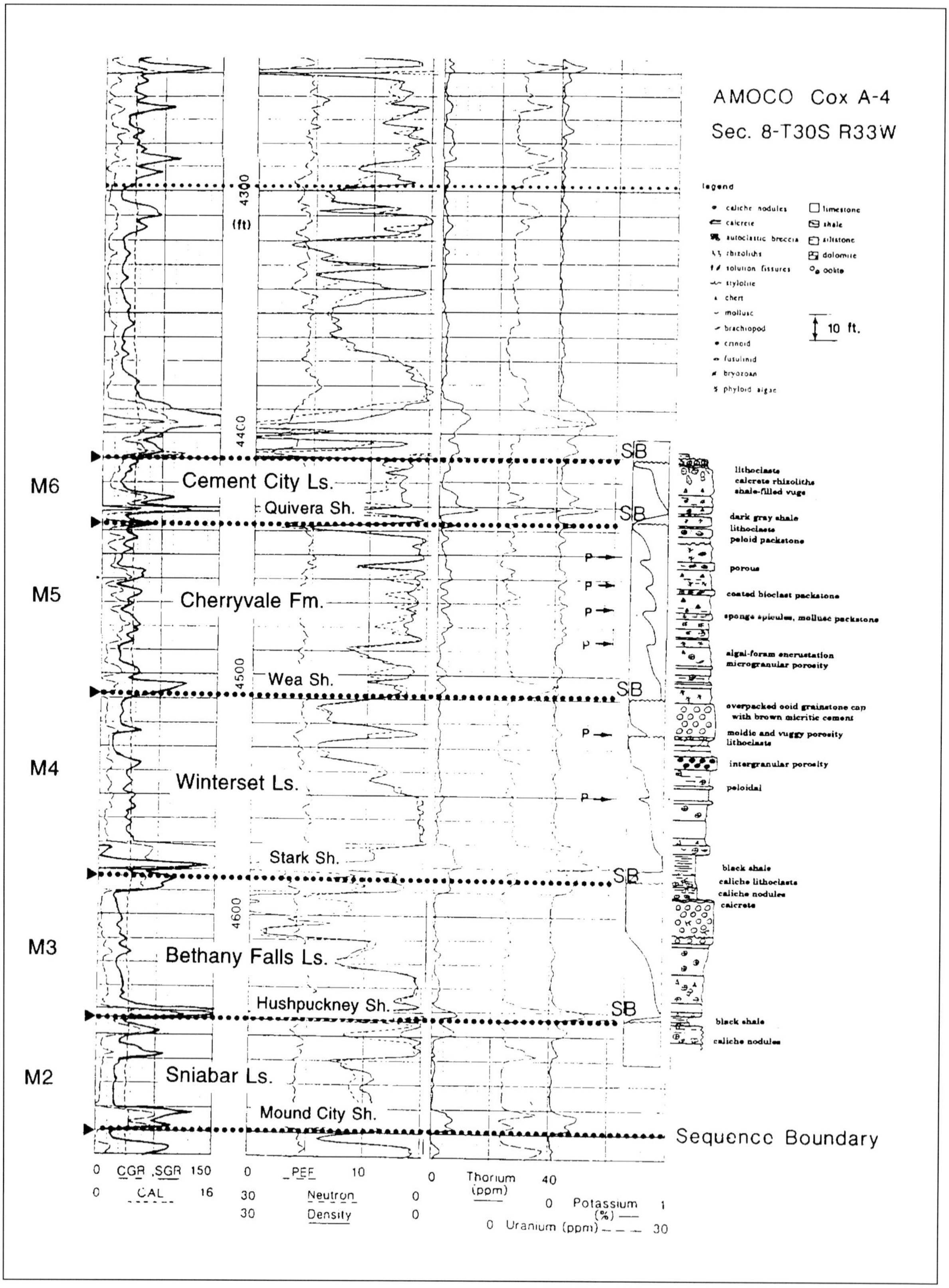

FIGURE 11 — Gamma ray, neutron, density, uranium, thorium, and potassium well logs, and graphic core descriptions of M2, M3, M4, M5, and M6 depositional sequences in the Amoco Cox No. A-4 well, Haskell County, southwestern Kansas. Well is located on lower shelf. Sequence boundaries (subaerial exposure surfaces) are indicated by heavy dotted lines. Minor cycles and parasequences are noted with P and arrow.

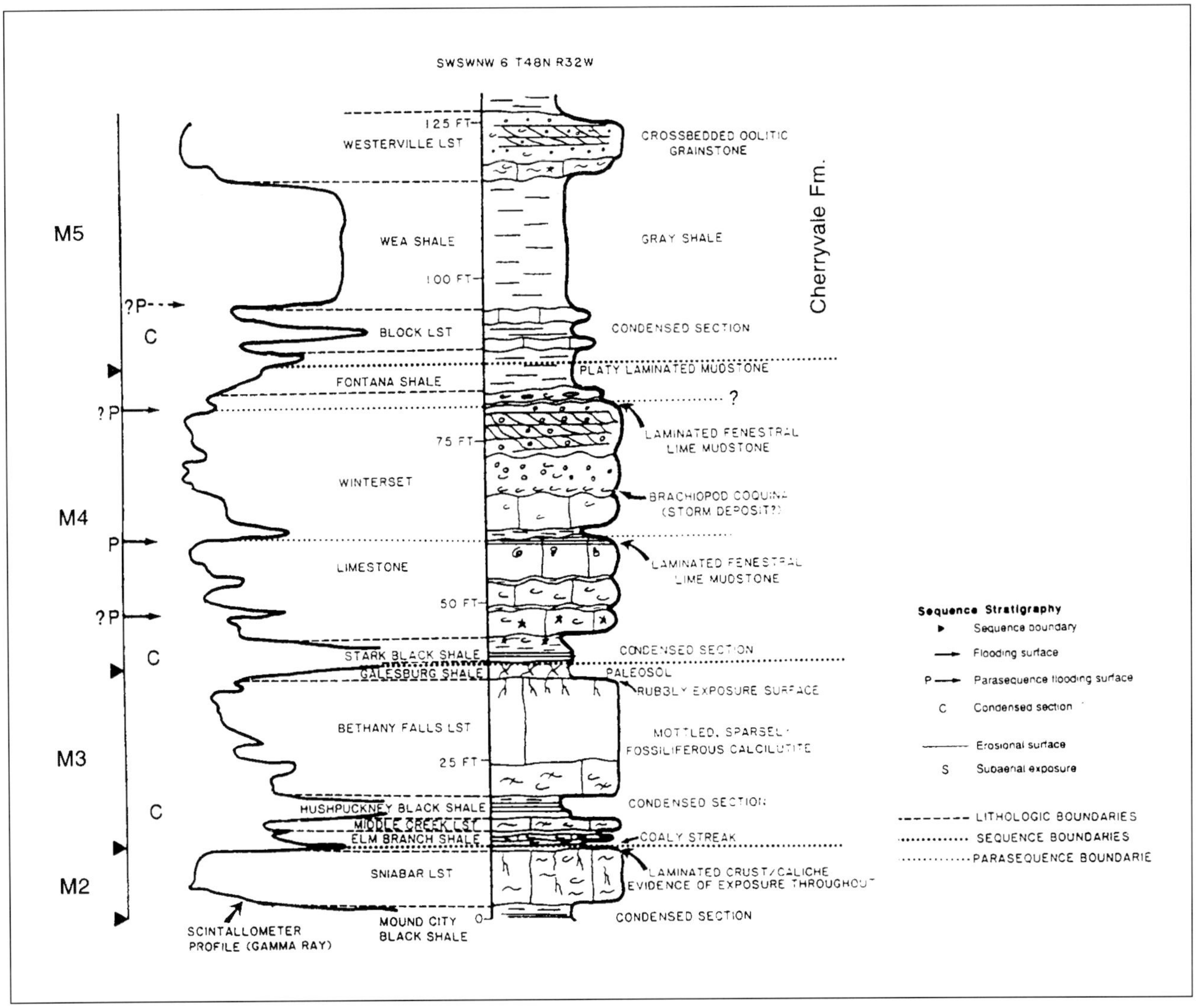

FIGURE 12 — Measured section of M2, M3, M4, and portion of M5 sequences in the Upper Pennsylvanian Kansas City Group at a surface exposure in Kansas City, Missouri (SW SW NW, Sec. 6 T48N-R32W) (from Watney et al., 1989). Lithologic description accompanied by gamma ray (GR) scintillometer profile (recording every 1 to 2 ft). Gamma ray units are not calibrated. Higher GR is to the right.

serve as proxies for the depositional sequences and sequence sets since local rock data are lacking to delimit subaerial unconformities.

Each prominent set of GSUs recognized shows notable stratigraphic re-organization from the previous set. Three significant carbonate packages representing three sets of GSUs are recognized: the Excello, Nuyaka Creek, and Muncie Creek genetic sets (with the names based on the condensed sections of the same name). Each genetic set contains five to seven primary GSUs. Each carbonate succession apparently downlaps into the basin in a manner similar to the near surface example discussed above. The south-facing carbonate platforms formed by these three sets of GSUs each underwent successive landward backstepping or retrogradation. The Stark Shale is associated with a smaller backstepping episode within the Nuyaka Creek set of GSUs. Subaerial exposure surfaces, although present and bounding each genetic set, are not practical boundaries for subsurface mapping because they are not as clearly defined as the maximum flooding surfaces (condensed sections) on the well logs (Galloway, 1989a, b).

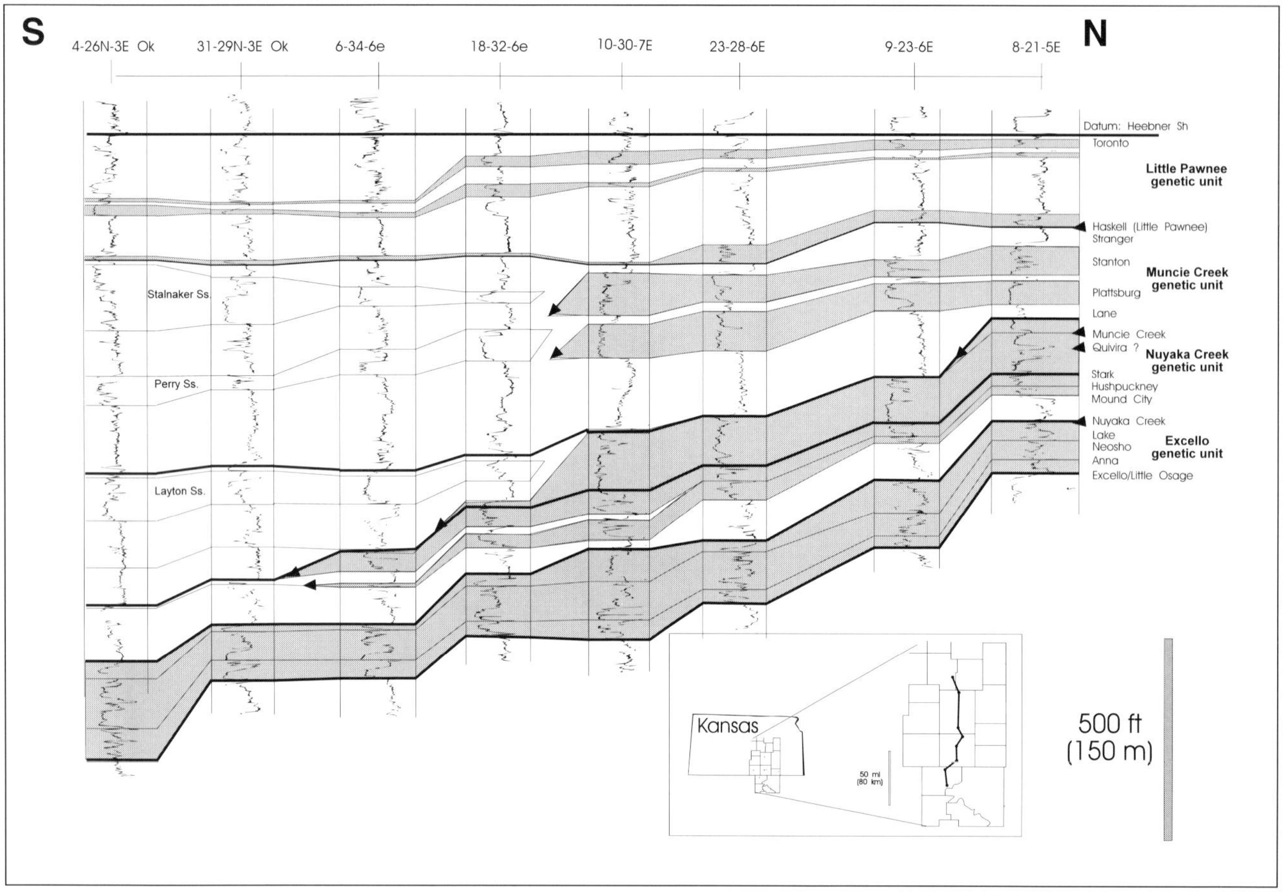

FIGURE 13 — North-to-south regional stratigraphic cross section datumed on the Heebner Shale for the Middle and Upper Pennsylvanian interval running approximately 125 mi (200 km) from central Kansas to northern Oklahoma. Index map included in figure. Gamma ray log traces and correlations of wells selected from regional stratigraphic data base showing shelf (right) to basin (left) transition. Carbonate-dominated sections on shelf (shaded intervals) pass laterally to a clastic-rich section in the basin. Major throughgoing marine markers (arrow heads located along right margin) bound the third-order genetic sets. Sandstones are interpreted as part of lowstand deposits. Dark screened strata = limestone; light screened strata = sandstone; white areas = predominantly shale and thin sandstone.

Each of the sets of GSUs ranges from 300 ft (90 m) to in excess of 500 ft (150 m) in thickness on the northern shelf. The generally abrupt backstepping of limestones associated within these genetic sets is interpreted as rapid (greater than normal) deepening and drowning of the carbonate shelf. The thin, organic-rich shales associated with these drowning events are also the condensed sections of individual basal primary genetic units in each genetic set. As previously mentioned, similar clastic-dominated GSUs with similar patterns and equivalent condensed sections are developed in flysch deposits that infill the southern Arkoma basin in Oklahoma, suggesting a common mechanism for their formation on opposite sides of the basin in contrasting depositional settings (Bennison [1985], p. 22 *in* Watney et al., 1989).

Successively younger primary GSUs within a set of GSUs generally step basinward or aggrade as suggested in detail in Fig. 9. This forward stepping in the Nuyaka Creek genetic set is described in three dimensions later in the section on mapping.

The sets of GSUs coincide closely with formal stratigraphic groups. The Marmaton Group, which occurs at the base of the cross section between the Excello Shale and the Nuyaka Creek Shale, consists of six distinctive primary GSUs (which comprise the Nuyaka Creek genetic set) (Fig. 2). The Marmaton Group was deposited on a broad carbonate shelf across the area included in this cross section. The shelf margin during Marmaton Group deposition was another 50 mi (80 km) south of the southern end of the cross section (Fig. 14).

In contrast, during accumulation of the Nuyaka Creek set of GSUs extending from the Nuyaka Creek Shale to the Muncie Creek Shale, the depositional shelf margin abruptly shifted landward (northward, Fig. 15) to within this line of section (Fig. 13). This abrupt landward shift of the carbonate shelf margin at the end of the Excello genetic set deposition is attributed to drowning of the lower portion of the shelf, proximal to the Arkoma and Anadarko basins in southern Kansas and northern Oklahoma during a relative rise in sea level (Fig. 15, Rascoe, 1962). Drowning rather than clastic poisoning of the carbonate margin is thought to be the mechanism for this landward shift, because general sediment starvation occurs at the beginning of the Nuyaka Creek genetic set over a broad area in southeastern Kansas and northeast Oklahoma (Watney et al., 1989). Comparing the lithofacies maps of the Marmaton (closely equivalent to the Excello genetic set) and Kansas City groups (closely equivalent to the Nuyaka Creek genetic set) in Figs. 15 and 16 indicates the backstepping nature of the Kansas City Group relative to the Marmaton Group in central Oklahoma. However, farther west in the Texas Panhandle, the backstepping of the Kansas City Group over the Marmaton Group is replaced by an apparent forward stepping (Hentz, 1994). This change suggests contrasting processes acting between these two areas, e.g., greater subsidence rates in the eastern region may account for the

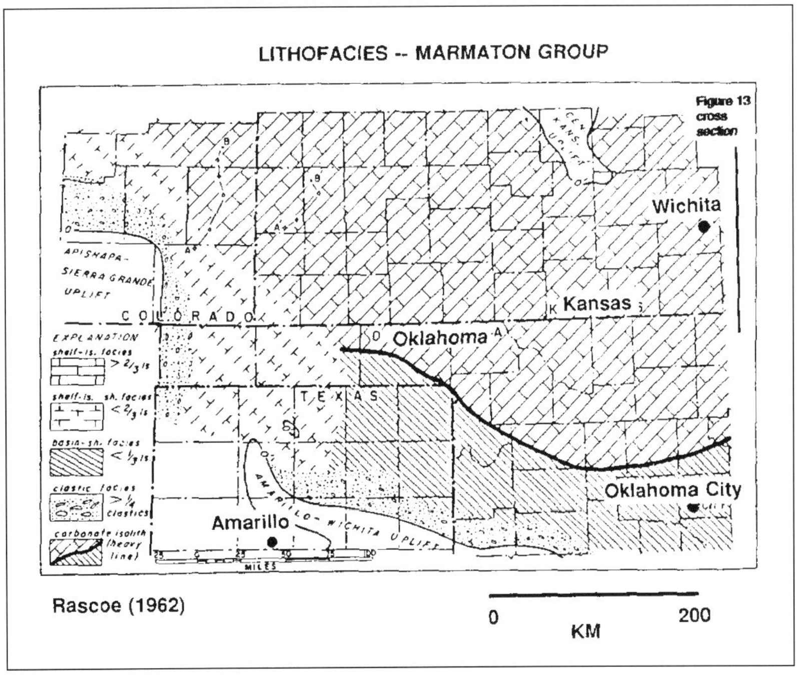

FIGURE 14 — Lithofacies map of Marmaton Group (modified from Rascoe, 1962).

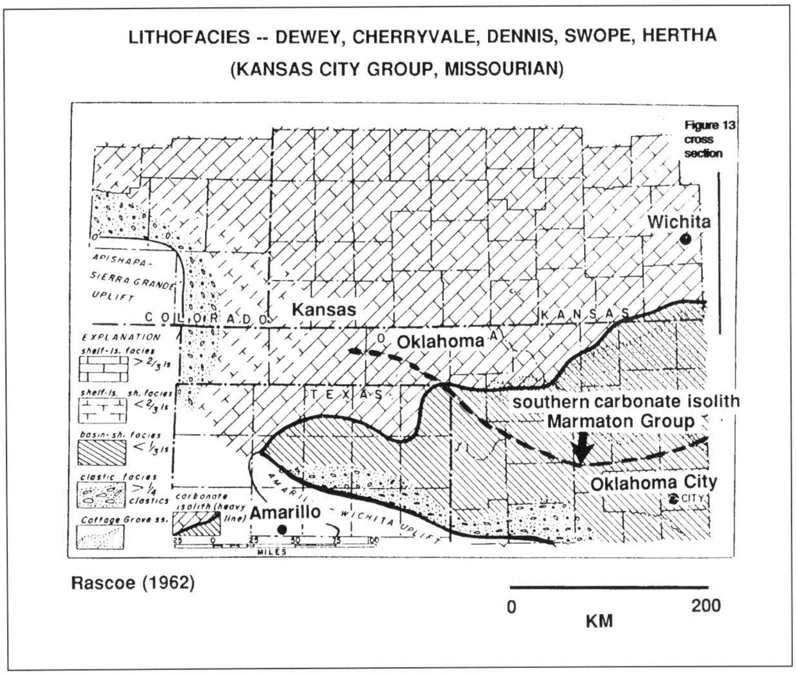

FIGURE 15 — Lithofacies map of the Kansas City Group (modified from Rascoe, 1962).

drowning and backstepping, or the west may have been able to keep up with an increase in carbonate production because of changes in climate or water conditions. Calculated subsidence rates were slightly less in western Kansas than in the east. Less clastics in the west may have allowed maintenance of carbonate productivity for longer times in each eustatic cycle in the west.

Clastics were intermittently introduced during apparent lowstands into some primary GSUs along the lower shelf and basin. One example is the clastics in the Hushpuckney GSU (containing the lower Layton sandstones), which occurs between the Stark and Hushpuckney shales (Fig. 13). However, most of the clastic infilling (e.g., the upper Layton and Stalnaker sandstones) comes near the end of deposition of the sets of GSUs followed by subsequent drowning. These later episodes of relatively voluminous clastic influx may be related to the extended periods of lowered sea level that are inferred to have occurred late in the deposition of the set of GSUs (see Fig. 16, by Bennison [1985], p. 22 *in* Watney et al., 1989; Bennison, this volume).

A key element about the boundary of the genetic sets is that the "clastic sediment delivery system" was optimal at this boundary and certain shelf locations previously carbonate dominated were transformed to clastic-dominated conditions, because carbonate-dominated cycles persist nearby in higher shelf locations and in areas isolated from the clastics. One explanation is that a long-term falling sea level might help to deliver significant new clastics to the shelf margin and basin (eastern Kansas and Arkoma basin) in a progressive, steady manner.

Isopach Maps of Primary GSUs in the Nuyaka Creek Genetic Set

Isopach maps of a succession of individual primary GSUs (proxies for depositional sequences) that constitute the Nuyaka Creek set of GSUs provide a visualization of the evolving nature of sediment accumulation of the northern shelf (Figs. 16 through 18). Each of these primary GSUs is bounded by condensed sections easily recognized on well logs. These surfaces usually directly overlie the subaerial exposure surface over much of the shelf. Thus, the maps essentially act as a series of snapshots with views of the changing high-resolution paleogeographic setting. These maps and supporting data show a complex pattern of deposition on the shelf, which is the framework for attempts at obtaining a more refined model of Pennsylvanian sedimentation.

The Nuyaka Creek genetic set begins with a drowning of the middle to lower shelf areas formed previously by the Excello genetic set in southeastern and south-central Kansas. Successive depositional episodes delineated by these overlying GSUs record the episodic infilling (lateral accretion) of this shelf.

Nuyaka Creek GSU

A thick wedge of marine siliciclastics up to 200 ft (61 m) thick was deposited in eastern Kansas during development of the Nuyaka Creek GSU (Fig. 16). The mapped interval extends from the base of the Nuyaka Creek Shale to the Mound City Shale (as proxy for the Pleasanton sequence). Decompacting the sediment may add up to 40% more section, further increasing the topographic relief developed on the lower shelf. The clastic unit abruptly ends in southern Kansas at an inferred constructional margin of the deposit adjoining a sediment-starved lower shelf. This shale-dominated unit, consisting primarily of fluvial deltaic sediments, was most likely sourced from the northeast (Howe, 1982). The sequence thins both southward and westward. Cores and well logs of cycles from the Central Kansas uplift suggest a positive shelf setting as revealed by consistent thinning of cycles, loss of beds, loss of black shale facies, in combination with more peritidal deposits, and more substantial paleosols (Watney, 1985; Watney et al., 1989). The thick clastic Nuyaka Creek GSU in eastern Kansas changes

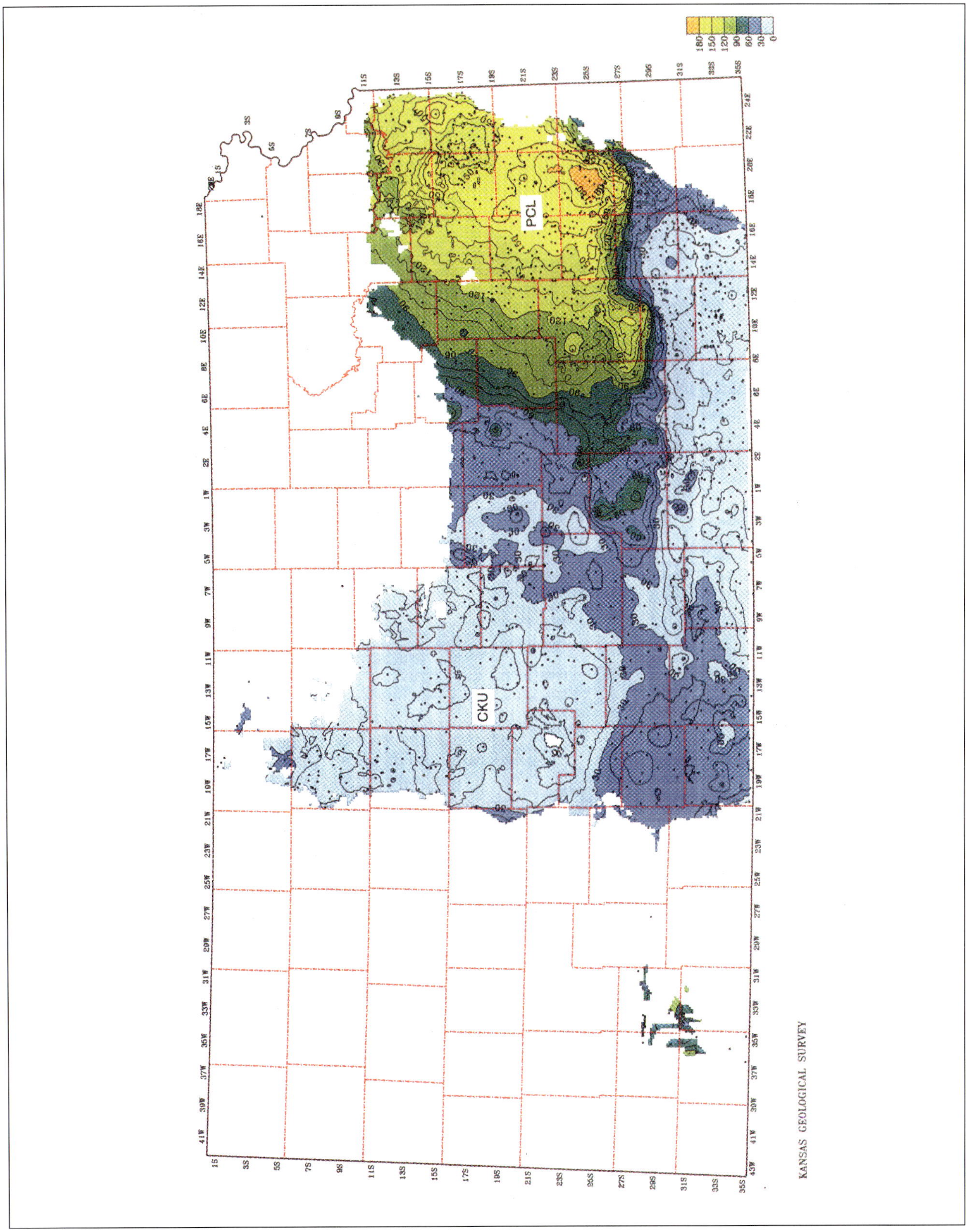

FIGURE 16 — Isopach map of Nuyaka Creek GSU. Contour interval is 10 ft; each shaded interval is 30 ft. County outlines are shown. Every other township and range is labeled along edges of map. Distance between them is 12 mi (19 km). Gridded and contoured using kriging option with SURFACE III.

to a cycle in western Kansas that is little different from the other adjoining sequences. The thick clastic wedge in the east probably owes its unusual thickness to several factors, including the relatively lower shelf elevation of eastern Kansas compared to the west, the proximity to a viable clastic sediment source, and added accommodation space created at the beginning of the Nuyaka Creek set of GSUs.

The clastic-dominated Cleveland Formation in the Texas Panhandle roughly occupies the same interval as the Nuyaka Creek GSU, with the common base of the Nuyaka Creek Shale (Hentz, 1994). The Cleveland Formation in this region represents a backstepping event like the Pleasanton. The Cleveland Formation contains two sequences with marine condensed sections developed in the basinward direction. These may be equivalent to minor cycles that have been observed within the Nuyaka Creek GSU in eastern Kansas (Watney et al., 1989).

Mound City GSU

The Mound City GSU is predominantly carbonate except in extreme eastern Kansas. On most of the shelf, the unit ranges from about 15 to 40 ft (4.6 to 12.2 m) in thickness. Areas of similar thickness of this GSU include the area over the crest of the underlying clastic lobe (PCL) of the Pleasanton Group and over the axial crest of the Central Kansas uplift (CKU). A narrow band of abrupt thickening is developed about 20 to 30 mi (32 to 48 km) north of the Oklahoma border in southeastern Kansas. Core and outcrop work reveal that this 78-mi (125-km) long band, up to 100 ft (30 m) thick and some 20 mi (32 km) wide, is a large carbonate bank complex consisting largely of phylloid algal wackestone (Watney et al., 1989). This bank developed on the abrupt basinward slope of the clastic wedge of the Pleasanton Group apparently in a desirable location for high carbonate production and increased accommodation space during the appropriate time of sea-level fall during deposition of the Sniabar Limestone. This represents a form of lateral accretion (Fig. 17) of the shelf that extended it basinward some 20 mi (32 km). Core and outcrop studies suggest that the internal framework of the carbonate bank is an amalgamation of smaller banks. The area south of the bank is the sediment-starved area that represents the depositional basin on the lower shelf. Local studies also indicate that large tracts of oolitic grainstone are developed in southwestern Kansas, expressed here as areas in which the Mound City genetic stratigraphic is relatively thick, rimming the western edge of the Central Kansas uplift and wrapping around the central region of the Hugoton embayment.

Hushpuckney GSU

The overlying Hushpuckney GSU resulted in additional accretion of the southern shelf margin and is also associated with an influx of clastics (Ladore Shale) that partially filled the lower shelf in southeastern Kansas. The clastics were probably derived from the Ouachita Mountains as sediments overfilled the Arkoma basin (Fig. 18) (Bennison, 1985). Relatively thick oolitic grainstones accumulated in the Hugoton embayment as a complex of porous carbonate lobes, while thicker oolitic grainstone is limited to the southern edge of the shelf forming a narrow rim in eastern Kansas (Fig. 19). The shelf was apparently steep enough over a broad area in western Kansas to sustain currents and waves, which generated a succession of oolitic grainstone tracts. Any clastics, which probably terminated oolite deposition in the east, were apparently too distant from this portion of the shelf. Local studies indicate that the grainstones developed in the east and west consist of separate stacked stratal units often separated by subaerial exposure surfaces (French and Watney, 1993). These minor cycles with exposure surfaces also punctuate the upper part of the Hushpuckney GSU in the area of mixed clastic/carbonate accumulation along the lower shelf in southeastern Kansas (Fig. 6: Watney et al., 1989).

Several trends of thicker porous carbonate (oolite) lobes are recognized in western Kansas, each

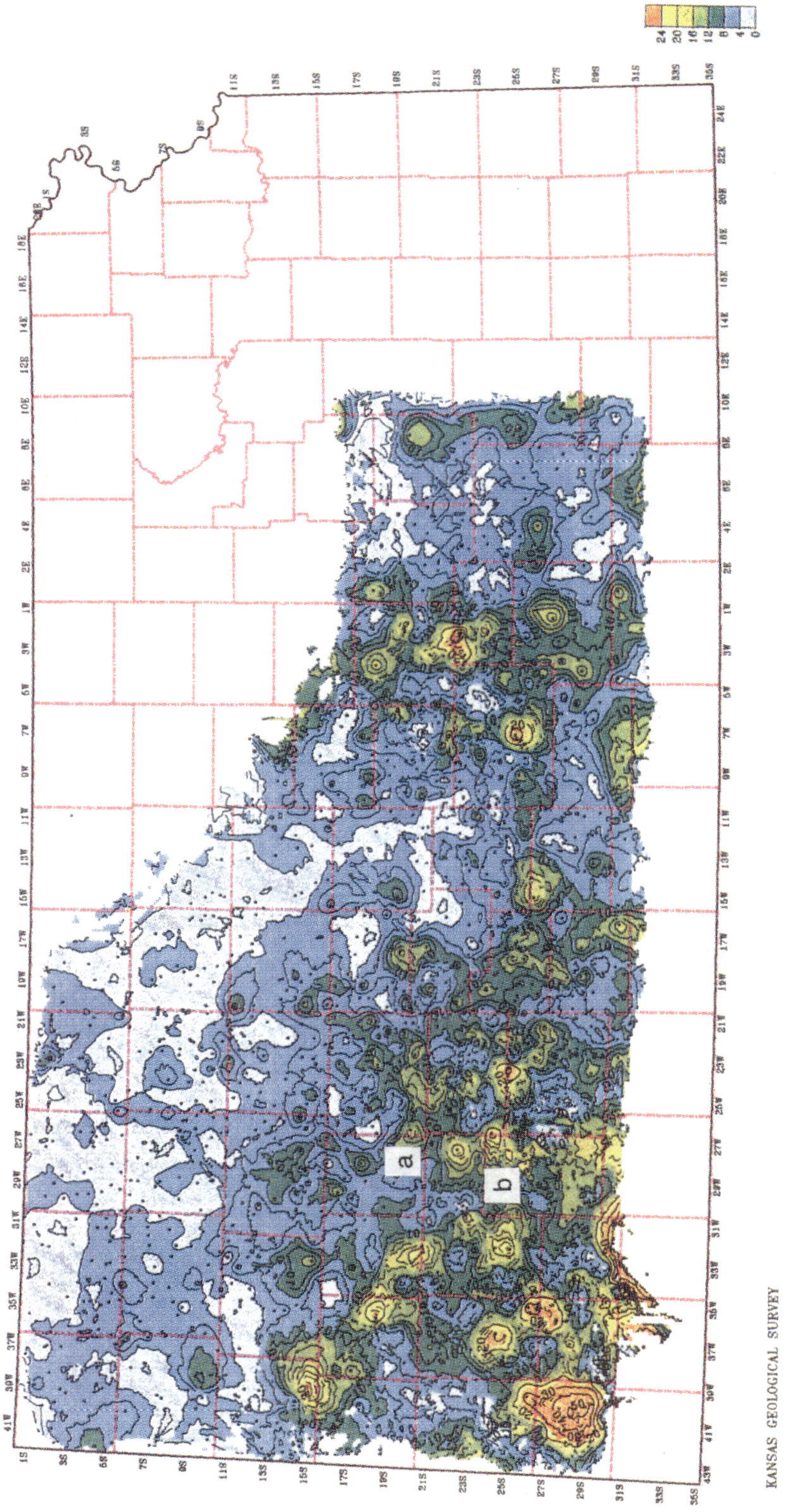

FIGURE 17 — Isopach map of Mound City GSU. Contour interval is 10 ft; each shaded interval is 20 ft. County outlines are shown. Every other township and range is labeled along borders. Distance between them is 12 mi (19 km). Gridded and contoured using kriging option with SURFACE III.

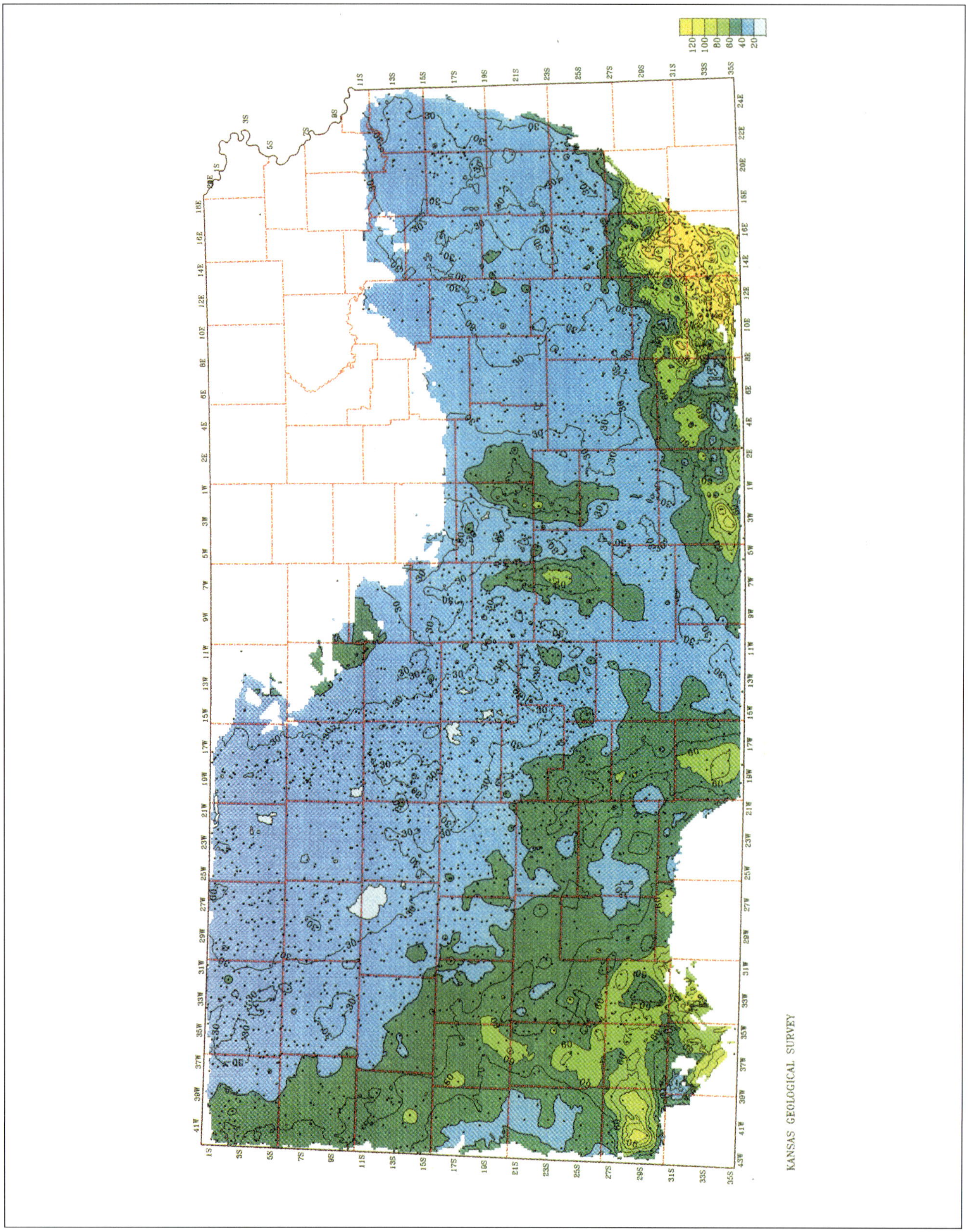

FIGURE 18 — Isopach map of Hushpuckney GSU. Contour interval is 10 ft. Shaded interval is 20 ft. County outlines are shown. Every other township and range is labeled along borders. Distance between them is 12 mi (19 km). Gridded and contoured using kriging option with SURFACE III.

trend paralleling the general strike of the shelf (identified in Fig. 19 as a and b). An upper shelf trend (a) wraps around the Hugoton embayment and west flank of the Central Kansas uplift. The mid-shelf trend (b) is characterized by thick porous lobes of carbonate that wrap around an apparently reactivated Cimmaron arch in extreme southwestern Kansas. This mid-shelf trend (b) then extends eastward across the Hugoton embayment and farther eastward across the southern margin of the Central Kansas uplift. The stacked oolites and minor cycles in southeastern Kansas are developed on a relatively steep slope with demonstrable relief (French and Watney, 1993).

The isopach thick in southeastern Kansas (Fig. 18) includes lowstand clastics (Ladore Shale) which were deposited basinward of the carbonate margin of the Bethany Falls Limestone. Clastics were deposited in low shelf positions while apparently bypassing high shelf locations occupied by paleosol development and, locally, minor marine shelf incision (Slamal, pers. comm., 1991). This filling episode suggests that the relief originally created on the basinward side of the clastic wedge of the Pleasanton Group was greatly reduced at this time, setting up a contrasting shelf setting for the next sequence.

Stark GSU

The isopach map of the succeeding Stark GSU reveals thickening toward the basin (Fig. 20). Thicker carbonate buildups (oolite and phylloid algal mudbanks) developed along the southern shelf margin as in the preceding cycle, but define a return of the shelf margin to near that of the previous sequence. Sediment-starved conditions in southeastern Kansas returned again to the lower shelf in Kansas bordering the Arkoma basin. This new carbonate margin was displaced approximately 20 mi (32 km) northward (landward) relative to the underlying Hushpuckney GSU, which includes the Ladore Shale, suggesting a moderate backstepping event. This is also observed in the cross section (Fig. 13). An isopach map of thickness of porous carbonate in the Winterset regressive limestone of the Stark GSU distinguishes lobes of oolitic grainstone up to 60 ft (18 m) thick with areal dimensions of tens of km (Fig. 21). Several tracts of thickened porous carbonate lobes that parallel strike are recognized in western Kansas — (a) upper northwest to southeasterly tract, and (b) a thicker more west-to-east trend that wraps around the southern Central Kansas uplift. Core and surface descriptions attest to the prominent development of minor sequences and parasequences in the Dennis (Watney et al., 1989; Felton and Heckel, 1992). The recurring marine inundations of the shelf, albeit minor, during the waning stages of regression may have encouraged the repeated occupation of a site by oolite shoal environments leading to stacking and widespread, thick grainstone accumulations.

Wea and Quivera GSUs

Overlying the Stark GSU are the Wea and Quivera GSUs, the uppermost units of the Nuyaka Creek genetic set. Isopachs of these two GSUs reveal infilling of the lower shelf and resultant loss of apparent topographic relief in Kansas (Figs. 22 and 23). Also, the thickness of these GSUs over much of the shelf in general is less than the underlying units, suggesting the loss of accommodation space due to infilling by sediment of previous cycles.

The features of these GSUs correspond with the close of the Nuyaka Creek genetic set. The overall stratigraphic pattern through this set is shelf-wide regression and basin margin filling. Evidence for the loss of sediment accommodation space in the Wea GSU includes the lapout of the marine portion of this unit along the upper shelf (zero contour line) in northwestern Kansas. Cores and logs trace the pinchout and joining of paleosols of the Wea and Stark GSUs (northwest Kansas, Fig. 22). Also, the lack of a distinctive condensed section in the Wea unit (except occasionally on the southern shelf) suggests shallower water during maximum flooding when the photic zone may still have been devel-

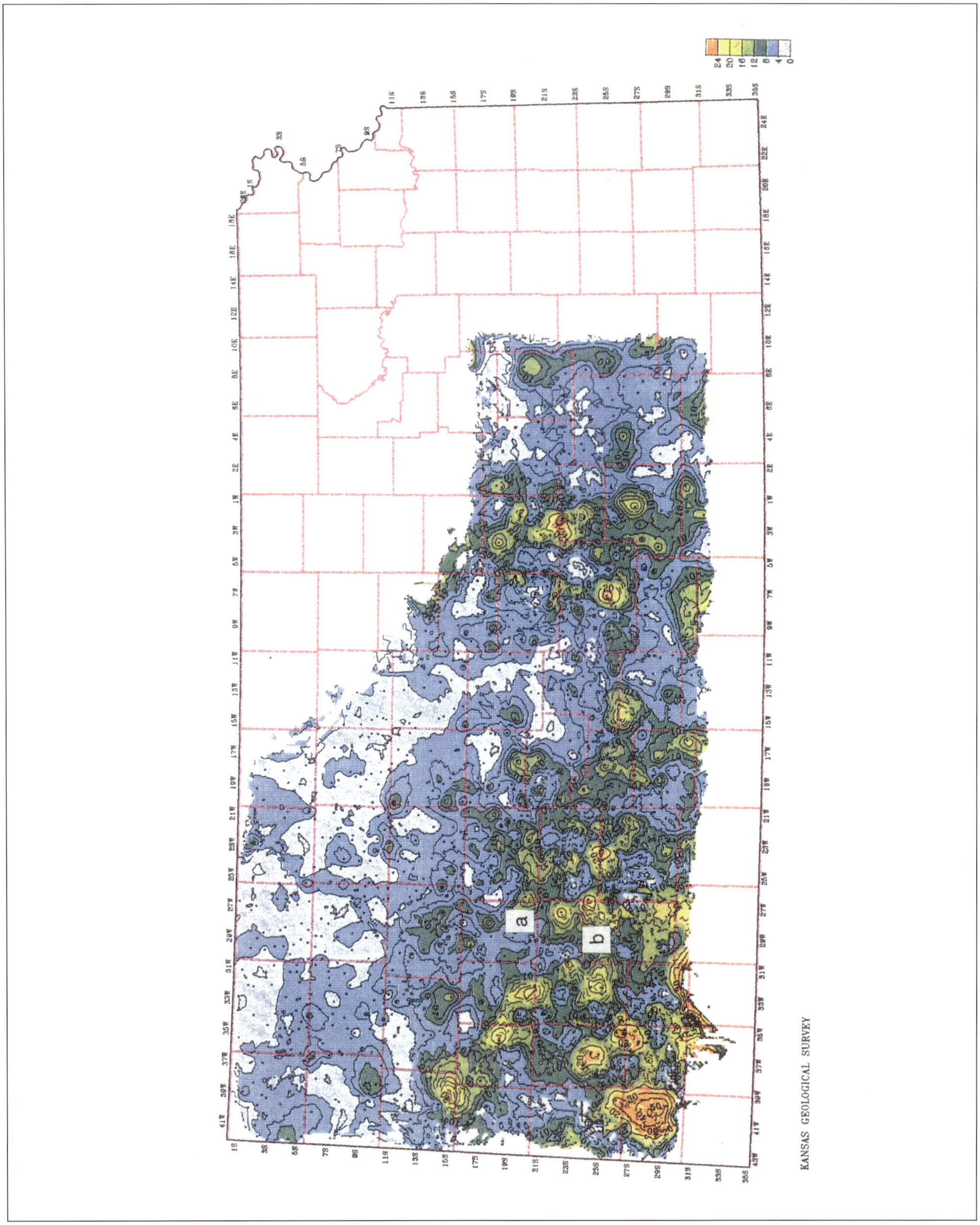

FIGURE 19 — Isochore map of thickness of porous carbonate above 8% porosity in the Hushpuckney GSU (Bethany Falls Limestone). Contour interval is 2 ft. Shaded interval is 4 ft. County outlines are shown. Every other township and range is labeled along borders. Distance between them is 12 mi (19 km). Gridded and contoured using kriging option with SURFACE III.

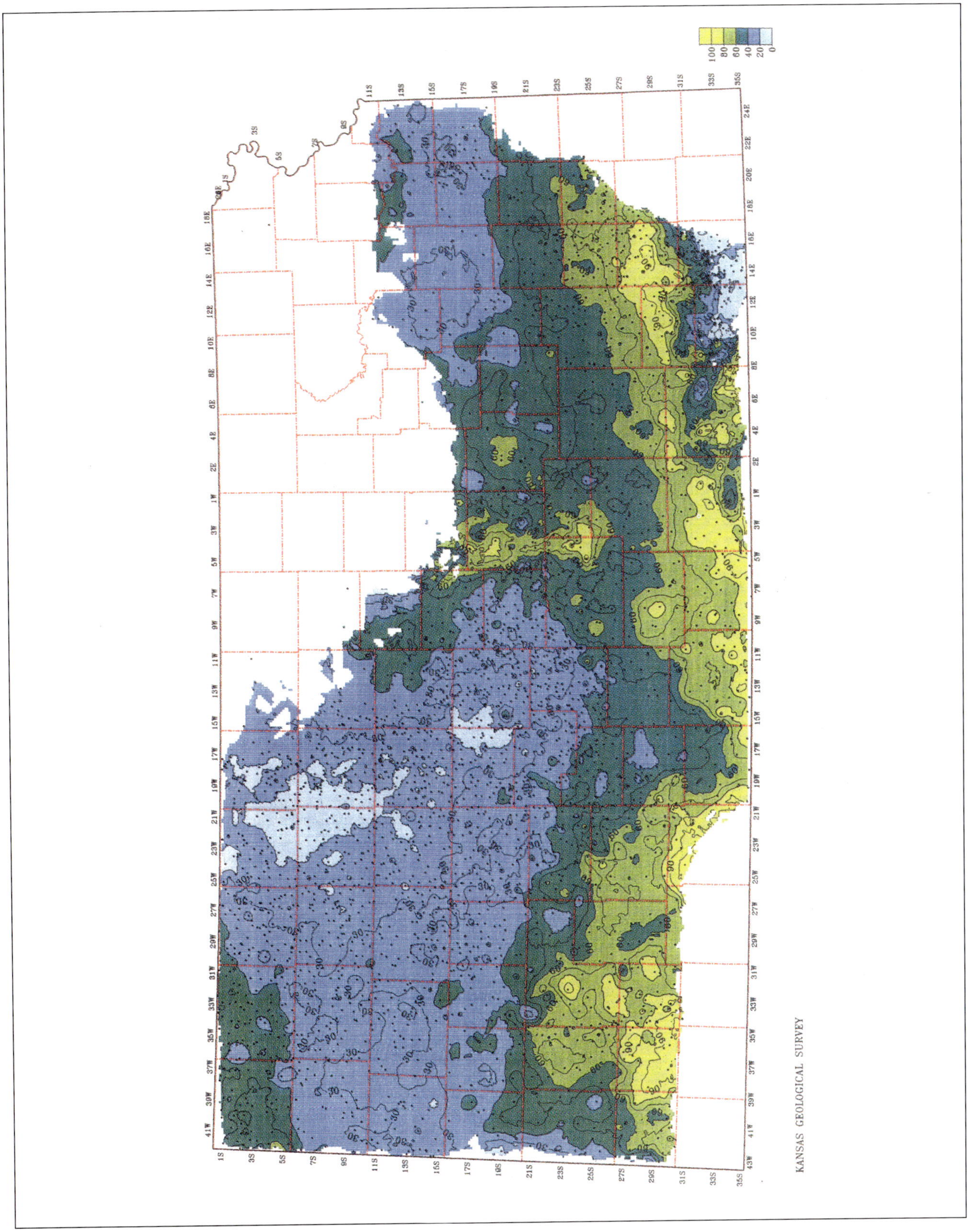

FIGURE 20 — Isopach map of Stark GSU. Contour interval is 10 ft. Shaded interval is 20 ft. County outlines are shown. Every other township and range is labeled along borders. Distance between them is 12 mi (19 km). Gridded and contoured using kriging option with SURFACE III.

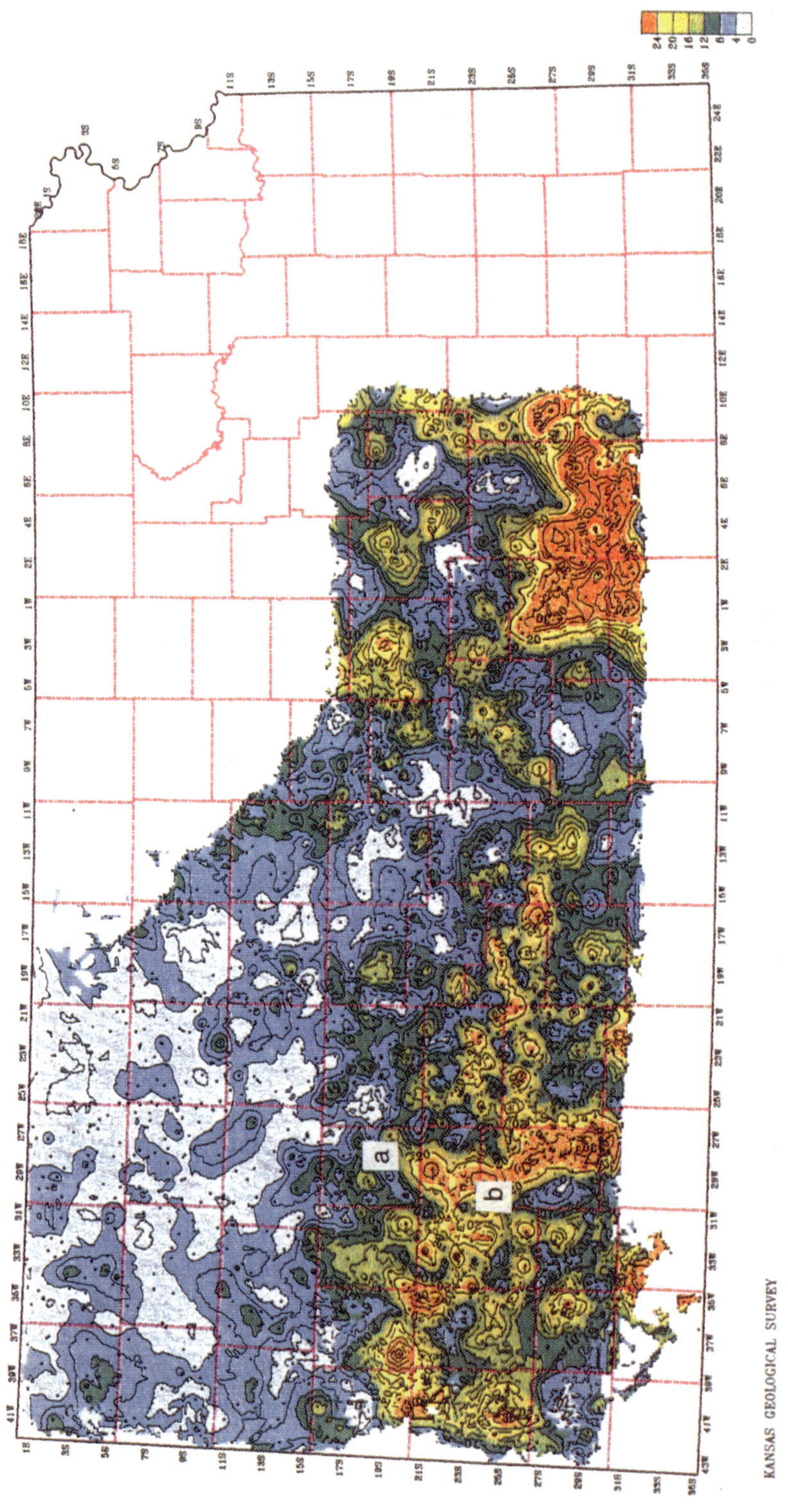

FIGURE 21 — Isochore of thickness of porous carbonate in GSU (Winterset Limestone) with porosity in excess of 8%. Contour interval is 2 ft. Shaded interval is 4 ft. County outlines are shown. Every other township and range is labeled along borders. Distance between them is 12 mi (19 km). Gridded and contoured using kriging option with SURFACE III. (a) updip lobate porosity; and (b) secondary line of lobate porosity discussed in text.

oped over a large part of the shelf. Thick lobate accumulations of oolitic grainstone in the Wea GSU are limited to the lower shelf such as in southeastern Kansas (40-ft [12-m] oolite bank described by Feldman and Franseen, 1991; Franseen et al., in press). Sediment accommodation space was apparently sufficient to permit accumulation of the oolite. The elevation and configuration (slope) of the shelf here were apparently suited to permit extended shallow, energetic sea-level conditions.

The Quivera GSU exhibits less than 20% of the range in thickness variation of the underlying sequences, attesting to a significant overall loss of sediment accommodation space across Kansas by this time (Fig. 23). A condensed section is developed over large areas of the shelf unlike the Wea unit, but the overall thickness of the cycle is reduced. This suggests that maximum flooding was deep enough to take most of the seafloor below the photic zone, but time available during regression in shallow marine conditions may have been reduced by more even topography during the Quivera GSU. Thick grainstones are also restricted to the lowermost shelf in the Quivera unit, resembling the geometry of the preceding Wea unit.

The last phase of accumulation of the Quivera GSU is a clastic deposit that is thickest on the lower shelf in southeastern Kansas. The low shelf setting for this deposit suggests a lowstand setting analogous to the Hushpuckney GSU. Also associated with these clastics is the formation of incised valleys higher on the shelf. Significant paleosols are also developed on the regressive carbonate unit across higher reaches of the shelf where clastic influx is minimal (e.g., multigeneration calcrete from Quivera GSU on Central Kansas uplift shown in Fig. 24). Moore et al. (1944) described sandstone-filled channels originating in the Chanute that cut down to as low as the Quivera Shale near the base of the Quivera GSU. Sandstones range in thickness from 40 ft (12 m) to as much as 100 ft (30 m). These sandstones locally contain limestone conglomerates at their base. The area of thick clastics continues southward into the Arkoma basin and is referred to as the Layton and Cottage Grove sandstones. These deposits have been interpreted as large fluvial deltaic deposits shed from the Ouachita (Wanless and Wright, 1978). Apparently, these clastic lobes onlapped the northern carbonate shelf during lowstand conditions. This lowstand setting for these clastics is also suggested in the regional cross section (Fig. 13)

The succeeding Muncie Creek genetic set begins with a significant landward (northward) backstepping (43 mi [70 km]) of the carbonate shelf margin at the horizon of the Muncie Creek Shale, another condensed section representing maximum flooding (Fig. 13). While the Muncie Creek Shale is also a condensed section of a primary GSU, it also heralds the onset of another significant longer term stratigraphic reorganization of the shelf, much like that of the Nuyaka Creek Shale. The carbonate margin previously residing in southern Kansas near the Oklahoma line abruptly retreated northward to central Kansas. This new carbonate margin steps progressively southward during successive primary GSUs in a manner similar to those within the underlying genetic sets. Siliciclastics were episodically deposited and partially filled the drowned portion of the shelf onto which succeeding carbonate cycles prograded.

There were two episodes of coarse-grained siliciclastic sedimentation on the lower shelf in the Muncie Creek genetic set including the Perry and Stalnaker sandstones. The latter and most prominent sandstone-rich interval was deposited at the close of this genetic set, below the Little Pawnee Shale which is the condensed section associated with the Haskell Limestone (Fig. 13). Large expanses of these two sandstone-rich clastic intervals accumulated on the lower shelf and basin at times of lowstand. These deposits are believed to be equivalent to the Lane and Stranger formations, respectively (Fig. 13). Both of these clastic-rich intervals include large clastic wedges and incised valley development on the shelf with the incised valleys in the Stranger Formation of the Douglas Group (Archer et al., 1994b) the most prominent. Western Kansas has no significant clastic units in the Douglas Group, much like the Pleasanton sequence. The strata again are attenuated in this area; it is con-

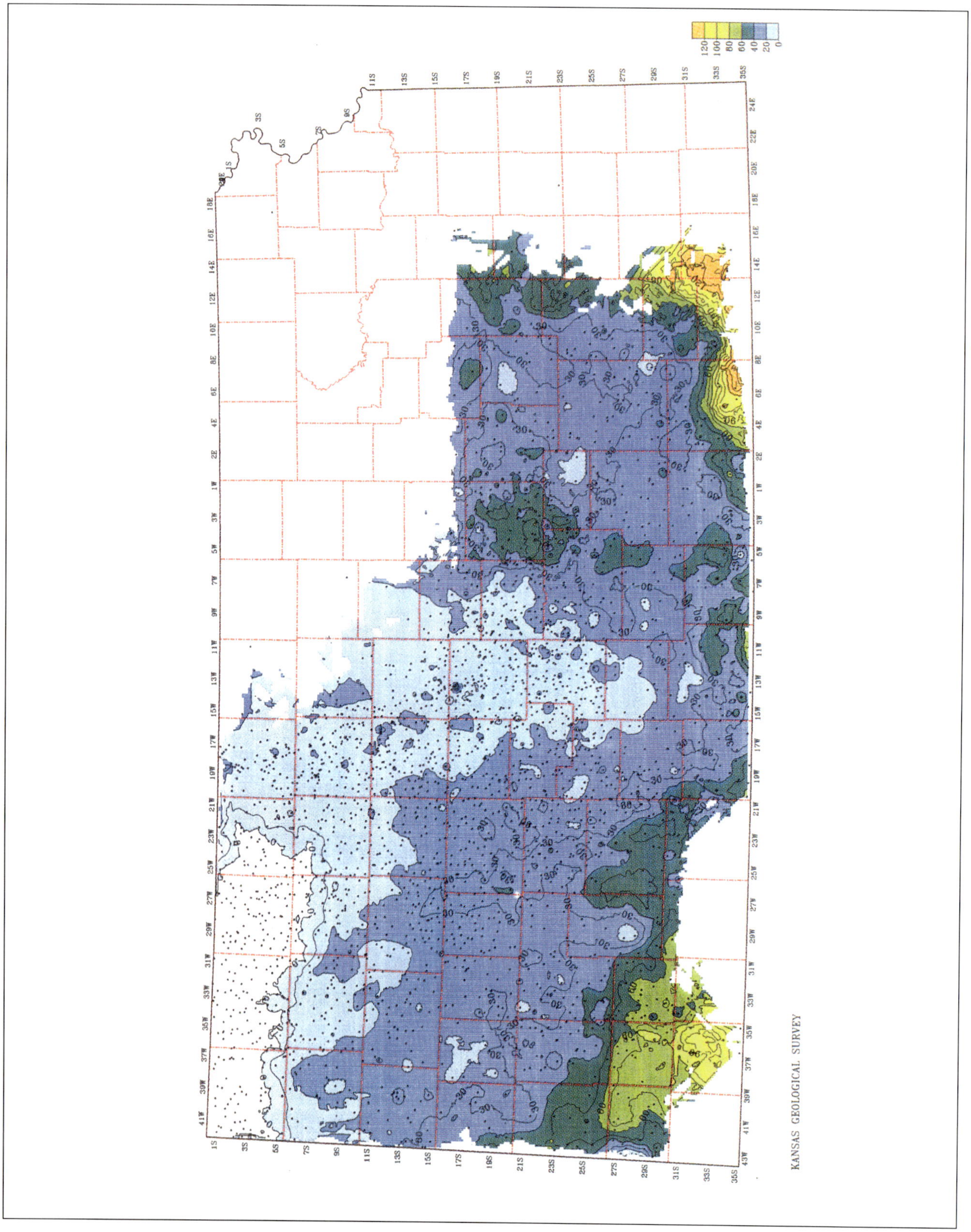

FIGURE 22 — Isopach map of Wea GSU. Contour interval is 10 ft. Shaded interval is 20 ft. County outlines are shown. Every other township and range is labeled along borders. Distance between them is 12 mi (19 km). Gridded and contoured using kriging option with SURFACE III.

cluded that western Kansas was again a positive area on the shelf relative to eastern Kansas, and the Central Kansas uplift acted as a barrier to clastics originating from the east. The Haskell Limestone is a regional marine marker bed that transcends the shelf and basin (including western Kansas) and represents the initial phase of deposition of another genetic set.

Recent studies by Archer et al. (1994b) and Feldman et al. (1995) described incised valleys up to 120 ft (41 m) deep and in excess of 7 mi (11 km) wide that have been cut into pre-Haskell Limestone strata (primarily the Stranger Formation) in the Douglas Group in northeastern Kansas. The streams that cut these valleys ran generally southward through eastern Kansas, paralleling presumed regional dip and passing onto the shelf margin in southeastern Kansas. Siliciclastic sediment apparently bypassed the shelf through these valleys, ultimately being deposited during a lowered sea-level stand that was established late in the deposition of the genetic set.

The lowstand sandstone and associated strata in the Stalnaker deposited along the lower shelf are both deltaic (Walton and Griffith, 1985) and estuarine (Archer et al., 1994b). This environment is consistent with lowstand conditions. The upward-coarsening succession of the Stalnaker and deposition over large areas restricted to the basin are consistent with a progressive lowering of relative sea level resulting ultimately in bypassing and incision of the upper shelf. The general pattern associated with the Muncie Creek genetic set is similar to the underlying sequence set: (1) rapid inundation and realignment of the shelf deposits and drowning of the lower shelf, (2) carbonate bank margin accretion and gradual loss of sediment accommodation space on the shelf, and (3) notable loss of accommodation space with restricted areas of clastic wedges and filling of lower shelf by clastics, accompanied incised valley formation shelfward.

Evidence of Relief on Shelf During Deposition of the Primary GSU

Individual GSUs exhibit notable thinning over the northern shelf and structurally positive areas (such as the Central Kansas uplift), with corresponding changes in facies and diagenetic effects. The interpretation of relief and topography on the shelf is very critical to evaluate mechanisms for sedimentation. Some of the evidence that such areas were topographically higher during deposition is described below.

(1) Loss of black shales over the crest of the Central Kansas uplift in all cycles. Where these condensed sections are lacking, very thin phosphatic and skeletal lag deposits occur locally at some of the flooding surfaces; in other cases (e.g., Wea GSU), the boundary between the transgressive and regressive carbonates exhibits no evidence of an interruption of sedimentation. The Stark and Quivera GSUs and the underlying Hushpuckney GSU all contain widespread black shales, each of which is locally missing over the Central Kansas uplift, and all of which grade northward to gray shale.

(2) Relatively shallow-water carbonate facies are more common and often occupy a greater proportion of the cycles over the Central Kansas uplift and upper shelf.

(3) Intensified subaerial exposure across the upper shelf is pervasive. Paleosols that developed along a single bounding surface exhibit features interpreted as increased weathering and duration of emergence northward (landward) across Kansas into southern Nebraska. Evidence includes depth of penetration of solution features (pipes), superimposed soil processes and translocation and alteration of clay minerals, and development of significant solution features including thick autoclastic breccias and mixed pebble carbonate conglomerate (Watney, 1980).

(4) The basal portions of regressive carbonates in the Hushpuckney and Stark GSUs on the lower shelf in southwestern Kansas contain dark, silty, pyritic, organic-rich lithofacies that contain a

dysoxic fauna. Transition from deposition of condensed section to carbonate may have been more episodic with intermittent return to reducing sediment-starved conditions on the seafloor. This appears to be related to position of the lower shelf and was most likely due to depth of the seafloor, upwelling, or simply organic productivity versus preservation of organic matter due to anoxia.

(5) The reason for downlapping of the carbonates into the basin is inferred lower elevation of the seafloor on the lower shelf. Reduced time in the photic zone led to less carbonate accumulation until water was too deep for the photic zone to penetrate and carbonate was not deposited. Greater subsidence may have helped to maintain the lower elevation and relief. Also, short-term accelerated subsidence may have helped to produce the relative rise in sea level to initiate the drowning, but this cannot be determined. Only longer term subsidence can be calculated from backstripping due to limited time resolution. However, subsidence along the lower shelf is deemed to have been sufficiently high to play an important role in maintaining the depositional topography along the shelf margin: over the duration of five primary GSUs, 350-k.y. duration each, total change in elevation attributed to subsidence during accumulation of a genetic set is 0.11 ft/k.y. (0.035 m/k.y.) (subsidence for shelf margin from Fig. 8) = 200 ft (61 m).

Third-Order Cycle Control on Carbonate Reservoir Development

The carbonate petroleum reservoirs that occur within primary (fourth-order) GSUs tend to exhibit predictable patterns of thickness, facies composition, and stratal complexity that correspond to their relative position in the third-order genetic sets. This pattern depends on shelf position and general accommodation space available. Primary GSUs that occur low in the third-order genetic sets generally were deposited in association with greater depositional topography and greater variability in thickness than the overlying primary GSUs constituting a third-order genetic set, e.g., the higher variable thickness associated with the oolite tracts in the Hushpuckney and Stark GSUs (Figs. 19 and 21). The carbonate grainstones in these lower sequences also are generally thicker and exhibit more thickness variability than their counterparts in the sequences near the top of the third-order genetic sets. These grainstones are frequently more oolitic than higher in the third-order genetic set and, consequently, subject to more severe diagenetic modification by percolating undersaturated meteoric waters that invaded the unit soon after deposition during late regression. As a result, these grainstones have erratic permeability due to their complex pore fabrics (Fig. 25, diagenetic sequence developed by R. Carlson), which poses difficulties in interpreting reservoir parameters from wireline logs (Guy and Doveton, 1993; Watney et al., 1993). Minor cycles commonly punctuate the thicker grainstones, adding additional heterogeneity in the form of thin, muddy, relatively impermeable wackestone layers or thin impermeable paleosols that may have their own early diagenetic overprinting separate from such features associated with the "master" paleosols that typically cap each primary depositional sequence (French and Watney, 1993).

Stratal successions from other third-order genetic sets from other locations on the shelf are nearly identical to those observed in the Muncie Creek genetic set (in thickness of primary GSUs and lithofacies successions). The underlying Excello genetic set that is developed basinward (southward) often resembles the Muncie Creek genetic set that is located more landward (northward). The similarities of these third-order genetic sets suggest comparable accommodation histories at different shelf locations and different times, i.e., similar magnitude primary cycles but overall deeper water during the Munice Creek genetic set.

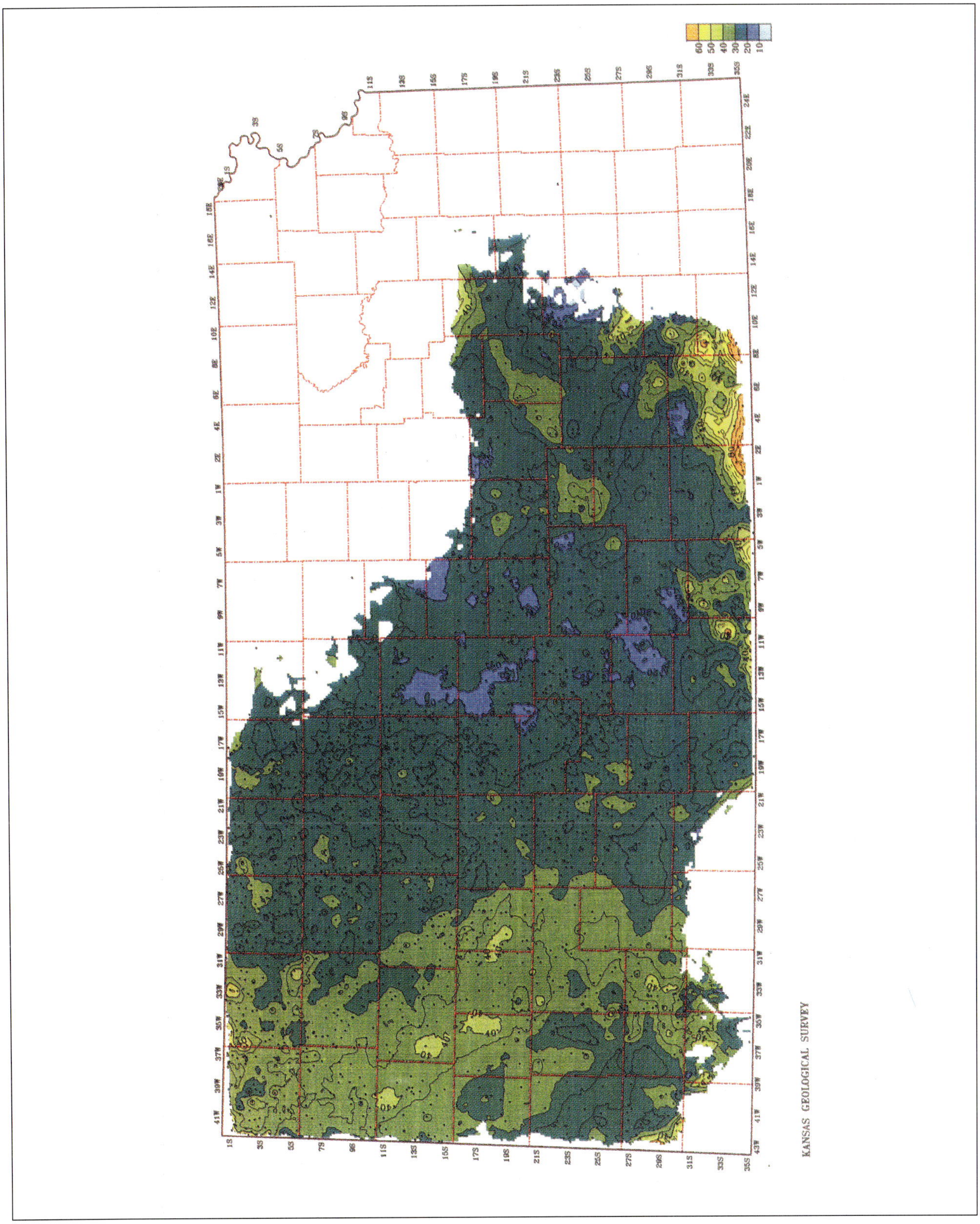

FIGURE 23 — Isopach map of Quivera GSU. Contour interval is 10 ft. Shaded inteval is 10 ft. County outlines are shown. Every other township and range is labeled along borders. Distance between them is 12 mi (19 km). Gridded and contoured using kriging option with SURFACE III. Shelf configuration is interpreted to be much more uniform than previous sequence in Kansas City third-order cycle.

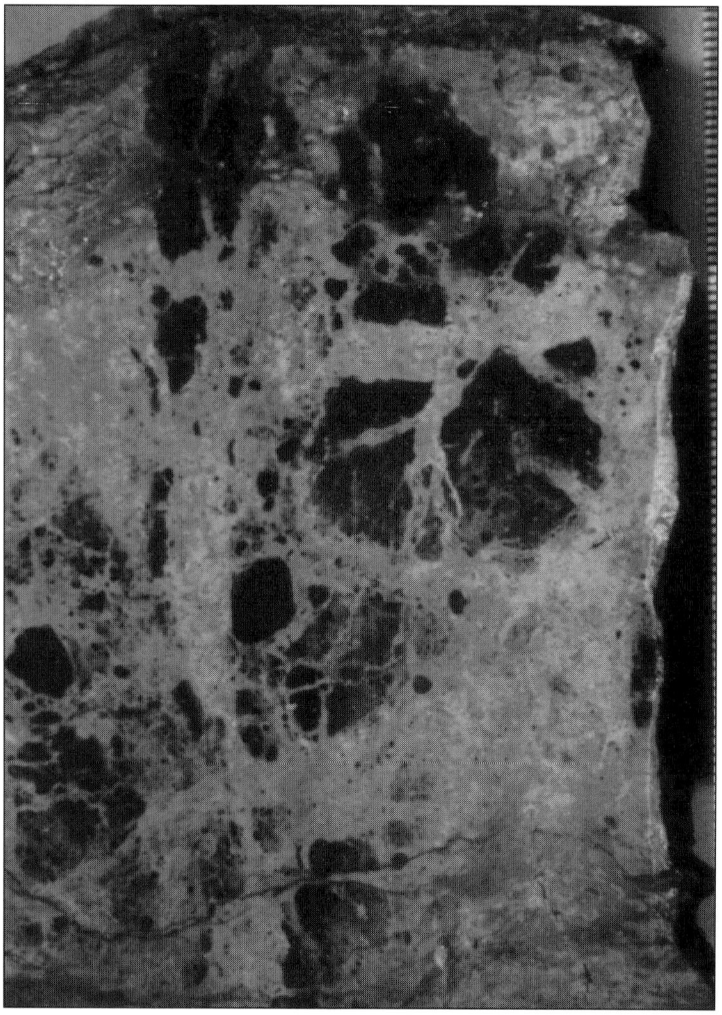

FIGURE 24 — Paleosol developed at the top of the Quivera GSU in core from the Pan American No. B-6 Ohlson well in NE SE NE, Sec. 28 T14S-R14W, Russell County in central Kansas. Slab photo shows multigenerational caliche clasts (dark) in matrix of microcrystalline calcite and siltstone. Clasts consist of mulitiple generations of circumgranular cracking. Scattered rhizoliths are also present. Paleosol developed at end of third-order Nuyaka Creek genetic set. Scale is millimeters. Depth of slab is 3097.5 ft.

Relative Sea-Level Curve for the Pennsylvanian

Ross and Ross (1987) constructed a relative sea-level (coastal onlap) curve for the Pennsylvanian based mainly on stages of fusulinid evolution. In the current study, a similar diagrammatic relative sea-level curve was developed for the Mid-Continent Pennsylvanian based on evidence provided by the distribution of third- and fourth-order cycles (Fig. 26). This curve reflects the third- and fourth-order cycles extending from the Atokan through the early Virgilian. The boundaries of the higher third-order genetic units in the mid-Desmoinesian through lower Virgilian are identified by the flooding units (Excello, Nuyaka Creek, Muncie Creek, and Little Pawnee shales).

A second-order rise in relative sea level, based on shelf margin stratal geometries and interpretations of modified Fischer (1964) plots (Figs. 27 and 28) is also incorporated into the relative sea-level diagram (Fig. 26). Modified Fischer plots were generated for four locations on the shelf with contrasting accommodation and subsidence histories. The interval of interest extends from the Middle Pennsylvanian Excello Shale to the earliest Virgilian Little Pawnee Shale (Haskell Limestone). Each primary (fourth-order) cycle was plotted using a constant cycle period of 350 k.y. Averaged backstripped subsidence values for that particular stratigraphic interval at that specific shelf location were used to construct the slope of the hypotenuse of the triangle. The thickness of each fourth-order depositional sequence is the vertical component of the triangles.

The second-order, sea-level rise coincides with the initiation of the Absaroka cratonic sequence (or supersequence) of Sloss at the beginning of the Pennsylvanian. The Fischer plots indicate a long-term increasing accommodation space across the shelf between the Middle Pennsylvanian Marmaton Group to the beginning of the Virgilian.

The trends of sediment accommodation on the Fischer plots change near the boundaries of the third-order genetic sets. However, the rates of accommodation change, as much as can be inferred from thickness among the fourth-order GSUs, are not consistent across the shelf (Fig. 27). This is, in part, probably related to excess sediment accommodation during which time minimal sediment was deposit-

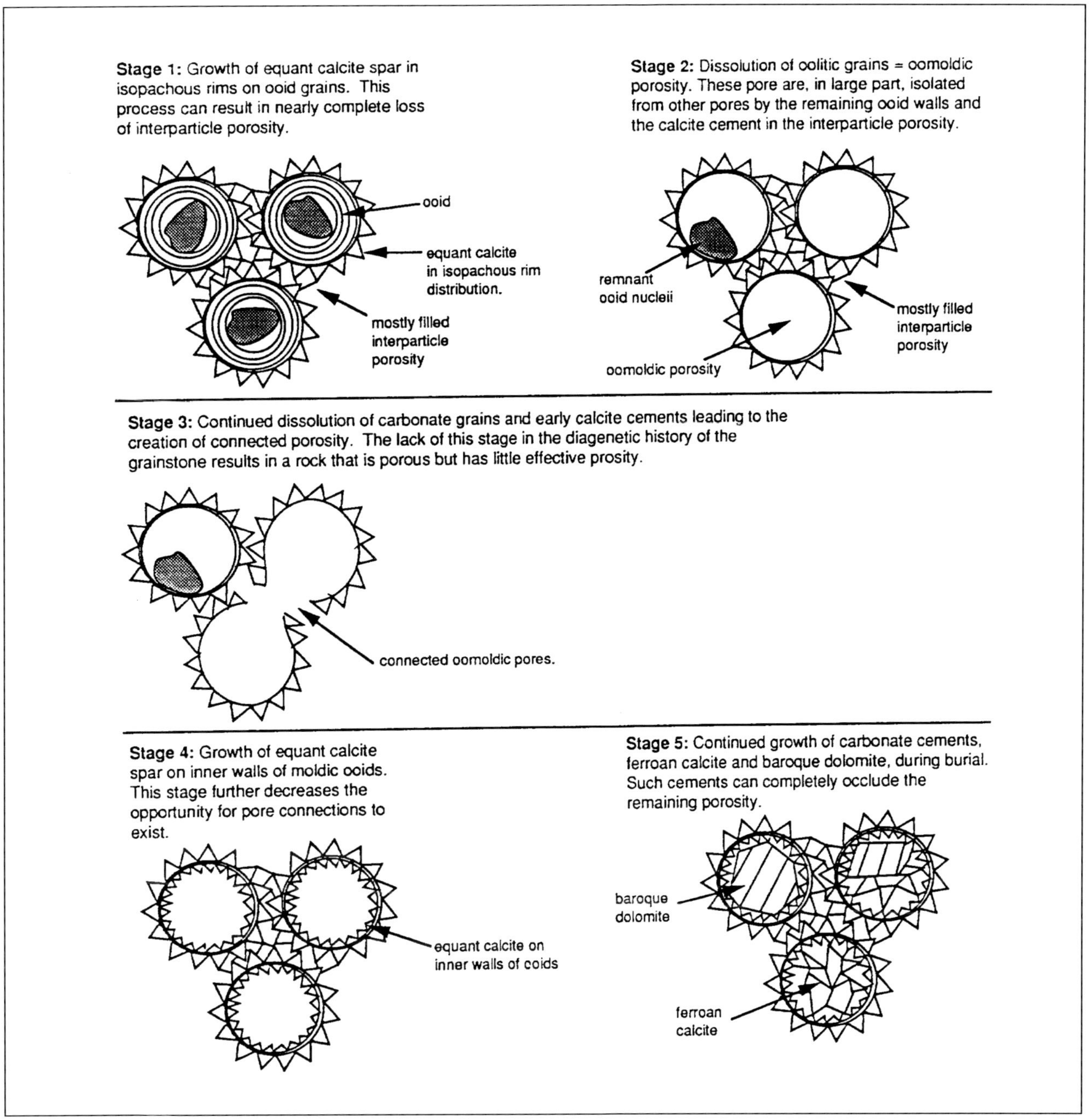

FIGURE 25 — Paragenesis of oolitic grainstones in the Lansing-Kansas City groups were divided into five major stages (R. Carlson *in* Watney et al., 1993). Meteoric waters, introduced during early subaerial exposure, were active during stages 1 to 3. Stages 4 and 5 occurred during burial. Permeable reservoir rock consists of connected oomoldic pores created by meteoric dissolution. Later crushing during early burial is thought to have occurred to enhance the connection of the oomolds in the weakened moldic fabric.

ed, e.g., during development of condensed section. Alternatively, the amount of sediment that accumulated in shallow water before sea level dropped to expose the shelf to extended nondeposition or erosion would also affect overall thickness. Furthermore, deviations from the estimated subsidence rates within individual genetic sets may limit the comparison of different shelf locations. Excess accommodation is inferred from the interpretation that the maximum water depth that occurred during deposition of

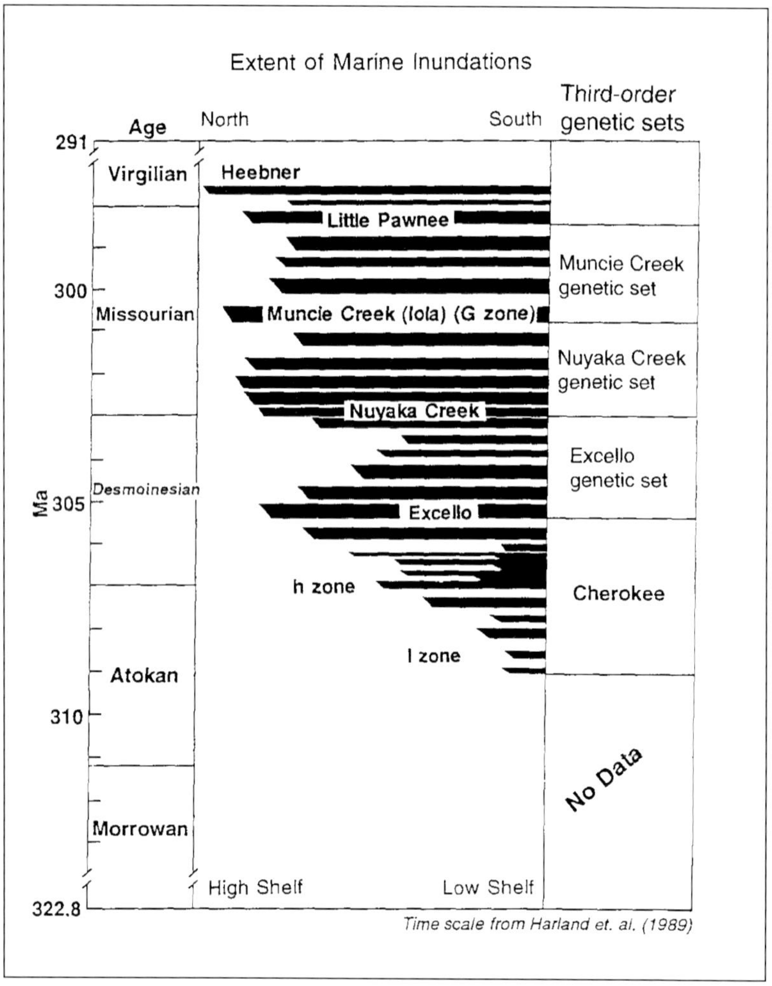

FIGURE 26 — Relative sea-level history of the Middle and Upper Pennsylvanian of the upper Mid-Continent. Letter scheme for mid to lower Cherokee based on Youle et al. (1994) who characterized series of fourth-order sequences along the margin of Hugoton embayment in southwest Kansas.

a given primary genetic depositional unit greatly exceeded the thickness of the genetic depositional unit (e.g., Heckel, 1977) with perhaps 330- to 660-ft (100- to 200-m) depths for black shales. The black shales common to these cycles suggest water depths exceeded depths of typical photic zones of today. Furthermore, other studies suggest similar water depths based on modeling extent of ice cover on Gondwana and resultant glacio-eustasy during the Late Paleozoic (Crowley and Baum, 1991). The greater water depths prevalent at lower shelf locations probably also lead to reduced carbonate sedimentation rates, because of less sea-bottom time within the optimal photic zone for carbonate production, resulting in relatively thin primary GSUs. In contrast, the same units deposited in higher shelf settings may be much thicker, because of longer sea-bottom times within the optimal zone and greater rates of carbonate production. Excess accommodation limits the extent to which relative sea-level history can be inferred from individual Fischer plots. However, the Fischer plots are believed to be able to provide useful information about longer term (e.g., second-order) accommodation history. Regional isopach maps illustrate the hazards of using one-dimensional thickness information to infer accommodation history, because the thickness within individual GSUs can vary so much laterally.

Third-order, seismic-scale cycles have been recognized in late Desmoinesian and Missourian strata in the Horseshoe atoll of the Midland basin in west Texas using seismic sequence analysis calibrated with examination of logs, samples, and fusulinids (Waite, 1993). The Desmoinesian-Missourian boundary (near the Nuyaka Creek Shale) corresponds to a sequence boundary with the drowning of a large area of the "reef." This is apparently equivalent to the prominent backstepping above the upper Desmoinesian Marmaton Group along the northern carbonate shelf margin in Kansas and Oklahoma associated with the base of the Muncie Creek genetic set. This same drowning event is also described by Hentz (1994) in the Texas Panhandle at the top of the Marmaton. It is not clear where the other boundaries are located at the top of the Kansas City and Lansing groups.

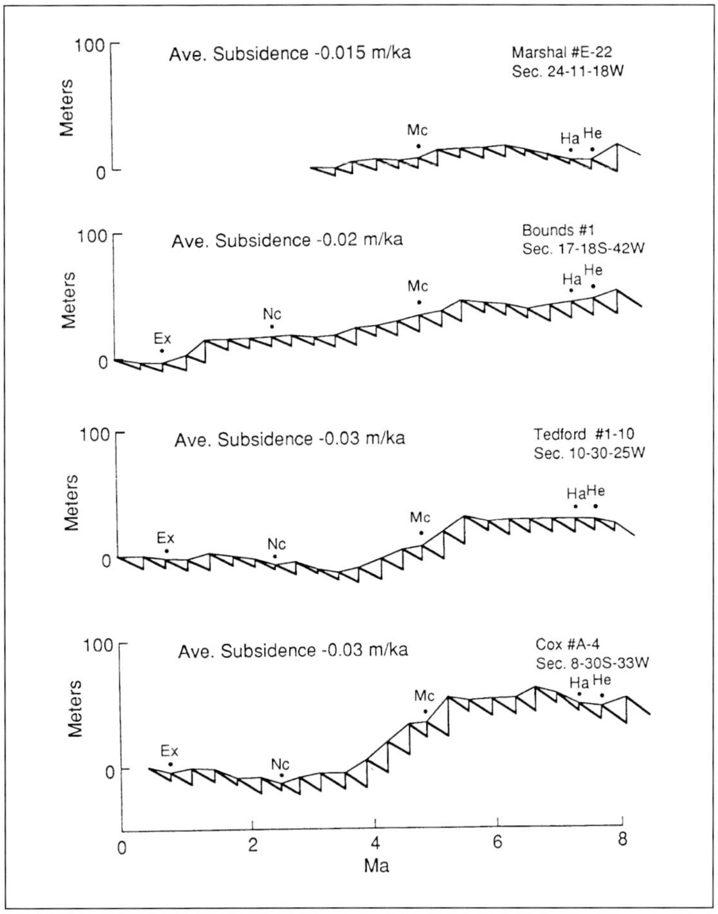

FIGURE 27 — Modified Fischer plots for selected wells located in western Kansas. Letters represent shales marking base of each third-order cycle. Plots provide estimates of sediment accommodation versus time. Upper surface of connected triangle is a measure of relative elevation of the depositional surface. Triangles are plotted for each fourth-order cycle, evenly spaced along time axis (350 k.y. each in duration). Vertical axis of each triangle is measured thickness from well log. Diagonal line of each triangle is calculated subsidence for that particular location on shelf. Well location in Fig. 28. Plot technique originally developed by Fischer (1964).

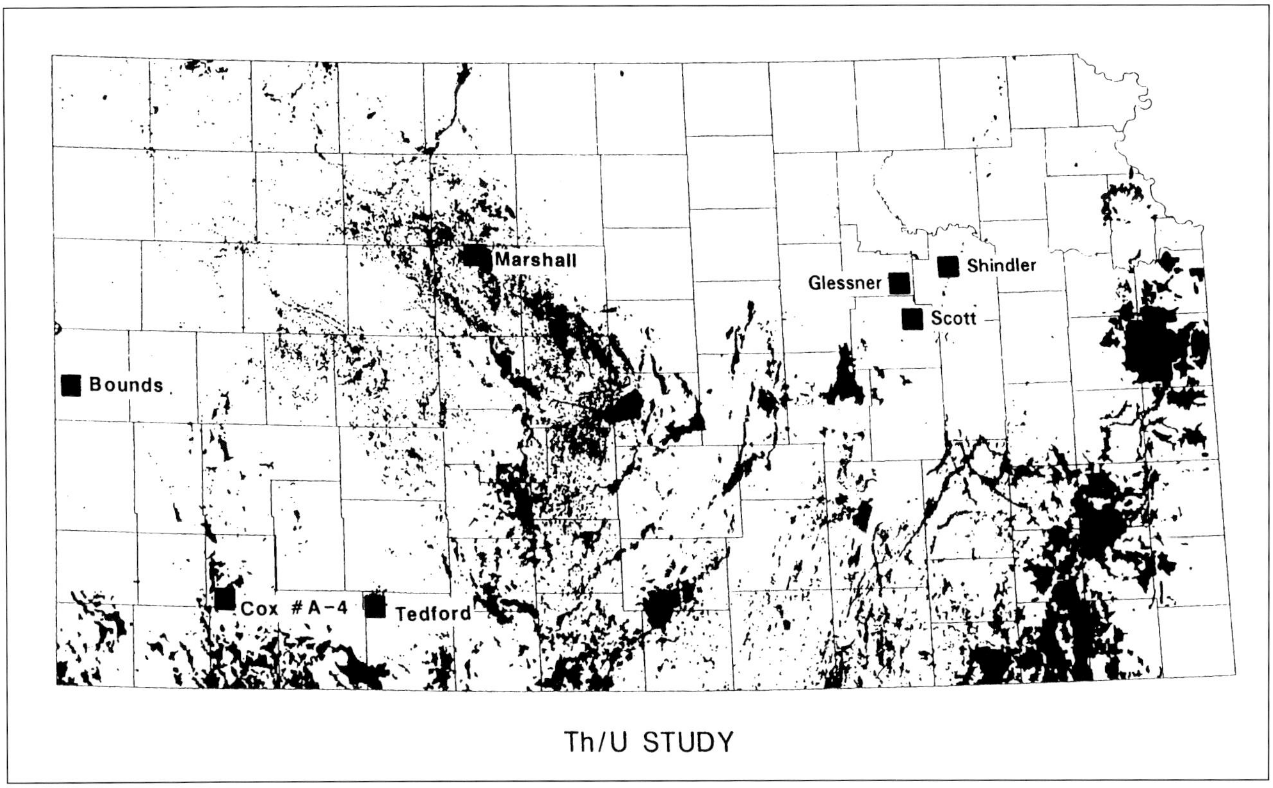

FIGURE 28 — Base map showing location of wells used for Fischer plots (Fig. 27) and study of Th:U ratio.

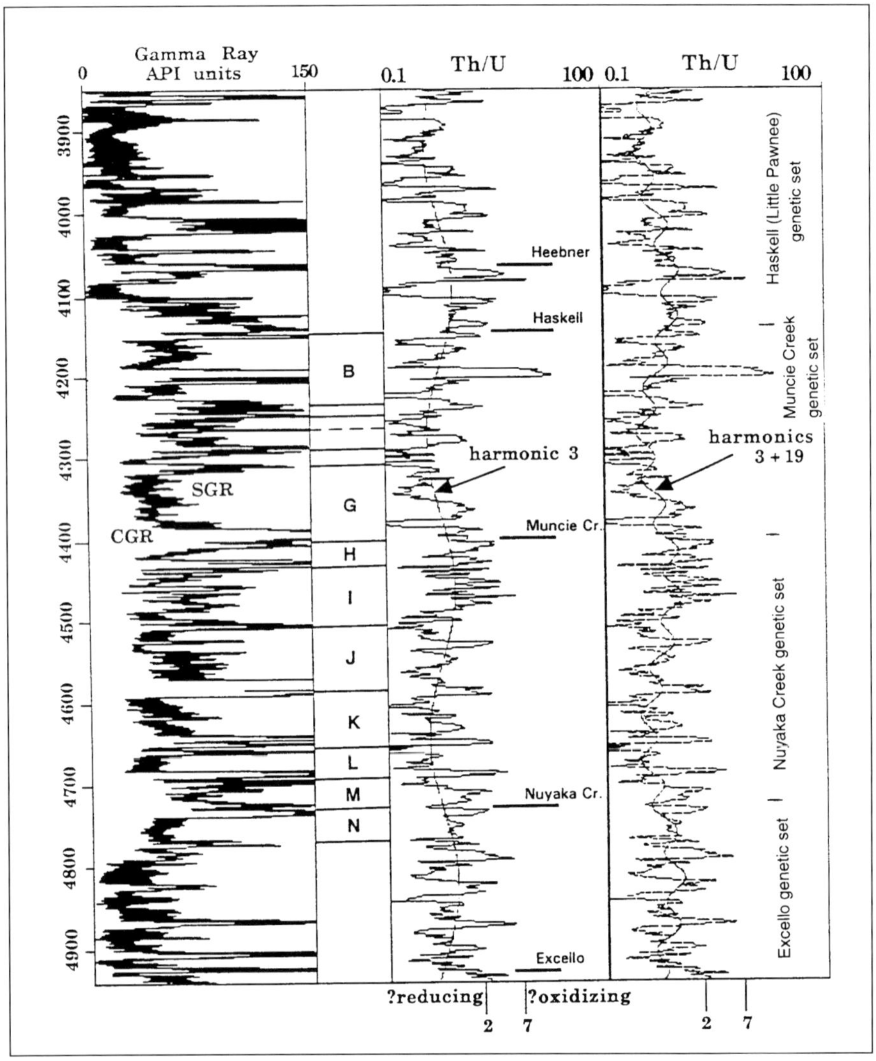

FIGURE 29 — Gamma ray, Th:U ratio, and two Fourier harmonics (third- and fourth-order) of Th:U ratio for Amoco Cox No. A-4 core include three third-order genetic sets: Excello, Nuyaka Creek, Muncie Creek, and Haskell (Little Pawnee). Top of Muncie Creek genetic set initiates at Haskell (Little Pawnee) and continues through the Heebner Shale. Both correspond with drowning and backstepping events.

PROFILE OF TH:U RATIO: INDICATOR OF RELATIVE SEA LEVEL FROM THE SHELF INTERIOR

The determination of relative sea-level change on the shelf margin is accomplished through the recognition and interpretation of sequences, sequence sets, their stratal geometries, and their analogous GSUs. However, in the shelf interior where local relief is much more limited, relative sea-level history (rates, duration, and magnitude of events) is much more difficult to reconstruct. Variations in accumulation rates, time available for sedimentation, and sediment accommodation potential across the shelf complicate the use of thickness and deposited lithofacies to infer relative sea-level history. Local variations in stratigraphy can create considerably different interpretations of the sea level — buildups of shoal-water facies such as oolites and algal banks, prograding clastic wedges, and differential structural movement (subsidence or uplift). Also, a considerable portion of time is probably not represented by any sediment, with resulting hiatus varying across the shelf according to elevation and local sediment supply. The net result is a high potential for complex stacking of stratal units with specific sites providing only limited information with which to interpret relative sea-level history and temporal changes. Independent methods are needed in these shelfal areas to assist in delineating cycles in general, and to help establish relative sea-level history.

The use of depth profile plots of the thorium:uranium Th:U ratio has been examined to help characterize the relative changes among sequence components, to establish a hierarchy of cycles, and to infer relative sea-level history (Watney et al., 1990, 1992; Doveton, 1991). The depth profile of the Th:U ratio was found to correlate with the stacking geometry observed along the shelf margin (Fig. 29) with the suggestion that redox profile and extent of marine inundation are correlatable.

COMPUTER MODELING

A two-dimensional (2-D) stratigraphic computer simulation (Fig. 30) of the shelf-margin cross section (Fig. 13) was used to experiment with the parameters developed in this study and similar analogues (Enos, 1991). The model is KANMOD programmed by J. French (French and Watney, 1990; Watney et al., in review). Parameters include eustasy, subsidence, and rates of sediment accumulation. The model simulates carbonates and condensed sections, but not clastic sedimentation. Sedimentation rates decrease with increasing water depth to a constant low value (0.03 ft/k.y. [0.01 m/k.y.]) below 115 ft (35 m), a rough estimate of the base of the euphotic zone in the Mid-Continent sea. Carbonate accumulation effectively stops below this depth. In modern oceans, the euphotic zone extends from under 80 ft (25 m) to over 200 ft (60 m) (Schlager, 1992). Carbonate sediments deposited within defined depth ranges (e.g., 49 to 66 ft [15 to 20 m]) are assigned separate colors, as are subaerial exposure surfaces. Condensed sections are similarly identified by color.

Rates of carbonate sedimentation vary with climate and water conditions. Perturbations in these and other factors are expressed through sea-level fluctuations and lead to reductions in the maximum expected rates. Some models factor in changing rates of sea level (such as a lag time) (Goldhammer et al., 1991) or reduced rates linked to rate of rise (SEDPAK, South Carolina Modeling Group). In KANMOD, rapid rises in sea level will simply overcome carbonate sedimentation and will lead to drowning of the shelf, once the depth of the euphotic zone is exceeded.

In KANMOD, the shelf is drowned each time a primary (fourth-order) depositional sequence is generated where the eustatic amplitude is set at 360 ft (110 m) based on arguments in Heckel (1994).

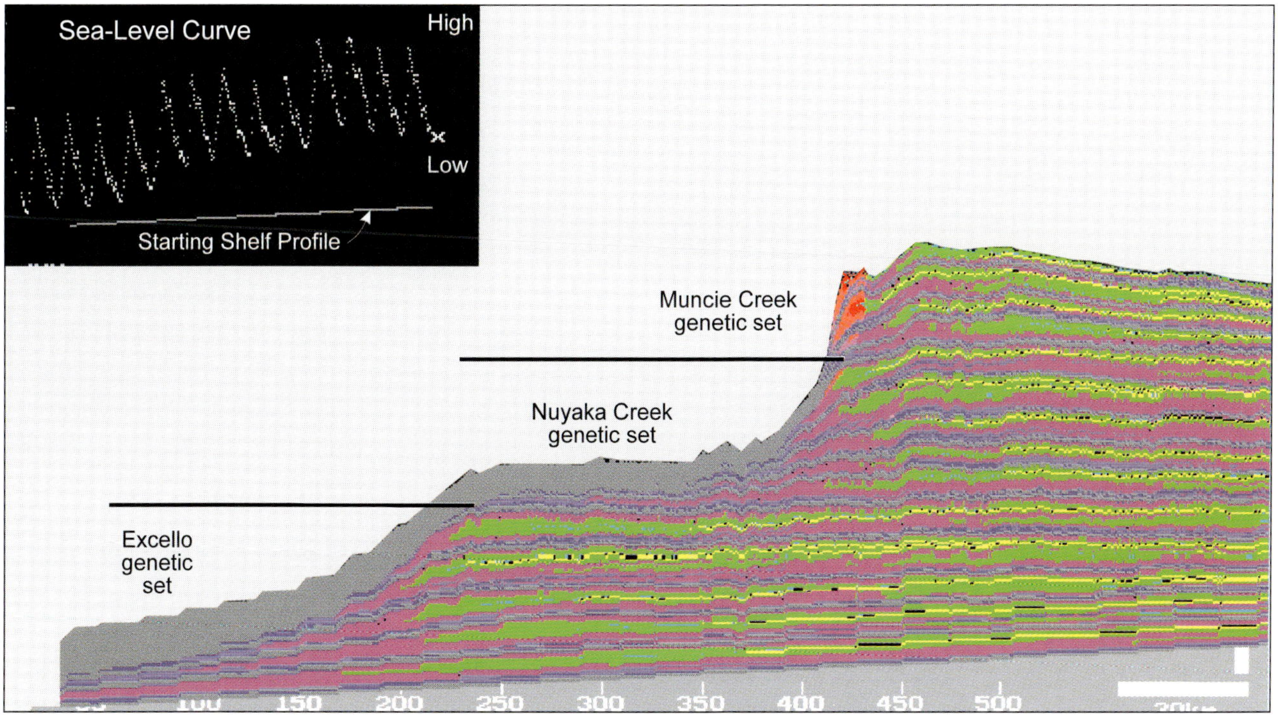

FIGURE 30 — The 2-D simulation of Middle to Upper Pennsylvanian shelf-to-basin carbonate system in northern Mid-Continent, bordering the Arkoma basin. Length of the profile is approximately 400 mi (650 km) and maximum height equals 980 ft (300 m). Run length is 3.56 m.y. Fourth-order GSUs (350-k.y. duration) are highlighted by gray-colored condensed sections. Paleosols are yellow, with increasingly deeper water facies expressed as purple to green colors. Third-order genetic sets (Excello, Nuyaka Creek, and Muncie Creek) are characterized by backstepping (approximating the Marmaton, Kansas City, and Lansing groups as labeled in this figure). This model uses an initial uniform slope, linear differential subsidence, with third- and fourth-order cyclicity generated from a relative sea-level curve in upper left and Fig. 31.

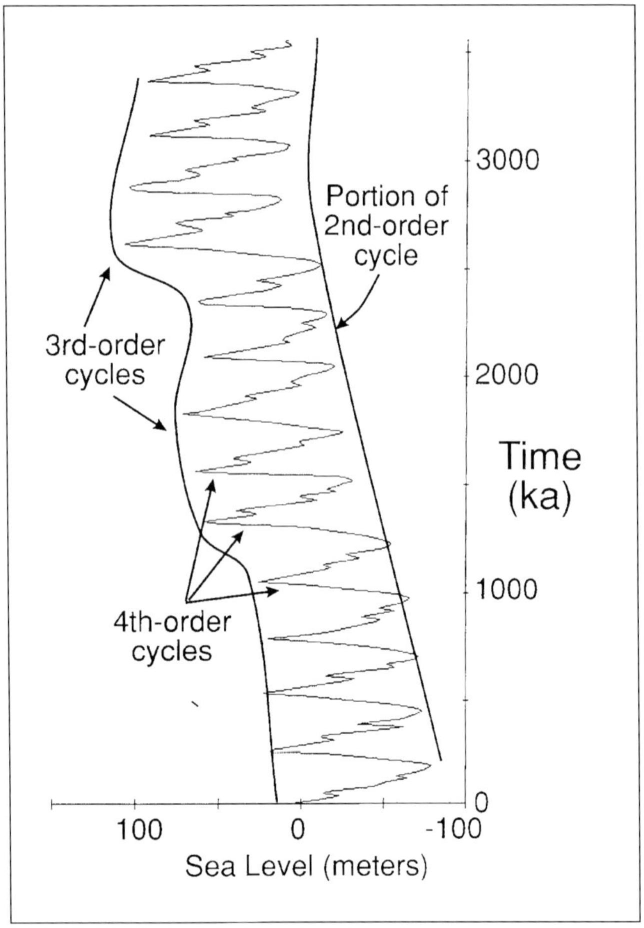

FIGURE 31 — Relative sea-level curve used in the 2-D stratigraphic simulation model in Fig. 30.

Rates of rise exceed 33 ft/k.y. (10 m/k.y.) during these rapid glacial eustatic rises (Heckel, 1994). This is followed by slower, staggered falls in sea level to complete a cycle over a 350-k.y. period (Fig. 31). These rates and amplitudes closely resemble Pleistocene eustasy (Denton and Hughes, 1983). Periods are longer than the Pleistocene, reflecting the probable dominance of the eccentricity orbital parameter during Pennsylvanian glaciation (Crowley et al., 1987). The length of this particular simulation is just over 3.5 m.y. and includes 14 fourth-order cycles.

The magnitude of the fourth-order, sea-level changes is at least 200 ft (about 70 m), based on stacking patterns and lithofacies along the shelf margin (Fig. 13) and water depths needed to exceed the photic zone and result in prolonged periods of shelf anoxia. Other workers have made similar estimates of eustatic change (100 to 200 m, Heckel, 1977, 1986, 1994 and Ross and Ross, 1987; 70 m, Adlis et al., 1988; 60 m ±15 m, up to ±175 m, Crowley and Baum, 1991). Crowley and Baum (1991) applied area-volume relationships developed for Quaternary ice sheets to estimate size of glacio-eustasy in Gondwana during the Carboniferous.

Eustatic rises in sea level during the Pleistocene were often high enough to drown viable carbonate reefs with rates now estimated to range up to 148 ft/k.y. (45 m/k.y.) (Blanchon and Shaw, 1995). Recent studies of the ice dynamics and ocean circulation by Bond et al. (1993) suggested that sharp spikes in sea level (Heinrich events) are much more common than once thought. These findings have strong implications for rates of climate change in analogous periods of geologic time, such as the Late Paleozoic. Shorter term, fifth-order cyclicity has been randomly incorporated into the relative sea-level curve. Actually each sequence should have its own characteristic set of minor cycles based on the Pleistocene model (Denton and Hughes, 1983) and the evidence for high-frequency, sea-level events in general during the Pleistocene.

Third-order cycles are modeled using an asymmetric (rapid rise and slower fall), 66-ft (20-m) amplitude, sea-level fluctuation (Fig. 31). This driver could have been the result of either changes in eustatic parameters or increased tectonic subsidence associated with development of the adjoining foreland basin, but is included in the relative sea-level curve since it affects the entire area that is modeled. The amplitude of these third-order fluctuations is based on the estimated relief of the drowned shelf. The range of sea-level change associated with third-order cyclicity is estimated at 66 ft (20 m) based on the lateral distance that the carbonate shelf backstepped coupled with the estimated slope of the shelf. Episodes of landward backstepping of the carbonate shelf margin were calculated to be 62 mi (100 km) and 31 mi (50 km) for the Kansas City and Lansing third-order sequences, respectively. Minimum estimates of the regional slope range from 0.53 to 1.10 ft/mi (0.1 to 0.2 m/km) based on regional thickness trends (see Watney, 1985b). Reduced rates of subsidence and/or eustasy coupled with sediment progradation were needed to re-establish sedimentation basinward on the drowned shelf.

A second-order cycle is also inferred from the record (Fischer plots) and stacking pattern along the shelf margin (Figs. 26 and 27). The magnitude of change in sea level of 230 ft (70 m) during the second-order rise is estimated from the modified Fischer plots. This value is in the range offered by Goldhammer et al. (1993). Other workers also recognize the rising sea level associated with this second-order cycle extending from Early to Late Pennsylvanian (Vail et al., 1977; Haq et al., 1987; Ross and Ross, 1987).

Subsidence increases linearly toward the basin (southward) and includes an optional flexural hinge. The sea-level curve used in the model represents a combination of a eustatic curve and subsidence that deviates from the linear constant rate (e.g., possible subsidence contributions to the third-order cyclicity). Long-term subsidence rates were acquired from backstripped subsidence modeling for various locations on the shelf (Fig. 8).

A number of runs using various combinations of parameters were used to examine the sensitivity of the model. Values were changed within constraints of the data to produce the results that most closely resemble the observed stratal geometries. The experiments suggest that on a simple low-relief ramp with linear, basinward-increasing subsidence, drowning of the lower shelf is generated at the initiation of third-order sequence sets. The first third-order rise generated the largest backstepping event, with about 40 mi (65 km) of landward displacement of facies tracts. This is similar to the 62-mi (100-km) backstep observed at the end of the Excello third-order genetic set (Fig. 13). The fourth-order cycles within this third-order genetic set reveal an upward trend of shallower facies in each cycle, with successively younger subaerial exposure surfaces stepping progressively farther basinward. This similarity is observed in cross sections, maps, and Th:U ratio profiles.

The minor fifth-order cycles resulted in the generation of subaerial surfaces that are shingled (en echelon style) toward the basin, forming toplap. The exposure surfaces dip basinward within individual fourth-order cycles until they reach conformity. At this point and basinward, the minor cycle can actually be referred to as a parasequence. The parasequences are recognized by repeated upward-shallowing events and may not have revealed enough information to suggest elements of forced regression. Thus, minor cycles and parasequences may be part of the same genetic fourth-order unit, the difference being mainly elevation on the shelf and whether or not subaerial exposure occurred.

The thickness and resolution of fourth-order depositional sequences are maximized in each third-order sequence set at shelf locations that experienced the most time under shallow-water conditions that were optimal for carbonate production. In contrast, the fidelity of the sedimentary record is reduced both basinward and landward as more time is spent either under conditions of sediment starvation (basinward) or subaerial exposure (landward). Thickness is not a close proxy for either water depth or time of sedimentation. The more accurate story comes from an integration of sequence stratigraphy and regional 3-D characterization of the depositional setting.

CONCLUSIONS

This study has attempted to demonstrate and classify GSUs and depositional sequences, and compare them to cyclothems. The study also attempted to recognize a cycle hierarchy in the Middle and Upper Pennsylvanian of the Mid-Continent. An integrated approach using cores, surface exposures, well logs, regional mapping, and cross sections was used to characterize second- and third-order cycles (genetic sets and sequence sets) containing higher frequency fourth-order (primary)

GSUs and depositional sequences and minor fifth-order cycles. Regional cross sections and maps of individual primary GSUs reveal details of the stacking geometries and relative sea-level history.

The 2-D stratigraphic computer simulation modeling was used to integrate parameters relating to generation of the observed sequence stratigraphic framework. The experiments in modeling provide an interactive perspective to address the dynamic conditions under which the GSUs and depositional sequences were possibly generated. An optimum realization was obtained by using parameters constrained by geologic study. The similarity to the observed geology indicates a possible convergence toward a quantitative understanding of the processes. It is hoped this work will aid in developing improved geologic process models for use in making better predictions for exploration and development of stratigraphic traps in the Mid-Continent.

ACKNOWLEDGMENTS

We acknowledge stimulating discussions and contributions of Evan Franseen, Howard Feldman, Eugene Rankey, Ross Black, and Tim Carr. Phil Heckel and Al Bennison are thanked for many years of fruitful discussions. J.C. Wong developed the database management system. Ajit Verma provided programming support for computer mapping and assisted in conducting the Fourier analyses. Appreciation is extended for reviews of the manuscript by Bob Scott, Phil Heckel, and Howard Feldman. This work was partially supported under DOE Grant No. DE-FG22-90BC14434.

SEQUENCES AND CYCLOTHEMS IN UPPER PORTION OF CHEROKEE GROUP (MIDDLE PENNSYLVANIAN, DESMOINESIAN), MID-CONTINENT, U.S.A.

Robert L. Brenner
Department of Geology
University of Iowa
Iowa City, IA 52242-1379

ABSTRACT

An integrated sedimentologic analysis of outcrops, cores, and geophysical well logs has delineated lithofacies and determined their stratigraphic relationships in siliciclastic cyclothems of the upper portion of the Cherokee Group in eastern Kansas and northeastern Oklahoma. Coals that represent coastal swamp settings maintained by rising sea levels mark the initiation of transgression in nonmarine interfluve settings (transgressive systems tract = TST). Gray shales represent marine shelf margin environments (shelf margin systems tract = SMST) and prodeltaic, progradational muds (regressive systems tract = RST). Deltaic complexes prograded across the margins of the Middle Pennsylvanian epeiric sea during times of regression and eustatic lowstand in response to increased amounts of siliciclastics supplied as fluvial valleys were incised (RST). Black, phosphatic clay shales represent marine maxima that can be laterally traced in outcrops, in cores, and with geophysical logs across the entire Mid-Continent (highstand systems tract = HST). Sandstones consist of both shoestring-shaped and thin sheet-like units within thicker siliciclastic sedimentary lobes. Shoestring-shaped sandstones formed as (1) incised fluvial valleys that were filled during sea-level rises and, therefore, represent TST in fluvial valleys, and (2) delta-distributary channel sands that were deposited as parts of deltaic progradational complexes (RST). Sheet-like sandstone bodies represent the reworked deltaic lobes that were reworked during marine transgressions and, therefore, represent a marine portion of the TST. Seat rocks beneath coals represent paleosols that mark emergent portions of shoreline complexes and represent maximum regression (lowstand systems tract = LST) within traditionally defined cyclothems. These emergent units constitute sequence boundaries. These cyclothems, which record glacio-eustatic sea-level changes that resulted from global climatic changes, are "allocyclothems," or allocyclic sequences based on sequence stratigraphic concepts.

All or parts of four sequences occur within the interval studied and can be traced across the study area along the outcrop belt and in the subsurface. These sequences range in thickness from about 2.5 to 20 m and represent time spans that approximate 100,000 years each. Sequence 1 consists of the Verdigris Formation, which is underlain by an exposure surface represented by a paleosol beneath the Croweburg Coal. Sequence 2 overlies a paleosol and consists of the Bevier Coal and a thin marine shale overlain by a regressive interval of shale and sandstone. Sequence 3 overlies a paleosol and consists of the Iron Post Coal in northern Oklahoma and equivalent strata northward in Kansas, including the Breezy Hill Limestone and a widespread shale that contains thin sandstone lenses capped by a paleosol underlying the Mulky Coal. Sequence 4 consists of the Mulky Coal, the overlying Excello Shale, and the Blackjack Creek Limestone members of the Fort Scott Formation. Because the base of the widespread, easily recognized Excello Shale is considered to be the base of the Marmaton Group, the Cherokee-Marmaton contact is at the base of a maximum flooding surface (MFS); according to the traditional (Exxon) sequence stratigraphy concepts, this surface is a parasequence boundary contained within a sequence. The base of the Oakley black shale of the Verdigris Formation and that of the Excello black shale would be considered sequence boundaries, according to "genetic stratigraphy" concepts. Although genetic sequences can more easily be traced across much of the Mid-Continent, traditional sequences offer the possibility of a higher resolution of stratigraphic correlations.

The extent of each sea-level change and concomitant climatic changes were major controlling factors that determined the distributions of lithofacies associations. Along the eastern margin of the seaway, deltaic lobes prograded southwestward in response to minor sea-level drops and concomitant fluvial valley incision. As individual delta lobes shifted and were abandoned, waves and currents winnowed their upper portions, leaving thin sheet-like lenses of sand. These reworked sands and overlying marine muds constitute autocyclothems that are relatively local in extent in eastern Kansas and northern Oklahoma, and are embedded within the regressive portions of allocyclic sequences.

INTRODUCTION

The Cherokee Group (Desmoinesian) in the Mid-Continent region east of the Nemaha uplift consists predominantly of siliciclastic mudrocks with lesser amounts of lenticular siliciclastic sandstones, thin limestones, and coals. These lithologies occur in repetitive sequences that are exposed in a belt that trends southwestward through the study area from western Missouri, through southeastern Kansas, and into northeastern Oklahoma (Figs. 1 and 2). The stratigraphic interval studied in detail consists of the upper portion of the Cherokee Group between the top of the Oakley Shale of the Verdigris Formation and the base of the similar Excello Shale of the Fort Scott Formation (Fig. 3). For this study, the author reviewed field and published information about the interval below the Oakley Shale down to the seat rock beneath the widespread Croweburg Coal and its correlative units. This expanded interval (i.e., from the Croweburg seat rock at the base to the Excello Shale at the top) is referred to in this study as the *upper portion of the Cherokee Group* (Fig 3).

Several workers (Aden, 1982; Reinholtz, 1982; Lardner, 1984; Denesen, 1985; Nelson, 1985) collected data and contributed interpretations from work done in local areas within the region (Fig. 2). Data and interpretations generated by these workers from detailed descriptions of outcrops, cores, and thin sections are available in theses at the University of Iowa and are on open file at the Kansas Geological Survey. Summaries of these data and interpretations have been published by Brenner (1989), and are used extensively in this study.

Purposes and Scope of Investigation

The purposes of this investigation were to (1) work out the regional stratigraphic relationships with-

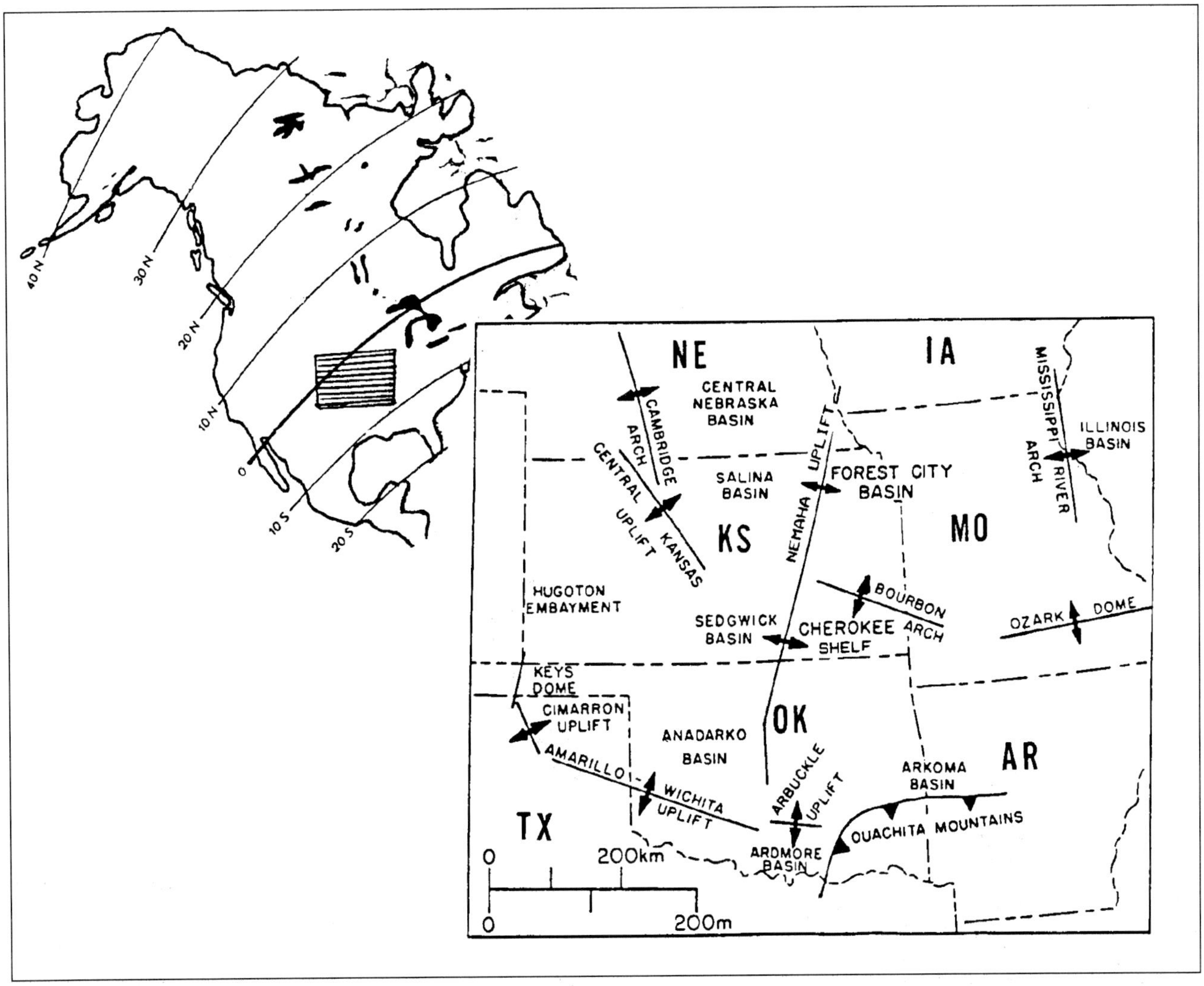

FIGURE 1 — Index map of North America showing Pennsylvanian tectonic elements in the Mid-Continent region of the U.S. (modified from Merriam, 1963; Lardner, 1984).

in the upper portion of the Cherokee Group, (2) reconstruct the successive paleogeographies of eastern Kansas and adjoining areas during this part of the Desmoinesian, (3) describe the petrographic characteristics and potential reservoir characteristics of sandstone units within the upper Cherokee, and (4) develop concepts that could be used to predict the occurrences and general characteristics of sandstone bodies that may be potential petroleum reservoirs. In this study, the author focuses on the relations between the concepts and realities of allocyclothems, autocyclothems, and *fifth-order* (tens of thousands of years) depositional sequences as they are preserved in the upper part of the Cherokee Group.

Tectonic Setting and Regional Structure

Desmoinesian strata were deposited on a tectonically stable depositional platform, bounded on the west by the Nemaha uplift, on the east by the Ozark dome, and on the south by the Arkoma basin (Fig. 1). During Early Pennsylvanian time, the shelf was divided into two depositional elements: the Cherokee basin in southeastern Kansas and northeastern Oklahoma, and the Forest City basin in

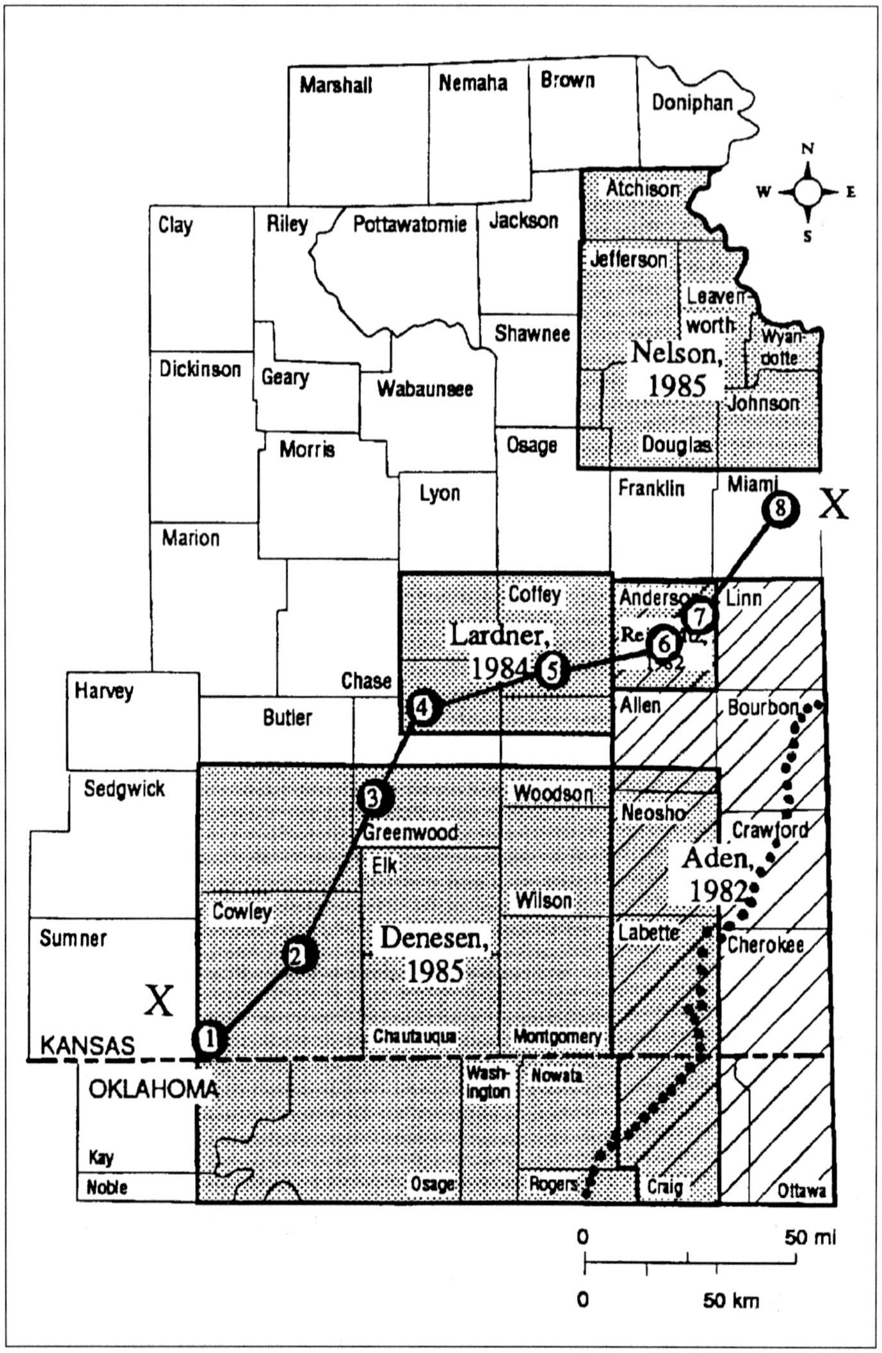

FIGURE 2 — Map of eastern Kansas and adjacent areas showing western limit of Cherokee Group outcrop belt (dotted line through southeastern corner of map area), locations of contributing worker's study areas, and locations of stratigraphic cross section X-X' and wells 1 to 8 that are shown in Fig. 9.

northeastern Kansas, and adjacent parts of Nebraska, Missouri, and Iowa. These two basins were separated from each other by the low-relief Bourbon arch, which trends northwest-southeast through east-central Kansas into Missouri (Merriam, 1963). By the time of upper Cherokee deposition, the Cherokee and Forest City basins were filled to the extent that the Bourbon arch was no longer a depositional barrier, and the region acted as a broad shelf area open to the deeper Anadarko basin to the south (Brenner, 1989).

The Ozark dome, which lies to the east of the Forest City-Cherokee shelf, persisted as a positive feature during the Paleozoic Era. The latest pre-upper Cherokee tectonic movements in the Ozark area seem to have begun prior to the deposition of upper Cherokee sediments (Branson and Huffman, 1965). Post-Desmoinesian erosion between the study areas and the Ozark dome has removed any evidence that this uplift might have influenced Cherokee sedimentation.

Tectonic subsidence in the study area increased southward and southwestward toward the foreland Arkoma and Arbuckle basins of Oklahoma. Siliciclastic sediment sources lay to the east, with west-flowing fluvial systems supplying sediments all along the eastern margin of the Desmoinesian seaway. Because accommodation space is created by subsidence, sea-level rises, and sediment compaction, and because the filling of this space resulted from sediment influx, the southern portion of the study area, with its greater accommodation space, has a thicker preserved sedimentary section.

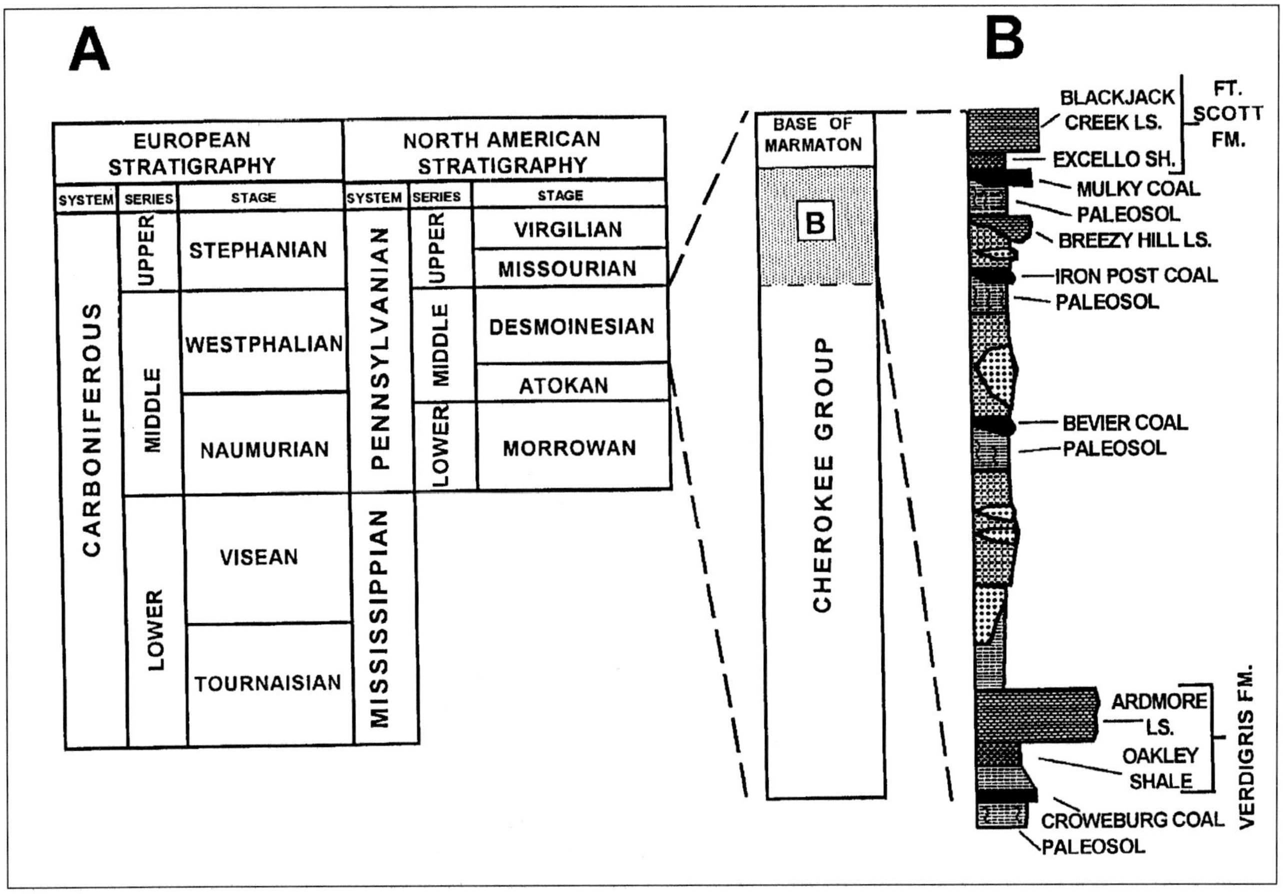

FIGURE 3 — (A) Position of Cherokee Group in North American Mid-Continent and European time-rock succession (modified from Zeller, 1968). (B) Position of upper Cherokee and basal Marmaton stratigraphic units discussed in this study.

STRATIGRAPHIC OVERVIEW

Cherokee Group Stratigraphy

Although vertical sections through the Cherokee Group appear to be repetitious and cyclic, lateral tracing of individual lithologies is difficult due to poor exposure and lateral facies changes. However, thin (<2 m), black, phosphatic clay shales have been traced both in outcrop and subsurface well logs over large portions of the Mid-Continent region. These high gamma count units appear to represent the deposition of suspended load and chemical sediments during times of maximum highstand when deeper parts of the region were covered by anoxic marine water (Heckel, 1977, 1984, 1986, 1994). Because of the relative ease of tracing the radioactive black shales in the subsurface, as well as along the outcrop belt, the stratigraphic interval analyzed in this study was defined by two widely traced black, phosphatic clay shales, the Oakley and Excello (Fig. 3). For this study, the author is including the Croweburg Coal and underlying seat rock which are stratigraphically below the Oakley Shale, but genetically related to the study interval.

A formal stratigraphic nomenclature for the Cherokee Group has not been agreed on by workers in the states represented in the study area. For this reason, this author will refer to the study interval between the Oakley and Excello shales as the *upper portion of the Cherokee Group* (or for the sake of brevity, the *upper Cherokee*), and use previously applied terms informally (Fig. 3B): *Ardmore Limestone* is a widespread limestone unit near the base of the study interval; the Verdigris Formation consists of the *Ardmore Limestone* and *Oakley Shale*. It is underlain by the *Croweburg Coal* and its seat rock. The *Breezy Hill Limestone* is a unit near the top of the interval in the southern part of the study area. In addition, informal but well-used terms will be noted to refer to several coal beds: the *Bevier Coal* which caps a paleosol above the *Ardmore*, the *Iron Post Coal* which is below the *Breezy Hill Limestone*, and the *Mulky Coal* which caps a paleosol above the *Breezy Hill Limestone*. These coal units and their underlying seat rocks are recognized across the study area (Howe, 1956), and they mark reversals in sea-level directions from regression (development of paleosols) to transgression (development of coastal swamps, followed by marine flooding) that are traceable using closely spaced well logs. For this reason, terms such as *Croweburg paleosol*, *Bevier paleosol*, *Iron Post paleosol,* and *Mulky paleosol* are used informally to mark exposed horizons of low sea-level stands below the respective coals.

Sequence Stratigraphic Concepts

Sedimentary sequences are deposited in response to the creation of accommodation space. The principles of sequence stratigraphy were first developed on continental margins and were related to eustatic sea-level changes and sediment influx rates (Mitchum et al., 1977; Van Wagoner et al., 1990). To apply these principles in other settings, one must understand these principles and then tailor the sequence model to account for local factors such as sediment flux and local tectonic effects (Posamentier and James, 1993; French and Heckel, 1994). In cratonic basins, slow subsidence rates and relatively shallow water depths result in reduced accommodation space for much of a basin's history. For this reason, sequences are thin relative to those formed on continental margins, and their unconformities are greatly subdued (Lindsay et al., 1993). Sequence boundaries and systems tracts in the Cherokee and Forest City basins can be identified by delineating erosion-bounded facies and tracing these unconformities across the region. Rapid changes in shoreline positions caused by sea-level changes over the low depositional slopes of the Cherokee and Forest City basins resulted in poorly developed *lowstand systems tract* (LST) deposits. Somewhat greater subsidence rates in the southern portion of the Cherokee basin provided accommodation space for development of LST deposits in Oklahoma (Walton, pers. comm., 1995). Areally restricted incised valley deposits that were cut during lowstands and filled during subsequent transgressions form fluvial valley *transgressive systems tract* (TST) deposits, while coals represent interfluve nonmarine TST deposits. The volumetric bulk of upper Cherokee sequences consists of progradation deposits, overlain by thin marine TST and *highstand systems tract* (HST) deposits. Due to the volumetric dominance of progradational sediments deposited during sea-level drop in these cyclothems, lumping them with HST deposits would not be appropriate. French and Heckel (1994) suggested using the term *regressive systems tract* (RST) deposits for these progradational deposits. This term is used in this study.

The widespread, black, phosphatic Oakley and Excello shales represent condensed intervals deposited at maximum sea-level highstands, and are therefore within sequences rather than bounding them. Delineating depositional sequence boundaries in an unambiguous manner in a region where most of the data come from geophysical well logs is a difficult task due to the subtle nature of unconformities in intracratonic settings (Lindsay et al., 1993). Another approach would be to adopt the genetic stratigraphy concepts of Galloway (1989a,b) in which stratigraphic intervals are subdivided at condensed sections (CS) or maximum flooding surfaces (MFS). Posamentier and James (1993) argued on philosophical grounds that as *sequence* implies a continuum and that since CS and MFS are always within continuums, they should not serve as boundaries. However, in the Pennsylvanian Mid-Continent the widespread black shales represent not only the most easily traced surfaces, but also major changes in depositional styles. Therefore, the approach in this study has been to use whatever is available to construct a stratigraphic framework with the highest possible temporal resolution. Using genetic stratigraphy on the upper portion of the Cherokee Group essentially reproduces the Kansas-type major cyclothems of Heckel (1977). However, by adding to this a modified version of sequence stratigraphic concepts, a higher resolution can be obtained in that minor (*fifth-order*) cycles can be recognized within genetic sequences bounded by the region-wide Oakley and Excello black phosphatic shales. The first step in constructing this stratigraphic framework was to undertake a detailed lithofacies analysis (Brenner, 1989), summarized and reevaluated below.

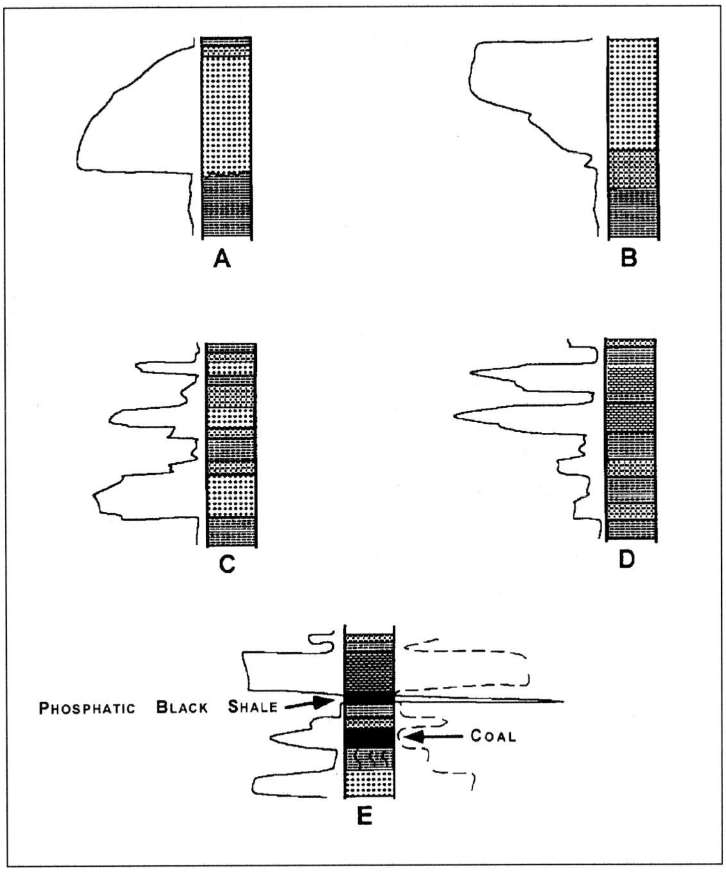

FIGURE 4 — Characteristic gamma ray log patterns and their lithologic interpretations: (A) upward increase in gamma counts forming a bell-shaped pattern representing an upward-fining sequence of sandstone to shale lying with sharp contact over shale; (B) upward decrease in gamma counts forming a funnel-shaped pattern representing an upward-coarsening shale to sandstone sequence with gradational base and sharp upper contact; (C) serrate pattern formed by interstratified sandstone and shale; (D) spiked pattern produced by thin, calcite-cemented sandstone beds; and (E) gamma ray and neutron log (dashed line) signatures of rock intervals containing coal and phosphatic black shale units.

LITHOFACIES ANALYSIS

Techniques Employed

The upper Cherokee outcrop belt provides but one limited northeast-southwest view of the sedimentary complexes that are present on the Cherokee shelf. This view has many interruptions caused by extensive present-day soil development and the low relief in the area. As a result, geophysical well logs from 1231 wells that penetrated the study interval in eastern Kansas and northeastern Oklahoma were examined (see Brenner, 1989, for geographic distribution). The gamma ray and neutron logs are the most useful logs for determining the lithologies penetrated, as confirmed by comparison of these logs with lithologies observed in same-well cores. Supplemented by density logs, they were used to determine lithologies in wells without cores. Electrical logs, which have a lower vertical resolution compared to radioactivity logs, were used in wells where radioactivity logs were not recorded. Figure 4 illustrates gamma ray curve signatures for various rock sequences encountered and confirmed by cores in the upper portion of the Cherokee Group.

Ten facies were identified from outcrops and well cores: (1) shale-chip sandstone conglomerate,

(2) homogeneous and horizontally bedded sandstone, (3) trough crossbedded sandstone, (4) interstratified ripple-laminated sandstone-siltstone-shale, (5) gray shale, (6) calcite-cemented, fossiliferous sandstone, (7) fossiliferous limestone, (8) black phosphatic shale, (9) coal, and (10) blocky, mottled, carbonaceous mudstone (Table 1).

Shale-Chip Conglomerate Sandstone Facies

Description. Conglomerate units were observed only in cores from northeastern Kansas (Leavenworth and Anderson counties). In northernmost cores (Leavenworth County), these units consist of medium- to coarse-grained sandstone with gray shale chips up to 20 mm in diameter. Chips are roughly disk-shaped, and lie parallel to stratification. In this area, this facies underlies trough crossbedded sandstone units, has sharp, erosional basal contacts, and overlies either sandstone or siltstone-shale units. Southwestward in Anderson County, the shale-chip conglomerate consists of fine-grained sandstone with shale chips having diameters less than 10 mm. Here, the conglomeratic facies forms units localized at the bases of homogeneous, fine-grained, micaceous sandstones that overlie interstratified sandstones and shales.

Where cores are present, the well log signature of conglomeratic sandstones consists of a blocky pattern of gamma ray and neutron log curves, having an irregular outer edge. It will be shown later that this well log signature is restricted to sediment and sandstone thick areas that form elongate patterns on isopach and sandstone isolith maps. Because conglomeratic units generally occur at the base of or beneath other sandstone units, they cannot be distinguished on well logs without the aid of cores or cutting descriptions (Fig. 5A).

Interpretation. The shale-chip conglomeratic sandstone facies represents channel lag deposits formed as channel currents cut into previously deposited clay-rich muds as sea levels dropped to lowstand positions. Conglomeratic units comprise the basal portions of sandstone channel complexes that were deposited during valley incision and as accommodation space created within northeast-southwest trending valleys as the sea level began to rise. This facies is considered as part of the LST and perhaps the base of the TST (Table 1).

Homogeneous and Horizontally Bedded Sandstone Facies

Description. Homogeneous and horizontally bedded sandstone units were studied on the outcrop and in cores from throughout the study area. In the cores from Leavenworth and Coffey counties, these units are oil-stained, fine- to medium-grained sublitharenites and litharenites that occur immediately above shale-chip conglomerate units (Figs. 5A, B). In Leavenworth County, these sandstones have the best sorting, and the highest porosity and permeability; thus, they provide the best petroleum reservoirs (Nelson, 1985). In the cores studied from Coffey County, this lithofacies is finer grained and more quartzitic than those observed in the Leavenworth County cores. On the outcrop belt, massive and horizontally bedded sandstones are fine to very fine-grained sublitharenites that directly overlie dark gray shales. Contacts with underlying shale units are sharp, but not erosional as evidenced by the lack of shale-chip conglomerates and the ubiquitous continuity and downward deformation of shale laminae where sandstone units form downward embayments into them without truncation of underlying strata.

LITHOFACIES	MAJOR CHARACTERISTICS	SYSTEMS TRACTS
Shale-chip conglomerate sandstone	Med.- to coarse-gr. SS with gray shale chips	LST, TST
Homogeneous-horizontally bedded sandstone	Fine- to med.-gr. SS with high porosity	RST, TST
Trough crossbedded sandstone	Fine- to med.-gr. x-bed. SS with high porosity	TST
Interstratified rippled sandstone-siltstone-shale	V. fine-gr. SS and mudrocks with horizontal wavy lam.	RST, TST SMST
Gray shale	Sparsely fossil., illitic	RST, SMST
Calcite-cemented fossiliferous sandstone	Fine- to med.-gr., well-sorted, found capping sandstone-rich intervals	TST
Fossiliferous limestone	Wackestone-packstone with marine fossils	TST, HST
Black phosphatic shale	Fissile, oblate spheroidal nod's.; conodonts	HST, MFS
Coal	Resinous-glassy, overlies mottled mudstone	TST
Blocky, mottled mudstone	Mottled, silty-clay, carbonaceous fragments	LST

TABLE 1 — Lithofacies recognized in the upper Cherokee interval. (HST = highstand systems tract; LST = lowstand systems tract; MFS = maximum flooding surface; RST = regressive systems tract; SMST = shelf margin systems tract; TST = transgressive systems tract.)

Interpretation. The homogeneous and horizontally bedded sandstone lithofacies represents rapid sand deposition in two different sedimentologic situations. Its position above shale conglomerates and erosional contacts in cores suggests that it was deposited during the waning phases of high-energy events and represents a transitional flow regime following a high flow regime event that caused the scouring of underlying muds. The sequence depicted in Fig. 5 (A-C) shows that this lithofacies grades upward into trough cross-stratified sandstones (discussed in the next section) that may represent deposition in the upper portion of the low flow regime of Harms and Fahnestock (1965). In fluvial channel deposits that were observed exclusively in the subsurface, this sequence records the backfilling of incised channels during the sea-level rises and, therefore, is the fluvial valley portion of the TST (Table 1). In outcrops, the homogeneous and horizontally bedded sandstone facies form units that are generally less than 1 m thick and form load structures into the underlying shales in some exposures. Here the sand appears to have been rapidly deposited as the result of a relatively high-energy event in a normally low-energy environment. This situation existed in crevasse splays that were formed when channel levees broke during floods, and in prodeltaic environments when sand-laden plumes extended seaward during floods. Because these prograding deposits were formed as the sea level was falling and landward valleys were being incised, they represent part of the RST (Table 1).

Trough Crossbedded Sandstone Facies

Description. Trough crossbedded sandstone units are found in isolated outcrops in northeastern Oklahoma which appear to be restricted to the lower one-half of the study interval. Other observations of this lithology were in cores from wells drilled into sandstones in Leavenworth, Anderson, and Coffey counties, Kansas. Sets are mostly less than 0.5 m thick and consist of medium- to fine-grained sublitharenites. Medium-grained sandstones are restricted to the northernmost cores; elsewhere this crossbedded facies consists of well-sorted, fine-grained litharenite. Reconstruction of core pieces indicates that cross sets are trough-shaped and unimodal (Brenner, 1989). In the cores from Leavenworth County, Kansas, this lithofacies overlies the homogeneous to horizontally bedded sandstone facies (Figs. 5B,C), while farther southwest in Anderson County, it lies directly above thin units of shale-chip conglomeratic sandstones.

Interpretation. The size and sorting of the trough crossbedded sandstone facies, along with their occurrences in elongated sandstone thicks, indicate that these sandstones represent low-amplitude dunes formed under conditions characteristic of the upper portion of the low-flow regime. The apparently unimodal dip of crossbeds within a particular core indicates that deposition took place under current directions that were constant at each locality. Crossbedded sandstone units in cores were always observed in upward-fining intervals that overlie the evidence of scouring (Lardner, 1984; Nelson, 1985). This sequential position (along with the other characteristics described above) suggests that this lithofacies represents subaqueous dune bedforms, possibly point bars in channel sequences. Having been formed in river valleys as rising sea levels created accommodation space, this facies represents a portion of the TST (Table 1).

Ripple-Laminated Sandstone-Siltstone-Shale Facies

Description. Ripple-laminated sandstone units are found throughout the outcrop belt from western Missouri to northeastern Oklahoma. In Missouri and Kansas outcrops, these units are restricted to the upper one-half of the study interval. In Oklahoma, rippled sandstones have been reported by Denesen (1985) in the lower as well as the upper portion of the study interval. Rippled units form slabby ledges that consist of 10- to 30-mm-thick sets of very fine-grained sublitharenites, commonly separated by 1 to 2 mm of medium gray shale. Individual ripple forms are usually either symmetrical or slightly asymmetrical; many are mottled, nearly obliterating external geometries and internal stratification. Where sedimentary dip directions are visible, they are unimodal within a unit, but often polymodal between units within an exposure. Ripple-stratified sandstone ledges invariably have sharp basal contacts with shales and often have gradational upper contacts with shales. Flaser beds and starved ripples are found above and below the rippled sandstones in some Crawford County, Kansas, outcrops.

Ripple-laminated sandstone units, similar to those found in outcrops, are found in all cores studied. In Leavenworth County, they are found directly above larger scale cross-stratified sandstone units within upward-fining sequences (Fig. 5D), as well as isolated units within shales. In cores where upper contacts of rippled units within upward-fining intervals are preserved, they consist of upward progressions from rippled sandstone, to flaser-bedded sandstone, to interstratified wavy sandstone and silty shale, and finally to shale. Upward-fining sandstone intervals with sharp bases are recognized on well logs by a gamma ray signature where gamma counts are low at the base, and gradually increase upward (Fig. 4A). Coupled with neutron or density logs, this signature forms a bell-shaped pattern. Some upward-coarsening sandstone-shale sequences that probably involve ripple-laminated sandstones have been interpreted using well logs. These sequences with gradational bases have gamma ray log signatures where gamma counts gradually increase upward with serrated lower portions that represent shale interbeds. Coupled with either a neutron or density log, this signature forms a funnel-shaped pattern (Fig. 4B). Isolated units within shale intervals have sharp basal contacts, and either sharp or gradational upper contacts that are commonly mottled and mixed with overlying muds.

Interpretation. The widespread lateral and vertical occurrences of ripple-laminated sandstones, as isolated units, within upward-fining sequences, and within upward-coarsening sequences, indicate that this lithofacies formed in a variety of depositional settings. The common link between these settings is that they were low-energy environments relative to the previously discussed sandstone environments, and that waves, currents, or both combined were strong enough only to move very fine-grained sand and smaller particles.

In Leavenworth County cores, this facies represents (1) upper portions of incised fluvial valley fills, (2) reworked sand that was deposited over channel bars under low flow-regime conditions, (3) splay deposits formed over interdistributary and flood basin muds during times of bank overflow, and (4) possibly the reworking of sand by small wind-driven waves on short-lived lakes that formed as channels were abandoned. In the first case, preserved ripple forms represent flow conditions in channels created by the tidal flow as rising sea levels formed estuarine settings. Rippled sandstones grade upward into mudstones, reflecting the infilling of estuaries and lakes with silts and clays. Splays in this setting consisted of ripple-laminated sands that were deposited over muds. These sands were later bioturbated as subaqueous organisms established or re-established their activities.

Shale units above and below rippled sandstone units along the outcrop belt contain low-diversity marine fossil fauna, such as foraminifers, brachiopods, and crinoids (Aden, 1982). Gibling et al. (1993) and Archer et al. (1994a,b) observed similar characteristics in strata associated with the Tonganoxie Sandstone of the Upper Pennsylvanian Douglas Group in Kansas. They interpret much of the channel fillings as tidally influenced estuarine deposits. The similarities in both physical characteristics and position within cyclothemic intervals suggest that much of the sediment of the ripple-laminated sandstone-siltstone-shale facies may have been deposited in stenotopic marine environments, such as tidal flats that exist in prodeltaic and interdistributary settings associated with tidally influenced, fluvially dominated delta complexes (Brenner, 1994). Sand deposition took place in these settings during times of high flood and ebb tidal flows, while suspension deposition of mud and organic activities dominated as velocities slackened at high tides. Where high rates of biological activity reworked sands prior to the next high-energy event, stratification was obliterated, creating slabby homogeneous sandstones similar to those described above in the homogeneous to horizontally bedded sandstone facies. Finer grain sizes and remnant burrows distinguish biologically reworked homogeneous sandstones formed by tidal currents from those rapidly formed under higher energy fluvial or marine conditions. The variety of stratigraphic positions where this facies occurs suggests that it represents portions of several systems tracts — RST, SMST, and possibly TST (Table 1).

Gray Shale Facies

Description. Along the Cherokee outcrop belt, the most abundant lithology found is gray, sparsely fossiliferous shale. In both the outcrop belt and shale intervals in well cores, shales along the northern and eastern margins of the study area and intervals containing sandstone units are siltier, richer in kaolinite, lower in mixed layered clay content, and contain illites characterized by X-ray diffraction patterns that indicate low crystallinity index values with broad, low-intensity 001 peaks (Aden, 1982).

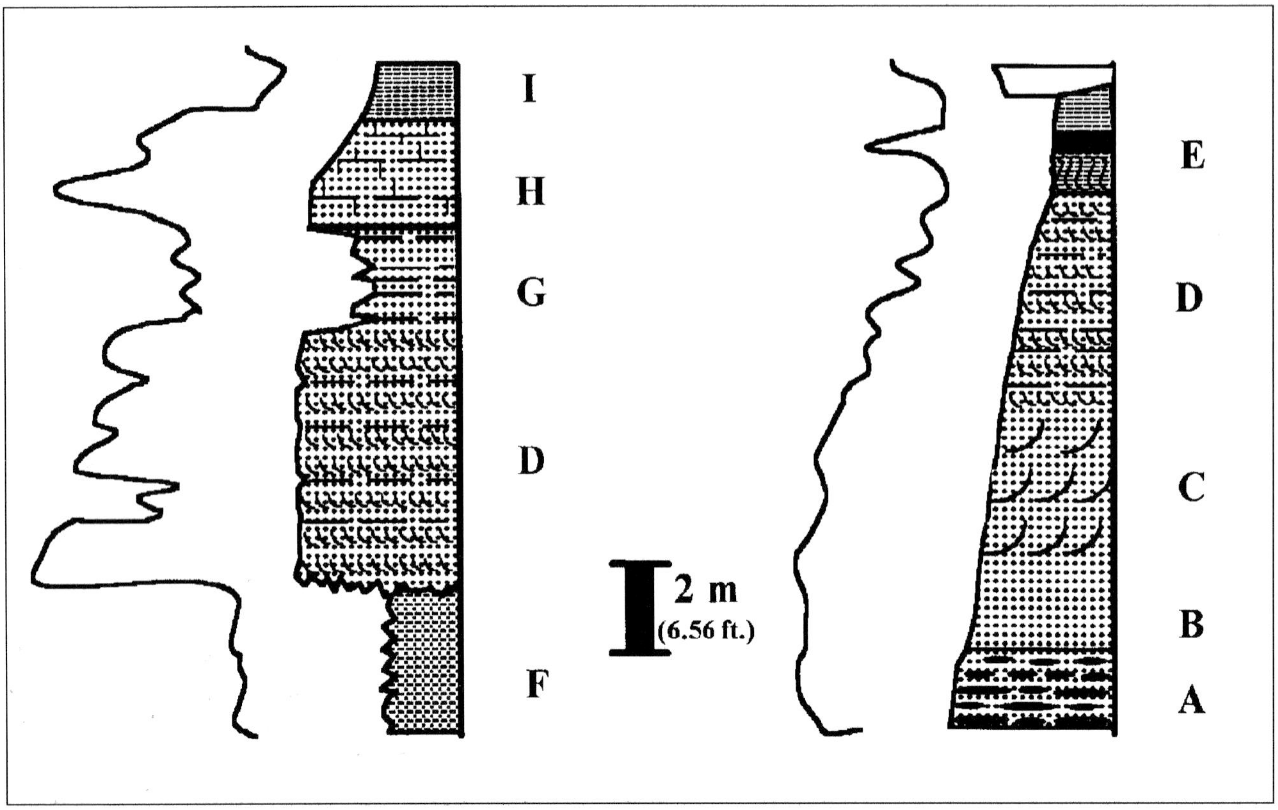

FIGURE 5 — Sandstone-rich rock sequences recognized in core samples from Leavenworth and Coffey counties, Kansas, and deduced from well log signatures in the study area. Lithologic units: (A) shale-chip conglomerate; (B) horizontally stratified sandstone; (C) unimodal cross-stratified sandstone; (D) ripple-laminated sandstone; (E) coal and interlaminated carbonaceous siltstone and shale; (F) interstratified siltstone and shale; (G) wavy stratified to mottled very fine-grained sandstone, siltstone, and shale; (H) fossiliferous, calcite-cemented sandstone; and (I) fossiliferous, dark gray clay shale (modified from Lardner, 1984, and Nelson, 1985).

In contrast, gray shales in the southwestern portion of the study area and within sand-free intervals contain higher clay percentages, have higher mixed layered clay contents, and produce X-ray diffraction patterns indicating higher illite crystallinity index values with sharper, more intense 001 peaks. As mentioned above, fossils are sparse in these shales, consisting primarily of a few species each of brachiopods and crinoids. However, both the diversity and abundance of marine fossils are higher in the more clay-rich shales than in the siltier, more kaolin-rich clays found associated with sandstones (Aden, 1982).

Interpretation. The gray shale lithofacies records the suspension deposition in low-energy environments. The great abundance and widespread distribution of this lithofacies suggest that it was deposited in subtidal shelf settings. The siltier nature of the shales, along with their richness in kaolinite, and low-illite crystallinity values along the eastern and northern margins of the study area suggest that these shales were deposited closer to siliciclastic point sources (Aden, 1982). They were deposited in delta-front, prodelta, interdistributary, and other shoreline-associated environments during periods of regression as muds were supplied from landward valleys that were being incised. These gray shales represent a portion of the RST (Table 1). High crystallinity indices of illite, lower kaolinite content, and scarcity of silt in the southwestern portions of the area indicate slower rates of deposition, farther from synchronous point sources (Aden, 1982). These observations along with variations in fossil abundance and diversity between these silty, low-illite crystallinity and clay-rich, high-illite crystallinity gray shale types suggest that the latter shales were deposited in more open marine shelf environments, and represent a portion of the SMST (Table 1).

Calcite-Cemented Fossiliferous Sandstone Facies

Description. Calcite-cemented fossiliferous sandstone units were observed only in cores from Coffey County, Kansas, by Lardner (1984). They consist of 0.3- to 0.6-m thick, very light gray, ferroan calcite-cemented, fossiliferous, fine- to medium-grained, well-sorted subarkoses. Fossils consist of abraded fragments of brachiopods, crinoids, and foraminifers. In the cores, this sandstone facies caps 3- to 4-m thick sequences of ripple-laminated sandstone and thin shales (Figs. 5G, H), and are in turn overlain by fossiliferous gray shales (Fig. 5I). These thin, pervasively cemented sandstones have well log signatures characterized by low gamma count spikes that are accompanied by symmetrical spikes on both neutron and density log curves (Fig. 4D). Analysis of well logs indicates that this lithology is widespread across most of the study area and occurs at several different stratigraphic horizons within the study interval, usually above signatures representing a sandstone or sand-rich mudstone interval.

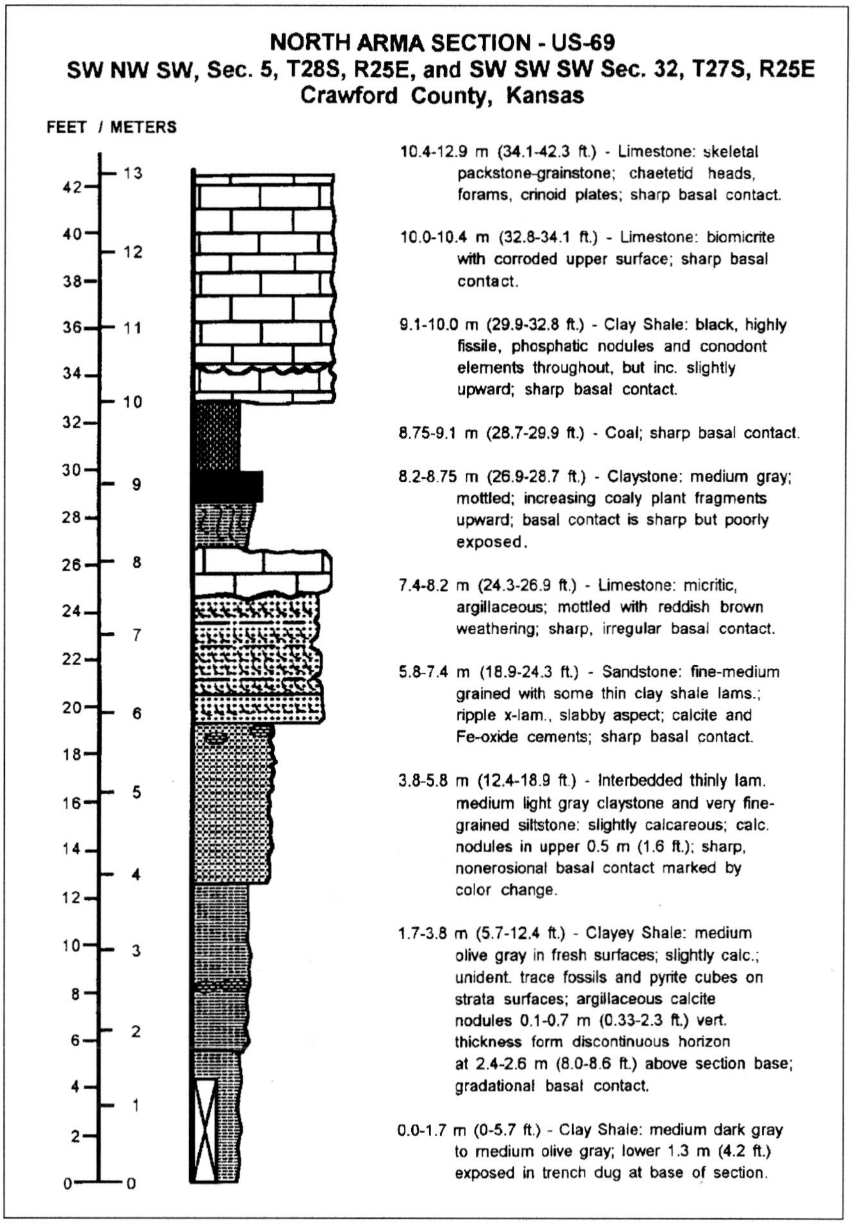

FIGURE 6 — Description of the uppermost portion of the upper Cherokee interval exposed in a road cut of U.S. 69 north of Arma, Crawford County, Kansas (modified from Brenner, 1989).

Interpretation. The coarser grain sizes, good sorting of siliciclastic grains, and abraded nature of marine fossil fragments indicate that the calcite-cemented fossiliferous sandstone units were deposited in relatively high-energy marine environments. The stratigraphic position of this lithofacies above unfossiliferous, ripple-laminated sandstones and below low-energy marine shales suggests that they represent sands that were wave-reworked during marine transgressions, either local or widespread (Lardner, 1984). The ripple-laminated to calcite-cemented fossiliferous sandstone sequence may represent wave-reworked sand bars or sheets such as those that form from the reworking of deltaic lobes as they subside along transgressive or retrogressive shorelines. This facies represents the basal portion of the marine TST (Table 1).

Fossiliferous Limestone Facies

Description. There are two stratigraphically and petrographically different fossiliferous limestone units within the study interval. The Ardmore Limestone is near the base of the study interval, below the 10 siliciclastic facies described above. It consists of a dense skeletal calcilutite containing marine brachiopods, bryozoans, echinoderm debris, corals, and algal fragments. Its thickness varies from 0.6 to 4.3 m (Denesen, 1985). The Breezy Hill Limestone (near the top of the study interval) is a skeletal calcilutite to calcarenite from eastern Crawford County, Kansas, southwestward into Oklahoma where it attains 3 m in thickness and lies below the Excello black shale. Near its northern limit in Crawford County, Kansas, it lies below a prominent mudrock and coalbed sequence (Fig. 6), whence it extends northward as a thinner and more mottled and argillaceous unfossiliferous calcilutite.

Interpretation. The Ardmore Limestone represents a widespread, low-energy carbonate shelf that covered most if not all of the study area during a time that siliciclastic point sources were far enough away that they had no influence. Carbonate sedimentation ceased as siliciclastic depositional systems prograded across the shelf. The lower portion of this unit may represent a thin TST that developed as the sea level rose, and the upper part represents an HST that covered the study area prior to a major regression (Table 1). The Breezy Hill Limestone represents a somewhat higher energy environment that transgressed the southern portion of the study area toward the end of the Cherokee deposition and may represent an HST formed after a minor transgression (Table 1). In Oklahoma, this unit grades upward into a regressive shelf deposit that resembles some of the marine carbonates that occur higher in the Pennsylvanian section. Northward, the Breezy Hill grades through a pedogenized rhizolitic shoreline deposit into a caliche (Knight, 1985).

Black Phosphatic Shale Facies

Description. Both the Oakley and Excello shales are traced across the entire study area and beyond (e.g., Heckel, 1977, 1994). Each consists of very dark gray to black fissile shale that typically both overlies and underlies dark gray shale. The black shale facies contains conodonts and oblate spheroidal, dark gray phosphate nodules, that have maximum diameters of less than 1 cm. Calcareous concretions are concentrated in layers in some Excello outcrops. The Oakley ranges from 0.5 to 1.8 m thick; the Excello shale is 1.5 to 2.0 m thick in northeastern Oklahoma and up to 2.5 m in southeastern Kansas.

Black phosphatic shales are represented on well logs by extremely high gamma ray spikes and accompanying low neutron spikes (Fig. 4E). This signature is so distinctive and the gamma ray counts are so high that these black shales can be readily identified, even where the black shale thickness drops slightly below the gamma ray log resolution limit of about 0.6 m.

Interpretation. These black shale units compose the most laterally extensive siliciclastic units in the Mid-Continent, having been traced from Oklahoma to Iowa and into the Illinois basin (Heckel, 1986;

Ravn et al., 1984). They represent slow deposition far from siliciclastic sources, in low-energy, anoxic marine environments that were created during times of major marine transgressions over the Mid-Continent region (Ece, 1987; Heckel, 1977, 1984, 1986, 1994). This facies was deposited while sea level was at its highest position and represents both the MFS and HST (Table 1). In some localities black shales lie directly over terrestrial facies such as coals (Fig. 6). This situation resulted from rapid sea-level rise relative to sediment accumulation, simultaneously removing siliciclastic point sources from the area and developing chemical conditions that precluded carbonate accumulations (Heckel, 1977, 1984; Brenner, 1989).

Coal Facies

Description. Each of the coals that were observed in cores and in outcrops rests on blocky, mottled light to medium greenish gray mudstone units. Mudstone-coal intervals occur on top of sandstone-rich intervals and upward-fining sandstone units. Mudstones appear to be laterally extensive although none of the coals can be unequivocally recognized in all well logs. Work along the outcrop belt by Howe (1956) indicates that many Cherokee coals are distributed over the entire study area. Coal beds in excess of 0.6 m thick have diagnostic well log signatures consisting of low gamma ray spikes that resemble those formed by sandstone and thin carbonate units. However, coal beds are represented on neutron and density logs by low neutron-density spikes that mimic rather than mirror the gamma ray spike (Fig. 4E).

Interpretation. Coals are the carbonized remains of plants that grew in swamps that covered low-lying coastal areas. To form and maintain extensive coal-producing swamps, the ground surface must be continuously wet, requiring a high water table (Cecil, 1990). The extensive Cherokee coals (e.g., the Croweburg, Bevier, Iron Post, and Mulky) formed as the water tables rose in response to the rising sea levels, and therefore represent the initiation of transgression across the region and form the base of the TST in nonmarine interfluve settings (Table 1). Locally distributed coals, such as the one described in the subsurface of northeastern Kansas by Nelson (1985), may have formed on delta lobes as abandonment and subsidence induced local transgressions flooded the delta complexes. Many more coals may have begun to form in this manner only to be destroyed during subsequent transgressions.

Blocky, Mottled Mudstone Facies

Description. Mudstone units vary considerably in composition from sandy to clay-rich, with varying amounts of carbonates (mostly ferroan calcite and siderite). All have blocky appearances in outcrops and contain carbonaceous plant fragments and root-shaped mottles. They underlie the Croweburg, Bevier, Iron Post, and Mulky coals, and serve as seat rocks everywhere that these coals were observed.

Interpretation. Mottled mudstones represent paleosols formed by plant growth and weathering of residual sediments. Because this facies requires subaerial exposure and withdrawal of the sea, it represents erosion/nondeposition and, therefore, marks sequence boundaries according to sequence stratigraphic concepts. This facies represents the LST in nonmarine interfluve settings (Table 1).

STRATIGRAPHIC ANALYSIS

Thickness Variations

The entire study interval between the widespread Oakley and Excello shales ranges in thickness from 12.2 m in Kay County, Oklahoma, to a thickness of more than 36.5 m in Montgomery County, Kansas (Fig. 7), where the base of a sandstone channel that cuts down through the Ardmore and underlying units is used as the interval base. Elsewhere in eastern Kansas, the upper Cherokee interval has a thickness range of 24.4 to 33.5 m.

A sandstone isolith map of the upper Cherokee in the study area shows high sandstone concentrations (Fig. 8) that generally correspond to the thicks on the interval isopach map (Fig. 7). Thus, sandstone percentage is greater in the north and east. Some sandstone highs do not directly relate to isopach thicks. One such case is in the northeasternmost part of the study area. The sandstone here seems to be part of a channel system that extends southwestward from southern Nebraska. Due to a low subsidence rate in this anorogenic cratonic setting, channels cut out previously deposited sediments, reducing the total thickness of upper Cherokee rock preserved in this area.

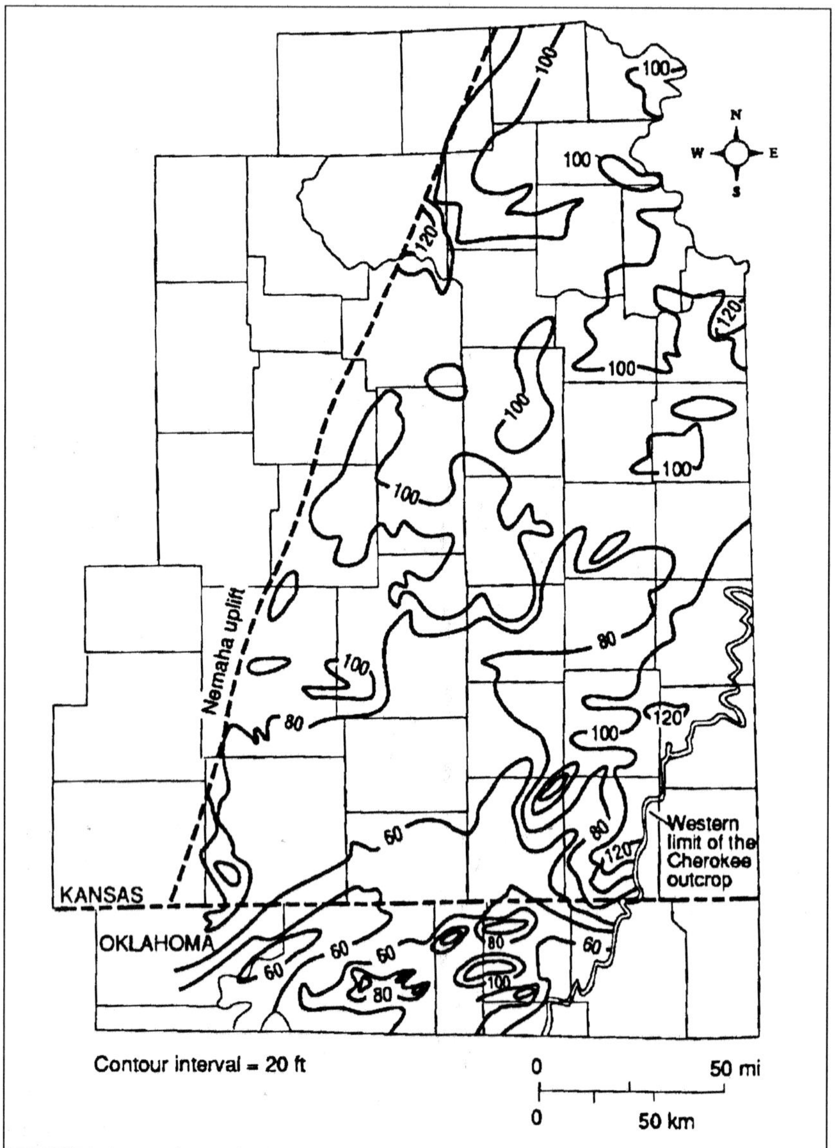

FIGURE 7 — Isopach map of upper Cherokee study interval, consisting of strata between top of the Oakley Shale and base of the Excello Shale (see Fig. 2 for location of counties) (modified from Brenner, 1989).

In northeastern Kansas, Nelson (1985) divided the sandstone units of the study interval into two genetically related sets. In the northern portion of his study area, all gamma ray/neutron well logs show a coal bed near the middle of the formation. The coal bed is not present in well logs and cores south of a line through the northwest corner of Wyandotte County through Leavenworth, Jefferson, and Douglas counties and into the central portion of Shawnee County (see Fig. 2 for county locations). The coal is replaced southward by a nearshore marine mudstone, perhaps as a result of a greater amount of compaction southward due to less sand in the section, keeping the land surface covered with sea water and preventing coal swamps from forming.

Lardner (1984) also divided the upper Cherokee into upper and lower sandstone-rich intervals in east-central Kansas. Because the coal bed observed to the northeast is absent, he divided the sandy intervals using an extensive clay-rich shale that may be partially equivalent to the coal bed used by Nelson (1985). The sandstones of the lower interval have a lobate pattern that appears to have prograded from an easterly source (Lardner, 1984).

In the southeastern Kansas and adjacent Oklahoma, Denesen (1985) placed the sandstone bodies into intervals based on their positions relative to coal beds. Because some of the coal beds are discontinuous, while others thin to below the resolution of geophysical well logs, Denesen (1985) traced coal beds to their resolution limits and then used well log signatures of units that he judged to be laterally equivalent.

The interval between the top of the Ardmore Limestone and the base of the Iron Post Coal contains thick "shoestring" sandstone bodies that trend northeast-southwest. Some of these sandstone bodies cut down through the Ardmore Limestone and underlying Oakley Shale in northern Oklahoma.

Stratigraphic Distribution of Sandstone

Individual sandstone bodies cannot be traced over areas more than a few tens of kilometers. Cross section X-X' (Fig. 9) extends along the trend of a sandstone-rich thick from northeast (Miami County) to Cowley County, Kansas, near the Oklahoma border in the southwest (see Fig. 2 for locations). Although most sandstone units are traceable for only small portions of the 270-km length of this cross section, they appear to occur within definite stratigraphic intervals.

Thicker sandstone units tend

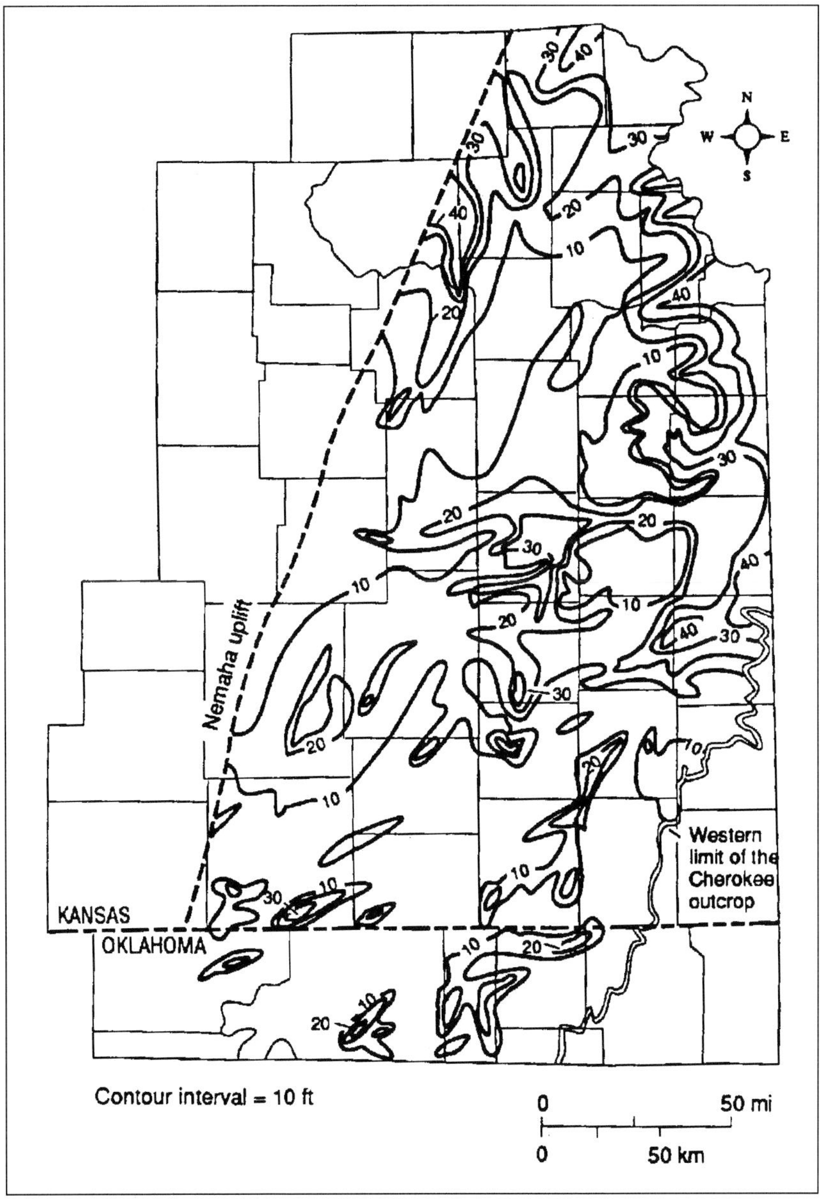

FIGURE 8 — Sandstone-isolith map of upper Cherokee interval (see Fig. 2 for location of counties) (modified from Brenner, 1989).

to have sharp basal contacts and either bell-shaped or blocky log signatures. Thinner units have either spike or funnel-shaped log signatures. When complex sequences of log signatures occur, spike or funnel-shaped signatures occur overlying blocky or bell-shaped signatures. Where sandstone units are laterally traced downdip, the signatures grade into irregular serrated curve patterns. Blocky and bell-shaped signatures change abruptly in strike-oriented cross sections and gradually in dip-oriented sections (Brenner, 1989).

These vertical and lateral patterns of well log signatures suggest that the thick, northeast-southwest-trending sandstones represent sharp-based channel complexes. The thinner, more extensive sandstone units with spike and funnel-shaped signatures seem to be upward-coarsening bar complexes. Irregular well log patterns, into which the sandstone signatures tend to laterally grade, represent interstratified sandstone and mudrocks. Cored intervals in Leavenworth, Anderson, and Coffey counties, Kansas, confirm these interpretations (Brenner, 1989).

DEPOSITIONAL SYSTEMS

The depositional systems within which lithofacies formed were delineated by integrating these observations and interpretations with the subsurface analysis discussed above. Sand deposition primarily took place in narrow trends that are spatially associated with lobate sediment thicks (Fig. 7) and high sandstone concentrations (Fig. 8). Most lobes seem to be areas where sand and mud entered the seaway as river systems reached the seaway margins. The lateral and vertical sequences of lithofacies are similar to those found within fluvially dominated delta systems and associated environments, and were depicted on two paleogeographic maps by Brenner (1989).

Incised Lower Fluvial Valleys

During times of sea-level lowstand, rivers incised underlying strata. The resulting incised paleovalleys were filled with sediments during subsequent rises of the sea level. These incised paleovalleys in the upper Cherokee are not exposed in the study area, but are detected by well logs in northeastern Kansas and northeastern Oklahoma (Brenner, 1989, 1994). Only the Kansas paleovalley fills were cored and studied. They consist of a vertical sequence of shale-chip conglomerate, overlain by horizontally stratified sandstone, unimodal cross-stratified sandstone, and ripple-laminated sandstone lithofacies overlying an erosional surface (Fig. 5A-D; Nelson, 1985; Brenner, 1989).

A re-evaluation of these sandstone bodies was prompted by recent studies of Upper Pennsylvanian sandstones, such as the Tonganoxie Sandstone of the Douglas Group in northeastern Kansas, by Gibling et al. (1993) and Archer et al. (1994b). They interpreted the Tonganoxie and similar sandstone units as estuarine paleovalley fills based on the vertical and lateral lithofacies relationships and their similarities with modern-day estuarine valley fills. The vertical sequence of lithofacies that they described are similar to that of the upper Cherokee units described by Nelson (1985) and Brenner (1989) in northeastern Kansas. The filling of coastal plain alluvial valleys included the formation of estuarine settings just prior to the time that marine transgression carried the shoreline landward over the filled incised valleys. Although the density of subsurface data in northeastern Kansas is not adequate to delineate lateral facies relationships in the upper Cherokee with the same precision that the Tonganoxie lithofacies were delineated, the linear trends of upper Cherokee sandstone bodies and their vertical succession of facies indicate that they may also represent estuarine valley fills. These linear sandstone trends form the proximal portions of lobate sandstone thicks shown on Fig. 8 in northeastern Kansas and northeastern Oklahoma.

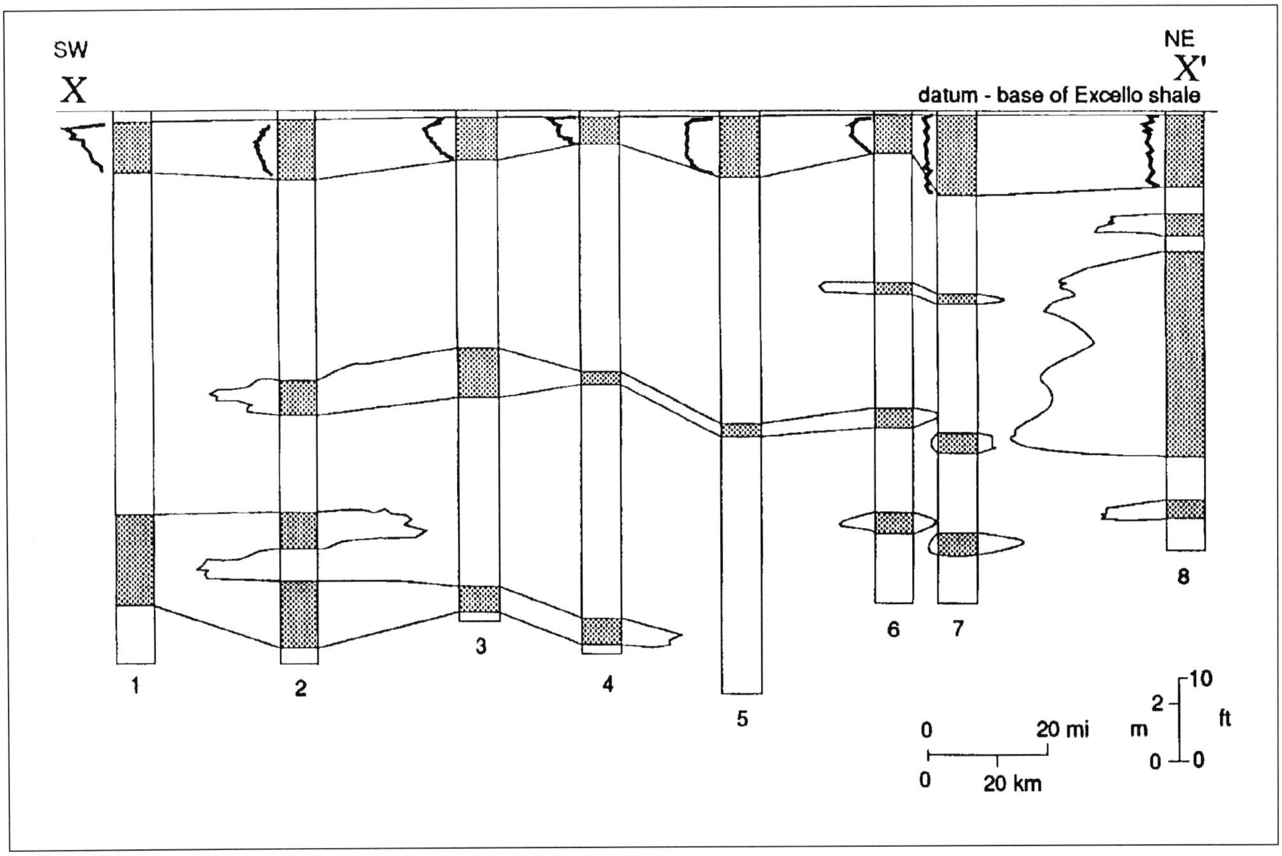

FIGURE 9 — Cross section X-X', northeast-southwest view of upper Cherokee sandstone bodies between Miami and Cowley counties, Kansas, showing stratigraphic positions of sandstone bodies in the study interval (see Fig. 2 for location of cross section and wells 1 to 8) (from Brenner, 1989).

Deltaic Complexes

Upward-fining sequences of sandstones and silty mudstones characterize the central portions of lobate sandstone thicks. These have sharp basal contacts, and core samples show that they consist of ripple-laminated sandstone overlain by fine-grained sandstone interstratified with siltstone and shale (G overlying D on Fig. 5; Lardner, 1984; Denesen, 1985; Nelson, 1985; Brenner, 1989). Off the sandstone trends, sandstones abruptly thin and grade into interstratified sandstone, siltstone, and shale. The upper portions of sandstones are often associated with or overlain by coal beds (Brenner, 1989; Denesen, 1985). This pattern is repeated in four stratigraphic positions within the study interval over the eastern and northwestern margins of the study area (Fig. 9).

Each lobate pattern on the sandstone isolith map represents one or more complexes of elongate delta lobes that formed during times of falling sea level in the study area. The northeast-southwest trends of relatively thick sandstone bodies within lobate thicks represent main distributary-channel complexes of delta systems. The characteristics of the sandstone and associated lithologies within these lobes that were described above suggest that the delta complexes were fluvially dominated, but had significant tidal influence. Smaller divergent trends of upward-fining sandstones represent complexes of minor distributary channels and crevasse splays.

Thicknesses of delta-lobe complexes are less than 9 m, reflecting the relatively low rates of subsidence in the Forest City basin and the Cherokee shelf compared to present-day settings along continental margins. In northeastern Kansas, delta complexes partially overlap the distal portions of incised lower fluvial valleys. Local rates of deposition probably exceeded accumulation rates by wide margins in this portion of the study area. Shifting channels resulted in a form of sedimentary cannibalism, where prograding channels eroded preexisting deposits and redeposited them in more basinward positions (Brenner, 1989).

Coal beds and underlying blocky, mottled, light-gray to tan mudstone units (interpreted as paleosols) cap some deltaic complexes along the eastern margin of the study area. These local paleosols and coals may represent local areas of emergence where deltaic sediments accumulated as fast or faster than they could be accommodated by subsidence, sea-level rise, or compaction. Rising sea levels resulted in the reworking of upper portions of the delta complexes, and redeposition of sand as shelf sand sheets. These are discussed below.

Prodelta and Interdistributary Environments

Between and adjacent to sandstone thicks, the upper Cherokee section primarily consists of mudrock with thin strata of siltstone and very fine-grained sandstone. The lithologic characteristics and positions relative to interpreted deltaic lobes, as well as the clay mineralogies of sampled mudrocks and the distributions of fossil assemblages, all indicate prodelta and interdistributary, tidally influenced depositional environments.

Thin, lenticular siltstone and sandstone beds encased within the mudrock sequences resulted from short periods of high-energy discharges of silt and sand from delta-distributary mouths into adjacent muddy-shelf areas. Evidence for this includes the sharp, nonerosional bases and upward fining of thin sandstones observed in cores and outcrops. Such events probably included spring tides and fluvial floods caused by intense storm systems, similar to typhoons that rake equatorial areas today during monsoon seasons. A short-term increase in a deltaic plane-jet flow could cause sands and silts, which were deposited earlier as distributary mouth-bar and tidal flat deposits, to be swept seaward into normally low-energy prodeltaic settings.

Crevasse-splay deposits also may have been deposited into interdistributary bays during floods. However, most of these deposits were later incorporated into the delta-lobe complex as distributary channels shifted within lobes. Only where a splay formed along the outer edge of a lobe, and thus occupied an extreme position relative to lateral lobe shifting, would it be identifiable as a splay in a delta-margin setting with a low subsidence rate.

Marine Shelf Environments

Portions of the upper Cherokee essentially lack sandstone and seem to represent low-energy shelf environments. For the most part, these intervals consist of clay shale containing marine fossils and some calcareous units. The most extensive of these sandstone-free intervals are in the upper part of the study interval toward the south. Characteristics of these intervals, such as more diverse marine fossils, higher illite crystallinity indices, and less kaolinite than mudrocks associated with deltaic complexes, show that they are more marine influenced and are located at a greater distance from siliciclastic point sources. Energy levels were low, with most sediment settling from suspension.

The Breezy Hill Limestone, a marine carbonate that lacks significant siliciclastic components, representing strand-line to shelf environments inaccessible to siliciclastic deposition, was deposited during or following a minor marine transgressive encroachment from Oklahoma into southeastern Kansas. On withdrawal of the sea, a lens of meteoric water diagenetically altered the Breezy Hill Limestone in the north to a silty calcilutite facies that represents a nonmarine caliche and rhizolite horizon (Knight, 1985).

Lenses of upward-coarsening marine sandstone, generally less than 3 m thick, occur in shelf areas and seem to represent the reworked coarse-grained portions of previously deposited delta lobes. Thin delta lobes that prograded well onto the shelf during sea-level lowstands were thoroughly reworked by marine currents and waves during subsequent transgressions. These elliptical sand sheets are calcite cemented and resemble thin, porous limestones on well logs (Fig. 5H).

SEQUENCES AND ALLOCYCLOTHEMS

Control of Regressive Sandstone Bodies

Many transgressive-regressive sequences, or cyclothems, occur in the upper Cherokee. The regressive portions of these cyclothems commonly consist of fluvial-deltaic complexes that represent delta-distributary progradational sands and prodeltaic and shelf muds, respectively. The thinness of TSTs is due to low rates of sedimentation caused by the removal of siliciclastic point sources that resulted from the rapid shifting of shorelines great distances toward higher portions of the gently sloping craton. Similar sequences have been observed in the lower portion of the Cherokee Group and have been interpreted as parts of deltaic complexes (Visher et al., 1971).

Two mechanisms seem to control the distribution of upper Cherokee sandstone bodies in the study area. First, glacio-eustatic (allogenic) fluctuations in the sea level (Heckel, 1986, 1994; Veevers and Powell, 1987) determined the general positions of shorelines and siliciclastic point sources in the study area by how far ice buildup caused the shoreline to retreat. Only during periods of eustatic regression did shorelines and deltaic systems prograde into the study area as increased supplies of sediments, derived from valley incisions, were transported to the shoreline by fluvial systems. During times of eustatic transgressions, shorelines and deltas were displaced far to the north and east, except for the Breezy Hill minor transgression when the shoreline extended through southeastern Kansas.

The second mechanism that controlled sand-body distribution was delta-lobe shifting that took place during times of eustatic regression and lowstand (Brenner, 1989). As distributary channels overextended themselves seaward, they shifted laterally by the diversion of flow through crevasses formed during floods and storms (Brown, 1979). Abandoned portions of a delta system subsided as muddy lobate sediment bodies compacted. Local minor transgressions occurred as marine waters covered subsiding lobes. Cyclothems formed in this fashion were autocyclic because they resulted from intrabasinal processes.

Stratigraphic Distribution of Allocyclothems

An analysis of cross sections made from closely spaced well logs delineated several cyclothems (Brenner, 1989). These cyclothems were traced over the entire study area and are considered allocyclic, whereas those confined to local areas are considered autocyclothems. Sandstone bodies occur in four intervals within the upper Cherokee. In some places, such as in Leavenworth County, Kansas, a sandstone-filled incised channel complex in the uppermost interval has cut through lower intervals (Nelson, 1985; Brenner, 1989). In other places, such as in Osage County, Oklahoma, similar channels lower in the section cut through the lowest intervals (Denesen, 1985; Brenner, 1989). In

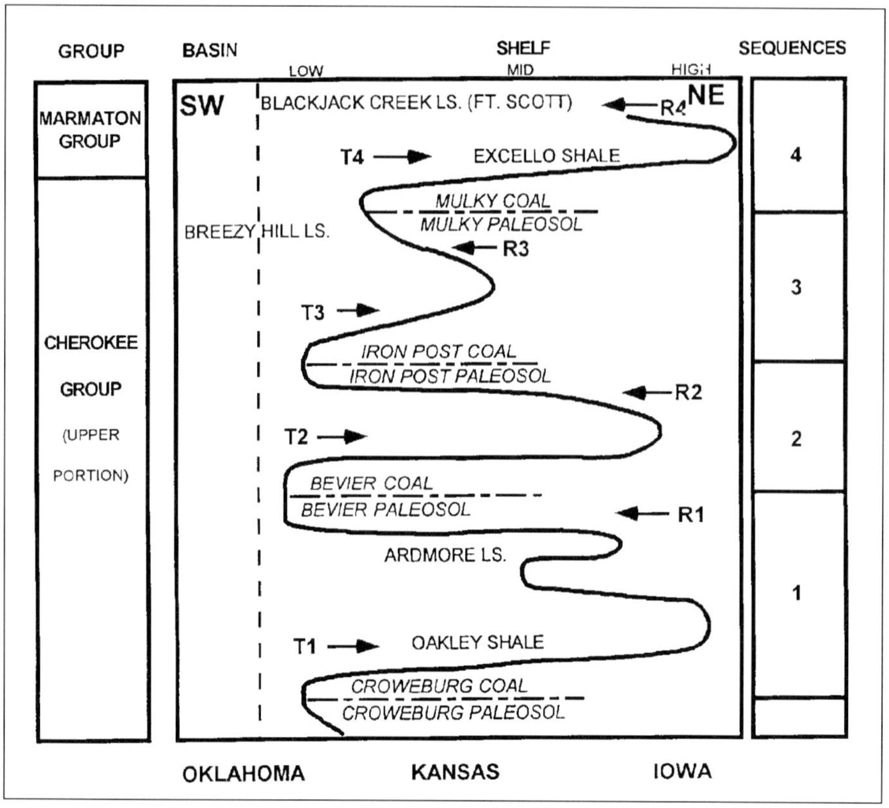

FIGURE 10. Stratigraphic positions of four depositional sequences in the upper Cherokee-basal Marmaton relative to a modification of eustatic sea-level curve published by Heckel (1986). T1-4 and R1-4 mark transgressions and regressions, respectively, with arrows showing sense of shoreline movement between Oklahoma and Iowa. Formal and informal rock unit terms are shown with the curve. Note that with exceptions for Oakley and Excello shales, lithic units cannot be individually traced across the entire study area, but rather a set of lithic units represents the horizons that reflect sea-level reversals.

places where discrete sandstone bodies are absent, gamma ray logs indicate four upward-coarsening shale-siltstone sequences within the study interval (Reinholtz, 1982; Lardner, 1984; Denesen, 1985; Brenner, 1989). These relationships are repeated in most of the well logs studied, but may be difficult to observe unless the logs are viewed in the context of strike- and dip-oriented cross sections, such as those presented by Brenner (1989), and illustrated by section X-X' (Fig. 9).

Heckel (1986) showed at least four minor allocyclothems between the Oakley and Excello shales. Addition of the information gathered in this study delineated these cyclothems as falling into all or parts of the four sequences depicted in Fig. 10. The base of Sequence 1 is marked by the Croweburg paleosol, and the sequence includes the Croweburg Coal and the overlying Verdigris Formation, including the Oakley Shale (T1 on Fig. 10), and regressive mudrocks and sandstones that overlie the Ardmore Limestone (R1 on Fig. 10). Sequence 2 overlies the Bevier paleosol and consists of the post-Bevier Coal transgression (T2 on Fig. 10) and the following regression terminated by the Iron Post paleosol (R2 on Fig. 10). Sequence 3 overlies the Iron Post paleosol and consists of the transgression that began with the Iron Post Coal (T3 on Fig. 10), including the Breezy Hill Limestone transgressive-regressive pulse, capped by the Mulky paleosol (R3 on Fig. 10). Sequence 4 consists of the Mulky Coal and equivalent strata, overlain by the Excello Shale (T4 on Fig. 10) followed by regression (R4 on Fig. 10) capped by a paleosol in the base of the Little Osage Shale Member of the overlying Fort Scott Formation in the Marmaton Group. These cyclothems are allocyclic because they can be traced at least across the region, and therefore formed as a result of extrabasinal processes of sea-level rise and fall.

These sequences fit within and overlap the boundaries of the genetic sequence (*sensu* Galloway, 1989a) bounded by the Oakley Shale (base) and Excello Shale (top). Heckel estimated that "major cycles" show a range of periods from 235 to 393 k.y. which approaches the long 413-k.y. Milankovitch eccentricity cycle (Heckel, 1986, p. 333). Assuming that this estimate is at least correct within a factor of 2, Heckel's "minor cycles," which correspond to the sequences defined above between the two major transgressions (T1 and T4 on Fig. 10), are equivalent to *fifth-order* cycles of Van Wagoner et al. (1990), with periods that approximate 100 k.y.

Relationship Between Autocyclothems and Allocyclothems

Within the widespread allocyclic sequences, many local regressive-transgressive cyclothems can be observed. These cyclothems correspond to the progradation, abandonment, and submergence of deltaic lobes and associated shoreline features. Thus these are autocyclothems, and they are restricted to times and places of basinward regression because of the availability of siliciclastic point sources during seaward displacement of the shoreline. This is in contrast to the removal of point sources from the study area — as the shorelines were displaced landward during regional transgressions, the only sediments available were thin highstand deposits of pelagic clays and organic matter. Regional regressions, accompanied by increased supplies of siliciclastic sediments produced by river valley incisions, allowed siliciclastic depositional systems associated with the shorelines and river mouths to re-establish themselves, and autocyclic processes to be resumed. The resulting siliciclastic sequences are dominantly progradational or upward-shoaling in character and are often capped by thin, marine-reworked sandstone sheets or hemipelagic shelf muds.

Although their individual characteristics were determined by local fluvial and deltaic processes, the stratigraphic distribution of autocyclothems was controlled by allogenic mechanisms. It is most likely that allocyclic transgressive-regressive sequences record the waxing and waning of Gondwanan ice sheets (Heckel, 1984, 1986, 1994; Veevers and Powell, 1987). An evaluation of all available evidence dealing with Mid-Continent Pennsylvanian marine cyclothems by Heckel (1994) showed that all alternative mechanisms that have been suggested fail to provide the timely dynamics required to cause the generation of regional sea-level changes needed to account for these allocyclothems.

SUMMARY AND CONCLUSIONS

Sedimentologic analysis of all available data shows that sands were predominantly deposited as northeast-southwestward-trending channels within incised lower fluvial valleys as sea levels rose, and in elongate delta-distributary channels of fluvially dominated deltaic lobes. These lobes of sediment prograded across the margins of the Pennsylvanian epeiric sea as river valleys were incised and siliciclastics were transported seaward during times of eustatic regression. Autocyclothems were formed as fluvial-deltaic channels laterally shifted. The allogenic regressive hemicyclothems and incorporated autocyclothems compose the *regressive systems tracts* within depositional sequences. During times of eustatic transgression, incised valleys filled with fluvial and estuarine sands and muds and siliciclastic point sources were shifted northward and eastward with the retreating shorelines. Thin shales and carbonate units were deposited in marine settings at slow rates during these times, resulting in thin *transgressive* and *highstand systems tracts*. Although transgressive limestones are considered part of the *TST*, the only extensively developed *HSTs* were formed during times of extensive marine flooding when deepwater anoxic conditions covered much of the Mid-Continent sea, resulting in the slow deposition of thin layers of phosphatic black muds (Heckel, 1977, 1984, 1986, 1994). These are represented by the Oakley and Excello black shales which bound the study interval; in the genetic stratigraphy scheme of Galloway (1989a, b), they form the genetic sequence boundaries.

Genetic sequences in the Cherokee Group probably represent about 400,000 years each; approximately equal to Heckel's (1986) major cyclothems and *fourth-order depositional sequences* of Youle et al. (1994). This approximates the highest resolution obtained through Pennsylvanian biostratigraphy. Depositional sequences (e.g., those depicted on Fig. 10), bounded by paleosols and equivalent strata, offer the possibility of a higher resolution of stratigraphic correlations, because they represent eustatic transgressive-regressive cycles with durations that approximate 100,000 years each. To conduct sequence stratigraphy at this scale of resolution, it is necessary to analyze closely spaced data over a broad area.

Because the tectonic setting for this portion of the Mid-Continent was one of relative stability during deposition of the upper Cherokee interval, glacio-eustatic fluctuation caused by the waxing and waning of Gondwanan ice sheets and consequential local climatic changes was most probably the main cause of fluctuations in sediment accumulation rates.

ACKNOWLEDGMENTS

Although I am solely responsible for the interpretations made in this manuscript, many people contributed data and ideas. Many of the data used in this study were collected and analyzed by Leon J. Aden, Stephen L. Denesen, James E. Lardner, Murray R. Nelson, and Philip N. Reinholtz while they were graduate students of mine at the University of Iowa. Professor Philip H. Heckel of the Geology Department at the University of Iowa first rekindled my interest in the Pennsylvanian of the Mid-Continent, and reviewed several versions of this manuscript. I also appreciate Tony Walton's thorough review of this manuscript which significantly aided in its final preparation. The Kansas Geological Survey supported most of the work done for this project. Additional support was supplied by the Oklahoma Geological Survey to some of my students. I am greatly indebted to all of these individuals and organizations.

FACIES AND STRATIGRAPHIC ANALYSIS OF CYCLOTHEMIC STRATA IN THE CHASE GROUP (PERMIAN, WOLFCAMPIAN), SOUTH-CENTRAL KANSAS

S.J. Mazzullo, C.S. Teal, and C.A. Burtnett
Department of Geology
Wichita State University
Wichita, KS 67260

ABSTRACT

Vertically recurrent strata of cyclic carbonate and siliciclastic facies compose the Chase Group (Permian, Wolfcampian) in south-central Kansas. These rocks were deposited in shallow to relatively deepwater environments on a broad ramp. A complex hierarchy of stratigraphic cyclicity is evident in these strata. The basic genetic stratigraphic unit recognized is the cyclothem, defined as a package of transgressive-regressive strata. Higher frequency, upward-shallowing subtidal and peritidal cycles are recognized within the seven cyclothems identified in the Chase Group. Terrestrial lowstand systems tracts (LSTs) with red shales are considered to be maximum-emergence lowstand deposits, the presence of which conceptually allows definition of six presumed third- and fourth-order depositional sequences. Accordingly, cyclothems are higher frequency third- and fourth-order units, and component cycles, most of which are not regionally correlative, are mostly of autogenic origin. Cyclothem development appears to have been affected by Milankovitch-forced glacio-eustasy and tectonism.

INTRODUCTION

Permian rocks, in particular those of Wolfcampian age, are found in several subsurface basins and are also exposed on outcrops throughout the Mid-Continent and southwestern United States (Peterson, 1980; Rascoe and Adler, 1983; Rascoe, 1988; Mazzullo, 1994; refer to additional papers in Fouch and Magathan, 1980 and Morgan and Babcock, 1988). Detailed, combined lithostratigraphic, paleontologic, and diagenetic studies of these rocks are imperative for evaluating long-term controls on deposition via sequence stratigraphic analysis. Important in this regard is the fact that Wolfcampian strata straddle the boundary between global "icehouse" conditions which existed in the Pennsylvanian and earliest Permian, and "greenhouse" conditions of the later Permian. Such scientific interest in Mid-Continent Wolfcampian rocks is equalled by economic interest in them because of the fact that

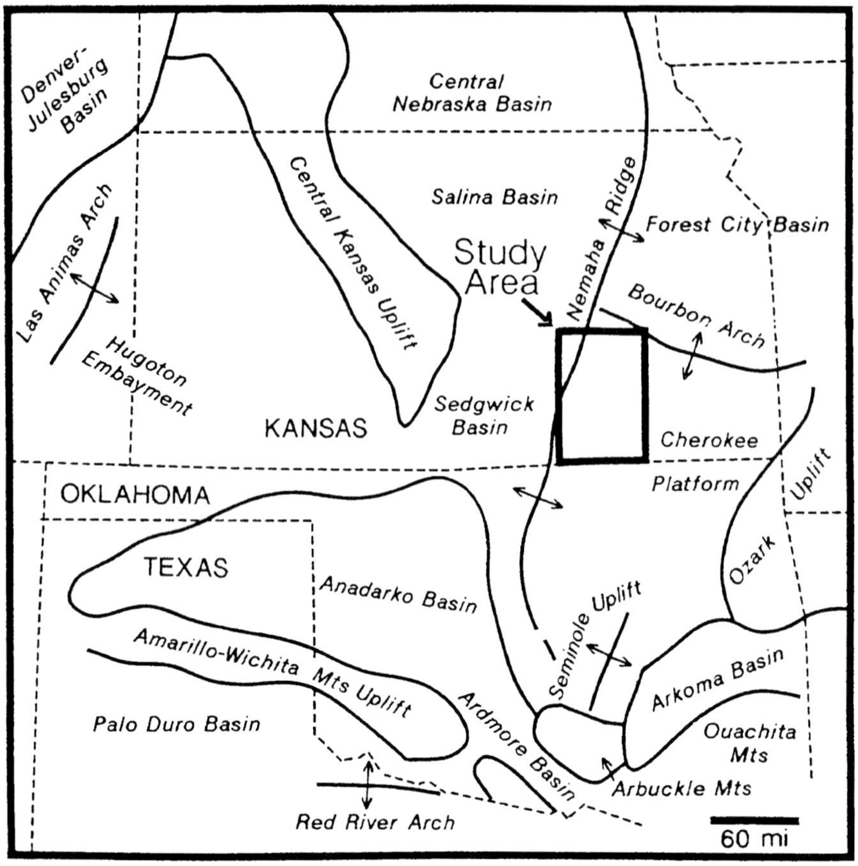

FIGURE 1 — Location of study area in south-central Kansas within the regional setting of Late Paleozoic deposition.

strata of this age are prolific hydrocarbon reservoirs in parts of Nebraska, Kansas, Oklahoma, Colorado, and Texas (Newell et al., 1987; Campbell et al., 1988; Mazzullo, 1994). A premier area in which to study Wolfcampian depositional facies in the sequence stratigraphic context is Kansas because, in contrast to some other areas of the Mid-Continent, deposition here occurred under conditions of relative long-term tectonic stability (Peterson, 1980; Rascoe and Adler, 1983; Rascoe, 1988). Deposition in such a relatively stable setting presumably fosters preservation in the rock record of composite eustatic signals with less "noise" than in strata deposited in more active areas. Therefore, combined sedimentologic and sequence stratigraphic analyses of these rocks in Kansas are likely to provide important concepts regarding fundamental controls on Wolfcampian deposition in the Mid-Continent.

This study focuses on outcrops of Wolfcampian strata of the Chase Group in Kansas, which include a complex mosaic of marine and terrestrial facies (Mazzullo and Teal, 1994). The Chase Group in Kansas is particularly important not only from the scientific viewpoint, but also from the economic standpoint because it is the major reservoir in the giant Hugoton gas field in the Hugoton embayment (Pippin, 1970), as well as in other fields located on the Central Kansas uplift (Fig. 1). Hence, information about facies types, extent, geometry, and depositional origin as determined first from detailed outcrop studies eventually can be applied by petroleum geologists to the analysis of subsurface hydrocarbon reservoirs. This study describes the sedimentary facies architecture, inferred sequence stratigraphy, and depositional origin of this section along its outcrop belt in south-central Kansas.

LOCATION, SETTING, AND STRATIGRAPHY

The Chase Group is exposed in an outcrop belt that extends from southeastern Nebraska, across eastern-central Kansas, and into northeastern-central Oklahoma (Fig. 1). In Kansas, these rocks were deposited on a relatively stable part of the craton which, during the Early Permian, included the slowly subsiding Cherokee platform, the Forest City, Salina and Sedgwick basins, and the buried Nemaha ridge and Bourbon arch (Fig. 1). The specific study area described in this work is located in south-central Kansas, and includes south-central Marion County and all of Butler and Cowley counties (Fig. 1). The Chase Group here consists of about 400 ft (122 m) of marine and marginal marine carbonate rocks interbedded with terrestrial to marine shales (Mazzullo and Teal, 1994). Ross (1963) assigned the Chase Group to the upper Wolfcampian, and most later workers have generally agreed with this assignment, correlating it to the uppermost Asselian to Sakmarian of Russia (Baars et al., 1994a, b). The Chase Group is overlain by evaporites and mostly terrestrial siliciclastics of the Wellington Formation (Sumner Group, Leonardian), and underlain by cyclic marine and terrestrial deposits of the Pennsylvanian-Permian Admire and Council Grove groups (Fig. 2). It is subdivided into seven formations, most of which include formally designated members (Fig. 2), that are regionally persistent on the surface and in the subsurface across Kansas, northern Oklahoma, and southern Nebraska. Although there are some regional variations in the thickness of individual formations and members of the Chase Group, its overall thickness varies very little (Moore, 1964; Chaplin, 1988), a fact that indicates relatively uniform subsidence during Wolfcampian time. The group thins regionally to the south, in north-central Oklahoma, to 300 to 334 ft (92 to 102 m), where it eventually passes from dominantly marine deposits to dominantly terrestrial siliciclastics (Chaplin, 1994). A similar change in facies is recognized in the subsurface along the west side of the Hugoton field in the Hugoton embayment (Pippin, 1970; Rascoe and Adler, 1983; Parham, 1993; Parham and Campbell, 1993), and may also have occurred along the erosionally backstepped, eastern side of the outcrop belt in Kansas (Mazzullo and Teal, 1994).

PREVIOUS STUDIES

Regional aspects of the stratigraphy of the Chase Group in Nebraska, Kansas, and northern Oklahoma were summarized by Condra and Upp (1931) and Zeller (1968). Cyclicity in the Chase Group in Kansas was first described by Jewett (1933) and Elias (1937), and later by Moore (1964) and Elias (1964). Hattin (1957), Cuffey (1967), Lutz-Garihan and Cuffey (1979), and others cited in this work studied selected units within the group, and based their depositional models mainly on paleontologic aspects of the rock record. More recent studies of depositional facies, cyclicity and, in some cases, inferred sequence stratigraphy, again of only selected units within the Chase Group, have been done in northern Oklahoma and parts of Kansas (Glossa, 1982; Busch et al., 1985; Toomey and Mitchell, 1986; Busch, 1988; Barrett, 1989; Twiss, 1991b; Toomey, 1992; Boardman and Nestell, 1993; Miller and West, 1993). Chaplin (1988, 1994) studied the stratigraphy and depositional facies of the entire Chase Group in northern Oklahoma (surface and subsurface), and Twiss (1991a) summarily described depositional facies in nearly the entire group from a core recovered in Riley County, Kansas. Studies of fusulinids from cores of the Chase Group in Oklahoma and Kansas were done by King (1988) and Sanderson and Verville (1970, 1988), respectively. Fusulinids from the Florence Member of the Barneston Formation currently are being re-examined by C.A. Ross.

A concerted effort to study regional aspects of the sedimentology, diagenesis, and sequence stratigraphy of the entire Chase Group in eastern-central Kansas (surface and subsurface) is presently in progress by the Wichita State University group (Mazzullo and Teal, 1994; Mazzullo et al., 1995a, b). The purposes of this long-term investigation are (1) to describe the Chase Group from a modern sedimentologic and stratigraphic perspective; (2) to understand the diagenesis of the component carbonate rocks; and in doing so (3) to ultimately provide a well-constrained outcrop model for the development of reservoir facies in the giant Hugoton field. However, the present study deals exclusively with

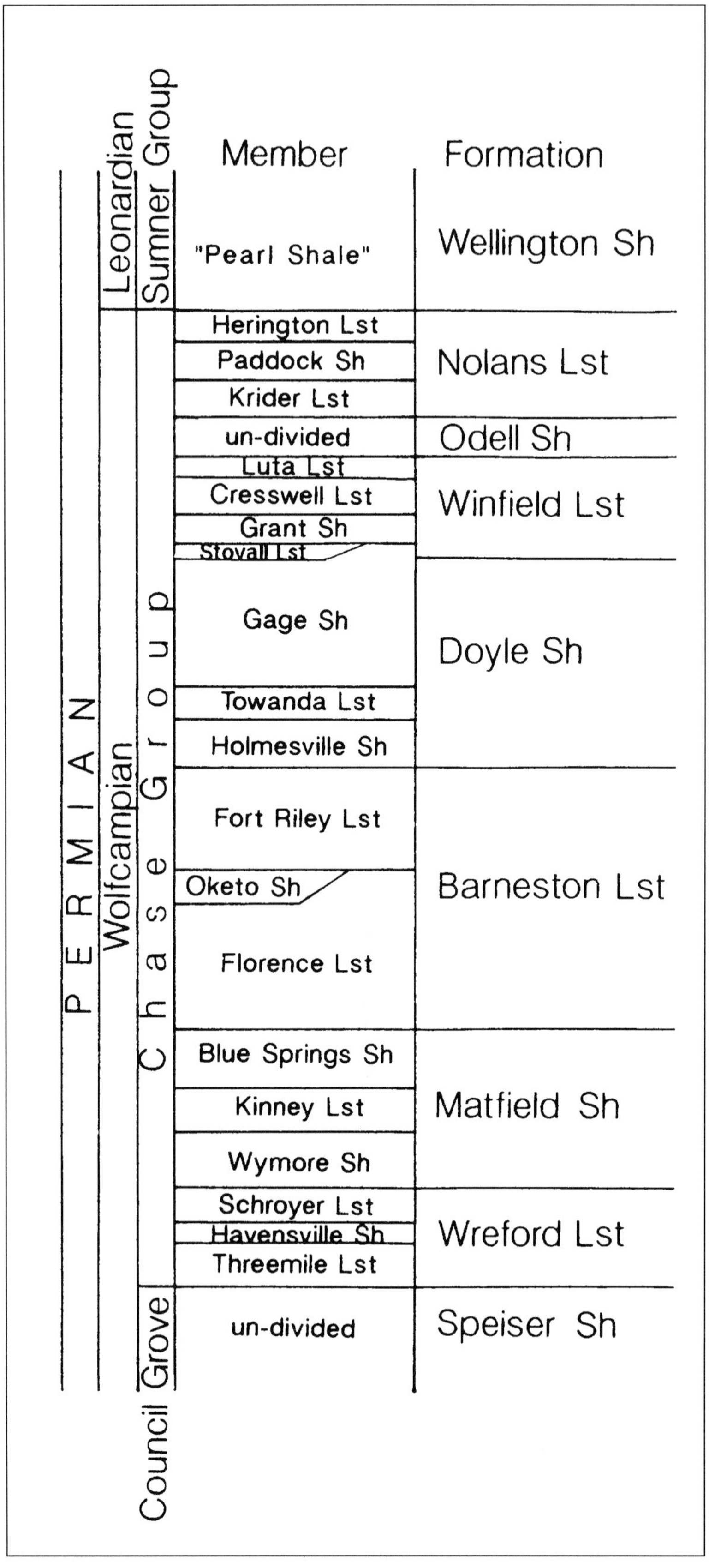

FIGURE 2 — Stratigraphy of the Chase Group in the study area.

outcrops of the Chase Group. These studies supplement detailed paleontologic studies of parts of the Chase Group by Darwin Boardman and his students and colleagues at Oklahoma State University (Boardman and Nestell, 1993).

APPROACH AND TERMINOLOGY

Cyclically deposited strata in Mid-Continent Upper Paleozoic rocks traditionally have been referred to as *cyclothems* (Heckel, 1986). Current concepts of sequence stratigraphy, on the other hand, recognize a hierarchy of stratigraphic units wherein the basic unit, the third- or fourth-order depositional sequence, is a composite eustatically controlled cycle of transgressive-regressive (T-R) strata. Allogenic (= allocyclic) and/or autogenic (= autocyclic), upward-shallowing deposits within sequences are, in turn, referred to as fourth-order or fifth-order cycles following the terminology of Goldhammer et al. (1991). Because of uncertainties regarding the duration of time represented by individual cyclothems and component cycles in the Chase Group, and often equivocal placement of sequence-bounding unconformities in ramp-type platforms such as that represented in the study area (Goldhammer et al., 1993), equivalence of cyclothems and cycles to such sequence stratigraphic units is not always readily obvious. Complicating the issue is the fact that the classification scheme of duration-based subdivision of Upper Paleozoic cyclic strata in Kansas (Busch et al., 1985; Heckel, 1986; Busch, 1988; Miller and

West, 1993; and others) is different from that proposed by others working elsewhere (Goldhammer et al., 1991, 1993).

However, cyclothems are clearly recognizable depositional entities in the Chase Group, as they are in Pennsylvanian strata. Accordingly, in the initial descriptions and interpretations presented below, cyclothems are considered to be the basic stratigraphic units in the section. The term *cyclothem* is used to refer to individual transgressive-regressive packages (Heckel, 1986), without initially assigning a hierarchal position to them based on duration. Boundaries between cyclothems and overlying lowstand deposits are apparently conformable to, in some instances, marked unconformities. Likewise, without initially advocating either origin (i.e., allogenic or autogenic) or time duration, the thinnest recognizable units composed of upward-shallowing subfacies within cyclothems are referred to below as *cycles.* By definition, cycles are bounded by either diastems or marine-flooding surfaces, and they variously compose either subtidal-types and peritidal-types (Goldhammer et al., 1989, 1993). In the following section, the cyclothems and component cycles in the Chase Group are described within the framework of inferred transgressive, highstand (stillstand), and regressive systems tracts as has traditionally been done in the Mid-Continent (Heckel, 1986). Later in this study, cycle and cyclothem development in the Chase Group are interpreted in terms of modern sequence stratigraphic terminology and inferred origin.

Representative localities in the study area were specifically chosen to describe and illustrate lithologies and inferred depositional environments in the Chase Group on the basis of relative completeness of exposed sections. These primary reference sections are supplemented by descriptions of outcrops from other localities so as to provide a more three-dimensional view of regional stratigraphy. These descriptions, the stratigraphic sections shown in the accompanying figures, and interpretations have been modified from those in earlier papers by Mazzullo and Teal (1994) and Mazzullo et al. (1995b). The present paper submits results of the ongoing studies, and as work continues, revisions and refinements of the interpretations presented herein are fully expected.

DESCRIPTION AND INTERPRETATION OF CYCLOTHEMS AND CYCLES

Cyclothems in the Chase Group are described below in ascending stratigraphic order. These rocks were deposited on a broad ramp that extended from eastern Kansas to southwestern Colorado, and from southeastern Nebraska to northern Oklahoma (Mazzullo and Teal, 1994). Major marine environmental zones on the ramp, as recognized in the study area and referred to in the following descriptions, are (1) *inner shelf facies*, which variously include (a) marginal marine, mostly siliciclastic-dominated deposits, (b) peritidal carbonate-dominated deposits, and locally siliciclastics, and (c) shallow subtidal, variously low-energy and high-energy deposits; (2) *middle shelf facies*, which mainly include carbonate deposits, and locally some shales representing low-energy, subtidal facies of moderate depth; and (3) *outer shelf facies*, mostly represented by cherty limestones. Terrestrial lowstand deposits are unfossiliferous shales. Strata in the study area were deposited in tropical marine environments that lay about 5° north of the equator (Scotese and McKerrow, 1990; Mazzullo, 1994), and climate became progressively more arid during Wolfcampian time (Patzkowsky et al., 1991).

Wreford: Threemile Cyclothem

This cyclothem includes part of the Speiser Formation (top of the underlying Council Grove Group), and the Threemile Member and lower one-half of the Havensville Shale Member of the Wreford Formation (Fig. 2). The reference section for this cyclothem (Fig. 3) is the roadside exposure along

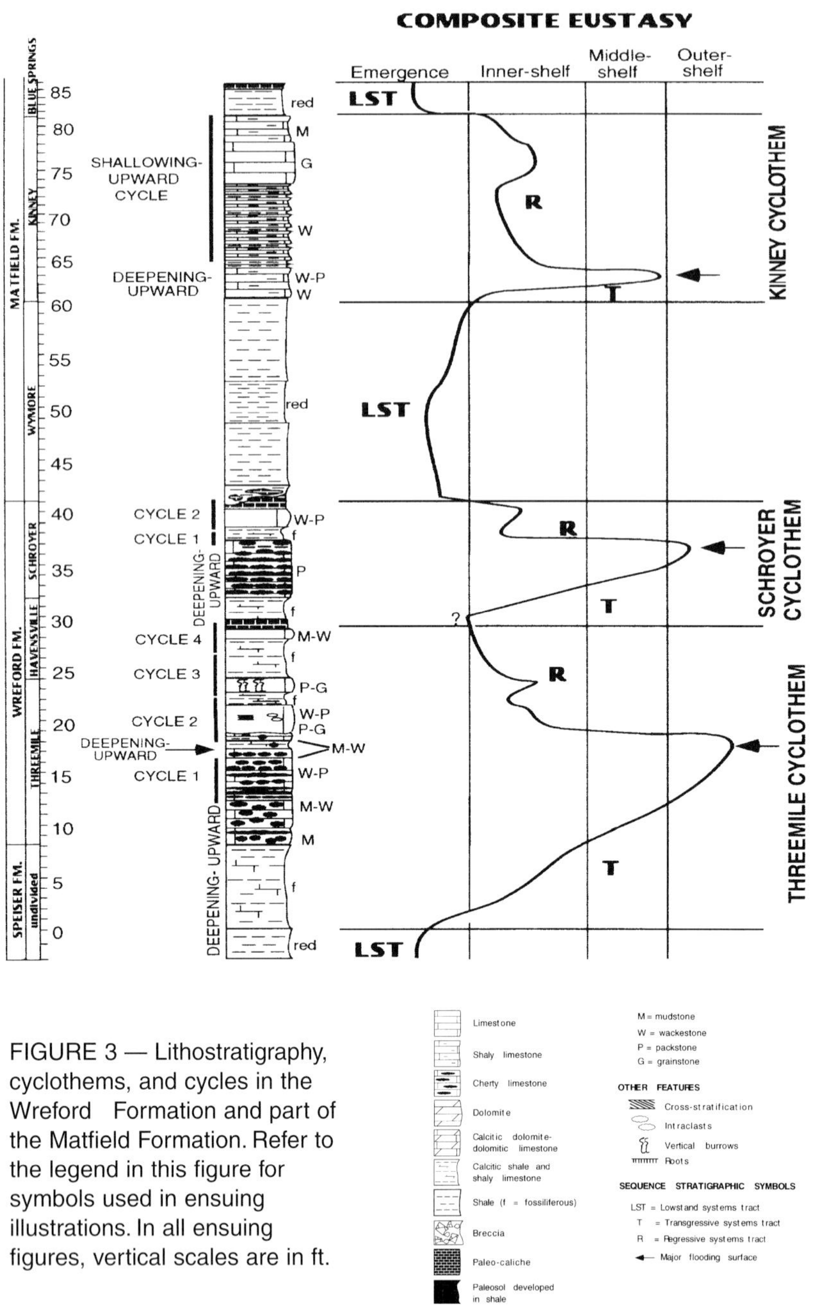

FIGURE 3 — Lithostratigraphy, cyclothems, and cycles in the Wreford Formation and part of the Matfield Formation. Refer to the legend in this figure for symbols used in ensuing illustrations. In all ensuing figures, vertical scales are in ft.

U.S. Highway 54 at "Lookout Point" in the Flint Hills, in Section 9 T26S-R8E, Butler County, Kansas. Supplemental sections are shown in Fig. 4.

The lower Speiser at this locality is a section of unfossiliferous red shale interpreted as terrestrial facies (Hattin, 1957). These beds are considered to be a lowstand systems tract (LST), the top of which is the base of the Threemile cyclothem. The overlying upper Speiser is 10 ft (3 m) of apparently noncyclic, gray, fossiliferous and bioturbated shale and shaly wackestone, with brachiopods, pelecypods, and crinoids. In sharp contact with the Speiser is about 4.3 ft (1.3 m) of cherty limestones in the lower part of the Threemile Member, which consists of lime mudstone to sparse biowackestone with fenestrate bryozoans, crinoids, siliceous spicules, and horizontal burrows (Fig. 3). This section, and the underlying fossiliferous Speiser, are interpreted as facies that deepen upward from marginal marine through inner shelf into middle shelf environments. Overlying cycle 1 consists of about 8.5 ft (2.6 m) of cherty limestones that slightly coarsens upward from lime mudstone and biowackestone to biowackestone and fine-grained biopackstone with similar fossils. This section is interpreted as proximal outer shelf facies. The succeeding 1.5 ft (0.5 m) of cherty lime mudstone to biowackestone is abruptly overlain by 0.3 ft (0.1 m) of gray, cherty, calcitic shale which contains a deepwater conodont fauna (D. Boardman, pers. comm., 1995). This section is interpreted as an upward-deepening facies deposited on the outer shelf, and the major flooding surface of this cyclothem is placed in the middle of the shale unit. This shale occurs rather high in the cherty portion of the Threemile section at this locality, whereas in southern Butler County and Cowley County, as well as in areas to the north (e.g., Chase, Morris, and Geary counties), presumably correlative shales occur stratigraphically lower in this

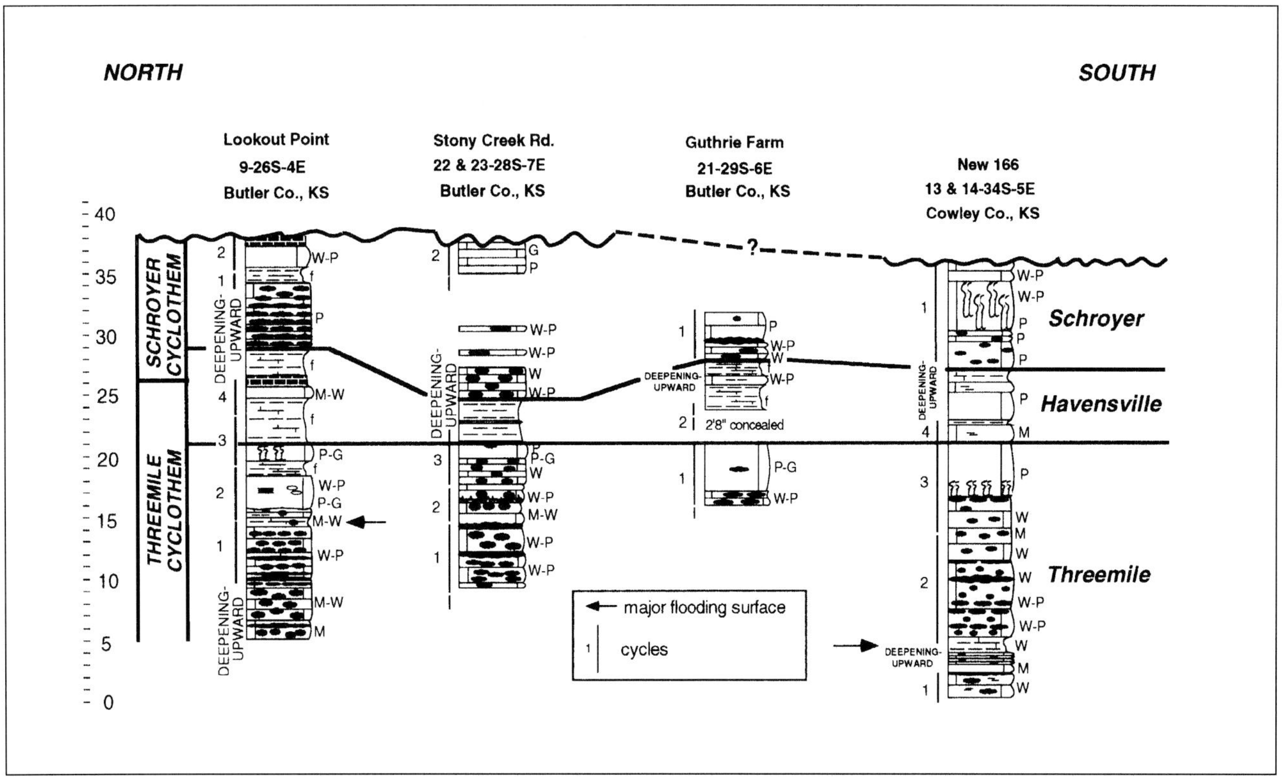

FIGURE 4 — North-south cross section and correlation of facies and cycles in the Threemile and Schroyer cyclothems. Section is hung on the base of the Havensville.

section (Fig. 4). Beds below this shale are included within the transgressive systems tract (TST) of the cyclothem.

Overlying cycles 2, 3, and 4 compose the regressive systems tract (RST) of the cyclothem. Cycle 2 includes the upper one-half of the calcitic shales and, in turn, (a) about 1.3 ft (0.4 m) of cherty, bioturbated lime mudstone to biowackestone at the top of the cherty part of the Threemile Member, which contains a slightly more diverse biotic assemblage of crinoids, brachiopods, and bryozoans than in underlying limestones; (b) 2.6 ft (0.8 m) of noncherty, cross-stratified, oncolitic, intraclastic, and locally porous limestone that grades upward from biopackstone to grainstone (this unit thins and changes facies along the outcrop to more micritic lithologies; the base of this unit is a minor scoured surface, presumably a marine erosional surface); and then (c) 0.7 ft (0.2 m) of sparsely fossiliferous, olive-green, calcitic shale, with subhorizontal burrows. This cycle is interpreted to represent abrupt shallowing to high-energy, inner shelf carbonate sands to marginal marine, inner shelf facies. Cycle 3 includes (a) the upper 0.7 ft (0.2 m) of the olive-green shale section, also interpreted as a marginal marine facies, overlain by (b) 1.2 ft (0.4 m) of nonporous biopackstone to grainstone (with pelecypods, crinoid fragments, foraminifers, brachiopods, and some oncolites) deposited in a moderate to high-energy, inner shelf environment; sharply overlain by (c) 2 ft (0.6 m) of sparsely fossiliferous, light brown and olive-green, calcitic shale (with brachiopods, crinoids, and bryozoans) interpreted as marginal marine deposits. Peritidal cycle 4 consists of (a) the remaining 2 ft (0.6 m) of sparsely fossiliferous, brown and olive-green shales; overlain by (b) a conspicuous 0.8-ft thick (0.2-m) bed of inferred inner shelf lime mudstone to biowackestone with brachiopods, pelecypods, and crinoids in the middle of the Havensville; abruptly overlain by (c) 1.0 ft (0.3 m) of unfossiliferous, yellowish green-brown, calcitic clay-shale with some pea- and marble-size calcite nodules that *may* be a poorly developed caliche soil. The top of the Threemile cyclothem is placed within this 1-ft thick (0.3-m) shale (Fig. 3).

There is not a lowstand terrestrial shale or readily recognizable unconformity between the Threemile and Schroyer cyclothems (i.e., within the Holmesville Shale) anywhere in the study area. Rather, the boundary between these cyclothems (as indicated above) is placed at the *possible* paleo-caliche at the top of cycle 4 in the upper Threemile (Fig. 3). Miller and West (1993) presumably found evidence of subaerial exposure at this approximate horizon in outcrops near Manhattan, Kansas. Tracing of the Havensville to the south of Lookout Point, from central and southern Butler County into southern Cowley County, however, has revealed no evidence whatsoever of an unconformity between the Threemile and Schroyer cyclothems. As shown in Fig. 4, for example, the Havensville exposed at a locality just north of the Butler-Cowley County line ("Guthrie Farm" outcrop) consists of fossiliferous, nearshore shales and shaly limestones. In contrast, in southern Cowley County ("New 166" outcrop), nearly the entire Havensville Member has changed facies to shallow water limestones presumably deposited slightly farther offshore, away from an influx of siliciclastics that characterize the more marginal marine facies of this unit to the north (e.g., Lookout Point: Fig. 3); no unconformity was found within this section at this "New 166" outcrop (Fig. 4). In such areas the boundary between cyclothems is placed at the point where inferred upward-shallowing facies are overlain by inferred upward-deepening facies (Figs. 3 and 4). In this manner, conformable cyclothem boundaries, like sequence boundaries, are placed within strata that presumably are correlative with unconformities farther updip.

Wreford: Schroyer Cyclothem

This cyclothem includes the remainder of the Havensville Member and the entire Schroyer Member of the Wreford Formation (Fig. 2). The reference section for this cyclothem is also the roadside exposures at "Lookout Point" (Fig. 3), and supplemental sections are shown in Fig. 4.

The upper 3 ft (0.9 m) of the Havensville consists of inferred marginal marine, sparsely fossiliferous (pelecypods, brachiopods, crinoids), light brownish-olive green, calcitic shale similar to that in the lower Havensville. The overlying 5.5 ft (1.7 m) of the lower Schroyer Member consists of cherty, fine-grained biopackstone that apparently is noncyclic. The basal 2 ft (0.6 m) of this section contains abundant brachiopods, whereas the upper 3.5 ft (1.1 m) contains fenestrate and ramose bryozoans, scattered brachiopods, spicules, and some crinoids; beds are very shaly in the top 0.5 ft (0.2 m) of this unit. The shales in the upper Havensville, and most of the 5.5 ft (1.7 m) of the lower Schroyer, are interpreted as an upward-deepening TST, with a major flooding surface of shaly limestone at the top of the section (i.e., just below the contact between overlying cycle 1 and the top of the underlying upward-deepening section: Fig. 3). The dominance of fine-grained packstones in this transgressive section suggests somewhat shallower depths of deposition than that of the more micritic cherty limestones in the Threemile cyclothem. Hence, a middle shelf depositional site is suggested for these rocks (Fig. 3).

The base of the RST includes the top few cm of the lower cherty Schroyer (distal middle shelf to possibly proximal outer shelf facies), which is abruptly overlain by about 1 ft (0.3 m) of olive-green, calcitic, and fossiliferous shale (commonly stained red on the outcrop) interpreted as marginal marine facies. The top of cycle 1 in this cyclothem is placed in the middle of this 1-ft thick shale. Overlying cycle 2 includes the remainder of this shale, and the uppermost bed of the Schroyer, whose thickness varies from 2.2 to 3.3 ft (0.7 to 1.0 m) along the outcrop here (Fig. 3). This bed is noncherty, nonporous biowackestone to packstone at the base, and biowackestone at the top, and includes oncolites, some bryozoans, crinoids, and foraminifers. It is interpreted as a moderate-energy, inner shelf facies. The top 1 ft (0.3 m) or so of this bed, and the basal 1.6 ft (0.5 m) of the overlying Wymore (described below), are replaced by unfossiliferous, light-colored micrite with root mottles, glaebules and incipient pisolites, and circumgranular cracks. Large vugs in the upper 1.5 ft (0.5 m) of the Schroyer are filled with this micrite, which is interpreted as a paleo-caliche. In a gully to the immediate south of Lookout Point, the top of the Schroyer likewise is a paleo-caliche, but it also contains vertical root casts and small vugs filled with green shale. Hence, the top of cycle 2 here is interpreted as having been truncated by subaerial exposure. In many exposures in southern Butler County and throughout Cowley County (Fig. 4), the noncherty beds of the upper Schroyer instead are grainstones, the top beds of which commonly have red and/or green shale infilling interparticle pores. Hence, throughout most of the study area, the top of the Schroyer is clearly a marked unconformity related to subaerial exposure. Therefore, the top of the Schroyer cyclothem is placed at the top of cycle 2 (= top of the Schroyer Member: Fig. 3). Subsequent deposition of the unfossiliferous green shales of the basal Wymore, which are interpreted as terrestrial, then covered this surface, and were themselves calichified. In this interpretation, the seeming single paleo-caliche that replaced the upper Schroyer and the basal Wymore at Lookout Point actually is two separate paleo-caliches. Other workers, who did not recognize the marked unconformity at the top of the Schroyer, instead placed the top of what the current authors refer to as the Schroyer cyclothem at the top of the paleo-caliche within the basal Wymore, where it is present (Hattin, 1957; Lutz-Garihan and Cuffey, 1979; Miller and West, 1993).

Lowstand Siliciclastic Facies — Overlying the paleo-caliche in the basal Wymore is 5.8 ft (1.8 m) of unfossiliferous green shale (Fig. 3) with scattered mm-thick layers of unfossiliferous micrite and discontinuous layers of pea- to marble-size, micritic carbonate nodules interpreted as incipient paleo-caliche layers. In turn, these beds are overlain by 4 ft (1.2 m) of unfossiliferous red shale in the middle Wymore. Both the green and red shales include apparent paleosols (Retallack, 1988), and are interpreted as terrestrial facies (Fig. 3). The overlying section is 8 ft (2.4 m) of regionally unfossiliferous green, yellow, and gray shale of the upper Wymore which may be entirely terrestrial as suggested by the sporadic occurrence of presumed paleosols throughout this section.

Kinney Cyclothem

This cyclothem is confined solely to the Kinney Member of the Matfield Formation (Fig. 2). The reference section for this cyclothem and associated beds is also the roadside exposures at "Lookout Point" (Fig. 3), and outcrops in adjoining fields to the immediate south of the road. Supplemental sections for the Kinney and Blue Springs are shown in Figs. 5 and 6.

Before proceeding with these descriptions, a brief discussion of the stratigraphy of the Kinney Member is in order. This member was formally named and described by Condra and Upp (1931) in reference to a section about 12 to 15 ft (3.7 to 4.6 m) thick that generally includes, in ascending order, (a) a lower section of fossiliferous limestone; (b) a middle section of fossiliferous shale, locally with interbedded limestones; and (c) an upper section of limestone, presumably with some interbedded shales, at least locally. Although some difficulty is found with their descriptions of the Kinney at some of their measured sections in Nebraska and Kansas, what is referred to as the Kinney at Lookout Point and elsewhere in the study area conforms both with the Condra and Upp (1931) tripartite definition of this member and its reported thickness

As so defined, the basal 2.6 ft (0.8 m) of the lower Kinney at Lookout Point is interpreted as the noncyclic, upward-deepening transgressive facies in this cyclothem. This section consists of shaly biowackestone to packstone with a brachiopod-dominated biota and accessory pelecypods and crinoids, and in its lower part, foraminifera-coated skeletal grains and high-spired gastropods; fenes-

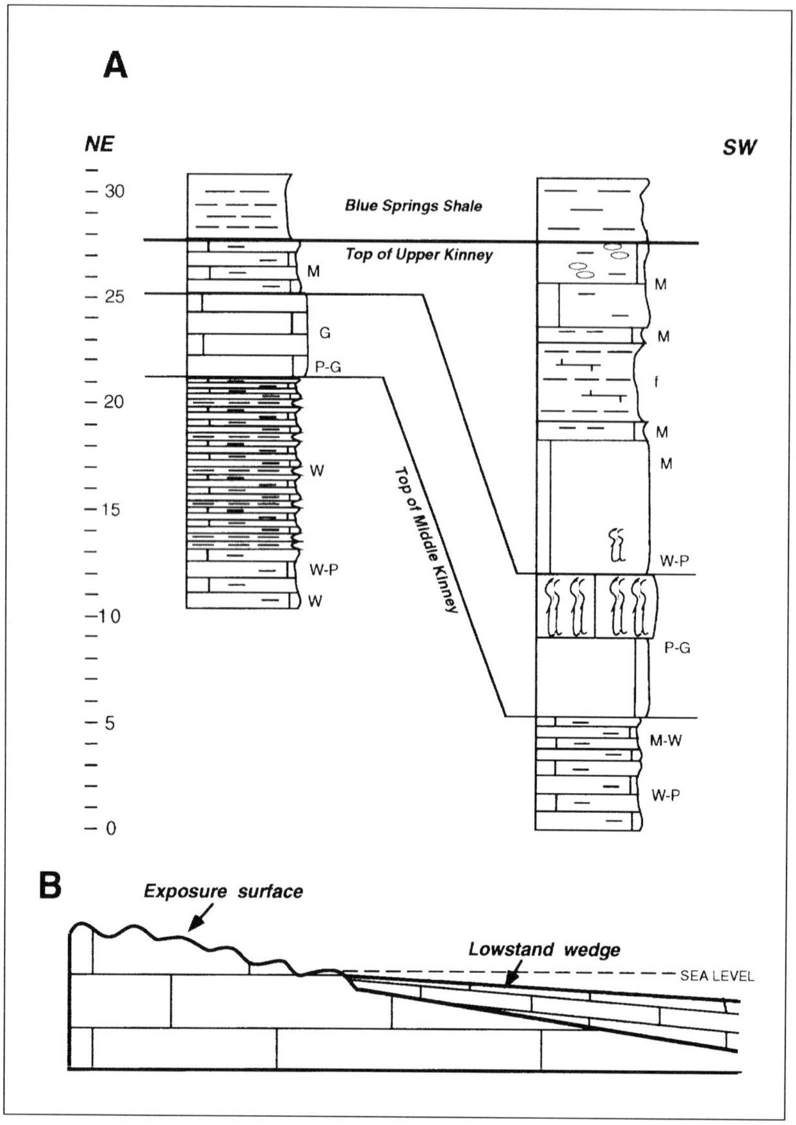

FIGURE 5 — (A) Correlation of the middle and upper parts of the Kinney from exposures at Lookout Point (from Fig. 3) southwest to the Bannon quarry; section is hung on the base of the Blue Springs Shale. (B) Interpretation of facies relationships between the sections at Lookout Point and Bannon quarry.

trate bryozoans are abundant at the top of this section. As shown in Fig. 6, this regionally persistent unit of the lower Kinney, which is correlated for a considerable distance south into Cowley County, locally is characterized by upward-fining sections of biopackstones and grainstones to shaly biowackestones. The lower Kinney is interpreted to be moderate to locally high-energy, inner shelf facies which rapidly deepen upward to middle shelf facies; the major flooding surface within the Kinney cyclothem likely occurs near the top of this section (Fig. 3).

Overlying regressive facies in the middle Kinney consist of (in ascending order) (a) about 2.5 ft (0.8 m) of fissile, sparsely fossiliferous black shale that grades upward to slightly more fossiliferous (brachiopod-dominated) gray, calcitic shale. The black shales in this section do not contain deep-water conodonts (D. Boardman, pers. comm., 1995); (b) 3 ft (0.9 m) of dark, fossiliferous and bioturbated, calcitic shale and shaly lime mudstone to biowackestone, with brachiopods, crinoids, and some pelecypods; and (c) 2.5 ft (0.8 m) of lighter colored (medium gray), lenticular bedded shale and shaly biowackestone, with abundant robust, burrowing pelecypods. This entire 8-ft (2.4-m) thick section is interpreted as an upward-shallowing low-energy, inner shelf facies. To the south, this unit thins and changes facies to include sparsely fossiliferous green shales, locally with lenses of brachiopod-rich, shaly limestones (Fig. 6). By these facies relationships, it appears that this section at Lookout Point was deposited slightly farther offshore than equivalent sections to the south, which are interpreted to have been deposited mostly in a marginal marine setting.

In turn, the middle Kinney section at Lookout Point is overlain by 4 ft (1.2 m) of the upper Kinney, which consists of medium bedded, light-colored limestones that coarsen upward from biopackstone-grainstone at the base, to biograinstone above (Fig. 3). At other locations in the study area (Fig. 5A), this grainstone is highly porous and commonly vertically burrowed. These rocks contain abundant foraminifers and accessory crinoids, ramose bryozoans, pelecypods, brachiopods, planispiral and

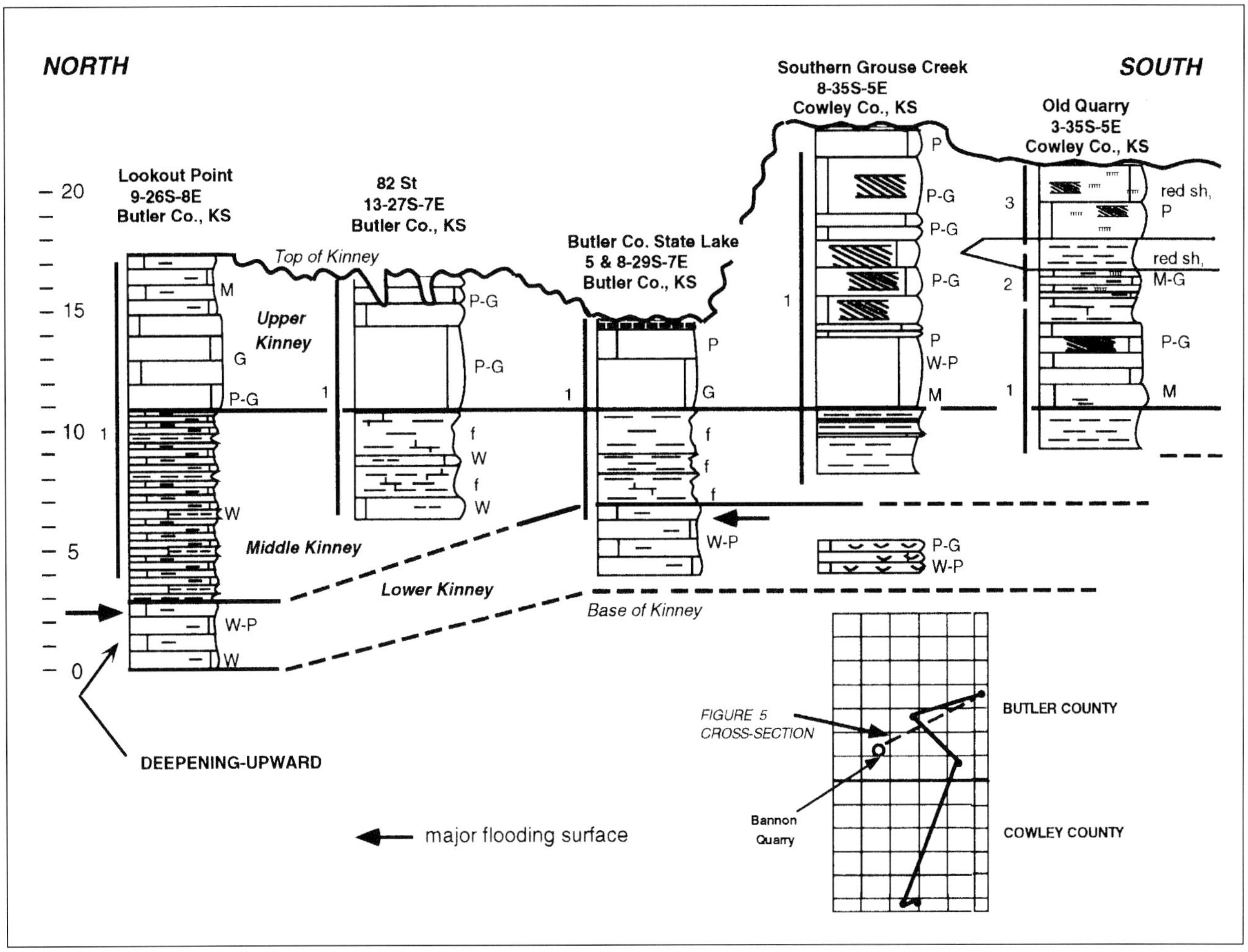

FIGURE 6 — North-south cross section of the Kinney cyclothem showing correlation of units and cycles. Section is hung on the base of the grainstone in the upper Kinney.

high-spired gastropods, and coated grains. They are interpreted as a sand shoal facies deposited in shallow, high-energy, inner shelf environments. To the south, this unit locally is cross-stratified, but eventually changes facies to intermixed red shale and fine-grained foraminiferal biograinstone near the Kansas-Oklahoma border (Fig. 6). The uppermost Kinney at Lookout Point is represented by a poorly exposed, 2.6-ft (0.8-m) thick section of thin-bedded, locally bioturbated lime mudstones, with scattered pelecypods and brachiopods, interpreted as low-energy, shallow, nearshore marine facies. Other than the long-term, upward-shallowing trend inferred for the RST in the Kinney cyclothem, higher frequency cycles could not be discerned in the section at Lookout Point (Fig. 3) or elsewhere in the study area (Fig. 6).

The middle and upper Kinney section exposed at Bannon (Southwest Butler) quarry in Section 29 T28S-R4E in Butler County (Fig. 5A) is most interesting, compared to that at other localities in the study area (Figs. 3 and 6), because of the much greater thickness of upper Kinney section here. At Lookout Point, for example, the biograinstones and overlying mudstones of the upper Kinney are 6.6 ft (2.0 m) thick (Fig. 3); in contrast, the correlative section at Bannon quarry is about 23 ft (7.0 m) thick (Fig. 5A). This thickened section includes about 7 ft (2.1 m) of biopackstones to grainstones at the base, which appears to correlate at least lithologically to the 4 ft (1.2 m) of biograinstone at Lookout Point, and is overlain by about 16 ft (4.9 m) of dominantly lime mudstones and some yellow shale (Fig.

5A). The lower part of this section likely correlates to the uppermost 2.6-ft (0.8-m) section of lime mudstones at Lookout Point (Fig. 3). Nowhere in the study area, from Marion and Chase counties south to the Oklahoma border, has a similarly thick Kinney section been seen. Of course, the question is why is this additional 16 ft (4.9 m) of section not present elsewhere in the study area? A possible answer to this question emerges on inspection of Fig. 6. Note on the inset map in this figure that the Bannon quarry section seemingly is the farthest downdip (to the west) outcrop of the Kinney within the outcrop belt of the Chase Group and, therefore, it likely represents the most seaward of the Kinney exposures examined in the study area. Although not found with certainty at Lookout Point because of incomplete exposures, the top of the Kinney at nearly all other localities in the study area is a major unconformity related to subaerial exposure, as evidenced by the common occurrence of paleo-caliche, small sinkholes and caves at the top of the Kinney (well displayed at the roadside outcrops in 13-27S-7E, Butler County: Fig. 6), eroded section, and red/green shale in interparticle pores within the grainstones; this unconformity variously extends from the top of the uppermost mudstone facies of the upper Kinney, downward to the top of the massive biograinstones (Fig. 6). All of the outcrops with such evidence of subaerial exposure lie east of the Bannon quarry section and, therefore, presumably are updip (landward) of it. It is possible that the "extra" section at the Bannon quarry was once present in areas to the east, but was removed by subaerial erosion prior to deposition of the Blue Springs section. This interpretation is not considered very likely because other than the Bannon quarry section, remnants of this thicker section were not found anywhere else in the study area. Instead, it seems more plausible that this extra section at the Bannon quarry may be a (nondetrital) lowstand carbonate wedge deposited in offlapping fashion downdip of a contiguous, subaerially exposed carbonate platform (Fig. 5B). By this interpretation, the seeming absence of a well-defined subaerial exposure surface at the top of biograinstone at the Bannon quarry is explained by this section being downdip of emergent platform deposits — that is, this area was still submerged (Fig. 5B). In all other areas, the top of the Kinney is considered to be the top of the Kinney cyclothem (Fig. 3).

Lowstand Siliciclastic Facies — The basal 3 ft (0.9 m) of the Blue Springs Shale Member at Lookout Point sharply overlies the Kinney, and is composed of unfossiliferous red shale interpreted as terrestrial facies (Fig. 3). Regionally correlative red shales similarly occur in the basal Blue Springs Member at the supplemental sections shown in Figs. 5A and 6. Although the thickness of these red shales varies somewhat throughout the study area, this section is capped by a regionally persistent, off-white to red micrite, about 1.7 ft (0.5 m) thick, with root mottles, incipient pisolites and glaebules, circumgranular cracks, and alveolar textures. The best exposure of this unit is in the roadside outcrops along Kansas Highway 96 in 27-27S-6E in Butler County. This micrite is interpreted as a paleo-caliche, and *may* be what Condra and Upp (1931) had referred to as the "Bruno Limestone." Overlying this inferred paleo-caliche as observed, for example, in road cuts along Kansas Highway 96 (27-27S-6E), is 8.5 ft (2.6 m) of unfossiliferous, greenish-gray and yellow-brown shale in the middle of the Blue Springs Member (shown in Fig. 7A), with presumed paleosols. All of these shales are interpreted as terrestrial facies.

Barneston: Florence-Fort Riley Cyclothem

This cyclothem is the thickest one in the Chase Group (about 93 ft: 28.4 m), and includes the uppermost Blue Springs Shale and all of the Florence and Fort Riley limestone members of the Barneston Formation. The composite reference section for this cyclothem (Figs. 7A and B) is based on the following outcrops: (1) for the Blue Springs and Florence section — roadside exposures along

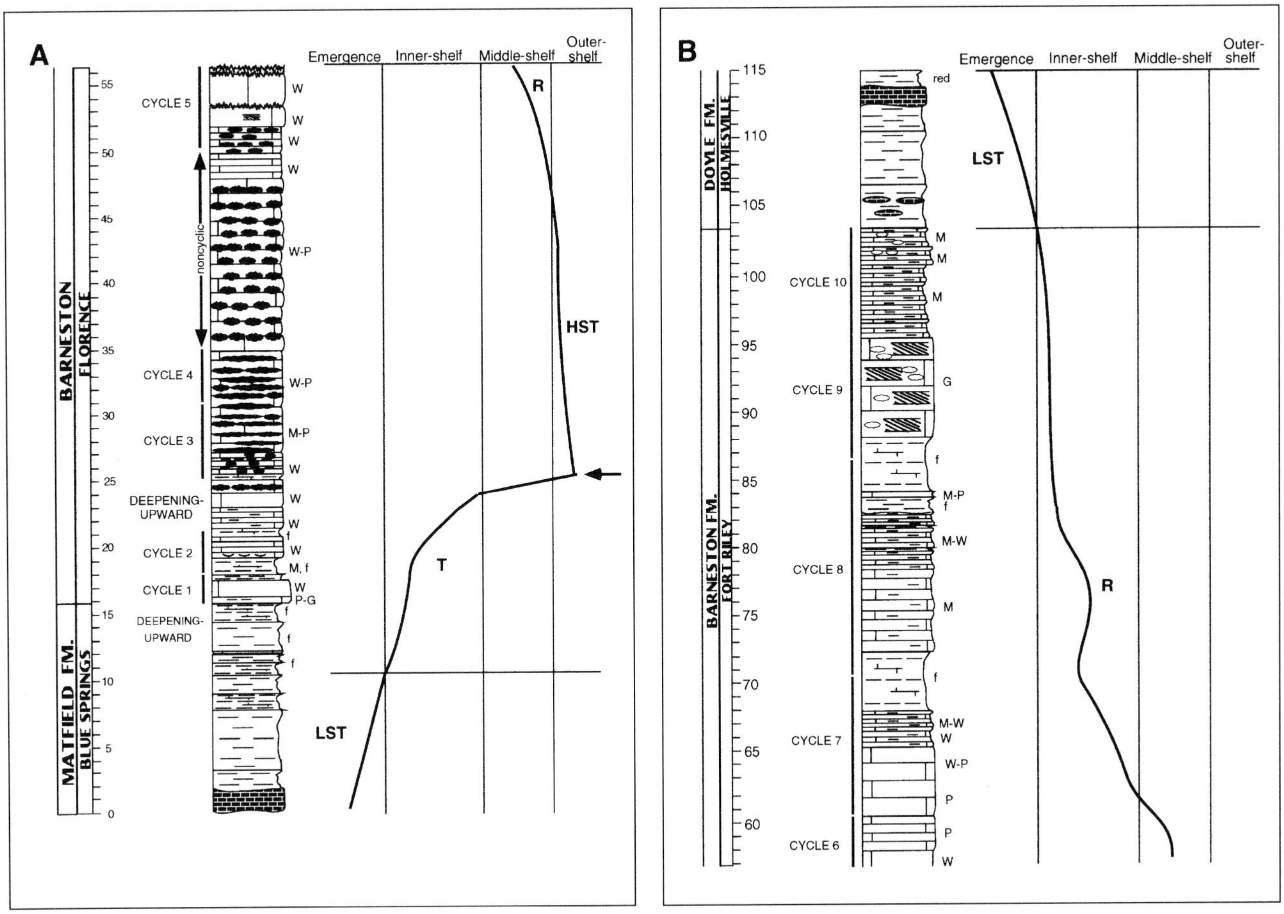

FIGURE 7 — (A) and (B) Lithostratigraphy and cycles in the Florence-Fort Riley cyclothem.

Kansas Highway 96 near the town of Leon (Section 27 T27S-R6E); the Florence Limestone along Cole Creek Road a few miles south of Leon (Sections 15 and 16 T28S-R6E); and the uppermost Florence and lower Fort Riley limestones exposed in road cuts along U.S. Highway 54 in Section 6 T26S-R6E (all of these outcrops are in Butler County); and (2) for the Fort Riley and lower Holmesville section — exposures in quarries in Section 10 T27S-R5E and in Section 9 T26S-R6E, and a roadside exposure ("Pumpjack Hill") along U.S. 54-77 in Section 27 T27S-R5E, all of which are in Butler County. Supplemental sections are illustrated in Figs. 8 and 9.

Two points concerning the usage of stratigraphic terminology must be discussed prior to describing this cyclothem. First, Prosser (1895) defined the Florence limestones from exposures near the town of Florence (particularly in the Jones Bros. quarry, now the Martin Marietta "Sunflower" quarry, in Section 6 T21S-R5E, Marion County, Kansas), but he did not formally describe these outcrops by providing a measured section. The field studies of the current authors indicate that the base of the Florence is not exposed at or for some distance around this type locality, and so contact relationships with the underlying Blue Springs Shale were never formally defined. In the past, workers have variously placed the base of the Florence either at the base of the cherty limestones of the Florence *sensu stricto*, as suggested by Prosser (1895) from exposures in the Fort Riley (Kansas) area, or at the base of the immediately underlying section of noncherty limestones, which overlie fossiliferous yellow shales (pelecypod rich) in the upper Blue Springs (Fig. 7A). The problem is that Prosser (1895, p. 773) maintained that the cherty Florence is underlain (in the Fort Riley area) by fossiliferous yellow shales of the uppermost Blue Springs; hence, according to him, the top of the Blue Springs is at the

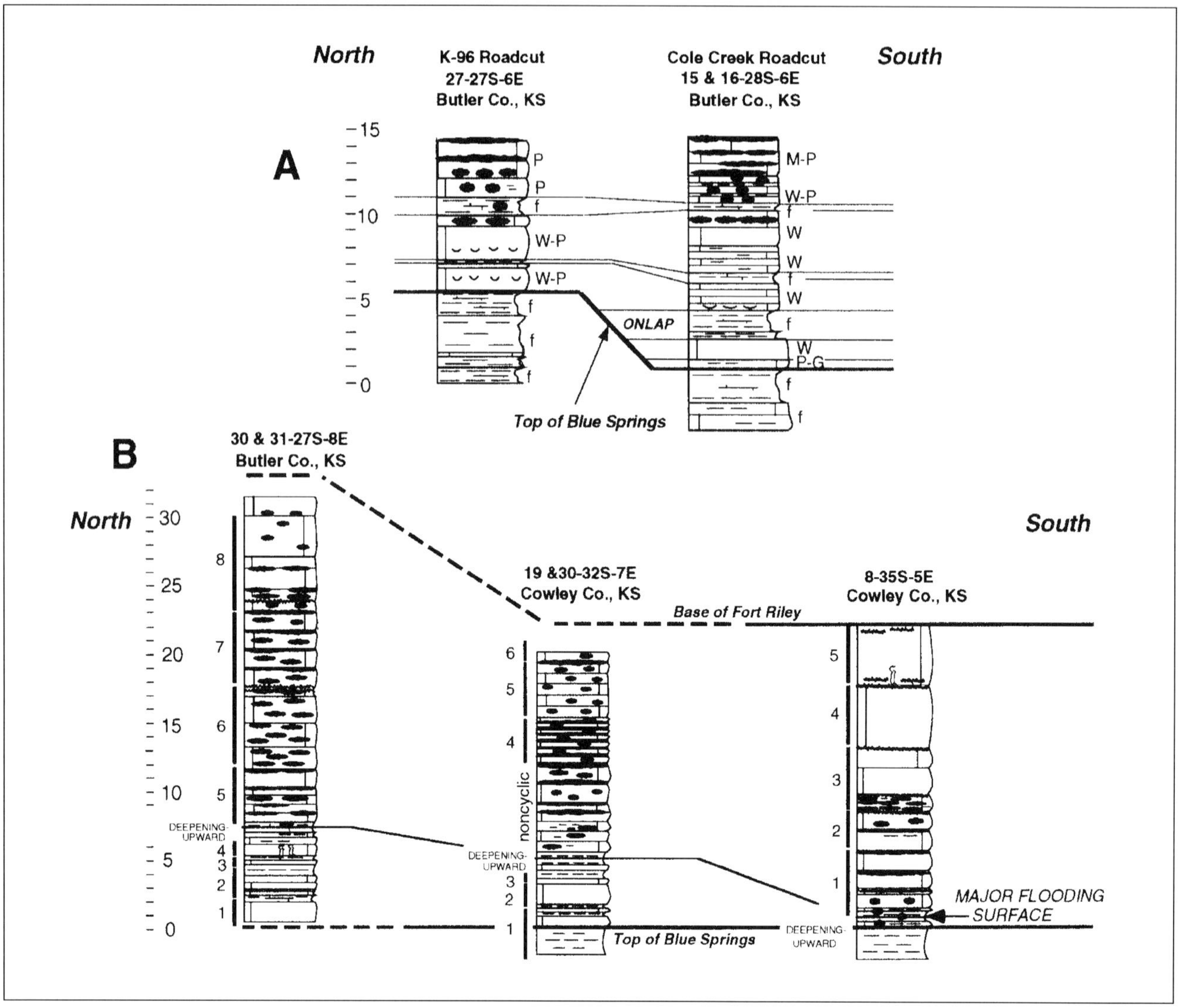

FIGURE 8 — (A) Correlation of the lower, noncherty Florence between two closely spaced outcrops showing onlapping relationships. (B) North-south cross section of units and cycles in the Florence portion of the Florence-Fort Riley cyclothem. Section is hung on the top of the Blue Springs Shale.

top of these shales. However, the authors have traced these yellow shales throughout their study area, and the shales are overlain everywhere (except in southernmost Cowley County) by a 4- to 7.2-ft (1.2- to 2.2-m) thick section of noncherty limestones, and some interbedded shales, directly beneath the main section of Florence cherty limestones (Figs. 7A and 8); it is this section of noncherty limestones that is the main issue here. Because the Blue Springs Member was defined by Condra and Upp (1931) as being dominantly composed of shales, and Prosser (1895) placed the top of the Blue Springs at the top of the fossiliferous yellow shales, in this study the authors are placing the base of the Florence at the regionally correlative contact between the fossiliferous yellow shales and the overlying noncherty limestones (Figs. 7A and 8). Second, where the Oketo Shale (which separates the Florence and Fort Riley members of the Barneston) is absent, as it is throughout the south-central Kansas study area south of Marion County, the contact between the Florence and Fort Riley limestones also is of concern, and most previous workers have not addressed this issue. In Butler and Cowley counties, the boundary between the Florence and Fort Riley is lithologically gradational (i.e., the rocks grade upward from cherty, bryozoan-rich limestones with occasional fusulinids [Florence *sensu stricto*]), through partly oncolitic, noncherty limestones with bryozoans [transitional beds], to dense oncolitic packstones [Fort Riley *sensu stricto*]. At most localities in the study area, down into northern Kay County, Oklahoma, the top of the transitional oncolitic facies between these members is a prominent stylolitic zone that the authors have chosen as the Florence-Fort Riley contact (Fig. 3).

The upper 5.4 ft (1.6 m) of the Blue Springs Member at the localities shown in Fig. 7A consists of fossiliferous and bioturbated, yellow calcitic shale, dominated by myalinid pelecypods, with accessory high-spired gastropods. These rocks are interpreted as marginal marine facies. The 7.2 ft (2.2 m) of the basal Florence Member is noncherty, and consists of two subtidal cycles (1 and 2), and part of an upward-deepening unit that persists into the lower part of the cherty Florence (Fig. 7A). The fossiliferous Blue Springs shales, together with cycles 1 and 2 and the overlying upward-deepening unit, are included within the TST of this cyclothem. Cycles 1 and 2 clearly exhibit an onlapping relationship as shown between two closely spaced roadside outcrops (Fig. 8A), one of which (in Sections 15 and 16 T28S-R4E) was used to compile Fig. 7A, and are interpreted as shallow water, marginal marine and inner shelf facies as suggested by subfacies lithologies. For example, cycle 1 consists of a 2-ft (0.6-m) thick section that fines upward from bioturbated biopackstone and grainstone through shaly biowackestone, to 5 in. (12.7 cm) of fossiliferous, fissile, black shale. The limestones in this cycle include pelecypods, crinoids, foraminifers, brachiopods, and small gastropods. The black shale contains finely comminuted bivalve fragments, and is interpreted as lagoonal rather than deepwater facies because of the absence therein of deepwater conodonts (D. Boardman, pers. comm., 1995). Cycle 2 consists of nearly 3 ft (0.9 m) of calcitic shale and shaly lime mudstone to biowackestone with a bed of coarse-grained, pelecypod biowackestone in the middle.

The upward-deepening unit consists of (in ascending order) (a) 2 ft (0.6 m) of the upper noncherty Florence section, which is thin-bedded shaly biowackestone and interbedded calcitic shale; (b) 2 ft (0.6 m) of biowackestone, cherty at the top; and (c) 5 in. (12.7 cm) of yellow-gray shale that contains deepwater conodonts (D. Boardman, pers. comm., 1995). This shale is regionally persistent throughout the study area (Fig. 8B), and is interpreted as the major flooding surface (placed in the middle of this unit) in this cyclothem. Overlying cycle 3 consists of the upper one-half of the yellow-gray shales, and then about 5.5 ft (1.7 m) of cherty mudstones, biowackestones, and fine-grained packstones with ramose and fenestrate bryozoans, fusulinids, and some brachiopods and crinoids (Fig. 7A). Overlying cycle 4 is 4.3 ft (1.3 m) of similar lithologies and biota, with 1-in. (2.5-cm) thick yellow shales at the top and base. The succeeding 15 ft (4.6 m) of the Florence consists of apparently noncyclic, cherty, biowackestone and fine-grained packstone with fenestrate and ramose bryozoans, crinoids, and brachiopods; fusulinids are present in only the basal 2 ft (0.6 m) of this section. Above the major flooding surface, cycles 3 and 4 and the overlying 15 ft of noncyclic section are interpreted as an extended highstand (stillstand) systems tract deposited in an open, outer shelf environment. Cycle 5 is a 6.2-ft (1.9-m) thick, upward-shallowing section that includes the basal 2 ft (0.6 m) of cherty biowackestone with abundant rugose corals (*Lophophyllidium* sp.) overlain by 4.2 ft (1.3 m) of stylolitic, noncherty, oncolitic biowackestone interpreted as part of the slowly RST deposited in a relatively shallow, middle shelf environment. The top of cycle 5 is the top of the Florence (Fig. 7A); the Oketo Shale is not present throughout most of the study area, and will not be discussed.

Cycle 6 constitutes the basal approximately 3 ft (0.9 m) of the Fort Riley (Fig. 7B), and consists of thick-bedded, dense oncolitic biopackstones wherein the size of oncolites decreases up-section. Cycle 7 is 9.9 ft (3.0 m) thick and includes (in ascending order) (a) 5 ft (1.5 m) of oncolitic biowackestone to packstone; (b) 2.6 ft (0.8 m) of shaly lime mudstone to biowackestone; and (c) 2.1 ft (of a total section that is 4.2 ft [1.3 m] thick: Fig. 7B) of dark to medium gray, calcitic shale with scattered brachiopods, pelecypods, and *Orbiculoides*. These shales do not contain a deepwater conodont fauna

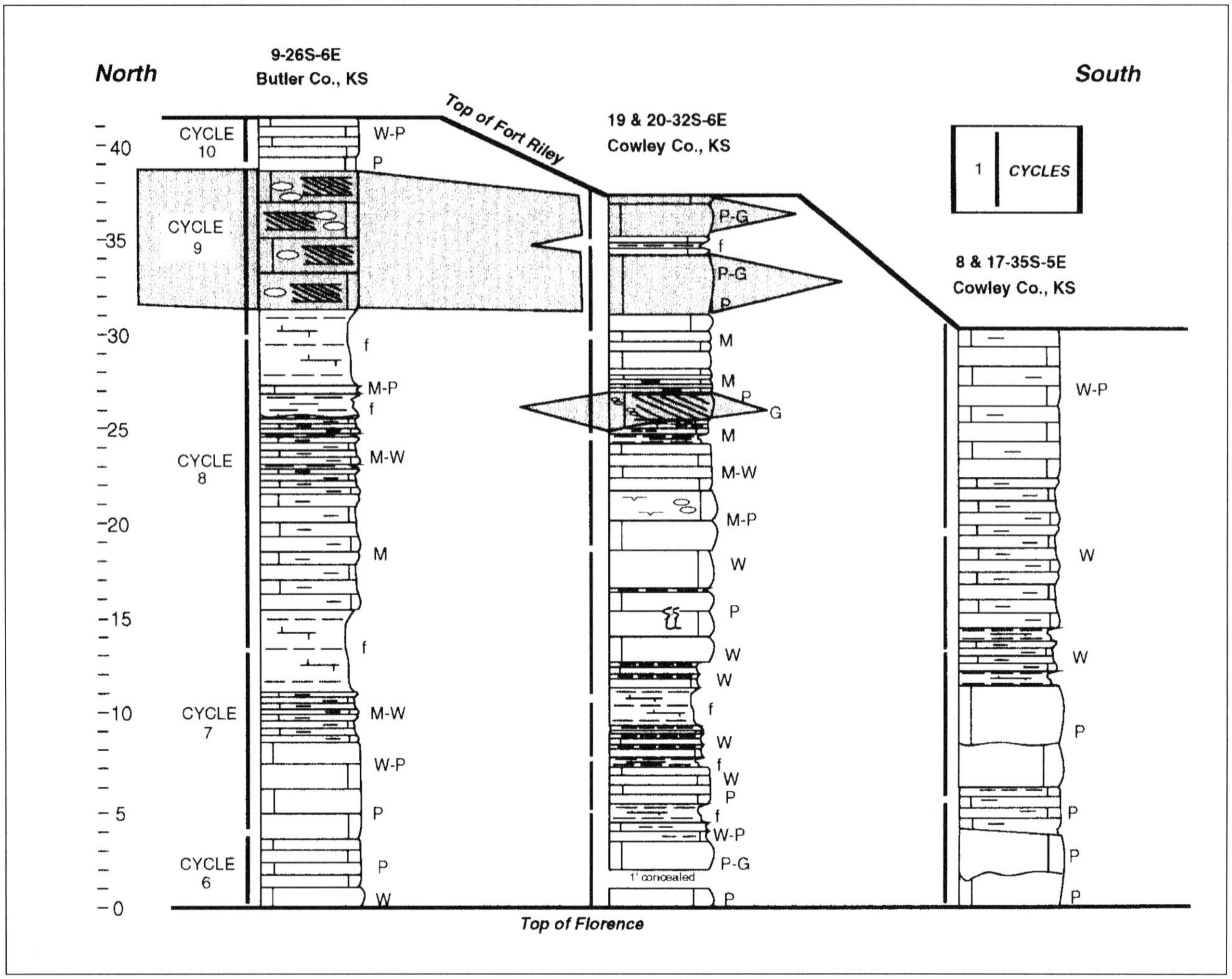

FIGURE 9 — North-south cross section showing lateral distribution of lithic units and cycles in the Fort Riley portion of the Florence-Fort Riley cyclothem. Section is hung on the top of the Florence.

(D. Boardman, pers. comm., 1995). The oncolitic limestones in cycles 6 and 7 persist from northernmost Butler County into southernmost Cowley County (Fig. 9) and into northern Kay County, Oklahoma (Mazzullo and Teal, 1994), but do not occur north from Marion County. In Geary and Riley counties, for example, equivalent strata include the "Fort Riley main ledge" of Swallow (1866). The overlying *Orbiculoides*-bearing shales shown in Fig. 7B are readily recognizable as far south as Winfield, Kansas (Fig. 9). Cycle 8 includes the upper 2.1 ft (0.6 m) of the *Orbiculoides*-bearing shale, and then about 14.5 ft (4.4 m) of interbedded calcitic shales and shaly lime mudstone to biowackestone with brachiopods, crinoids, and some pelecypods (Fig. 7B). In southern Marion County, and in the vicinity of Winfield to just north of Silverdale, Kansas, the middle part of this cycle locally includes several feet of cross-stratified biograinstone (Fig. 9). Cycles 6 through 8 are interpreted as an extended RST of middle shelf facies that grade upward to inner shelf facies. The cross-stratified biograinstones from around Winfield to Silverdale are interpreted as high-energy, inner shelf facies.

At the localities shown in Fig. 7B, cycle 9 is about 9.3 ft (2.8 m) thick, and consists of thick-bedded, relatively coarse-grained, porous biograinstones with lenses of desiccated lime mudstone and clasts of such. The cross-stratification here is "herringbone," with bimodal-bipolar dip directions. This facies, interpreted as having been deposited in a peritidal, inner shelf environment, is not present in southern Marion County, west-central Butler County, and southern Cowley County (Fig. 9); nor is it present to the north, for example, in Geary and Riley counties. Hence, it is a locally developed facies deposited on a portion of the buried Nemaha ridge (Fig. 2). Regional stratigraphic mapping has indicated that this grainstone is not a facies of the underlying shaly limestone section in the middle part of the Fort Riley (i.e., the grainstones are not developed at the expense of the middle Fort Riley, whose thickness is more or less uniform throughout the study area [Fig. 9]). There is no definitive evidence of subaerial exposure of the biograinstones at the localities shown in Figs. 7B and 9.

Cycle 10 is recognized as a separate cycle from cycle 9 because the latter is a peritidal facies, and so overlying beds should be terrestrial. Instead, cycle 10 (nearly 16 ft thick [4.9 m]) begins with 5.9 ft (1.8 m) of dark gray, shaly, laminated lime mudstones (with *Planolites, Orbiculoides*, and scattered brachiopods, pelecypods, and bellerophontid gastropods) interpreted as a shallow, nearshore lagoonal facies. This section is overlain by 2.6 ft (0.8 m) of cryptalgal (small hemispheroids) limestones (Fig. 7B). Cycles 9 and 10 are present in Butler and northern Cowley counties, but are not recognized with certainty in central and southern Cowley County or in northern Kay County, Oklahoma (Fig. 9). However, these cycles appear to be present in southern Marion County. The top of the Florence-Fort Riley cyclothem is placed at the top of cycle 10 (Fig. 7B).

Lowstand Siliciclastic Facies — At roadside outcrops along U.S. Highway 54-77 in Section 27 T27S-R5E ("Pumpjack Hill": Butler County), the Fort Riley is overlain by inferred terrestrial facies in the lower Holmesville. At this location (Fig. 7B) there are 11 ft (3.4 m) of unfossiliferous green, gray, orange, and yellow shale with calcite and some quartz-replaced anhydrite nodules similar to those described by Chowns and Elkins (1974) and Ulmer-Scholle and Scholle (1994), and horizons of brecciated micrite, with incipient pisolites, glaebules, and circumgranular cracks, interpreted as paleocaliches. The top 2 ft (0.6 m) of section exposed here consists of red soil believed to result from weathering of unfossiliferous red, terrestrial shale (Fig. 7B). The remainder of the overlying Holmesville section is poorly exposed in the study area, but regionally consists of shales (Condra and Upp, 1931; Zeller, 1968). Partly exposed in a road cut along U.S. Highway 77 in Section 3 T28S-R4E, Butler County is 9 ft (2.7 m: base not exposed) of unfossiliferous yellow-brown-orange shales also interpreted as terrestrial (Fig. 10).

Towanda Cyclothem

This cyclothem includes all of the Towanda Limestone Member of the Doyle Formation (Fig. 2). The reference section for this cyclothem (Fig. 10) is a composite of (a) roadside outcrops along U.S. Highway 77 in Section 3 T28S-R4E; (b) Towanda outcrops in Section 15 T27S-R4E (around Augusta Lake); and (c) along a county road in Section 17 T27S-R4E; all outcrops are in Butler County, Kansas. Supplemental sections are discussed below.

The basal 0.5 ft (0.2 m) of the Towanda section at the localities shown in Fig. 10 consists of porous, brecciated lime mudstone, with evaporite-nodule molds, inferred to be an evaporite dissolution-collapse breccia. This section is interpreted as peritidal sabkha deposits. The overlying section consists of about 5 ft (1.5 m) of unfossiliferous shale which is dark green-gray in the lower part and light green-gray above. Conceptually, somewhere within this section, which is interpreted as marginal marine facies, is the major flooding surface of this peritidal-dominated cyclothem. Hence, the lower few feet of Towanda is an upward-deepening (from peritidal to marginal marine) TST. The upper part of this shale, and the overlying 3 ft (0.9 m) of brecciated lime mudstone, compose peritidal cycle 1 (Fig. 10). This breccia is interpreted as a supratidal sabkha, evaporite-dissolution breccia. Cycles 2

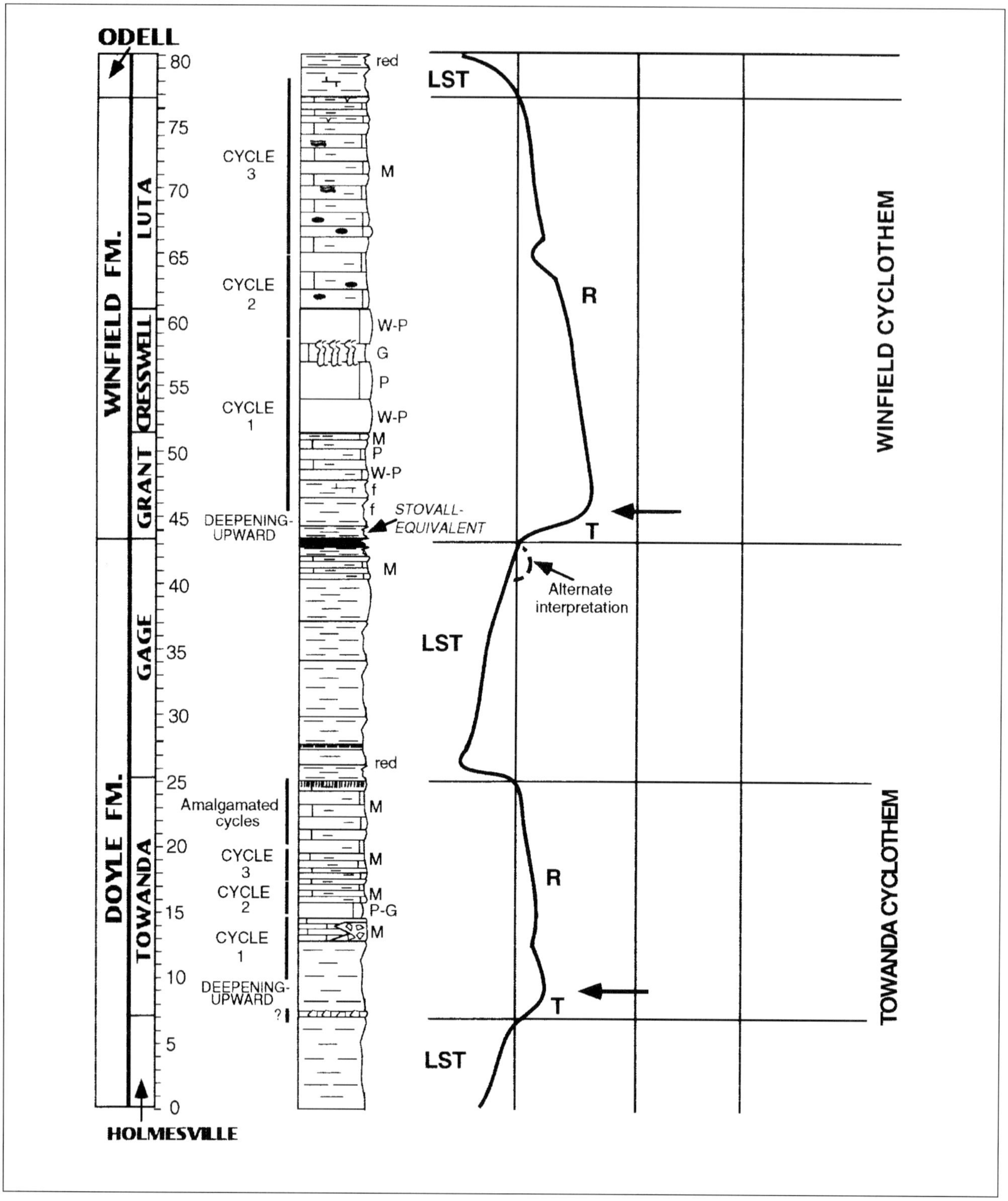

FIGURE 10 — Lithostratigraphy, cyclothems, and cycles in the Towanda and Winfield cyclothems.

and 3 are each about 3 ft (0.9 m) thick; cycle 2 is an upward-fining and upward-shallowing section from biopackstones and grainstones deposited in a shallow inner shelf environment, to shaly lime mudstones with abundant calcite geodes believed to be sabkha facies with replaced evaporite nodules. The upper beds of the Towanda are a section of thin, amalgamated peritidal cycles composed of shaly lime mudstones with calcite geodes; this cycle is capped by a thin lime mudstone layer with abundant vertical rootlets inferred to be an incipient paleosol (Fig. 10). Above the inferred major flooding surface, the Towanda section represents the peritidal-dominated RST. (Fig. 10).

The Towanda is the most lithologically variable cyclothem within the Chase Group. Facies in Cowley County variously include peritidal (sabkha) limestones at some localities, and intraclastic, oolitic biograinstones at others. Similar facies occur in southern and central Butler County, but at the type section indicated by Fath (1921), the section is dominantly peritidal with some fine-grained, foraminiferal wackestones and packstones. In contrast, the 10 ft (~3.1 m) of exposed Towanda section in southern Marion County at road cut exposures along U.S. Highway 77 in Section 18 T21S-R5E and in the Sunflower quarry (in Section 6 T21S-R5E) is dominated by foraminiferal packstones and grainstones and associated shallow subtidal lime mudstones and shales.

Lowstand Siliciclastic Facies — At roadside outcrops in Section 31 T30S-R4E in Cowley County, the approximately 5.5 ft (1.7 m) of the lower Gage Member of the Doyle Formation (Fig. 2) consists of unfossiliferous red shale with small, inferred caliche nodules at the top and possible paleosols throughout. These beds (Fig. 10) are interpreted as terrestrial facies. The remainder of the Gage Member exposed here, and in roadside outcrops in Sections 26 and 36 T28S-R4E (Butler County), consists of unfossiliferous yellow-gray and green-gray shale, with presumed paleosols, overlain by 1.5 ft (0.5 m) of laminated, unfossiliferous, shaly limestone characterized by laminae of recrystallized micrite and laminar pores that may have resulted from evaporite dissolution. These laminated carbonates, and the underlying shales, may entirely be terrestrial facies (Fig. 10). Alternatively, this entire section may record a vertical transition from terrestrial shales, to possibly peritidal carbonate sabkha facies (i.e., the laminated, recrystallized micrite) of a minor, short-lived transgression (Fig. 10). However, the top 1.5 ft (0.5 m) of the Gage at the locality shown in Fig. 10 is composed of unfossiliferous and blocky, black and organic matter-rich shale, with slickensides and antiforms, interpreted as a paleosol (Retallack, 1988). This presumed paleosol has also been recognized at the top of the Gage at other localities in the study area.

Winfield Cyclothem

The composite reference section for this cyclothem (Fig. 10) includes the following outcrops: (a) the Grant and Cresswell members exposed in road cuts along 170th St. in adjoining Sections 25 and 36 T28S-R4E in Butler County; (b) these units, plus the Stovall, exposed along Kansas Highway 15 in Section 31 T30S-R4E in Cowley County; and (c) the Luta and basal Odell exposed in the Martin Marietta "North Marion quarry" in Section 28 T19S-R4E in southeastern Marion County, Kansas. Although not the thickest cyclothem in the Chase Group, the marine phase of this cyclothem extends the farthest south (into Oklahoma) of all of the cyclothems in the Chase Group (J. Chaplin, pers. comm., 1994).

The authors have chosen to retain the name Luta to refer to the upper member of the Winfield Formation, above the Cresswell. Although the Kansas Geological Survey presently includes the beds designated as Luta within the Cresswell Member (Zeller, 1968), such a change has never been formally designated. The Luta is sufficiently distinct from the Cresswell, and also is regionally mappable to warrant its retention as a formal stratigraphic unit.

The Stovall Member of the Winfield Formation is not present everywhere in the study area. Where it is present, it consists of about 3 to 4 ft (0.9 to 1.2 m) of nonporous, shaly lime mudstone with rare pelecypods (and at some localities in Cowley County, oncolites). Bed thickness generally increases upward from laminae to thin beds. The Stovall is interpreted as a peritidal to extremely shallow water, inner shelf facies. Such a lithology is different from what has been considered "typical" Stovall, which at locations to the north of the study area is commonly cherty (Condra and Upp, 1931). Nevertheless, unless a new member is designated for this unit, it is regarded as the Stovall Member because of its position in the stratigraphic section (Fig. 10). At those localities where the Stovall seemingly is absent (e.g., in Sections 25 and 36), the lower 1.5 ft (0.5 m) of the Grant consists of laminated and unfossiliferous, calcitic shale to shaly lime mudstone that resemble tidal bundles or tidal rhythmites described by Archer (1991) and Archer et al. (1994); this section is stratigraphically equivalent to the Stovall (Fig. 10). The overlying approximately 3.5 ft (~1 m) of the Grant Member consists of yellow-gray, calcitic shale and shaly limestone, with brachiopods, pelecypods, bryozoans, and crinoids. A thin zone of dark gray, fossiliferous shale occurs in the middle of this unit, which is interpreted as the major flooding surface in this cyclothem. Hence, the Stovall (where present) and/or Stovall-equivalent basal strata in the Grant, together with the lower one-half of the 3.5-ft thick section of overlying Grant, are interpreted as an upward-deepening TST that includes, in ascending fashion, peritidal facies to moderate depth, inner shelf facies (Fig. 10)

Overlying cycles 1, 2, and 3 are included within the extended RST of this cyclothem, facies of which were deposited in progressively shallower inner shelf environments (Fig. 10). Cycle 1 includes the upper one-half of the 3.5-ft thick Grant section described above, which is overlain by (a) 4 ft (1.2 m) of the upper Grant Member, which consists of shaly biowackestone to packstone with abundant conspicuous, red-stained oncolites (referred to as the "pebbly zone" in older studies); and (b) a 6.4-ft (2.0-m) thick section in the Cresswell of biowackestone to packstone at the base (with brachiopods, bryozoans, and crinoids) overlain by porous, oncolite biograinstone (with accessory foraminifers, crinoids, and brachiopods). The top of this section is intensely vertically burrowed (Fig. 10), but has no evidence of subaerial exposure. Cycle 2 consists of (in ascending order) (a) 2.8 ft (0.9 m) of relatively coarse-grained oncolite biograinstone at the base, grading upward to finer grained oncolite wackestone to packstone with similar fossils (obvious evidence of subaerial exposure has not been found at the top of this unit, which is the top of the Cresswell [Fig. 10]); (b) about 4 ft (1.2 m) at the base of the Luta, of light gray, bioturbated, shaly lime mudstone, with some chert, crinoids, and brachiopods (this unit contains some soft and crumbly nodules that, based on XRD analyses, are composed of anhydrite that has been largely replaced by chert — these nodules may be of secondary, postdepositional origin); and (c) 0.7 ft (0.2 m) of similarly fossiliferous, light gray lime mudstone with burrows, desiccation cracks, and apparent molds of evaporite nodules. As so described, cycle 2 includes low-energy inner shelf facies that shallow upward to intertidal facies.

Peritidal cycle 3 is about 13 ft (4.0 m) thick, and grades upward from (a) about 9 ft (2.7 m) of thin-bedded, light gray, bioturbated, shaly and unfossiliferous lime mudstone, also with secondary(?) evaporites (silicified anhydrite), interpreted as shallow water to possibly peritidal, inner shelf facies; (b) 3 ft (0.9 m) of laminated, cryptalgal lime mudstone, also with silicified evaporite nodules, and large-diameter desiccation polygons, interpreted as peritidal sabkha facies (this unit is the top of the Luta); and (c) the lower approximately 1 ft (0.3 m) of a 2-ft (0.6-m) thick section in the basal Odell of green and light gray calcitic shale, and shaly lime mudstone with cryptalgal laminites and fenestrae, interpreted as marginal marine and peritidal facies. The top of the Winfield cyclothem is placed in the middle of this 2-ft thick shale (Fig. 10).

As shown in Fig.11, the Luta is recognized throughout the study area, but it is characterized by

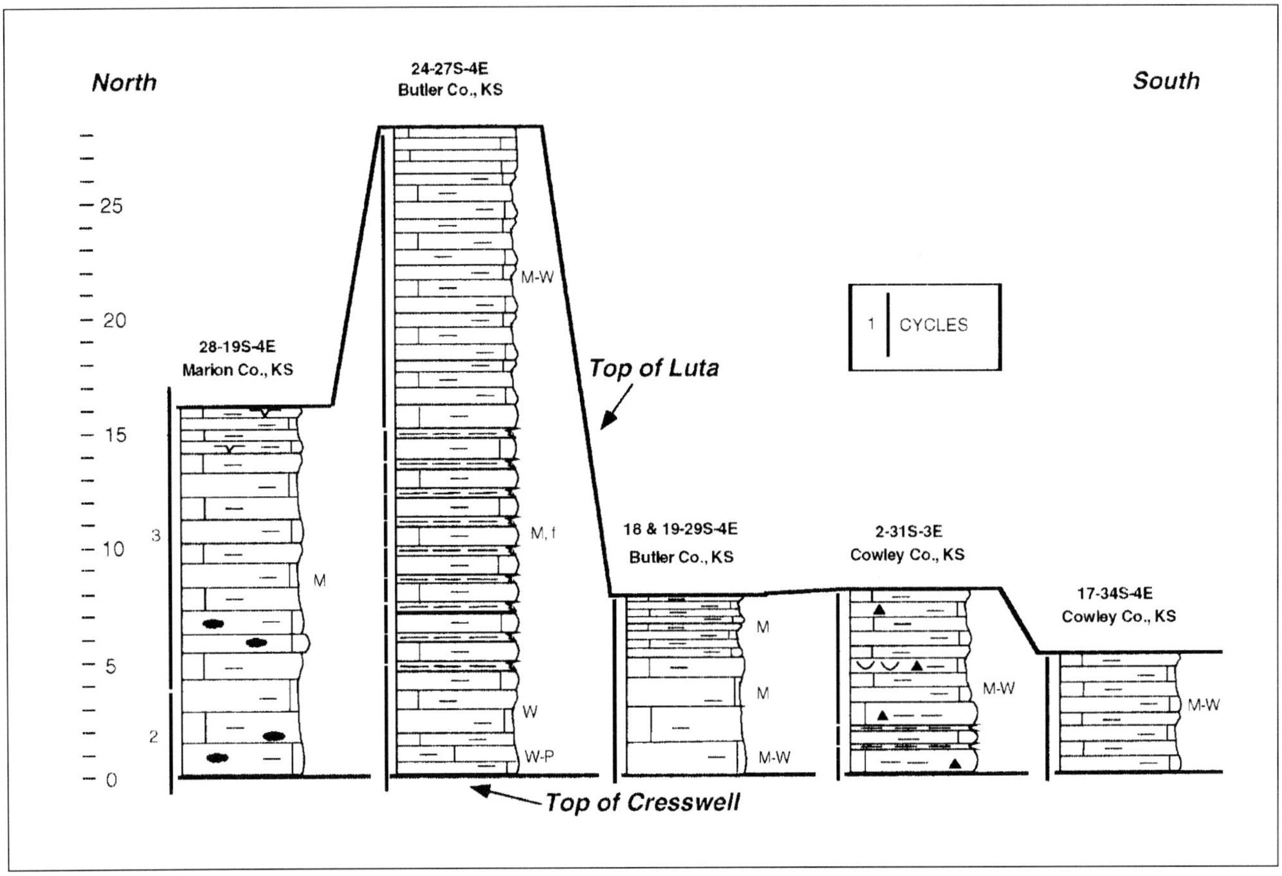

FIGURE 11 — North-south cross section showing cycles and thickness variations in the Luta Member of the Winfield Formation (upper part of the Winfield cyclothem). Section is hung on the top of the Cresswell Member of the Winfield Formation.

pronounced facies changes and extreme variations in thickness. With reference to Luta facies as illustrated in Fig. 10, correlative facies in Butler and Cowley counties instead are of more marine aspect in that they are composed of generally fossiliferous limestones. The Luta is about 16 ft thick (4.9 m) at the reference section shown in Fig. 10, 10 to 12 ft thick (3.1 to 3.7 m) at two closely spaced locations in central Butler County (Section 7 T27S-R4E and Sections 18 and 19 T29S-R4E), and 28 ft thick (8.5 m) at another nearby location in central Butler County (Section 24 T27S-R4E); it is 8 ft thick (2.4 m) in northern Cowley County (Section 2 T31S-R3E), and 5 ft thick (1.5 m) in central-southern Cowley County (Section 17 T34S-R4E) (Fig. 11).

Lowstand Siliciclastic Facies — The reference section for lowstand deposits represented by the Odell Formation is at the Spring Hill Municipal Golf Course in Arkansas City, in Section 18 T34S-R4E in Cowley County, Kansas (Fig. 12). Here, the Odell is about 30 ft (9.2 m) thick, and consists mostly of unfossiliferous green and gray shales with scattered paleosols. However, the top 3 ft (0.9 m) of this section here consists of fossiliferous shale and some shaly lime mudstones (Fig. 12). This unit changes facies to a bed of shaly, oncolitic limestone at "140 Foot Hill" just a few miles to the south (by the Quaker Haven Camp in 5-35S-4E), and is the unit that Toomey and Mitchell (1986) mistakenly included within the Krider Member of the Nolans Formation. At roadside exposures along Kansas Highway 15 in northern Cowley County (Section 3 T31S-R3E), just east of Udall, as well as throughout Butler County, the upper Odell is not similarly fossiliferous. Instead, it consists of unfossiliferous grayish-green and green shales, also with some paleosols.

Nolans: Krider-Paddock-Herington Cyclothem

This cyclothem (Fig. 12) includes the Krider, Paddock, and Herington members of the Nolans Formation, and the basal beds of the overlying Wellington Formation (Fig. 2). The reference sections for this cyclothem include (1) for the Krider, the road cut exposure along Kansas Highway 15 in Section 3 T31S-R3E in Cowley County; (2) for the Paddock, the exposure along a north-south county road in Section 3 T29S-R3E in Butler County; (3) for the Herington, the road cut exposure along Kansas Highway 15 in Section 34 T30S-R3E in Cowley County; and (4) for the uppermost Herington and lower Wellington, the exposures adjoining the Santa Fe Railroad trestle where it crosses U.S. Highway 54 in Sections 24 and 25 T27S-R3E in Butler County.

Cycle 1 and the base of this cyclothem begin within the uppermost Odell where it is fossiliferous (Fig. 12); where the Odell is not fossiliferous, the base of the cyclothem is defined at the base of the Krider Member of the Nolans Formation. Within cycle 1, the uppermost few feet of the fossiliferous Odell are interpreted as marginal marine facies. The overlying 2.5 ft (0.8 m) of the Krider is composed of shaly biowackestone to packstone at the base, grading upward to shaly biowackestone (both lithologies with pelecypods, brachiopods, crinoids, and relatively abundant high-spired gastropods), and capped by black, calcitic shale with abundant fenestrate bryozoans and accessory pelecypods, crinoids, and brachiopods (the black shales are about 1 ft [0.3 m] thick, and the top of cycle 1 is placed in the middle of them). These units in cycle 1 are interpreted as shallow, inner shelf facies (i.e., the limestones) that grade upward to marginal marine, lagoonal facies (i.e., the black shales) and which constitute the TST of this cyclothem (Fig. 12). The upper one-half of the fossiliferous black shale section, and near the top of the 2-ft (0.6-m) thick section of overlying limestone in the upper Krider (which is biowackestone to packstone at the base, biowackestone at the

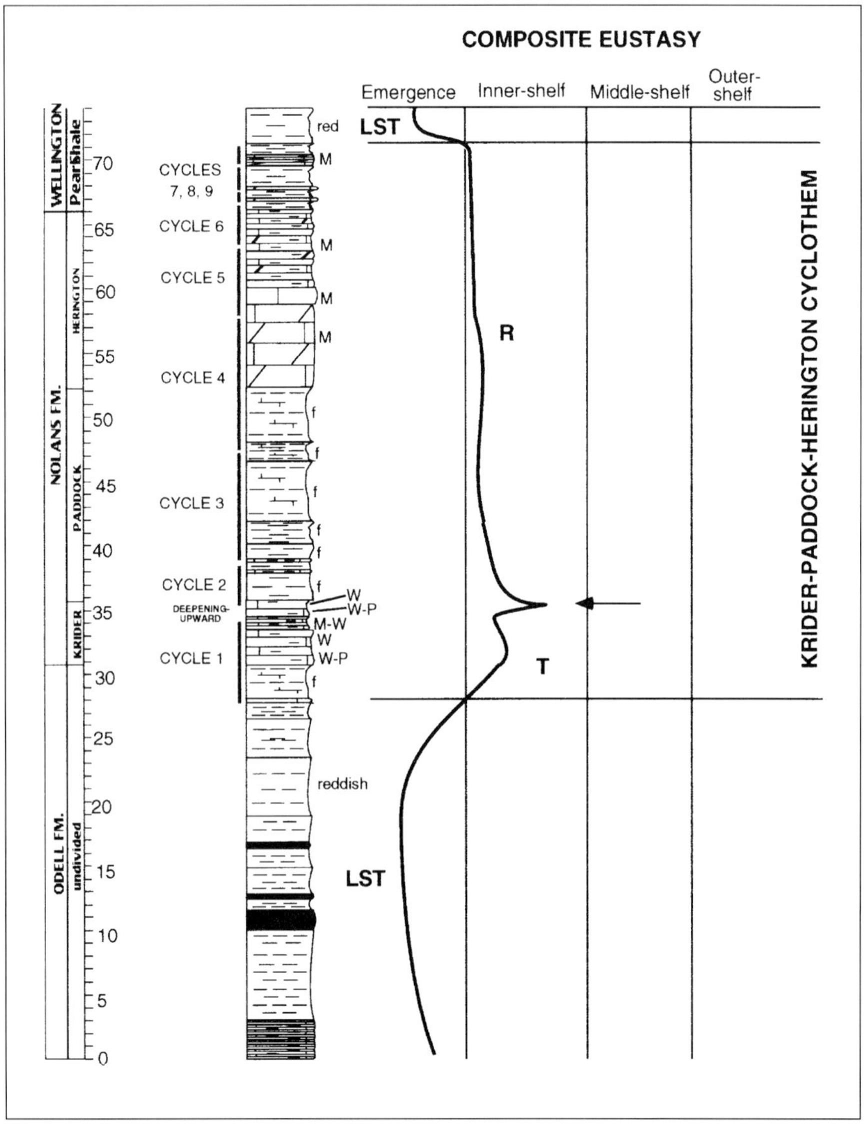

FIGURE 12 — Lithostratigraphy and cycles in the Odell lowstand tract and Krider-Paddock-Herington cyclothem.

top, with fenestrate bryozoans and accessory crinoids, pelecypods, and brachiopods), constitute an upward-deepening section in this cyclothem. Interpreted facies are marginal marine lagoonal deposits up through a presumed major flooding surface represented by middle inner shelf facies, respectively (Fig. 12).

Cycle 2 includes the uppermost few cm of the Krider and, in the basal Paddock, the overlying approximately 4 ft (1.2 m) of sparsely fossiliferous (dominated by pelecypods with some crinoids and bryozoans) yellow-gray shales capped by laminated cryptalgal deposits (Fig. 12). This cycle is interpreted as upward shallowing from inner shelf to marginal marine facies. Cycle 2 and successive cycles are part of an extended RST in this cyclothem (Fig. 12). Cycle 3 consists of about 6.4 ft (2.0 m) of similarly sparsely fossiliferous (pelecypods, brachiopods) yellow-gray shales (marginal marine facies) capped by shales with presumed calcite-replaced evaporite nodules, interpreted as peritidal sabkha facies. Cycle 4 includes (in ascending order) (a) the upper 5 ft (1.5 m) of the Paddock, which consists of yellow-gray calcitic shales, with scattered pelecypods and laminae of palisade calcite crystals presumably replacive of subaqueous evaporites (this section is interpreted as mixed peritidal and marginal marine facies); and (b) about 5 ft (1.5 m) in the basal Herington of thin-bedded, light gray, shaly and generally unfossiliferous, calcitic dolomites with abundant calcite nodules presumably replacive of evaporites (syndepositional or postdepositional?), interpreted as shallow subtidal to peritidal facies. Discontinuous lenses of crinoid-siliceous spicule wackestone to packstone occur in the basal 0.5 ft (0.15 m) of this section here (but nowhere else in the study area), and may be a lag deposit. Cycle 5 is about 5.5 ft (1.7 m) thick, and grades upward from dolomites as in the immediately underlying cycle (shallow subtidal facies), to laminated dolomites (laminations appear to have resulted from physical processes of deposition, e.g., tidal rhythmites?, rather than being of cryptalgal origin) interpreted as peritidal facies (Fig. 12). Calcite nodules that presumably are replacive of evaporites are not found in this section here, but are abundant in this same facies elsewhere in the study area. The upper 2.7 ft (0.8 m) of the Herington consists of similar unfossiliferous lime mudstone which does not contain calcite nodules here, but which is abundant in other areas. The basal 5.3 ft (1.6 m) of the lower Wellington consists of interbedded yellow-brown, unfossiliferous shale and shaly lime mudstone, both with presumed replaced evaporite nodules; near the top, light gray lime mudstone with palisade calcite crystals is presumably replacive of subaqueous evaporites. These upper 2.7 ft of the Herington, and the basal Wellington, are interpreted as four stacked peritidal cycles (6 through 9) of marginal marine to peritidal sabkha subfacies; the top of the cyclothem is placed at the top of cycle 9 in this section (Fig. 12).

Lowstand Siliciclastic Facies — Cycle 9 in the Krider-Paddock-Herington cyclothem is overlain by 2.8 ft (0.9 m) of unfossiliferous red shale in the Wellington, with conspicuous "cauliflower" chert nodules that likely replaced evaporite nodules (Chowns and Elkins, 1974; Ulmer-Scholle and Scholle, 1994). This section is interpreted as a terrestrial sabkha facies (Fig. 10).

INFERRED PALEO-WATER DEPTHS OF CYCLOTHEMS

Conceptual Foundation From Previous Studies

Carbonates — Assumed paleo-water depths of maximum transgressive carbonate units in Upper Paleozoic cyclothems in Kansas have been the subject of debate for decades. The two schools of thought on the matter, the shallow water versus deepwater origin, are largely based on paleontological arguments. Proponents of the deepwater origin of transgressive carbonate units include Elias (1937, 1964) and Hattin (1957). Elias invoked analogy with the then-known habitats of some modern large foraminifers in suggesting that fusulinid-bearing transgressive units in Kansas cyclothems were deposited at depths of 180 to 200 ft (55 to 61 m). Accordingly, cherty limestones in the Chase Group (the Threemile, Schroyer, and Florence), some of which contain fusulinids, were believed by these authors to have been deposited in the depth range of 110 to 200 ft (34 to 61 m). Later, Moore (1964) and Lutz-Garihan and Cuffey (1979) inferred considerably shallower depths for these units on the basis of paleontological data and regional stratigraphic relationships. According to Boardman et al. (1984) and Boardman and Nestell (1993), fusulinid-bearing rocks in Pennsylvanian and some Permian cyclothems were deposited at depths of less than 165 ft (50 m). Other proponents of the shallow water school (Tasch, 1957; Imbrie et al., 1959, 1964; Laporte, 1962; Laporte and Imbrie, 1964; McCrone, 1963, 1964) suggested that fusulinid-bearing rocks were deposited at even shallower depths of about 50 to 60 ft (15 to 18 m) or less. In fact, McCrone (1964) went so far as to suggest that most Kansas Wolfcampian cyclothems were probably deposited in water less than about 60 ft (18 m) deep, and an overall shallow marine setting, at least for the Chase Group, was also suggested by Twiss (1991a) and Miller and West (1993). Tasch (1957) suggested that fusulinids may have lived in water as shallow as 5 ft (1.5 m) deep.

Shales — The dark gray and typically phosphatic, conodont-rich, black shales in transgressive units in Kansas Pennsylvanian cyclothems are considered by many workers to be progressively deeper water deposits (Elias, 1964; Watney, 1985a; Heckel, 1986, 1994; Boardman and Nestell, 1993; and others). Heckel (1977, 1994) suggested that these black shales were deposited in anoxic environments at depths of at least 330 ft (100 m), whereas others have suggested somewhat shallower depths of 230 to 330 ft (70 to 100 m: see Boardman and Nestell, 1993). Black shales become less common in Kansas Lower Permian cyclothems relative to Pennsylvanian cyclothems, suggesting general long-term shallowing during the Late Paleozoic (Elias, 1964), and some have been interpreted to have been deposited at even shallower depths than those in the Pennsylvanian (Imbrie et al., 1964; Laporte and Imbrie, 1964; McCrone, 1964). According to Boardman et al. (1984) black shales in Lower Permian and some Upper Pennsylvanian cyclothems may have been deposited in depths slightly greater than 165 ft (50 m). In contrast, others argue for a lagoonal origin of some of these shales based on their geochemistry (Coveney et al., 1989).

Depths Inferred for the Chase Group

No fusulinids have been found in the cherty Threemile and Schroyer cyclothems in the study area, although some have been described from the basal part of the Threemile Member in Chase County, Kansas (Sanderson and Verville, 1970). Instead, bryozoans are the dominant biotic element in these cyclothems, neither of which contain black shales. Modern bryozoans are particularly common in open-shelf environments at depths of about 70 to 260 ft (21 to 80 m)(Ryland, 1967). Notwithstanding likely evolutionary changes in bryozoan habitat since the Permian, it is suggested that maximum paleo-water depth of the middle to outer shelf TST in the Threemile cyclothem may have been in this range. In contrast, cherty, bryozoan-rich, middle to outer shelf transgressive limestones in the Schroyer cyclothem are dominated by slightly coarser grained and less micritic lithologies (fine-grained packstones), the presence of which may suggest somewhat shallower depths of deposition, perhaps 70 to 100 ft (21 to 31 m). This contention is in accordance with paleo-depths postulated by McCrone (1964) and others (60 ft or less: 18 m) for bryozoan-rich carbonates in Upper Paleozoic cyclothems in

Kansas. Fusulinids are relatively abundant in the cherty, prolonged highstand systems tract (HST) in the Florence cyclothem, which also is bryozoan-rich. In correlative units in the Permian basin, fusulinids generally are not found below inferred depths of about 50 to 60 ft (15 to 18 m) unless they have been reworked (A.M. Reid, pers. comm., 1994). Hence, it is believed that water depths of cherty, fusulinid-bearing beds in the Florence may have been on the order of 50 to 60 ft (15 to 18 m), an inferred depth range that is in accordance with that suggested by the shallow water school of workers listed above. Associated bryozoan-rich wackestones and packstones that are devoid of fusulinids, however, may represent water depths like those in the Schroyer cyclothem. If such were the case, then many more thin cycles would be present than are indicated in Figure 7B; this aspect of the sedimentology of the Florence awaits further study.

The authors believe that the cherty limestones composing the Threemile and Schroyer cyclothems, and the Florence portion of the Florence-Fort Riley cyclothem, represent the deepest water facies of the carbonate deposits in this section; these rocks were deposited on the middle and outer shelves (Figs. 3 and 7A). In contrast, inferred water depths for the noncherty limestone cyclothems in the Chase Group (Kinney cyclothem, Fort Riley portion of the Florence-Fort Riley cyclothem, and the Towanda, Winfield and Krider-Paddock-Herington cyclothems) represent considerably shallower water depths, and all such limestones were deposited mainly on the inner shelf (Figs. 3, 7B, 10, and 12).

Black shales occur in the Kinney, Florence-Fort Riley, and Krider-Paddock-Herington cyclothems, and medium to dark gray, *Orbiculoides*-bearing shales occur in the Fort Riley. The lack of a deepwater conodont fauna in these shales, the absence of nonskeletal phosphate, and lack of evidence of syndepositional dissolution of included carbonate allochems in these shales are not consistent with their deposition in deep (~330 ft [100 m]), anoxic environments (Heckel, 1986, 1994). It is for these reasons that these shales have been interpreted as being mostly of lagoonal origin.

HIERARCHAL ORDERING OF CYCLES AND CYCLOTHEMS: SEQUENCE-STRATIGRAPHIC APPROACH

A composite vertical section of the seven cyclothems and intervening lowstand deposits within the Chase Group, compiled from Figs. 3, 7, 10, and 12, is shown in Fig. 13. The authors infer that red shales in lowstand systems tracts (LSTs) in this section indicate periods of maximum emergence. With this codicil, then the seven marine cyclothems can logically (and naturally) be grouped into six lower frequency units separated by lowstand siliciclastics which include maximum lowstand red shales. In ascending order, these units are (1) the Wreford, the only compound-cyclothemic unit within the Chase, which includes the Threemile and Schroyer cyclothems; (2) the single-cyclothemic Matfield (Kinney cyclothem); (3) Barneston (Florence-Fort Riley cyclothem); (4) Doyle (Towanda cyclothem); (5) Winfield (Winfield cyclothem); and (6) Nolans (Krider-Paddock-Herington cyclothem)(Fig. 13). The pertinent question is what do the cycles, cyclothems, and these lowest frequency units represent in terms of sequence stratigraphy? A possible answer to this question lies in analysis of the (1) regional extent of cyclothems and component cycles; (2) stacking patterns, thicknesses, and types of cycles within individual cyclothems; and (3) in the compound-cyclothemic Wreford unit, comparison of the thickness and relative extent of transgression of the individual cyclothems.

Regional Correlation of Cycles and Cyclothems

The cyclothems in the Chase Group incorporate formally designated formations and members that mostly persist from Nebraska, through Kansas, and into northern Oklahoma (Condra and Upp, 1931). Accordingly, the fact that cyclothems inherently are regionally widespread units suggests that their overall deposition was controlled, at least in large part, by allogenic processes. However, important exceptions to this generality exist: (1) the fact that the grainstones and overlying micritic lithologies in the upper part of the Fort Riley (i.e., cycles 9 and 10 in the Florence-Fort Riley cyclothem: Fig. 13) do not occur to the north of the study area, or in central and southern Cowley County (Fig. 9); and (2) the Luta Member of the Winfield Formation, which composes cycle 3 and part of underlying cycle 2 in the Winfield cyclothem (Fig. 13), is not present in northern Kansas and Nebraska (Condra and Upp, 1931). These facts may suggest that the development of at least part of the Florence-Fort Riley and Winfield cyclothems was controlled by factors other than allogenic processes.

The thicknesses and types of cycles (subtidal versus peritidal) in each of the cyclothems, from the representative localities illustrated in Figs. 3, 7, 10, and 12, are shown only for reference in the composite section in Fig. 13. In terms of their regional extent and correlation, however, the important illustrations are Figs. 4, 6, 8, 9, and 11. For example, cycles are *not* regionally correlative in the Threemile and Schroyer cyclothems (Fig. 4), the Florence-Fort Riley cyclothem (Figs. 8 and 9), and the upper part of the Winfield cyclothem (the uppermost Cresswell and Luta: Fig. 11). At most localities in the study area, the Kinney cyclothem seemingly is a single, upward-shallowing cycle. However, three cycles conceivably can be interpreted to exist at the old quarry locality in Section 3 T35S-R5E in southernmost Cowley County (Fig. 6), and together with the inferred lowstand carbonate wedge at the Bannon quarry in Butler County (Fig. 5) complicate interpretations of cycle correlation in this cyclothem. Cycles are only grossly regionally correlative in the Towanda cyclothem, the lower part of the Winfield cyclothem, and in the Krider-Paddock-Herington cyclothem (Figs. 6, 10, and 12). These observations suggest that both allogenic *and* autogenic controls affected cyclothem development in the Chase Group.

Cycle Stacking Patterns and Types

Stacking patterns and types of cycles within the TSTs and the HST/RSTs of the Chase cyclothems are shown in several illustrations. Relationships indicated in the TSTs of these cyclothems are described. (1) There is no consistent trend in cycle thickness in the Threemile and Schroyer cyclothems (Fig. 4). According to many workers, cycles inherently should thicken upward in response to increasing accommodation during transgression in dominantly allogenically controlled depositional systems (Goldhammer et al., 1989, 1993). However, cycles do deepen upward in these cyclothems in apparent response to rates of transgression that exceeded rates of sediment deposition. (2) Although cycles likewise deepen upward in the Florence-Fort Riley cyclothem, they also thin upward (Fig. 8), an observation that is in apparent contradiction to predicted depositional responses to increasing accommodation in allogenically dominated systems. Cycles also deepen upward in the Towanda and Winfield cyclothems (Fig. 10), and both thin upward and deepen upward in the Krider-Paddock-Herington cyclothem (Fig. 12). Cycles that both deepen upward and thin upward alternatively can be explained by progressively slower rates of deposition of increasingly deeper water facies during transgression. However, at least cycles 1, 2, and 3 in the Florence-Fort Riley cyclothem, and respective cycles in the Towanda, Winfield, and Krider-Paddock-Herington cyclothems, are not deepwater deposits, but rather are relatively shallow water, inner shelf facies. Hence, the presence of such cycles, which mostly are not regionally correlative, in a TST, suggests a strong component of autogenic control on their deposition.

Cycles in RSTs theoretically should thin upward and include progressively more peritidal-type facies in response to decreasing accommodation *if* cycle development was related to dominant allogenic controls on deposition (Goldhammer et al., 1989, 1993). Relationships indicated in the RSTs of the Chase cyclothems are (1) consistent upward-thinning cycles are not apparent in the Threemile, Schroyer, Florence-Fort Riley, and Winfield cyclothems, although a change from subtidal-type to peritidal-type cycles is noted in the latter three cyclothems (Figs. 4, 8, 9, and 10); and (2) component cycles both thin upward and change from subtidal-type to peritidal-type in the Towanda and Krider-Paddock-Herington cyclothems (Figs. 10 and 12). These relationships suggest mixed allogenic and autogenic controls on cycle development.

Compound-Cyclothemic Wreford Unit

The interesting aspects of cyclothem development in the Wreford Formation are (1) that the major flooding surface and the transgressive cherty limestones apparently represent greater depths of deposition in the Threemile cyclothem than they do in the Schroyer cyclothem; (2) cyclothem thickness decreases from the Threemile into the Schroyer; and (3) exclusively subtidal-type cycles occur in the Threemile cyclothem, whereas the Schroyer cyclothem is capped by a peritidal cycle (Fig. 13). These relationships suggest that *overall* development of the Threemile and Schroyer cyclothems was controlled by allogenic processes.

Origin of the Lowest Frequency, Lowstand-Bounded Units

The aforementioned hierarchal stratal characteristics of the cycles and cylothems in the Chase Group are consistent with both those features in allogenically controlled deposi-

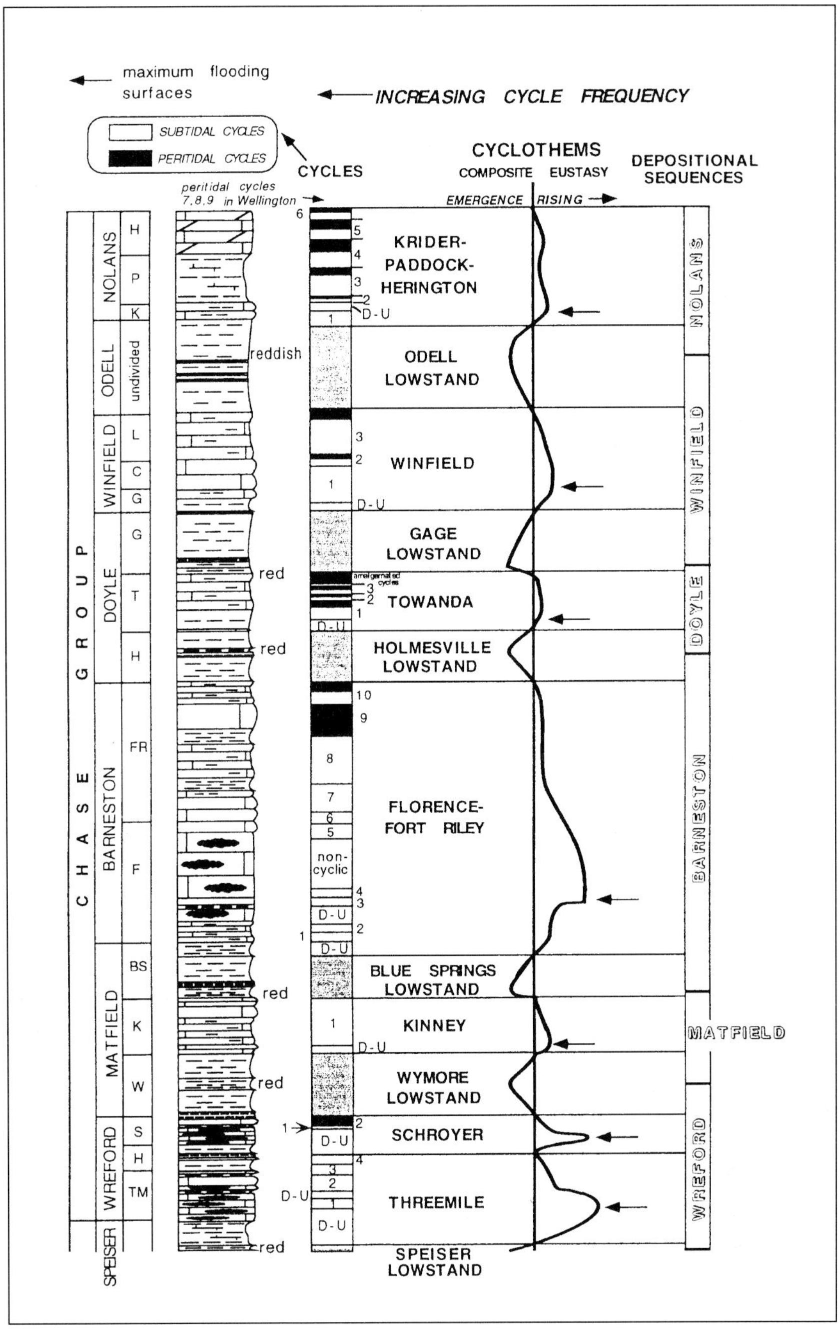

FIGURE 13 — Composite section of cyclothems, and inferred depositional sequences, in the Chase Group.

Sequences	Duration	Cyclothems	Duration
Nolans	1.26 m.y.	Lowstand	28 k.y.
		Krider-Paddock-Herington	625 k.y.
		Lowstand	612 k.y.
Winfield	988 k.y.	Lowstand	12 k.y.
		Winfield	588 k.y.
		Lowstand	385 k.y.
Doyle	644 k.y.	Lowstand	54 k.y.
		Towanda	381 k.y.
		Lowstand	210 k.y.
Barneston	2.07 m.y.	Lowstand	228 k.y.
		Florence-Fort Riley	1.6 m.y.
		Lowstand	207 k.y.
Matfield	625 k.y.	Lowstand	46 k.y.
		Kinney	389 k.y.
		Lowstand	189 k.y.
Wreford	917 k.y.	Lowstand	149 k.y.
		Schroyer	226 k.y.
		Threemile	447 k.y.
		Lowstand	95 k.y.

TABLE 1 —Thickness and compacted adjusted durations of sequences and cyclothems in the Chase Group, based on a total duration of 6.5 m.y. for the upper Asselian to Sakmarian Chase Group and 20% compaction for shales.

tional systems as described by Goldhammer et al. (1993), and in autogenically controlled depositional systems (Ginsburg, 1971). The cyclothems are grouped within six lowstand-bounded units (Wreford, Matfield, Barneston, Doyle, Winfield, and Nolans) which are separated by terrestrial red shales (Fig. 13). These lowest frequency units are interpreted as depositional sequences. Assigning absolute durations to these sequences is extremely hazardous because of basic uncertainties as to the total amount of time represented by this section of rocks, the completeness of the section (Hardie et al., 1986, 1991; Algeo and Wilkinson, 1988; Fischer and Bottjer, 1991), and in particular of the amount of time, rates of accumulation, and rates and degrees of compaction of the marine carbonate versus the lowstand siliciclastics in the section. With these severe reservations in mind, however, it is noted from several sources that the approximate duration of the Chase Group (uppermost Asselian to Sakmarian) is consistently about 6.5 m.y. (Harland et al., 1982; Palmer, 1983; Menning, 1989). It is assumed that the shales in the section must have compacted, perhaps more so than the carbonates. Shales compact about 40% at relatively shallow burial depths (< 500 m) (Reike and Chilingarian, 1974), although amount of compaction varies greatly depending on factors such as rate of deposition (of shales and overburden) and relative amounts of clay and coarser fractions in the shales (Meade, 1966). Because of these many uncertainties, the authors believe a modest estimate of 20% shale compaction is appropriate. Rates of deposition of carbonate and shale vary in different environments and through time and, therefore, are difficult to quantify in a section as paleoenvironmentally and texturally diverse as in the Chase Group. Hence, the authors are forced to assume grossly equal rates of deposition for each of the carbonate units, and different but equal rates for each of the lowstand siliciclastic units, in the Chase Group section. However, the authors do not believe that thin carbonate-dominated cyclothems or siliciclastic lowstand deposits represent the same amount of time as thicker sections of such. Likewise, thicker sequences are assumed to represent more time than thinner ones. With all of these wild-eyed assumptions, the authors have calculated thickness and compacted/adjusted durations of the sequences and cyclothems in the Chase Group merely for the sake of argument (Table 1), and can tentatively assign them a hierarchal ordering according to the classification of Goldhammer et al. (1991).

By this admittedly crude method, sequence durations range from 625 k.y. to about 2.1 m.y., which would suggest that they are third-order and fourth-order depositional sequences. Marine cyclothem durations range from 226 k.y. to 1.6 m.y.; hence, with the exception of the third-order Barneston (1.6 m.y. duration), they are fourth-order units (Table 1). Component cycles within the cyclothems likely are dominantly autogenic cycles because most are not regionally correlative; those that are grossly correlative on a regional basis may be fifth-order or higher frequency units. Heckel (1980, 1986, 1994), Busch and Rollins (1984), Busch et al. (1985), Busch (1988), Barrett (1989) and others suggested that Upper Pennsylvanian and some Lower Permian cyclothems in the Mid-Continent are fifth-order cycles (300 to 500 k.y. duration), and that their component cycles are sixth-order or even higher frequency cycles. The discrepancy in hierarchal ordering between these authors and the estimations of the current authors results mainly from the different classification schemes used, i.e., Busch and Rollins (1984), their Table 1, versus that of Goldhammer et al.(1991) which is used in this study.

DISCUSSION

Possible Controls on Deposition

The fundamental control on deposition of Mid-Continent Upper Paleozoic cyclothems has been attributed by many authors solely to climate-induced eustasy (Wanless and Shepard, 1936; Wheeler and Murray, 1957; Heckel, 1977, 1980, 1986, 1994; Ross and Ross, 1985, 1987; Boardman and Nestell, 1993). The more recent studies of Heckel (1977, 1980, 1986, 1994), Boardman and Heckel (1989) and Boardman and Nestell (1993) based their conclusion of fundamental glacio-eustatic control on comparison of inferred cycle periodicities in the Mid-Continent Pennsylvanian to Milankovitch orbital cyclicities. As discussed by DeBoer (1991), Goldhammer et al. (1993) and others, this approach, although sometimes unavoidable, is wrought with inherent problems. However, the inferred climate-forced eustatic model proposed by these workers did in fact operate at a time of known Southern Hemisphere glaciation and deglaciation (Crowell, 1978; Veevers and Powell, 1987), which strengthens their argument for at least some component of glacio-eustasy. Heckel (1986, 1994) not only argued against local tectonic controls, but also against any broad-based tectonic controls on Pennsylvanian cyclothem formation. Although he cited some studies of tectonic processes of relatively high periodicity (Cathles and Hallam, 1991), he believed that most broad-based, episodic tectonic processes presumably operate at considerably greater periodicities (> 2 m.y.), which contrasts to the relatively short time span of the Pennsylvanian. Furthermore, he believed that the corresponding extremely slow rates of sea-level rise induced by such tectonic activity (but see Klein and others, discussed below) could not account for the characteristic facies trends observed in typical marine-dominated, Mid-Continent Pennsylvanian cyclothems.

Other workers instead have argued either for a strictly tectonic control (Weller, 1930, 1956, 1958), or a combined tectonic-eustatic (but not necessarily a glacio-eustatic) control on Upper Paleozoic cyclothem formation (Elias, 1937; Moore, 1950, 1964; Watney, 1985a, b; Klein and Cloetingh, 1989; Klein and Willard, 1989; Klein, 1990b; Klein and Kupperman, 1992; Miller and West, 1993). According to the Klein group, the formation of Mid-Continent Pennsylvanian cyclothems resulted from the concurrent interplay between relatively long-term flexural deformation, resulting from plate tectonic activity along the eastern and southern margins of North America, and eustasy. In this model, the formation of Appalachian-type cyclothems (nonmarine-dominated) presumably was controlled largely by tectonic-induced sea-level rise due to proximity to active plate margins. In contrast, they suggested that marine-dominated, Kansas-type cyclothems largely owed their origin to eustasy, at times controlled by Milankovitch-induced glaciation that was superposed on moderate tectonic subsidence in Mid-Continent areas that were inboard of tectonically active margins. Illinois-type cyclothems would represent an intermediate type of deposit. In deference to Heckel (1986, 1994), Klein and Kupperman (1992) discussed several studies (Plafker and Savage, 1970; Watts et al., 1982) that presented evidence that broad-based, short-term tectonic pulses (periodicities < 10 k.y. to 1 m.y.) associated with relatively rapid sea-level changes indeed are known to occur within the range of Milankovitch periodicities. Whether or not such short-period mechanisms are viable mechanisms for cyclothem formation, however, is questionable.

The important point in consideration of the possible mechanism(s) responsible for the deposition of cyclothems and sequences in the Chase Group is the fact that as during the Pennsylvanian, there *was* both moderate tectonism in the Mid-Continent during the Early Permian due to plate-margin orogenesis (Kluth and Coney, 1981; Bond et al., 1984; Ross, 1986), *and* Southern Hemisphere glaciation (Crowell, 1978; Veevers and Powell, 1987). Regardless of whether or not the periodicities of episodic tectonic pulses are long or short, they must have overlapped eustatic cyclicity at times during the long period of time represented by the Mid-Continent Upper Paleozoic section! Conceivably, then, there should be tectonic signals, albeit subtle ones perhaps, within these cyclothems as there may also be in Pennsylvanian cyclothems (Klein and Kupperman, 1992). Perhaps such possible tectonic signals are recorded only in long-term subsidence and accommodation (Klein and Kupperman, 1992). With specific regard to the Chase Group, Elias (1937) and Moore (1950, 1964) suggested that eustatic fluctuations during the Wolfcampian may have resulted from pulses of tectonically induced change in the volume of ocean basins. Others solely invoke glacio-eustasy (Boardman and Nestell, 1993) or climatic changes (i.e., arid to humid, and vice versa) concurrent with eustasy and some tectonism (Miller and West, 1993) to explain cyclicity in the (lower) Chase Group. Regardless, the fundamental questions pertaining to the origin of cyclicity in the Chase Group, in reference to the different schools of thought discussed above, are (1) was cyclothem and cycle development indeed related to glacio-eustasy as an extension of seemingly genetically related Pennsylvanian cyclothems; and (2) are there recognizable tectonic signals amid this stratigraphy? These questions are considered below.

Analysis of Chase Group Cyclicity

Evidence for Glacio-Eustatic Control on Deposition — As discussed above, the stratigraphic hierarchy that *is* evident in cyclothem and component cycle development in the Chase Group is analogous to that in many ancient, cyclic platform carbonate strata whose deposition is believed to have been controlled largely by forced eustatic changes in accommodation (Goldhammer et al., 1990, 1991, 1993; Montanez and Osleger, 1993). There is departure from these general themes, however, as discussed earlier in this study, as evidenced by portions of cyclothems that are not regionally developed, and anomalous cycle stacking patterns. Such departures can be related to (a) local effects of differential rates of accommodation change and sediment accumulation, which may affect short-term, dominant autogenic or allogenic controls on sedimentation (Ginsburg, 1971; Hardie and Shinn, 1986; Goldhammer et al., 1993) and/or (b) tectonism (discussed below).

Inferred eustasy in many ancient cyclic deposits has been interpreted to have been glacio-eustatically driven via Milankovitch forcing, an interpretation that is all the more plausible if they were deposited during times of known glaciation (Mid-Continent Pennsylvanian cyclothems: Heckel, 1986, 1994 and others). In this regard, estimated durations of cyclothems in the Chase Group (Table 1), deposited during a period of glaciation, although clearly based on crude and equivocal calculations, show some dominant periodicities that indeed may reflect Milankovitch forcing. For example, durations of some of the cyclothems are multiples of eccentricity periodicities (Table 1). Glacio-eustasy likely must have controlled at least some of the development of cyclic deposits in the Chase Group.

Possible Evidence for Tectonic Control — As discussed above, many of the characteristics of cyclicity in the Chase Group are inconsistent with sea-level fluctuations and resulting changes in accommodation affected solely by Milankovitch-forced glacio-eustasy. Although none of the individual arguments presented below is unequivocal, together they caution one not to blindly invoke glacio-eustasy as the sole mechanism driving cyclothem formation even in known periods of glaciation. (1) Many of the cyclothem periodicities do not match those of Milankovitch cycles (Table 1). Missing sections due to erosion or missed beats, for example, can be two reasons for such mismatches (Hardie and Shinn, 1986). (2) For the same reasons, perhaps, typical "Schwarzacher bundles" (DeBoer, 1982; Schwarzacher and Fischer, 1982) are not identified within the Chase Group section (Fig. 13). (3) Heckel (1984, 1994) suggested that the asymmetrical shapes of inferred sea-level curves of Pennsylvanian cyclothems, which result from a typical stacking pattern of thin transgressive and thick regressive facies tracts, are analogous to those of Pleistocene glacio-eustatic curves. Of course, equating the thickness of facies tracts to elapsed time in ancient rocks is questionable practice, and one cannot justifiably compare such curves as Heckel (1986, 1994) has done to what in reality is a time-adjusted curve of Pleistocene eustasy. Disregarding this point for the sake of argument, however, Fig. 13 shows that whereas many of the inferred sea-level curves of the Chase Group cyclothems are asymmetrical, some are not (i.e., the Towanda and Krider-Paddock-Herington). With the exception of these cyclothems, those other cyclothems with presumed asymmetrical sea-level curves are characterized by cycle stacking and subfacies stacking patterns that are mostly inconsistent with an interpretation of forced eustatic control on deposition. (4) A long-term trend of shallowing from dominantly marine facies (Chase Group) to dominantly terrestrial facies (Wellington Formation), and a corresponding decrease in the extent of relative transgression from the Wreford up into the Nolans (Fig. 13), likely reflect a long-term decrease of tectonic subsidence and accommodation. Yet, several interesting aspects of cyclothem development in this section that depart from this long-term trend are (a) the interpretation that cyclothem thicknesses are erratic rather than incorporating progressively thinner cyclothems through time (Fig. 13); (b) the seemingly anomalous, rapid facies dislocations from deepwater, cherty limestones to shallow water, noncherty limestones in the Threemile and Schroyer cyclothems (Fig. 4); (c) the seeming anomalously thick sections of proximal inner shelf deposits (shales and shaly limestones) that are overlain by more distal inner shelf carbonate sands in the RSTs in the Kinney cyclothem (Fig. 3) and the Fort Riley portion of the Florence-Fort Riley cyclothem (Fig. 7B); (d) the restricted areal occurrence of grainstones and overlying mudstones (cycles 9 and 10) in the upper Fort Riley in the study area (Fig. 9); and (e) the highly variable thickness of the Luta in the Winfield cyclothem (Fig. 11).

Points a and c above may suggest periods of increased subsidence during long-term subsidence decrease which, of course, indeed could be related solely to complex cyclicity caused by superposed Milankovitch precessional, obliquity, and eccentricity periodicities. With regard to point b above, similar abrupt facies dislocations from deep to shallow water facies are not present in the RSTs of the other cyclothems; rather, such changes are gradual (Fig. 13). If these regressive units in the Threemile and Schroyer were siliciclastic rather than carbonate, then perhaps a strong case could be made for invoking periods of rapid siliciclastic input and consequent shallowing related to a climate-controlled model of deposition (Cecil, 1990). Alternatively, increased far-field tectonism could have resulted in increased terrigenous input. However, the anomalous facies patterns developed in these regressive *carbonates* instead could indicate rapid higher frequency eustatic fall, or conversely, minor tectonic uplift. Tectonism superimposed on glacio-eustasy could have affected facies development in the compound-cyclothemic Wreford sequence. However, point d above is much more difficult to explain via eustasy. The fact is that cycles 9 and 10 in the Florence-Fort Riley cyclothem are present only in Butler and northern Cowley counties, on the Nemaha ridge, which is seismically active even today (Steeples, 1987). Hence, the Nemaha ridge, although buried in Wolfcampian time, likely was periodically tectonically active during deposition of the Chase Group. A pulse of tectonic subsidence that resulted in formation of a local depocenter in which these cycles were deposited is a logical explanation for their distribution. In a related point (e, above), extreme thickness variations in the Luta must have involved local depocenters of increased subsidence during deposition of the Winfield cyclothem; interestingly, the thickest Luta section in the study area (in the quarry in Section 24 T27S-R4E, Butler County: Fig. 11) lies directly astride the Humboldt syncline, which defines the eastern edge of the Nemaha ridge!

Thus, it appears that cyclothem and sequence development in the Chase Group likely resulted from a complex interplay between Milankovitch-forced glacio-eustasy and tectonism of varying periodicity operating at both regional and local scales.

SUMMARY AND CONCLUSIONS

The seven cyclothems recognized in the Chase Group were deposited in inner shelf to outer shelf environments on a broad ramp-type platform. The cyclothems represent third- and fourth-order units, and component cycles may include fifth-order eustatic cycles (i.e., those that are grossly regionally correlative), or more likely are mainly of local autogenic origin. Cyclothems and intervening lowstand deposits are grouped into six lower frequency, third- and fourth-order marine depositional sequences. Sequence boundaries are conceptually defined by the occurrence of terrestrial LSTs that contain red shale, which are considered to represent periods of maximum subaerial emergence. An imprint of Milankovitch-forced glacio-eustatic control on cyclothem development is suggested by stacking patterns and types of cycles in some cyclothems, and the fact that deposition of the Chase Group occurred during a period of known Southern Hemisphere glaciation. However, anomalous aspects of cyclothem and cycle development in the Chase Group can be best explained by invoking a model of composite eustatic-controlled sea-level fluctuations that incorporate glacio-eustasy and tectonism.

INFLUENCE OF SEA-LEVEL FLUCTUATION ON RESERVOIR QUALITY OF THE UPPER MORROWAN SANDSTONES, NORTHWESTERN SHELF OF THE ANADARKO BASIN

Z. Al-Shaieb
J. Puckette
A. Abdalla
Oklahoma State University
Stillwater, OK 74078

ABSTRACT

Upper Morrowan reservoirs in the northwestern shelf of the Anadarko basin are primarily valley-fill sandstones that formed as part of transgressive-regressive cycles controlled by sea-level changes. The lowering of the sea level resulted in the basinward migration of the shoreline and the development of incised river valleys. Lag conglomerate composed of shale intraclasts was the major facies formed during this episode. On the other hand, valley-fill sediments composed of fluvial, estuarine, and flood-plain facies were deposited during major relative sea-level rises. The complex depositional pattern and resultant diverse lithofacies are the main factors that contributed to the heterogeneity of these reservoirs. Depositional style within these channels was the major factor that affected reservoir quality.

Fluvial dominated sandstones are typically coarser grained, contain less calcite cement, and are good-quality reservoirs. Marine-dominated sandstones contain abundant skeletal grains and carbonate cement that occluded porosity. Finer grained (tidal-dominated) estuarine sandstones are often poor-quality reservoirs due to biogenic modification that destroyed primary porosity.

GEOLOGIC SETTING

The northwestern shelf of the Anadarko basin (including the Hugoton embayment) is located in the Texas and Oklahoma panhandles, southeastern Colorado, and western Kansas (Fig. 1).

The paleogeography and positive elements during deposition of upper Morrowan rocks are shown in Fig. 2. These include the Apishapa-Sierra Grande uplift to the west, Ancestral Front Range and Transcontinental arch to the north, and the Wichita-Amarillo uplift to the south. The Central Kansas

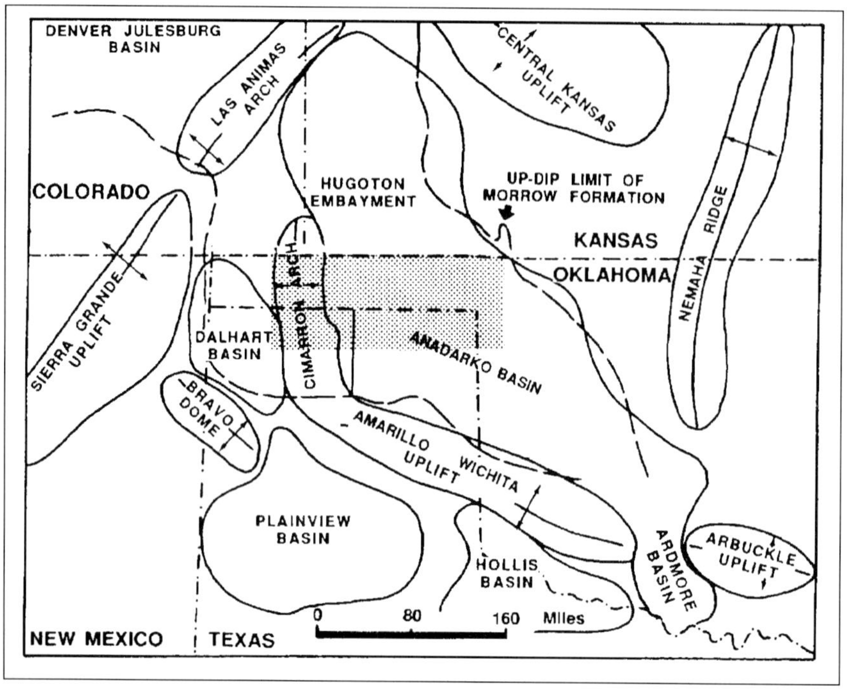

FIGURE 1 — Study area in northwestern Anadarko basin and Hugoton embayment (after Khaiwka, 1968).

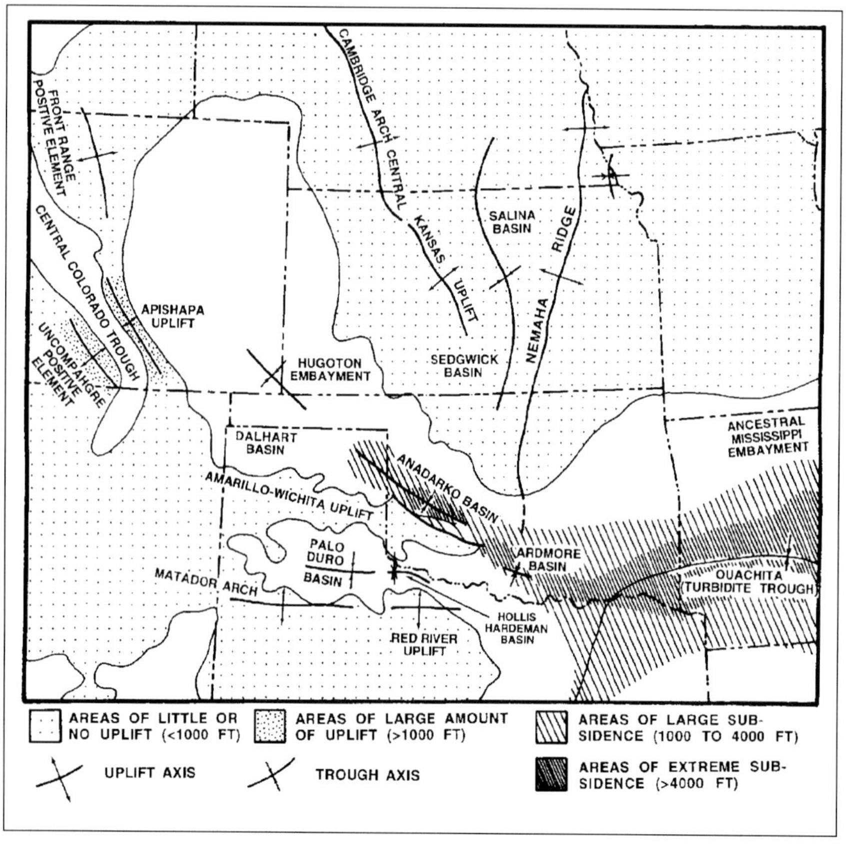

FIGURE 2 — Paleogeography and structural features during Morrowan time (Sonnenberg et al., 1990).

uplift was a relatively low-relief positive area northeast of the basin (Sonnenberg et al., 1990). These features influenced Morrowan depositional style and served as sediment sources.

The Morrowan Series is the lowermost Pennsylvanian series in North America and is located above the Mississippian Chesterian and below the Atokan (Bates and Jackson, 1987). In the Anadarko basin, strata that underlie the Thirteen Finger Limestone and overlie Mississippian strata are described by the operational name "Morrow Formation," an informal stratigraphic unit. In this study, strata of Morrowan age are divided into the "upper Morrowan" and "lower Morrowan," after the convention of earlier studies (including Rascoe and Adler, 1983). In general, the "Squaw Belly Limestone" is used to separate the upper and lower Morrowan intervals. However, in the panhandle region of Oklahoma, the division into upper and lower intervals is commonly made at the top of the more resistant (calcareous) shale shown in Fig. 3. The lower Morrowan consists primarily of marine sandstones that were deposited on the eroded Mississippian topography. They represent a variety of depositional facies including shoreface (Wheeler et al., 1990) and transgressive valley-fill sandstones (Gerken, 1992). Marine muds and fluvial sands were the dominant sediments in the upper Morrowan. In the

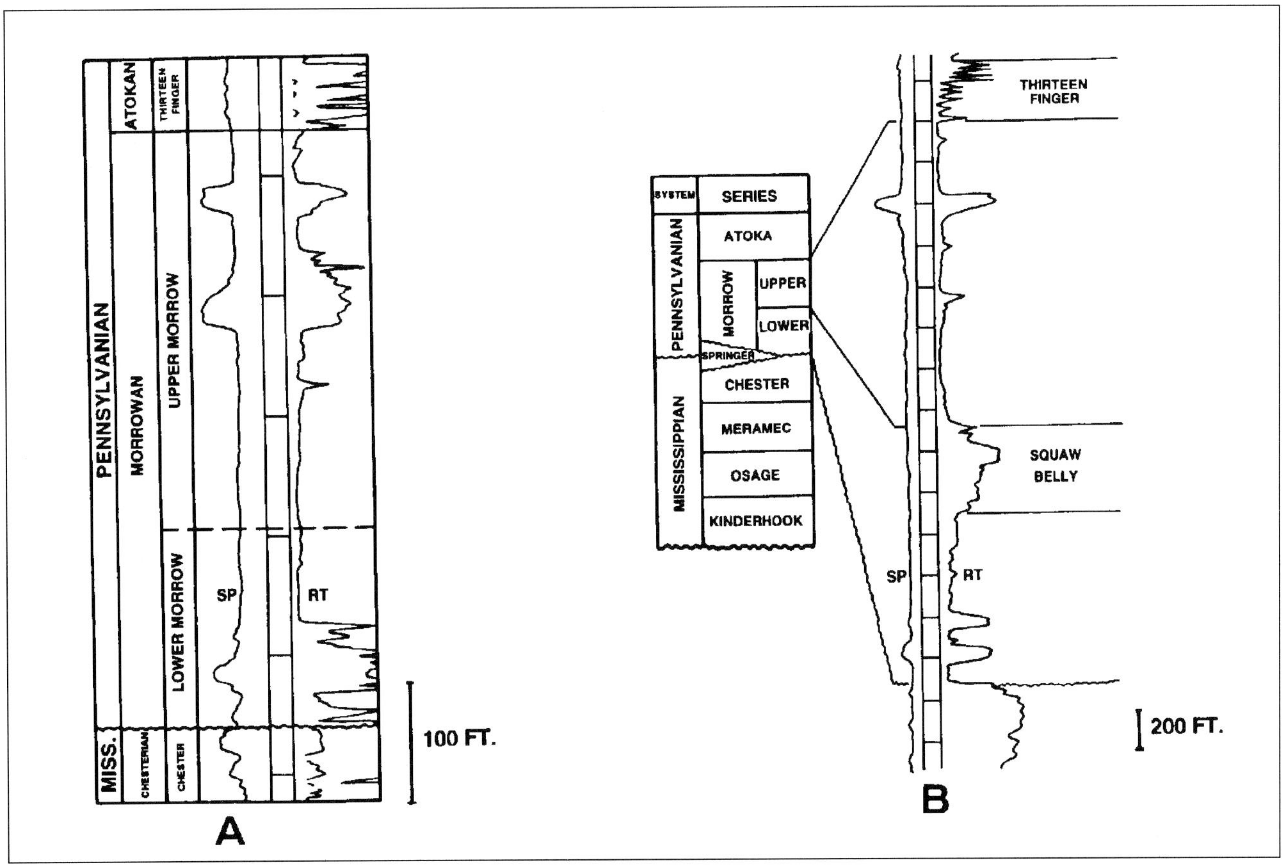

FIGURE 3 — Morrowan stratigraphy showing informal stratigraphic divisions: (A) Panhandle region of Oklahoma (Cornish, 1982); and (B) Central basin (Shelby, 1980).

Oklahoma panhandle, sediments were deposited on a very low-angle slope of approximately 1 ft/mile. Reported upper Morrowan stream gradients of approximately 0.4 to 0.9 ft/mi (Cornish, 1982) and <2.5 ft/mi (Swanson, 1979) also suggest flat terrain. The flat-lying topography made the shelf susceptible to alternating inundation and subaerial exposure as the sea level fluctuated in response to tectonic or glacio-eustatic cycles.

PREVIOUS INVESTIGATIONS

The economics of upper Morrowan reservoirs have made them primary drilling targets in the Oklahoma and Texas panhandles, western Kansas, and eastern Colorado. As a result, they have been the focus of many published and unpublished studies including Forgotson et al. (1966), Benton (1971), Swanson (1979), Cornish (1982), Rascoe and Adler (1983), Harrison (1990), Bowen et al. (1990), Sonnenberg et al. (1990), and Wheeler et al. (1990).

Benton (1971) established that the upper Morrowan sandstones in the Postle field (T5N-R13ECM, Texas County, Oklahoma) were deposited within a fluvial system that drained the eroded Morrowan paleotopography. Swanson (1979) suggested deltaic processes controlled their deposition. Subsequent studies by Krystinik and Blakeney (1990), Sonnenberg et al. (1990), and Wheeler et al. (1990) described valley-fill reservoirs in eastern Colorado and western Kansas where at least seven episodes of incision and valley fill have occurred.

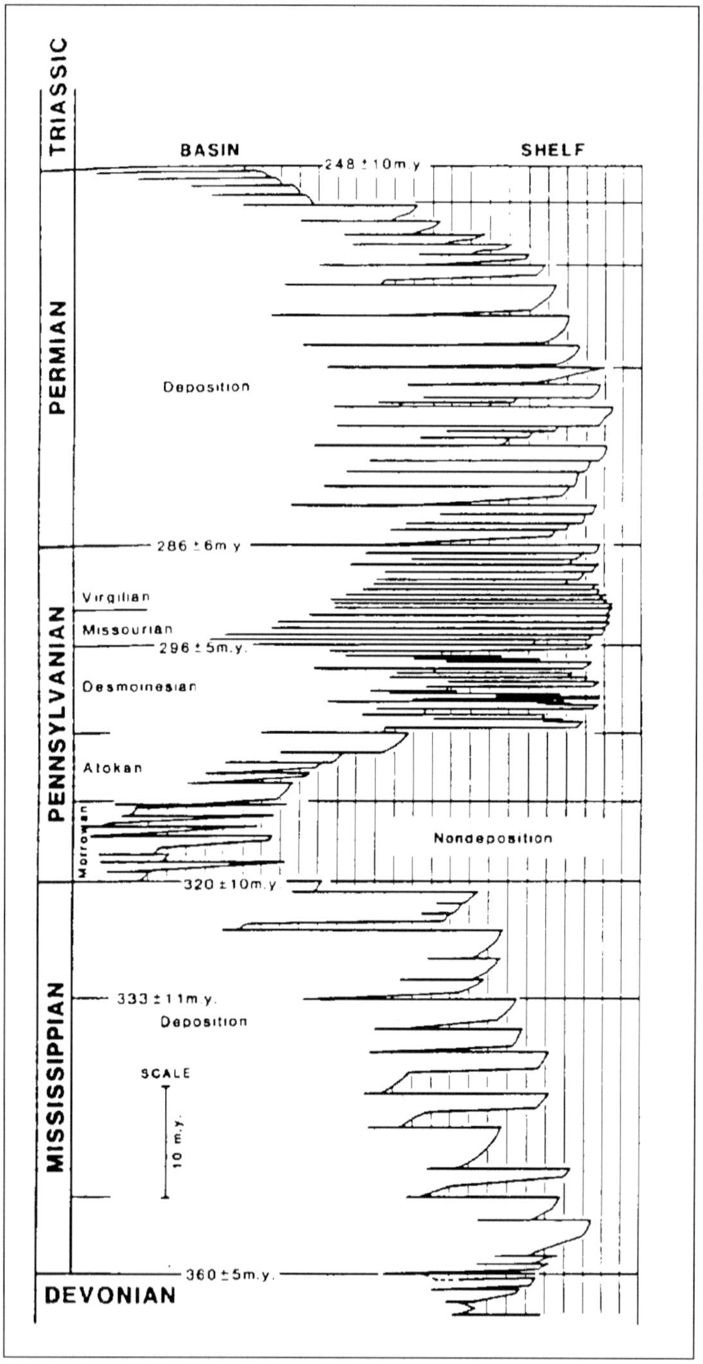

FIGURE 4 — Worldwide coastal onlap curve depicting relative sea-level lowstand and transgressive/regressive cycles during the Morrowan (Ross and Ross, 1988).

SEQUENCE STRATIGRAPHY OF THE MORROWAN

The application of the concepts of sequence stratigraphy has significantly contributed to the interpretation of siliciclastic sequences. These concepts emphasize the importance of sea-level fluctuation as a major control on the sedimentary style in the nearshore environment. According to these concepts, the upper Morrowan channel-fill deposits are interpreted as portions of two major processes: cutting of the valleys during episodes of regional lowering of the sea level and infilling of valleys during subsequent rises in the sea level. Models depicting the effects of sea-level changes on sedimentary processes by Weimer (1988), Sonnenberg (1990), Krystinik and Blakeney (1990), and Wheeler et al. (1990) provided insights into the depositional style of these sequences.

The worldwide coastal onlap curve (Fig. 4) from Ross and Ross (1988), demonstrated eustatic sea-level changes during the Carboniferous. These fluctuations generated the transgressive-regressive depositional cycles that were prevalent in Pennsylvanian and Permian shelf rocks in the Mid-Continent region. Stratigraphic sequences in these rocks have been defined on the basis of their bounding lowstand erosional surfaces (Youle et al., 1994; Weimer et al., 1988). Figure 5 (Sonnenberg et al., 1990) depicts a typical Morrowan stratigraphic sequence.

FIELD STUDIES

Several upper Morrowan fields in the Oklahoma and Texas panhandles were examined in detail to determine the depositional environments of these reservoirs.

The Northwest Eva field in T4N-R10ECM, Texas County, Oklahoma, was included in this study because of the availability of cores and other subsurface data. Harrison (1990) mapped the upper

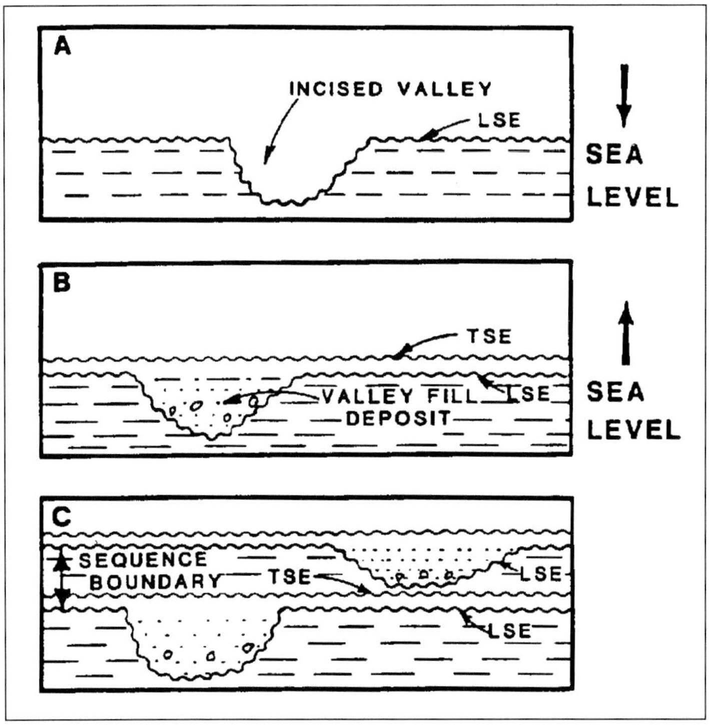

FIGURE 5 — Schematic diagram of a Morrowan stratigraphic sequence. (A) Lowstand Surface of Erosion (LSE) during sea-level lowstands. The LSE is a sequence boundary. (B) Formation of valley-fill deposits composed of fluvial and estuarine deposits during a sea-level rise. A Transgressive Surface of Erosion (TSE) forms when the shoreline migrates landward over the valley-fill and shelf deposits. (C) Deposition of blanketing marine shale. The sequence is delineated by the next sea-level lowstand. Sequence boundaries are the lowstand surfaces of erosion (LSE) in this model (Sonnenberg et al., 1990).

Morrowan Purdy Sandstone and delineated the reservoir trend that extends northeast to southwest through the field (Fig. 6). Cross section A-A' (Fig. 7) illustrates how the sandstone in this trend filled a valley that apparently was eroded into the underlying shale. Two cores were examined in the sand trend: the Gulf Oil Corp. Ferguson No. 1, in the NW SE Section 2 T4N-R10ECM, and the Gulf Oil Corp. Kelly No. 1, in the SW NW Section 1 T4N-R10ECM.

The Gulf Ferguson core contains 47 ft of shale, siltstone, and sandstone. The basal part of the cored interval (4453 to 4466 ft) consists of 13 ft of fossiliferous marine shales that are disconformably overlain by 16 ft of conglomeratic, upward-fining sandstone (4435 to 4451 ft) containing small- to medium-scale crossbeds and several small-scale upward-fining intervals. This sandstone is overlain by 4 ft of dark-gray silty shale (4431 to 4435 ft). The upper 13 ft of core (4418 to 4431 ft) consists of a second conglomeratic channel-fill sandstone (4424.6 to 4431 ft) and abandoned channel-fill siltstone, sandstone, and shale (4418 to 4424.6 ft). Portions of the cored interval are shown in Fig. 8.

The second core in the Northwest Eva field is the Gulf Oil Corp. Kelly No. 1 in the SW/4 NW/4 Section 1 T4N-R10ECM. The basal part of the cored interval (4434 to 4455 ft) consists of 21 ft of dark-gray fossiliferous shale. This marine shale is disconformably overlain by a medium- to coarse-grained sandstone that grades upward into carbonaceous shale with coaly stringers. The sandstone is 6 ft thick (4434 to 4428 ft). It contains flowage structures, small-scale crossbeds, and organic debris. The carbonaceous shale and siltstone (4428 to 4422 ft) contain plant debris. The overlying carbonaceous shale extends from 4422 to 4416 ft, where a disconformity separates it from a fossiliferous marine shale (4416 to 4415 ft). Photographs of this core are shown in Fig. 9.

Another core examined in this study is the Petroleum Inc. Hendrix No. 3, from the Carthage field in Texas County. The upper Morrowan interval in this core is composed of sandstone and shale. The core contains at least four significant genetic units. These are a fine-grained sandstone (4558 to 4568 ft), upward-fining, coarse- to medium-grained sandstone (4537 to 4558 ft), fossiliferous (marine) sandstone (4527 to 4537 ft), and marine shale (4526 to 4527 ft) (Fig. 10). Wireline logs suggest the basal sandstone is in sharp contact with an underlying (marine) shale that was not cored.

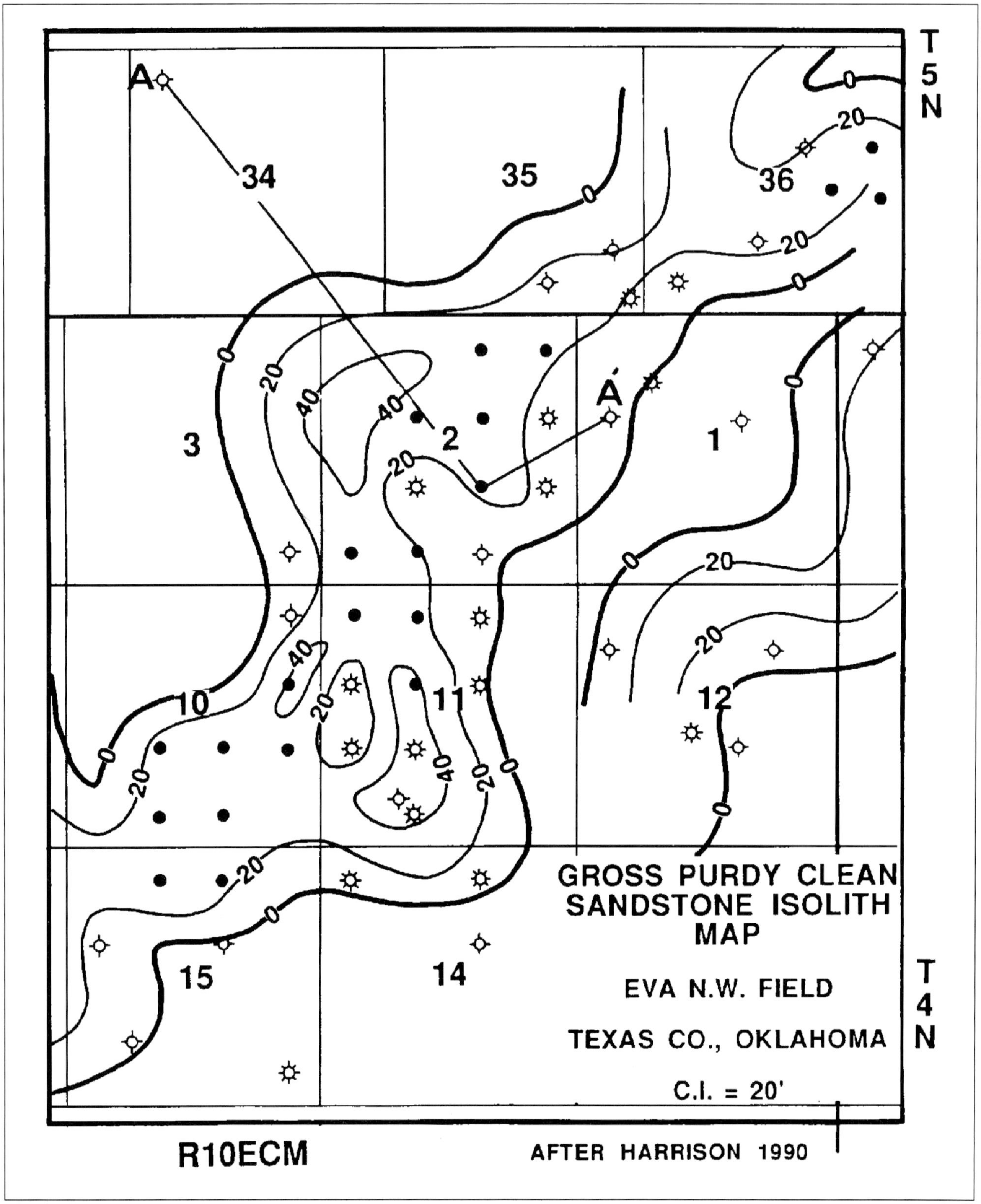

FIGURE 6 — Thickness map of the upper Morrowan Purdy Sandstone; Northwest Eva field, Texas County, Oklahoma (Harrison, 1990).

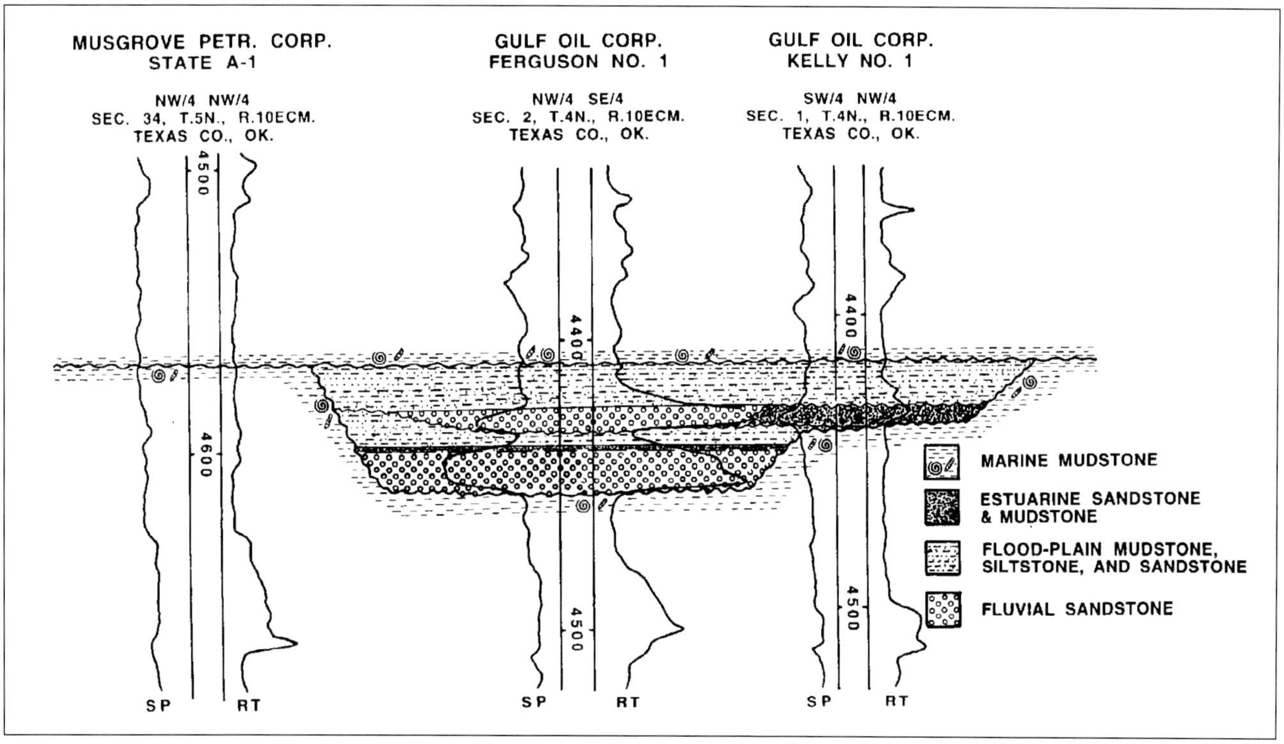

FIGURE 7 — Schematic cross section A-A' through the Northwest Eva field showing the apparent valley incision. The proposed valley-fill facies are identified (after Harrison, 1990). Location of cross section line A-A' is shown in Fig. 9.

DEPOSITIONAL FACIES

Six distinct depositional facies were recognized in the Morrowan cores. These are fossiliferous marine shale (MSH), clay-pebble conglomerate (CLC), clay- and silt-rich burrowed, fine-grained sandstones and interbedded sandstones and shales (BFS), conglomeratic to medium-grained, graded (upward fining) sandstones (CGS), fossiliferous, coarse- to fine-grained sandstone (MSS), and flora-rich shales, siltstones, and sandstones (FRS).

Fossiliferous Marine Shales (MSH)

Description

The MSH facies consists of dark gray, thin planar to laminated, fissile to platy mudstones (Fig. 8, unit F; Fig. 9, units A and E). They are sparsely burrowed and contain brachiopod and echinoderm bioclasts. This is the dominant facies in the section and forms the background sediment (Krystinik and Blakeney, 1990) beneath the coarser grained CLC and BFS facies. The MSH overlies MSS and FRS facies at the top of sand-rich intervals. Here, it commonly develops a fossil-rich conglomeratic zone at its base (Fig. 10, unit A). The MSH overlies older marine shale (MSH) facies in areas where channel-fill sediments are absent. In this setting, the contact is recognized by bioclast-rich conglomerate and/or an exposure surface (paleosol) on the lower shale.

Wireline logs indicate that the MSH facies is in sharp contact with overlying coarser grained facies. The lower contacts of MSH with sand-rich facies assemblages appear gradational. Contacts where no sandstone was deposited (marine shale with marine [MSH] or flood-plain [FRS] shale) are usually

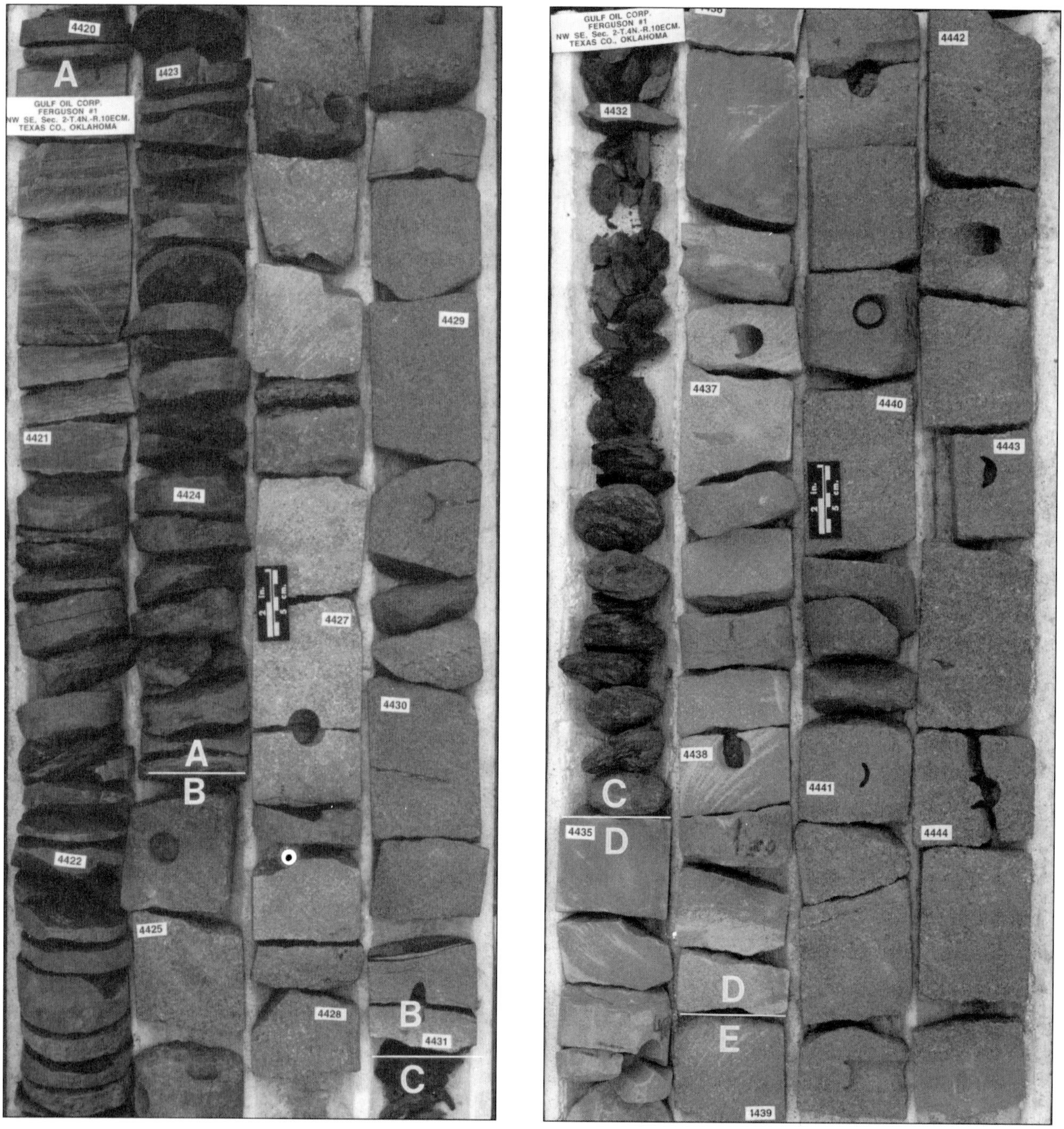

FIGURE 8 (above and opposite page) — Cored interval: Gulf Oil Corp. Ferguson No. 1. (A) Abandoned channel-fill siltstones and shales (4420 to 4424.6 ft) in sharp contact with (B) coarse-grained fluvial sandstone (4424.6 to 4431 ft). Note the erosional contact at 4431 ft between sandstone and (C) underlying dark shale (4431 to 4435 ft). (D) Fine- to medium-grained sandstone (4435 to 4438.7 ft). (E) Coarse-grained sandstone (4438.7 to 4451 ft) unconformably overlies (F) dark-gray marine shale (4451 to 4453 ft).

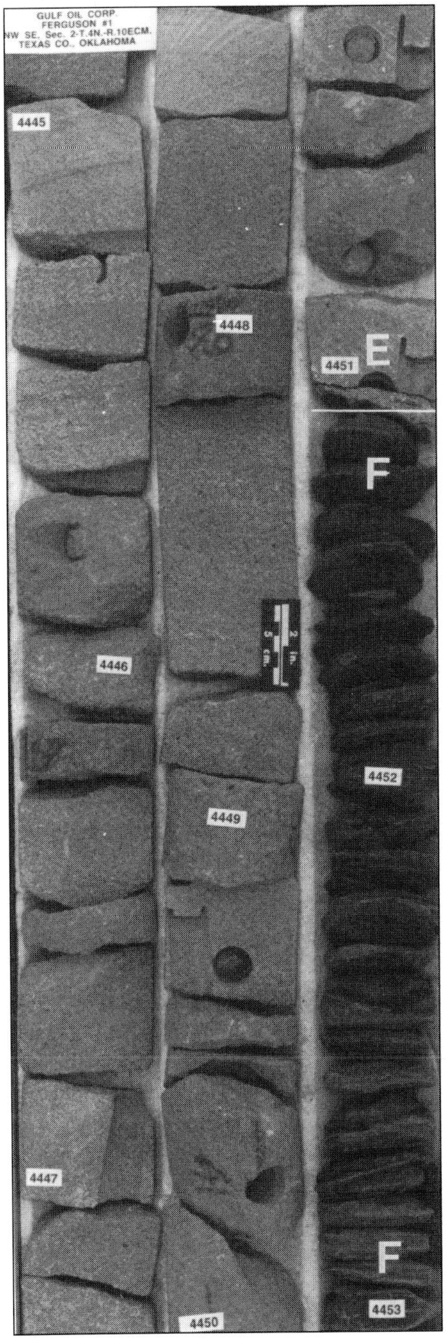

only detectable in cores.

Interpretation

Invertebrate fossils, absence of plant debris, and burrows suggest the MSH facies was deposited as shelf muds during maximum marine transgressions. The conglomeratic zone at the base of this facies represents transgressive lag deposited above a surface of erosion (ravinement surface).

Clay-Pebble Conglomerate (CLC)

Description

The CLC facies is composed of thin (<1 to 2 ft thick) pebble conglomerates that consist of claystone clasts in mud matrix. The clasts are not easily discernable on wireline logs due to their low permeability and shale-like signature.

Interpretation

Channel lag that formed during sea-level lowstands as streams eroded the channel floor and banks.

Burrowed Fine-Grained Sandstones and Interbedded Sandstones and Shales (BFS)

Description

The BFS facies is composed of silt- and clay-rich, burrowed fine-grained sandstones with thin, coarse sandstones and interlaminated sandstone/shale intervals (Fig. 10, unit D). This facies is typically 1 to 10 ft thick. Coarse-grained sandstones are <1.5 ft thick. Fine-grained sandstone intervals contain ripple laminae, are wavy to planar bedded, and often homogenized by biotic activity. They are interbedded with dark gray shale and mudstone.

Interpretation

The presence of burrowing coupled with the interbedded nature of this rock suggests that the BFS facies reflects the interaction of fluvial and marine processes. Horizontally laminated fine-grained sandstones suggest a tidal flat environment, whereas coarser grained sandstones and crossbedding imply higher energy (tidal channel) conditions. Biotic activity, sand, and mud imply a well-oxygenated mixed energy setting such as a tidal-dominated estuarine deposit.

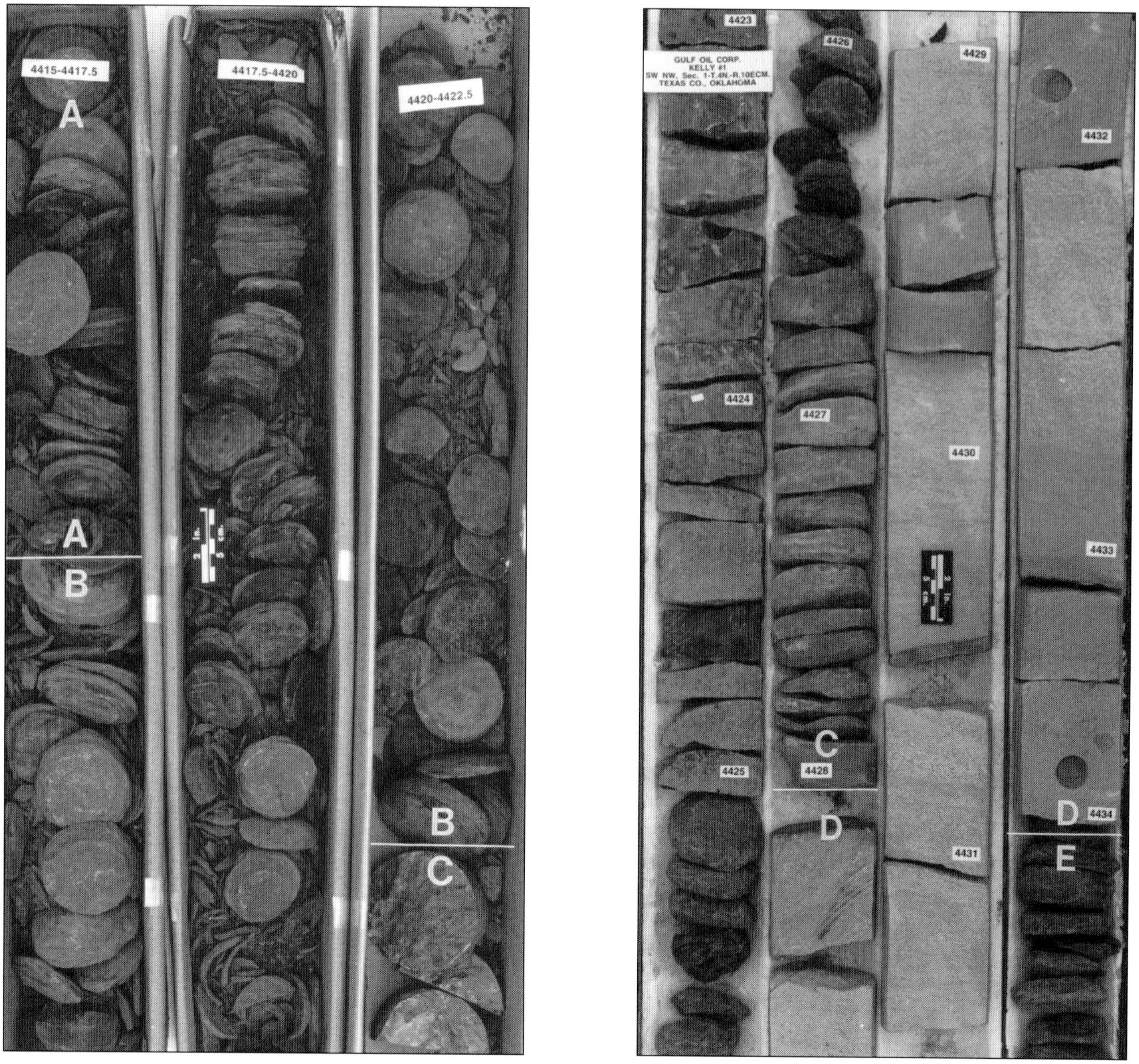

FIGURE 9 *(above and opposite page)* — Cored interval: Gulf Oil Corp. Kelly No. 1. (A) Dark-gray fissile marine shale (4415 to 4416 ft) and (B) gray, silty flora-rich shale (4416 to 4422 ft). (C) Shaly siltstone and sandstone (4422 to 4428 ft). (D) Medium- to coarse-grained sandstone (4428 to 4434 ft) in sharp contact with underlying (E) dark-gray marine shale (4434 to 4442.5 ft).

Conglomeratic, Coarse-Grained Sandstones (CGS)

Description

The CGS facies consists of conglomeratic, granule to medium-grained sandstones (Fig. 8, units B and E; Fig. 9, unit D). They are typically 2 to 30 ft thick, massive to planar crossbedded, poorly sorted, and composed of small-scale, upward-fining units. The CGS facies forms sharp contacts with underlying estuarine or marine sediments, but can be gradational with overlying estuarine sand-

stones. Wireline logs (gamma ray and porosity logs) infer a graded nature and porosity development within this facies. It may repeat and form stacked sandstones within a given valley-fill assemblage.

Interpretation

Coarse grain size, overall upward-fining character, and lack of burrowing identify the CGS facies as fluvial sandstone deposited by high-energy (braided) streams whose energy decreased as the base level rose.

Fossiliferous Sandstone (MSS)

Description

The MSS facies sandstones range from 5 to 10 ft thick and contain abundant echinoderm, bryozoa, and brachiopod fragments (Fig. 10, unit B). These planar crossbedded units grade upward from coarse- to medium-grained sandstone, become richer in clay and silt matrix, and are burrowed. The MSS facies is typically calcite cemented and low porosity. The MSS may have sharp or gradational contacts with underlying CGS facies. Upper contacts with shelf muds (MSH facies) are sharp and easily recognized, if transgressive lags were deposited.

Interpretation

Abundant abraded bioclastic debris and crossbedding suggest the MSS facies was deposited under marine-influenced (estuarine) conditions. Increasing matrix toward the top implies deepening water and increasing marine dominance within a drowned estuary.

Flora-Rich Shales, Siltstones, and Sandstones (FRS)

Description

The FRS facies is rich in plant fossils and ranges from 1 to 30 ft thick. Dominant lithofacies are light to dark-gray, planar laminated mudstone, gray siltstone, and very fine-grained sandstone (Fig. 9, unit B). Carbonaceous debris, including leaf and stick impressions, root traces, and coaly material, are common. Contacts with fluvial facies are typically sharp, whereas those with estuarine and marine (MSH) facies may be difficult to recognize unless transgressive lag or invertebrate fossils are identified.

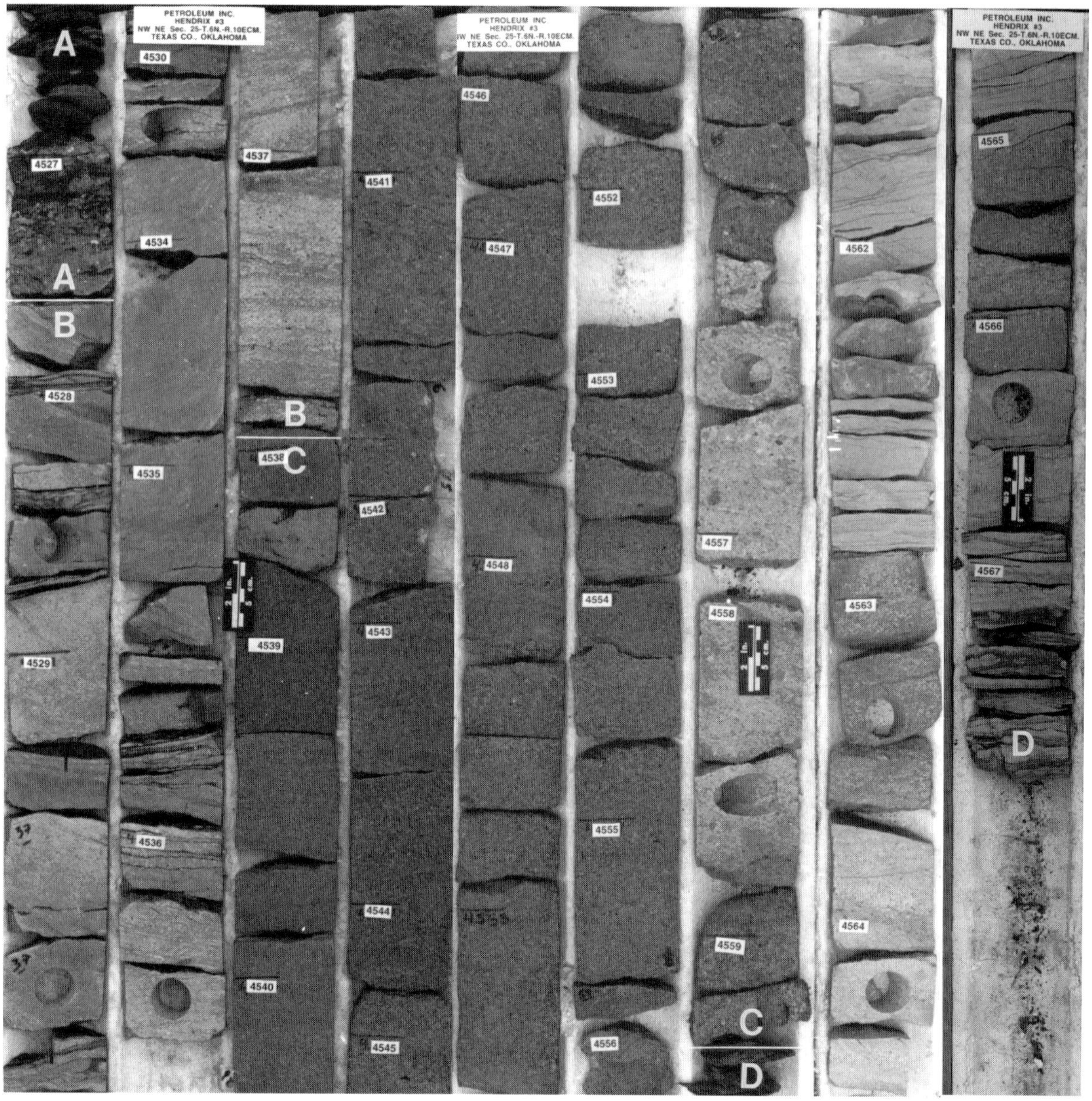

FIGURE 10 — Cored interval, Petroleum Inc. Hendrix No. 3. (A) Dark-gray marine shale with fossil hash zone at base (4526 to 4527.5 ft). (B) Medium- to coarse-grained sandstone with shale partings, burrowing, and abundant invertebrate fossils (4527.5 to 4538 ft). (C) Very coarse crossbedded sandstone (4538 to 4559.5 ft) with basal clay-pebble conglomerate in sharp contact with underlying dark shale. (D) Mixed interval of dark shale, small-scale upward-fining sandstones, and interbedded sandstone and shale (4559.5 to 4568 ft). Finer grained sandstones are burrowed and contain clay-rich laminae.

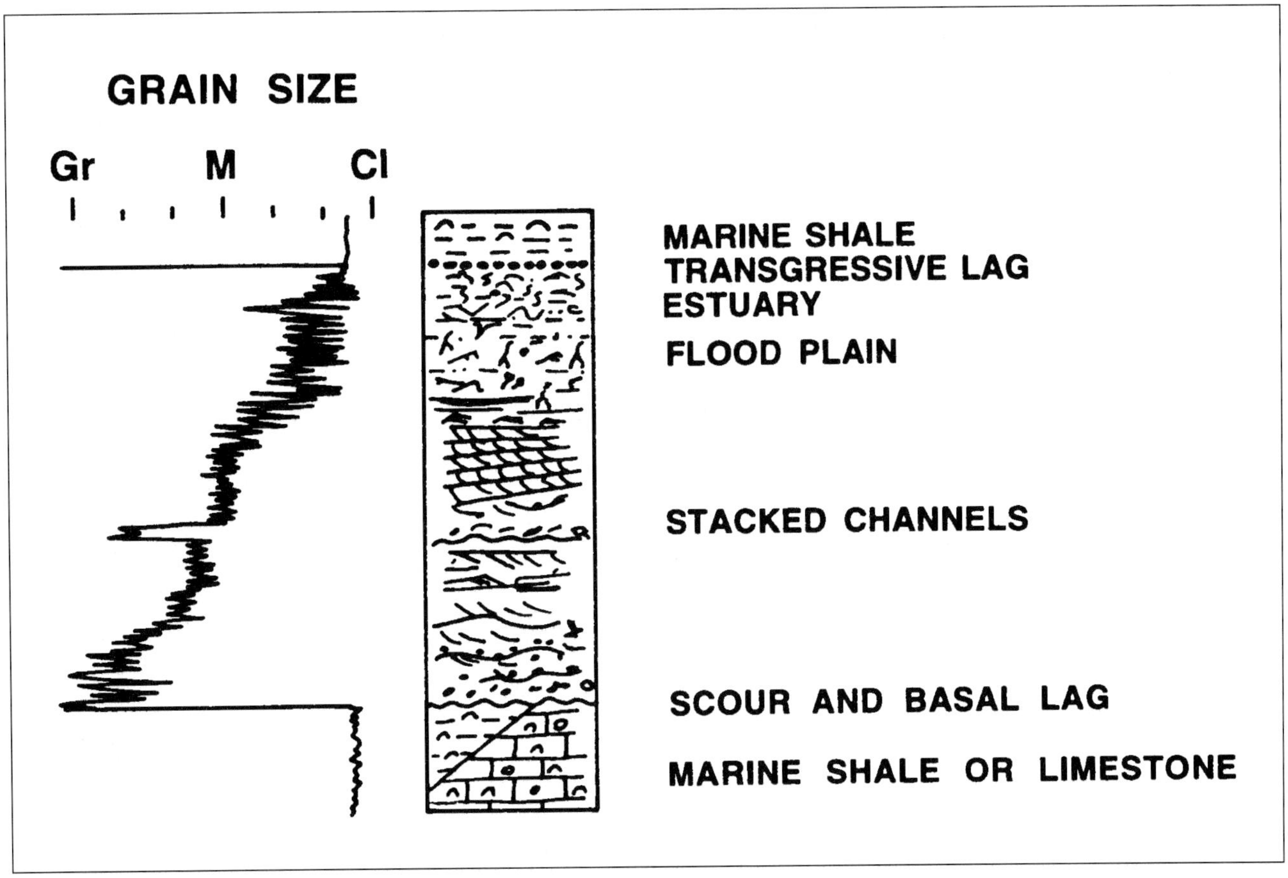

FIGURE 11 — Typical vertical profile through a valley-fill deposit (Krystinik and Blakeney, 1990).

Interpretation

Abundant plant debris, root traces, and coaly material suggest abandoned channel-fill and swamp deposits within the fluvial flood plain.

These facies are similar to the sediment categories proposed by Krystinik and Blakeney (1990) as a typical vertical profile in a Morrow valley-fill deposit (Fig. 11).

FACIES INTERPRETATION IN CORED INTERVALS

The Petroleum Inc. Hendrix core contains four of the six depositional facies. These are the burrowed estuarine (tidal flat) sandstone (BFS), fluvial sandstone (CGS)(Fig. 12), estuarine (MSS) sandstone (Fig. 13), and overlying marine mudstone (MSH). The marine mudstone is separated from the underlying sediments by a fossil hash deposit (Fig. 14) that typifies a Transgressive Surface of Erosion (TSE) (Weimer, 1988, 1984). The marine shale at the base of the valley fill and scour/unconfor-

FIGURE 12 *(above left)* — Coarse-grained, porous, and oil-stained fluvial sandstone with carbonate-cemented planar crossbeds and upward-fining graded intervals. Petroleum Inc. Hendrix No. 3. Depth 4544 ft.

FIGURE 13 *(above center)* — Medium-grained, carbonate-cemented estuarine sandstone with invertebrate grains. Petroleum Inc.Hendrix No. 3. Depth 4537 ft.

FIGURE 14 *(above right)* — Marine mudstone with fossil debris conglomerate at base. Fossil hash zone is typical of the transgressive surface of erosion (TSE) (Weimer, 1984, 1988) developed by a regional marine transgression. Petroleum Inc. Hendrix No. 3. Depth 4527 ft.

FIGURE 15 *(left)* — Plant fossils in flood-plain siltstone and shale. Ferguson No. 1. Depth 4420 ft.

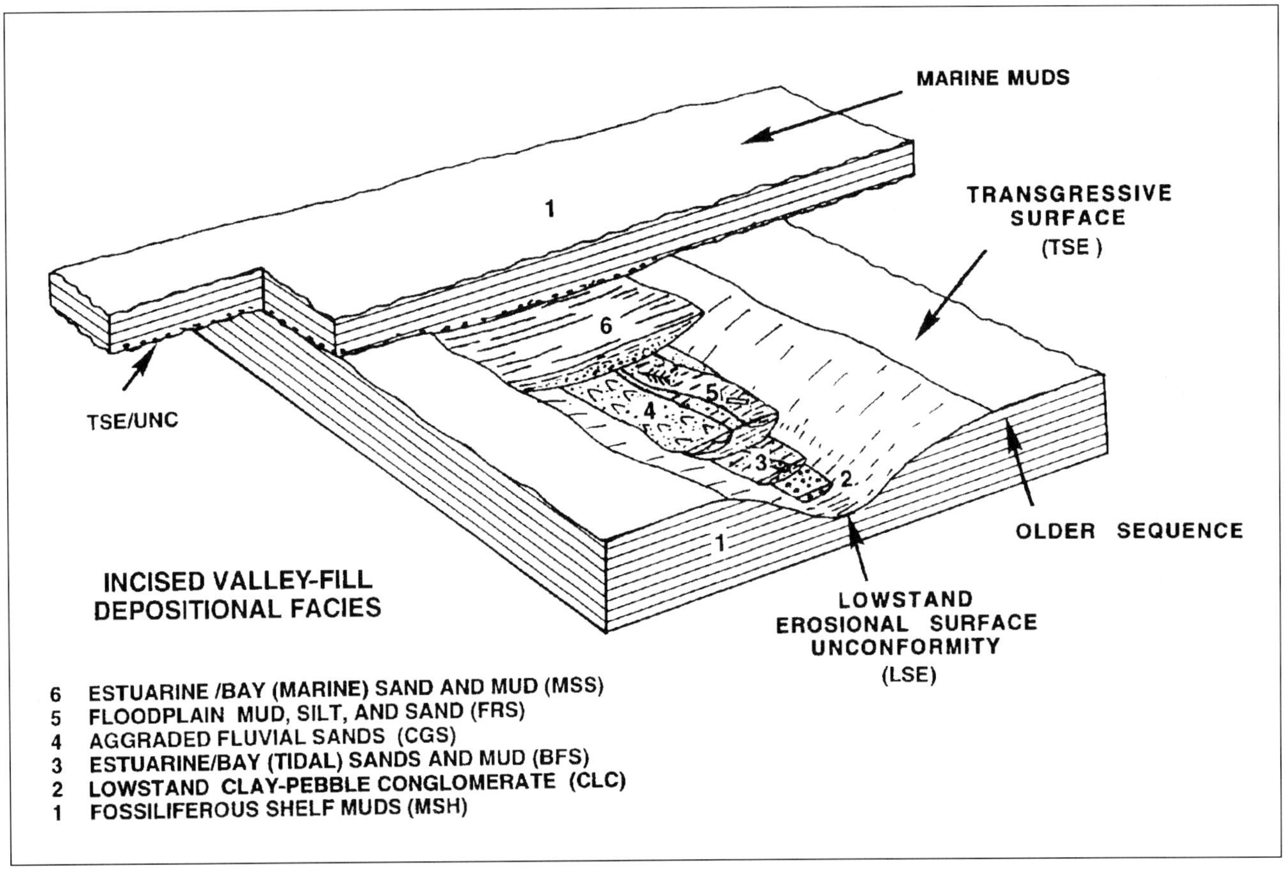

FIGURE 16 — Generalized schematic block diagram illustrating the deposition of valley-fill facies that could occur with sea-level rise following incision (after Wheeler et al., 1990).

mity (LSE) were not cored, but can be inferred from log signatures.

The Gulf Ferguson core contains fossiliferous marine shale (MSH), a sharp erosional contact (LSE), coarse-grained fluvial sandstone (CGS), dark gray flood-plain mudstone (FRS), a second fluvial sandstone (CGS), and flora-rich (FRS) flood-plain shales and siltstones (Fig. 15).

The Kelly core contains marine shale (MSH), an erosional contact (LSE), an estuarine sandstone (BFS) that grades upward into flood-plain (FRS) siltstones and carbonaceous mudstones with coaly material (Fig. 9), and highstand marine shale (MSH).

The cores represented above are a subset of 12 cores examined in the Oklahoma panhandle region. In addition to these cores, over 1000 wireline logs were also examined. These data suggest that upper Morrowan sandstone deposits are dominated by upward-fining, channel-fill assemblages that are enclosed in marine shales. Some of these sandstones represent point bar sandstones that developed along meanders in fluvial valleys (Swanson 1979; Cornish 1982).

DEPOSITIONAL MODEL

The processes that controlled the deposition of the various sedimentary facies were strongly influenced by changes in the sea level. Fluvial valleys eroded into existing shelf sediments during sea-level lowstands. Lowstand deposition occurred in the form of channel-lag conglomerates that contain abun-

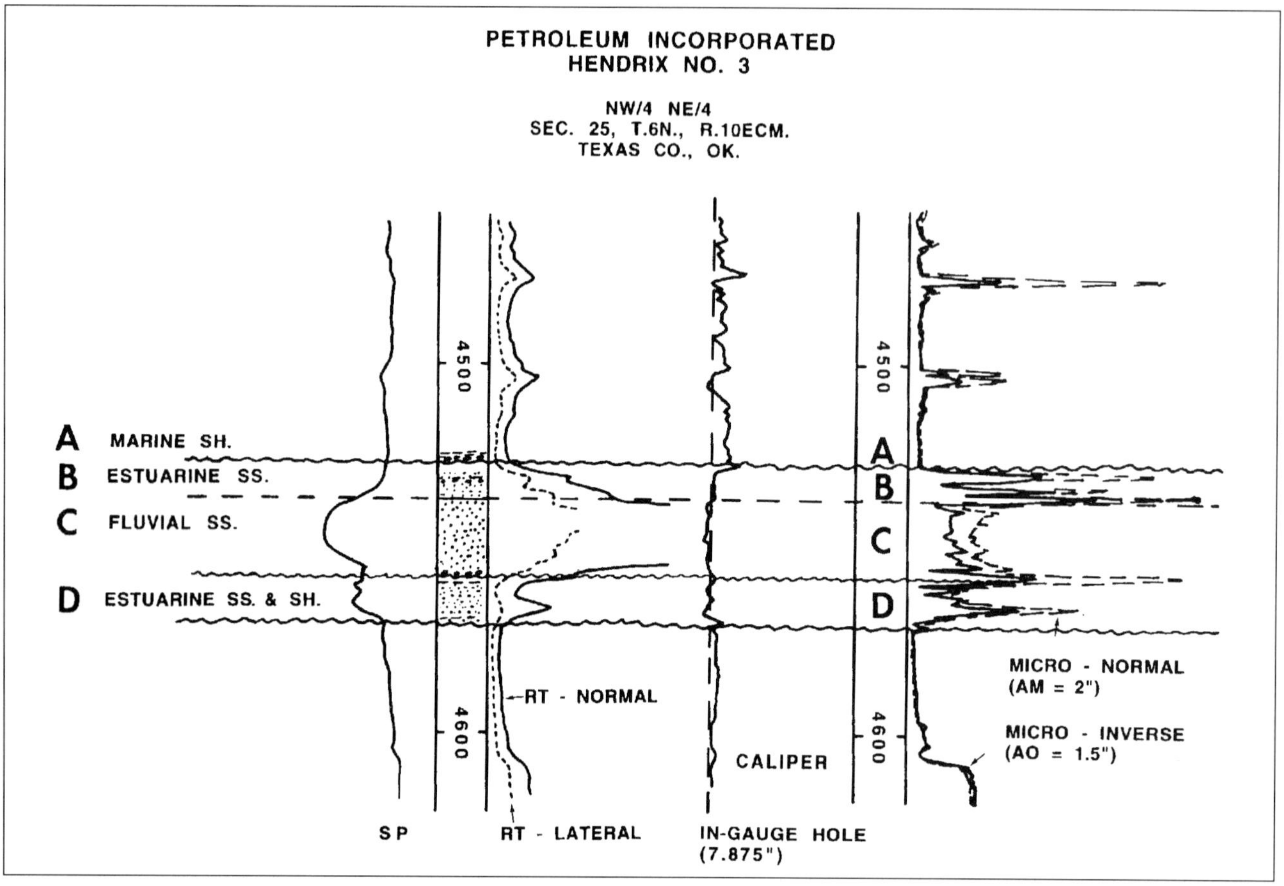

FIGURE 17 — Composite wireline log schematic diagram across the upper Morrowan sandstone interval in the Petroleum Inc. Hendrix No. 3. Positive microresistivity separation identifies porous reservoir in the upper 11 ft of the fluvial sandstone (C). Estuarine sandstones (B) and (D) exhibit low porosity (after Puckette, 1993).

dant clasts of the background sediment (CLC facies). Infilling the incised valleys occurred during the transgressive phase produced by the rising sea level. The initial flooding of valleys produced estuaries where finer grained sand, silt, and mud were deposited (BFS facies). In direct response to the rising base level, coarser grained fluvial sediments (CGS facies) were deposited in an updip direction from the interface between the fluvial and marine systems. As the sea level continued to rise, the fluvial sands were succeeded by additional estuarine sediments (MSS facies). Abandoned channels and marshes became the loci for the deposition and preservation of plant debris within the valley (FRS facies). A generalized schematic block diagram of a valley-fill depositional sequence is shown in Fig. 16. Substantial variations in grain size and shale content within valley trends can be attributed in part to minor sea-level oscillations during the overall transgression. The dropping sea level rejuvenated the system and caused incision of existing valley-fill sediments. Multiple episodes of incision and deposition generated a variety of facies successions, such as estuarine (BFS) succeeded by fluvial (CGS), formed stacked channel sequences, and contributed to reservoir heterogeneity.

SEQUENCE STRATIGRAPHY AND RESERVOIR QUALITY

A depositional sequence is a genetically related succession of strata bounded by unconformity surfaces or their correlative conformities (Mitchum, 1977). For the Morrow, the widely accepted low-

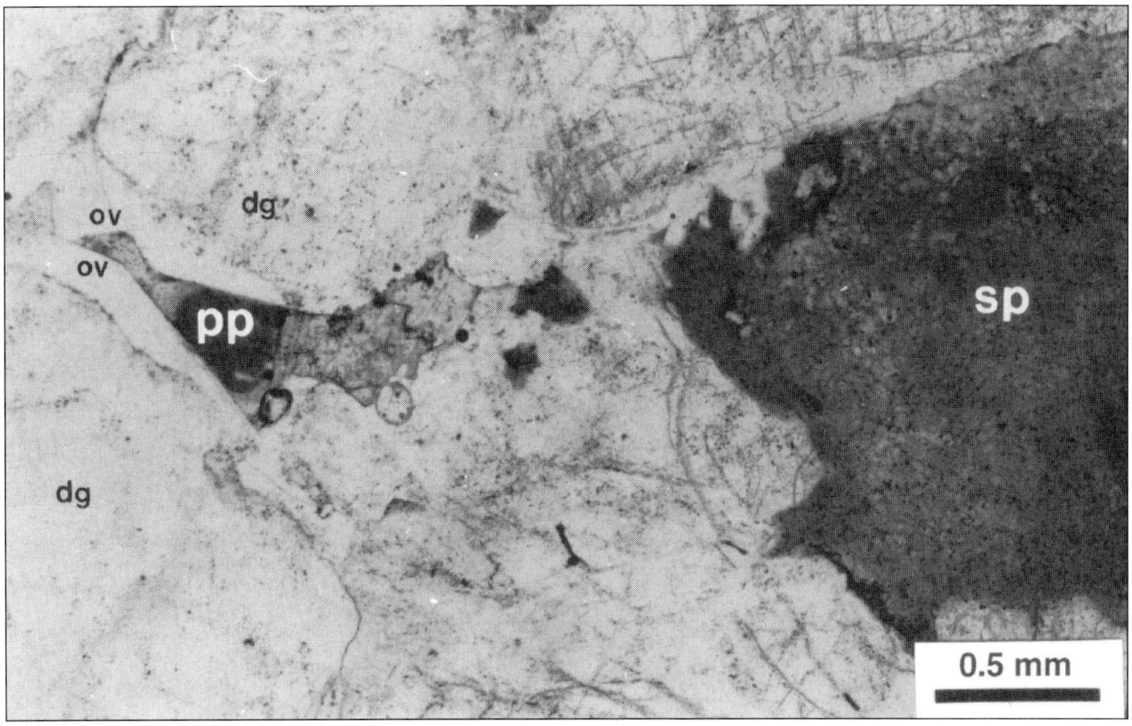

FIGURE 18 — Photomicrograph illustrating primary and secondary porosity in the Petroleum Inc. Hendrix No. 3. Primary porosity (pp) is framed by euhedral crystal faces of syntaxial quartz overgrowths (ov) that border inclusion-rich detrital grains (dg). Secondary porosity (sp) is represented by large oversized pore formed by the dissolution of a detrital feldspar grain. Depth 4560 ft. Plane-polarized light (ppl).

FIGURE 19 — Photomicrograph of secondary intragranular porosity formed by the dissolution of feldspar. Silica cement (SC) occludes some porosity. Petroleum Inc. Hendrix No. 3. Depth 4560 ft. Plane-polarized light (ppl).

FIGURE 20 — Photomicrograph of tidal-dominated estuarine sandstone with abundant clay matrix and low porosity. Petroleum Inc. Hendrix No .3. Depth 4573 ft (ppl .

stand erosional surfaces (LSE) are considered the sequence boundaries (Fig. 5), and sediments between these bounding surfaces constitute a genetically related succession of strata. Within a generalized application of sequence stratigraphy concepts, the CLC facies lags (Fig. 16, facies 2) are lowstand systems tract (LST) deposits. The valley-fill sediments (Fig. 16, facies 3, 4, 5, and 6) represent transgressive systems tract deposits. The marine shelf muds (Fig. 16, facies 1) represent the culmination of the transgressive and highstand systems tracts. Forced regression (Posamentier et al., 1992) and the LST are evidenced by the incision of shelf muds that marks the next sequence boundary. The major aim of this study is to apply the concepts of sequence stratigraphy to the construction of a precise depositional framework of the upper Morrowan reservoir rocks. However, this type of analysis has also contributed to the identification of reservoir quality lithofacies within the stacked facies of a valley-fill sediment assemblage.

The Petroleum Inc. Hendrix No. 3 core illustrates this application. The four depositional facies in the Hendrix core have distinct textures, compositions, and diagenetic histories. A composite log schematic for the Hendrix (Fig. 17) indicates that the porous reservoir in the cored interval is the fluvial sandstone (CGS) facies, whereas the underlying (tidal) BFS and overlying (marine) MSS estuarine sandstones are tightly cemented. Development and preservation of porosity in the sandstone facies are primarily functions of their depositional history and composition.

Porosity in the fluvial dominated sandstone is found in the middle and top of the genetic unit. Preserved primary porosity was recognized by the straight euhedral crystal faces on quartz overgrowths

FIGURE 21 — Photomicrograph of marine-dominated estuarine sandstone with abundant invertebrate grains and carbonate cement. Petroleum Inc. Hendrix No. 3. Depth 4543 ft (ppl).

that border pores (Fig. 18). Secondary porosity is evident as enlarged intergranular, intragranular, and microporosity. Enlarged intergranular porosity (Fig. 18) was generated by the dissolution of framework grains such as feldspars and granitic rock fragments. Intragranular porosity developed within framework grains as the result of preferential dissolution (Fig. 19). Microporosity was observed within some partially leached framework grains and more commonly between kaolinite booklets.

Porosity in the base of the fluvial dominated unit was occluded by carbonate cement and pseudomatrix. Pseudomatrix was formed by the ductile deformation of clay clasts. Carbonate cements are dolomite and calcite. Calcite has a coarse equant morphology. The dolomite is baroque type that apparently formed from calcite when hot fluids (>80°C) (Radke and Mathis, 1980) moved through the reservoir.

The basal estuarine sandstone failed to retain significant porosity due to the presence of a clayey detrital matrix (Fig. 20). This matrix was dispersed among the sand grains by burrowing animals.

The fossiliferous marine-dominated estuarine sandstone that immediately overlies the fluvial reservoir in the Hendrix core is tightly cemented by calcite cement (Fig. 21). This cement may have been derived from the stress-induced dissolution of fossil fragments during sediment compaction.

CONCLUSIONS

The upper Morrowan on the northwestern shelf of the Anadarko basin consists of both marine (shelf) shales and fluvial, flood-plain, and estuarine valley-fill sediments. Deposition of these sediments was controlled by changes in the relative sea level. As the relative sea level dropped, incised valleys eroded exposed shelf muds. During rises and marine transgressions, valleys were filled with fluvial, estuarine, and flood-plain sediments. With continued transgression, these deposits were blanketed with marine muds.

Reservoir quality was influenced by depositional facies. Fluvial sandstones are typically coarser grained, contain less calcite cement, and are better reservoirs than the estuarine sandstones. Marine-dominated estuarine sandstones contain abundant bioclasts and carbonate cement. Tidal-dominated estuarine sandstones are often matrix-rich, fine-grained, poor-quality reservoirs.

ACKNOWLEDGMENTS

The authors wish to express their appreciation to Darwin Boardman, Justin Evans, Catherine Price, Aaron Rice, and Gary Stewart for their helpful contributions. We thank J. W. Shelton, A. W. Walton, and Rick Erickson for their constructive criticisms and valuable suggested improvements.

EVIDENCE FOR SEA-LEVEL FLUCTUATION AND STRATIGRAPHIC SEQUENCES IN THE COUNCIL GROVE GROUP (LOWER PERMIAN), HUGOTON EMBAYMENT, SOUTHERN MID-CONTINENT

J. Puckette
D. R. Boardman II
Z. Al-Shaieb
School of Geology
Oklahoma State University
Stillwater, OK 74078

ABSTRACT

The Council Grove Group in the Hugoton embayment is formed by a series of sequences composed of shallow marine limestones, shales, and nonmarine red beds. The repetitive nature of these units documents glacial eustatic, cyclic sea-level oscillations.

The Council Grove contains two distinct types of sequences. Those in the lower part (Foraker through Grenola formations) consist of (1) a basal transgressive surface, (2) a shallow marine limestone with a marine condensed section that represents the transgressive and highstand systems tracts, (3) shallow water carbonates with a weathered upper surface (exposure surface) that represents forced regression, and (4) red caliche-bearing, blocky claystones and siltstones (red beds) that may correspond to the lowstand systems tract (LST) of the shelf margin. Sequences in the upper part (Beattie Limestone through Speiser Shale) are similar to those in the lower Council Grove, but lack the marine condensed section that represents a well-defined highstand systems tract (HST).

The lower sequence boundaries are characterized by thin (<1ft) transgressive lag deposits which separate the limestones from the underlying nonmarine red beds. The transgressive systems tracts (TSTs) consist of upward-deepening carbonate. Maximum flooding within the HST is represented by a carbonate marine condensed section containing glaucony and skeletal phosphate, including abundant conodonts. Forced regression is represented by upward-shoaling carbonates punctuated by minor flooding surfaces that delineate parasequences. The upper surfaces of the regressive carbonate units and overlying nonmarine claystones and siltstones exhibit evidence of subaerial exposure. The limestones developed weathered zones consisting of carbonate regolith in a red-green clay and silt matrix. The succeeding claystones are red and contain nodular caliche. These features indicate

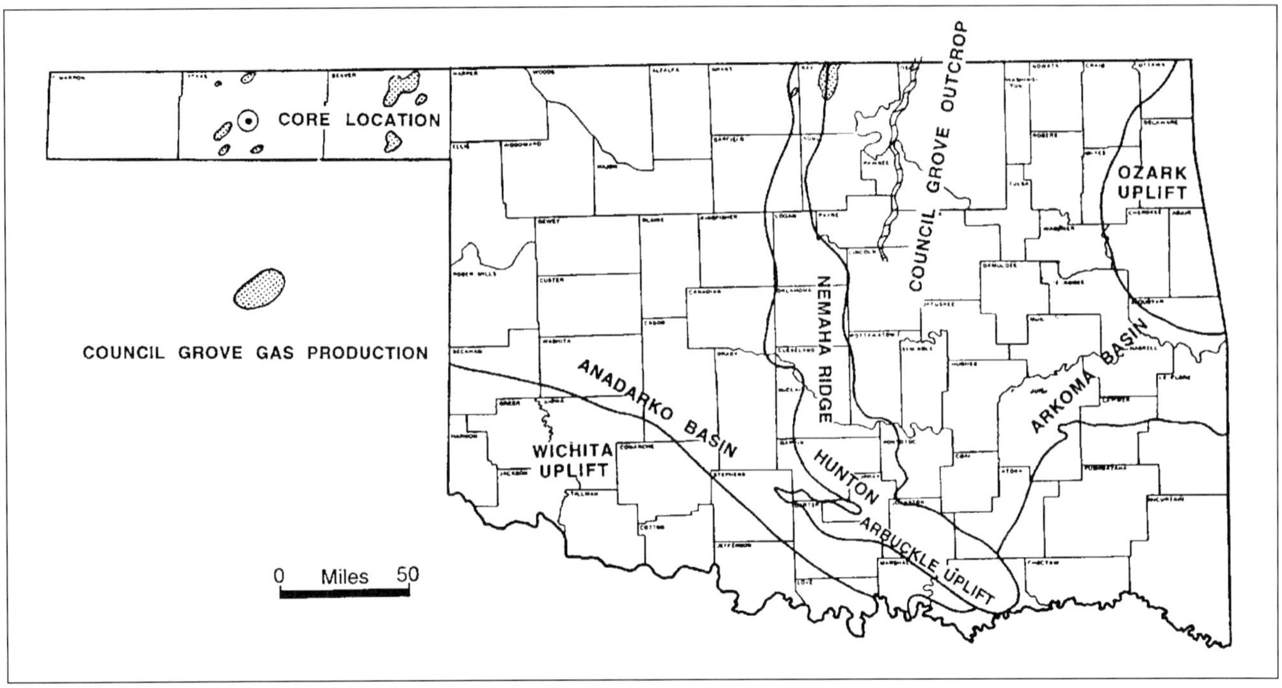

FIGURE 1 — Locations of Council Grove outcrops, core, and gas producing areas.

subaerially exposed rocks and paleosols. A variety of karstic features (such as vugs and solution channels) are common in the limestones. Some of these dissolution features are filled with red clay.

Porosity in the Council Grove carbonate units is secondary and formed by the dissolution of carbonate grains and cement. The highest porosity values (>10%) are found in upward-shoaling ooid and skeletal grainstones. Karsted rocks with vugs and channels filled by infiltrated sediment exhibit porosity of <10%.

INTRODUCTION

A number of carbonates in the Council Grove Group are gas reservoirs in the Hugoton embayment and northern Oklahoma platform. Whereas production is commonly associated with structural anomalies, porosity development appears to have been influenced by the depositional setting and facies. To interpret the evolution of these reservoirs, the vertical sequence of the Council Grove Group in the Hugoton embayment was analyzed by examining cores from the Cities Service Stonebraker A-80 well in T3N-R14ECM, Texas County, Oklahoma. Stratigraphic correlations, facies relationships, and reservoir characteristics were established and interpreted.

COUNCIL GROVE GROUP STRATIGRAPHY

The Council Grove Group was examined in the surface outcrops (Boardman and Nestell, 1994), and the core located on Fig. 1. Correlation of the subsurface and surface sections was primarily accomplished using faunal (conodont) biostratigraphy and lithofacies analyses, as well as wireline log signatures. Differentiation of the section into genetically related units was keyed to the recognition of lowstand exposure surfaces and subsequent transgressive flooding that indicated significant changes

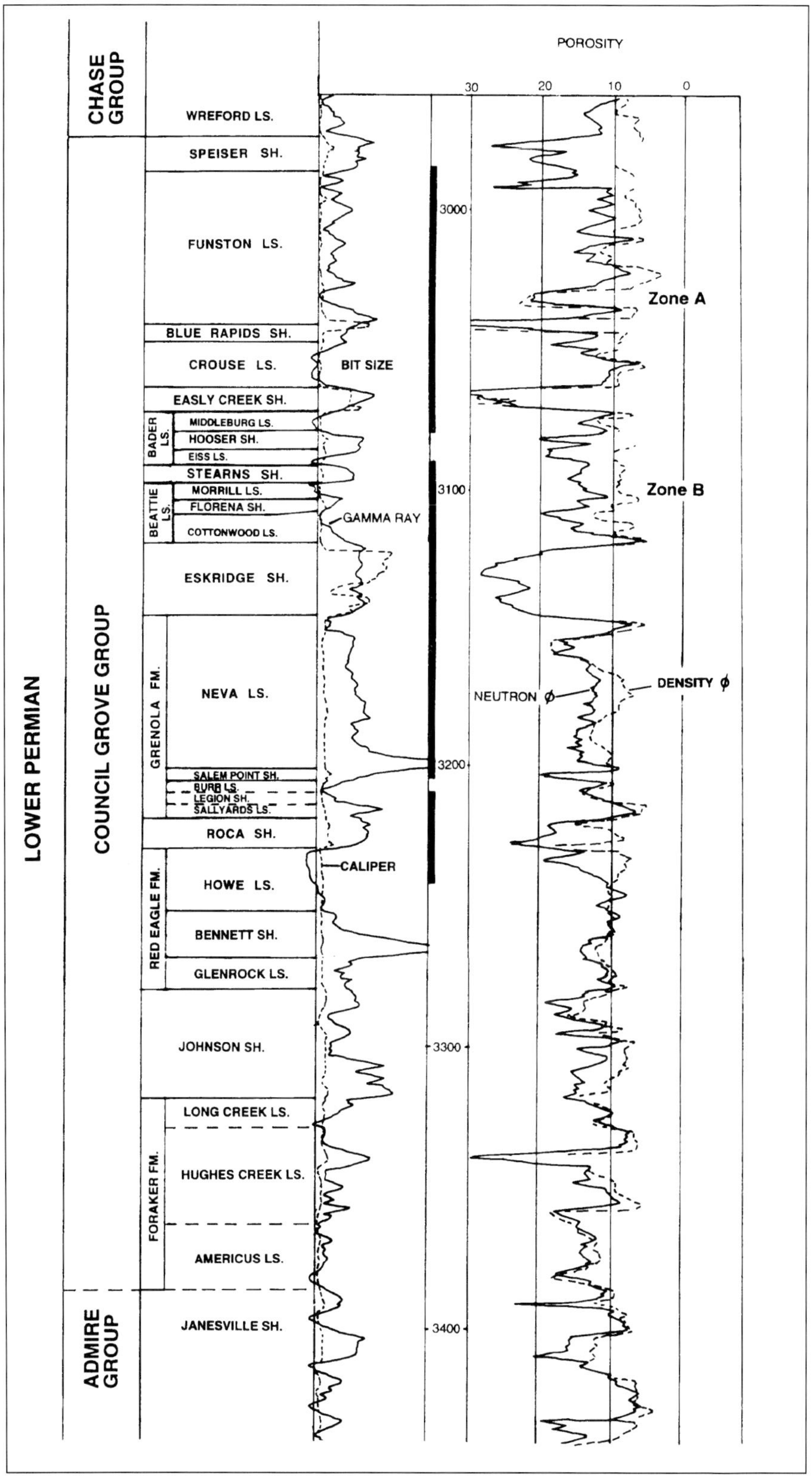

FIGURE 2 — Stratigraphic nomenclature and gamma ray, caliper, and porosity logs of Council Grove Group, Hugoton embayment, Texas County, Oklahoma. Cities Service Stonebraker A-80 well. Cored intervals are indicated by black strip. Radioactive minerals concentrated along a marine condensed section generated the high gamma ray response at 3198 ft in the base of the Neva Limestone.

in depositional setting.

The available core consisted of 221.5 ft from the upper 237 ft of the Council Grove Group, and included those units between the Red Eagle Formation and the Wreford Limestone of the overlying Chase Group (Fig. 2).

COUNCIL GROVE DEPOSITIONAL FACIES

The cored rocks are characterized by repetitive, shallow water (subtidal to lower intertidal) carbonates and terrestrial redbed claystones and siltstones. The noted absence of upper intertidal and supratidal facies suggests nondeposition of these facies or their subsequent erosion. The general absence of these facies from all cycles infers nondeposition, and thus a mechanism for the rapid change in sea level from a subtidal to terrestrial setting in a given area. Boardman and Nestell (1994) proposed that glacial eustatic cycles provided a repetitive mechanism that generated these rapid oscillatory sea-level fluctuations. These fourth-order eustatic cycles (Youle et al., 1994) forced relatively rapid regressions and transgressions of the Wolfcampian seas and generated the lithofacies relationships evident in these rocks.

NO.	SYMBOL	DESCRIPTION
1	L-WP	**Light gray lithoclastic wackestone to packstone**: Carbonate and siliciclastic lithoclasts and skeletal grains in silt and mudstone matrix. Basal carbonate unit that forms sharp contact with underlying red beds.
2	SW	**Light to medium gray skeletal wackestone:** Peloids and fossil grains in a mudstone matrix. Large fossil grains are disarticulated, abraded, and algal coated; small grains are whole.
3	PS-P	**Medium gray peloid and skeletal packstone:** Disarticulated, abraded, and commonly algal-coated pelmatozoan, foraminifer, ostracode, brachiopod, and bryozoan fragments with varying amounts of sparry cement and mud matrix.
4	S-SH	**Calcareous siltstone, silty mudstone, and medium and dark gray shale:** Finely laminated, dark, calcareous, silty shale without burrows and fossil grains that grades upward into lighter gray to green silty mudstone and shale with brachiopods and burrows. Also, medium to dark gray fossiliferous shale with disarticulated pelmatozoans and brachiopods, and whole fusulinids in a matrix rich in bryozoan, ostracode, and trilobite debris.
5	OB-PG	**Crossbedded ooid and bioclastic packstone to grainstone:** Medium to light gray packstones and light-colored grainstones. Dominant skeletal grains are disarticulated and abraded pelmatozoans, fusulinids, brachiopods, and foraminifers. Ooids are dominant nonskeletal grains. Algal coatings and grain micritization are common.
6	O-PG	**Oncoid packstone to grainstone:** Micritized composite grapestone clusters and single grains up to 12 mm. Cemented by anhydrite.
7	I-WP	**Gray to green intraclastic wackestone to packstone:** Intraclastic wackestone to packstone (weathered carbonate) that grades upward into red silty claystone with carbonate lithoclasts and caliche nodules and crusts.
8	RC	**Red claystone and clayey siltstone:** Nonfossiliferous red claystones and siltstones with nodular caliche and fracture-filling calcite (red beds).

TABLE 1 — Council Grove Group lithofacies.

LITHOFACIES

The principal lithofacies (Table 1) recognized include (1) light gray lithoclastic wackestone to packstone, (2) skeletal wackestones, (3) peloid and skeletal packstones, (4) calcareous siltstones, silty mudstones, and shales, (5) crossbedded ooid and bioclastic packstone-grainstones, (6) oncoid packstone-grainstones, (7) green to red lithoclastic wackestone to packstone, and (8) red claystones and siltstones.

Light Gray Lithoclastic Wackestone-Packstone Facies (L-WP)

Description

The L-WP facies consists of carbonate and siliciclastic lithoclasts and skeletal grains in a mud and silt matrix. The diameters of the lithoclasts average 5 mm and range from <1 to 20 mm. They are rounded to subrounded and consist of skeletal wackestone, mudstone, and abraded caliche nodules. The bioclasts are primarily broken echinoderm, brachiopod, and ostracode grains. The L-WP is typically <0.5 ft (15 cm) thick. The percentage and size of lithoclasts decrease upward, and the facies forms a gradational contact with the overlying skeletal wackestones and packstones. The L-WP has a sharp contact with the underlying redbed claystones and siltstones.

Interpretation

The L-WP facies was deposited during the marine transgression that flooded the exposed redbed units. The composition and rounding of the lithoclasts suggest they were derived in part from local red beds and in part from more distant carbonate sources. Bioclastic debris suggest that transgressing water supported a diverse marine fauna.

Light to Medium Gray Skeletal Wackestone Facies (SW)

Description

The SW facies consists of peloids, oncoids, pelmatozoan, brachiopod, bryozoan, foraminifer, and ostracode grains in a mudstone matrix. Small grains are whole, whereas larger bioclasts are disarticulated and abraded. Abraded fragments are commonly algal-coated and suggest reworking and transportation. The SW gradationally overlies the L-WP and may contain a few small (<1 mm) lithoclasts. Silt is common in this facies. The SW also occurs in sharp contact with underlying dark-gray shale or grain-rich packstone-grainstone facies when it occurs within the upper part of a carbonate unit.

Interpretation

The small intact grains and mud matrix indicate the SW facies was deposited below a normal effective wave base. The broken and abraded bioclasts and oncoids were likely transported into this environment by occasional storms. The SW represents a deepening of water after the initial marine flooding of subaerially exposed red beds or deepening associated with parasequence boundaries within carbonate units.

Medium Gray Peloid and Skeletal Packstone Facies (PS-P)

Description

The PS-P facies commonly forms a gradational contact with the underlying SW facies. The dominant grains are pelmatozoans, foraminifers, ostracodes, brachiopods, and bryozoans. The bioclasts are disarticulated, abraded, and commonly algal coated. Sedimentary structures are not apparent. The upper contact of PS-P is usually gradational with more grain-rich carbonates and sharp with dark shales or skeletal wackestones.

Interpretation

The decreasing mud content and increase in coated grains and bioclastic debris suggest an increase in wave energy, hence a decrease in water depth. The PS-P facies represents the initial upward-shoaling phase from the underlying SW facies. It was deposited below to slightly above normal wave base.

Calcareous Siltstone, Silty Mudstone, and Medium and Dark Gray Shale Facies (S-SH)

Description

The S-SH facies consists of dark, finely laminated calcareous shale without burrows or bioclasts that grades upward into lighter gray to green, burrowed, silty carbonate mudstones and shales, dark to medium gray calcareous siltstones and shales with large horizontal burrows and intact brachiopods, and silty, thin-bedded mudstone with large vertical burrows and bioclasts. The S-SH is in sharp contact with the underlying SW to PS-P facies and gradational with the overlying skeletal packstones or grainstones.

The finely laminated, dark gray, nonfossiliferous, calcareous shale occurs above a hardground surface that contains an accumulation of skeletal phosphate debris. Gamma ray logs indicate this zone forms a "hot" marker bed (Fig. 2).

The medium to dark gray fossiliferous shale phase of the S-SH occurs in several carbonate units. The dominant grains are disarticulated pelmatozoans and brachiopods and whole fusulinids in a matrix that is rich in bryozoan, ostracode, and trilobite debris. These grains are relatively unabraded and only sparsely coated. Skeletal phosphate also occurs in this facies.

Interpretation

The dark, finely laminated silty to calcareous shale that lacks burrows and bioclasts represents stratified dysoxic to anoxic conditions during maximum water depths. The accumulation of skeletal phosphate below this mudstone represents a marine condensed section. The "hot" gamma ray log signature may coincide at least in part with the phosphate layer.

With shallowing, conditions became more hospitable to marine fauna and burrowing became more common. Intact fossils suggest low-energy deposition below the normal wave base. An increase in macro-invertebrate fossil debris and carbonate matrix indicates continued shallowing and a return to normal marine conditions.

The occurrence of the medium to dark gray fossiliferous shale within carbonate units identifies periods of flooding and increased water depth. The bases of the shales may define parasequence boundaries.

Crossbedded Ooid and Bioclastic Packstone to Grainstone Facies (OB-PG)

Description

The OB-PG facies grades from medium- to light-gray packstones to light-tan to cream grainstones that contain low-angle crossbedding. The OB-PG forms gradational contacts with underlying PS-P facies. This grain-rich facies forms sharp upper and lower boundaries with mud-rich (S-SH and SW) shales and wackestones. The dominant skeletal grains are disarticulated and abraded pelmatozoan, fusulinid, brachiopod, and foraminifer bioclasts. Ooids are the dominant nonskeletal grains. Algal coatings and grain micritization are prevalent.

Interpretation

The low-angle crossbedding, ooids, and coated abraded skeletal grains indicate a high-energy environment. The OB-PG represents shoal deposition that culminates the upward-shallowing sequences within the carbonate intervals.

Oncoid Packstone to Grainstone Facies (O-PG)

Description

This tan to gray algal-influenced facies is present in only one sequence where it forms sharp contacts with overlying and underlying SW facies. The oncoids are completely micritized large composite grapestone clusters of grains (5 to 12 mm in diameter). The intergranular cement in the O-PG facies is anhydrite.

Interpretation

The accretionary nature and composite grapestone grains suggest that the O-PG facies was deposited in a shoal environment in which quiescence allowing grain-to-grain cementation alternated with turbulence (probably storms) causing partial breakup of the clusters. The anhydrite cement suggests that sometime after deposition and possibly dissolution, this rock was saturated with sulfate-enriched brines from a nearby evaporative environment.

Gray to Green Intraclastic Wackestone to Packstone Facies (I-WP)

Description

The I-WP facies is light-green to gray, intraclastic wackestone to packstone that grades upward into red silty claystone with carbonate lithoclasts, caliche crusts, and nodules. Circumgranular cracking and hematite staining are common. The intraclasts are ovoid to polygonal and are <1 mm to 5 cm long. Solution-seam boundaries between clasts are common. The I-WP has gradational contacts with the underlying facies (generally peloidal-skeletal packstone) and succeeding red beds.

Interpretation

The intraclasts, circumgranular cracking, hematite staining, and caliche indicate this facies has been altered by subaerial exposure. The lack of rhizoconcretions or root traces suggests that soil development was inhibited by arid conditions. Grain contact dissolution represents burial diagenesis.

Red Claystone and Clayey Siltstone Facies (RC)

Description

The RC facies consists of silty red claystone and clayey siltstones. These red beds lack marine fossils and laminated structure. This facies may be green-gray or variegated at the base or top. The RC contains nodular caliche and fracture-filling calcite. It lies gradationally above the I-WP facies and has a sharp (probably erosional) boundary with the overlying L-WP facies.

Interpretation

The caliche, red color, and absence of marine invertebrates indicate that this facies was deposited under subaerial conditions. The lack of obvious wet-climate paleosol features such as (rhizoconcretions, root traces, and horizonation) suggests deposition in an arid environment unsuitable for

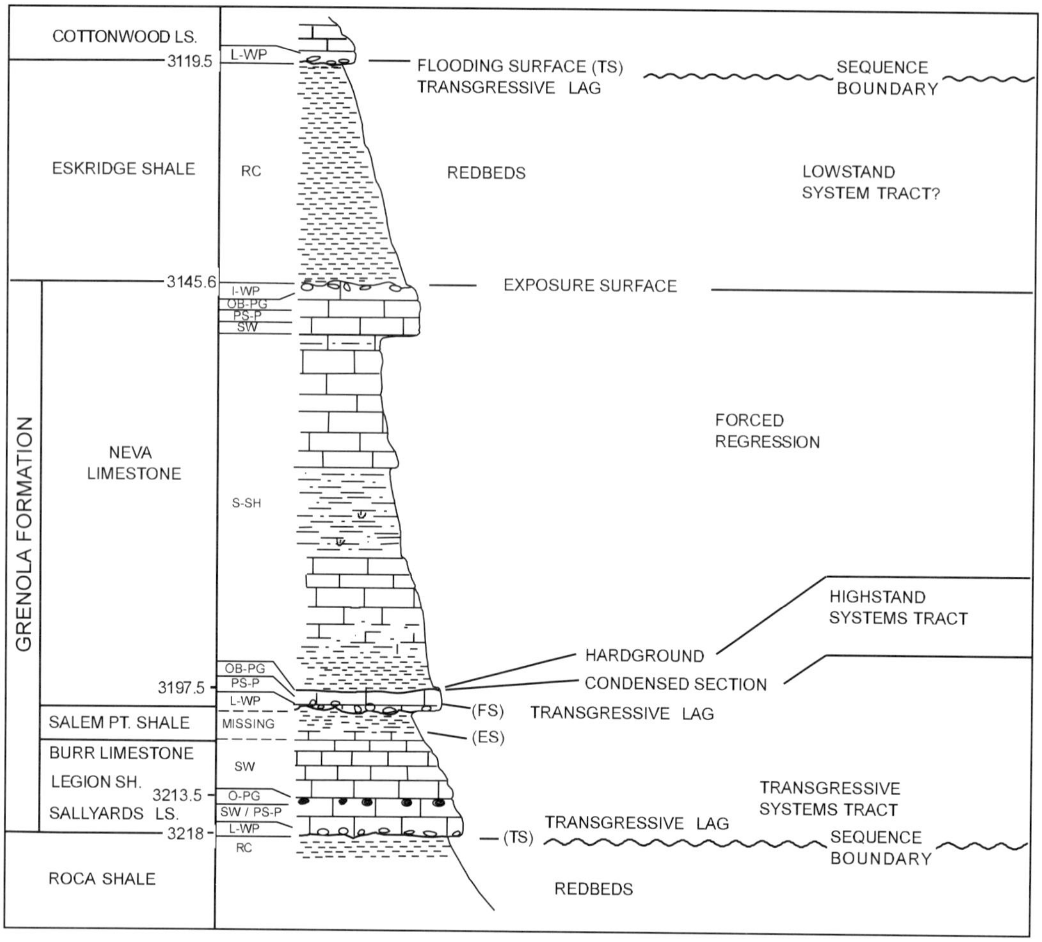

FIGURE 3 — Schematic diagram of facies succession in Grenola Formation depositional sequence. Weathering profile is based on lithic characteristics of the units and caliper log measurements.

much plant growth and/or preservation.

DEPOSITIONAL SEQUENCES

The Council Grove contains two distinct types of depositional sequences. Those in the lower part (Foraker to Grenola formations) contain (1) a basal transgressive surface and lithoclastic L-WP facies, (2) a shallow marine limestone (SW, PS-P, and O-PG facies) of a transgressive systems tract (TST) that ultimately culminated as a marine condensed section and dark calcareous shale (S-SH facies; basal unit) within the highstand systems tract (HST), (3) mixed carbonate siliciclastic units (S-SH facies) and upward-shoaling carbonates (OB-PG facies) of the forced regression (Posamentier et al., 1992) that culminate with a weathered upper exposure surface (ES) (I-WP facies), and (4) redbed claystone/siltstone units (RC facies) that correspond to the lowstand systems tract (LST) of the shelf margin.

Depositional sequences in the upper part (Beattie Limestone through Speiser Shale) are similar to those in the lower Council Grove, but lack a well-defined marine condensed section and HST. Within the upper part, the HST and the forced regression consist of upward-shoaling limestones that are punctuated by flooding surfaces or contain evidence of gradual deepening and shallowing events.

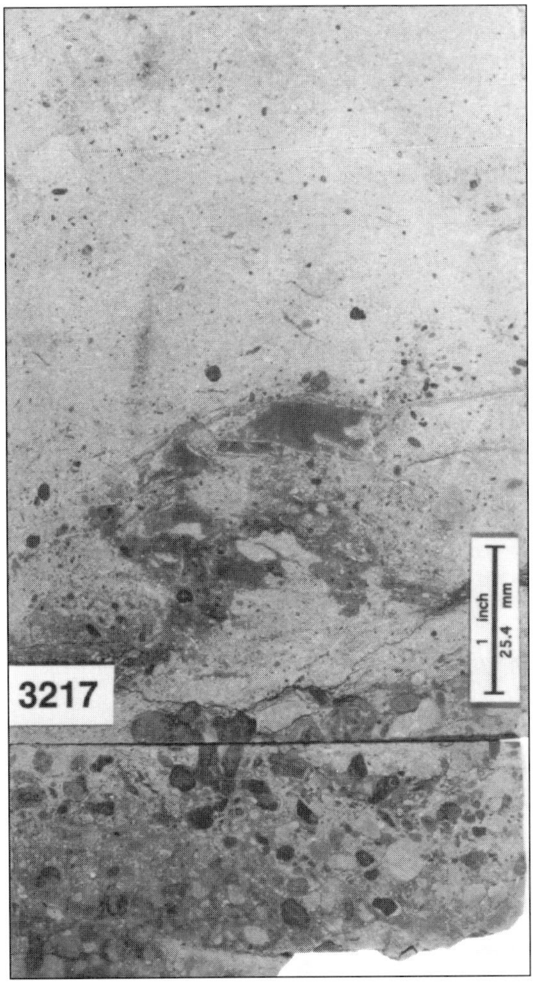

FIGURE 4 — Transgressive lag deposit (L- WP facies) at base of Sallyards Limestone. This transgressive surface (TS) separates overlying marine carbonates from underlying nonmarine red beds and defines the lower boundary of the Grenola sequence.

Lower Council Grove Depositional Sequences: Grenola Formation

Illustrating the type of sequence in the lower Council Grove, the Grenola Formation consists of the Sallyards Limestone, Legion Shale, Burr Limestone, Salem Point Shale, and Neva Limestone (Fig. 3). The lower sequence boundary is characterized by a thin (<1 ft) transgressive lag deposit (Fig. 4) (L-WP facies) that separates the basal limestone from the underlying nonmarine red claystones (RC facies) of the Roca Shale. The pebbles that compose the lag are bioclasts, marine limestones, red claystones and siltstones, and caliche (Fig. 5). The clasts are typically well rounded and coated with algal-produced micrite. They occur in a silt- and clay-rich mud matrix.

The wackestones-packstones (SW and PS-P facies) that characterize the TST in the Sallyards Limestone interval are relatively thin and suggest only a short-term deepening of water following the initial flooding of the underlying nonmarine Roca surface. In contrast, oncoid packstone-grainstone (O-PG facies) (Fig. 6) at 3213.5 ft indicates shallowing water. This unit may reflect a minor regression associated with the Sallyards Limestone and Legion Shale. The Burr Limestone (Fig. 3) contains an upward-shoaling interval followed by deepening conditions. A complete Burr section was not seen due to missing core. The Salem Point Shale is represented by a thin, brown, crumbly claystone (possibly a paleosol) that reflects lowstand conditions. The Salem Point Shale was then inundated by the Neva transgression. Maxi-

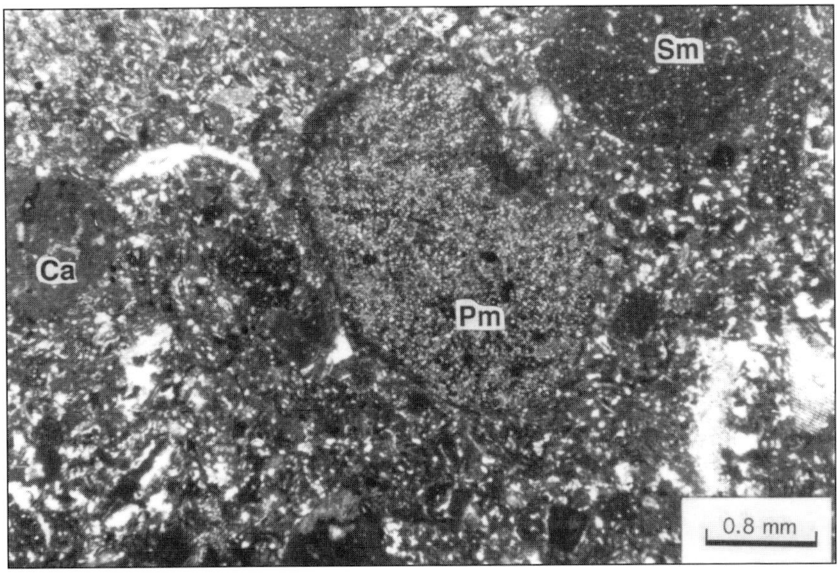

FIGURE 5 — Photomicrograph illustrating pebble morphology and composition in Sallyards transgressive lag (L-WP facies). Ca = caliche, Pm = peloid dolomudstone, and Sm = silty mudstone. Depth 3217 ft. Plane-polarized light.

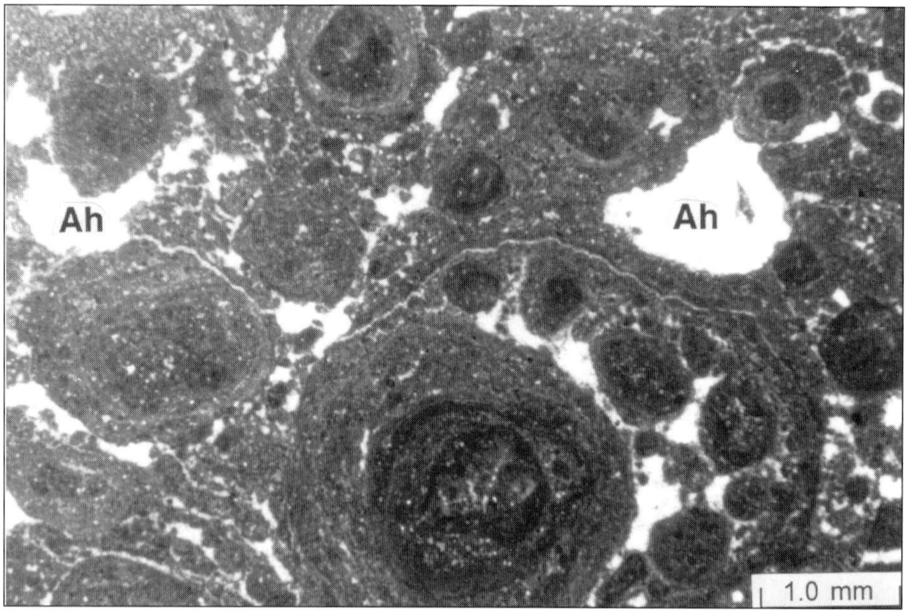

FIGURE 6 — Oncoid packstone-grainstone (O-PG facies) in Sallyards Limestone that indicates shallowing conditions following the marine transgression. Note grapestone grains and anhydrite cement (Ah). Depth 3213.5 ft Plane-polarized light.

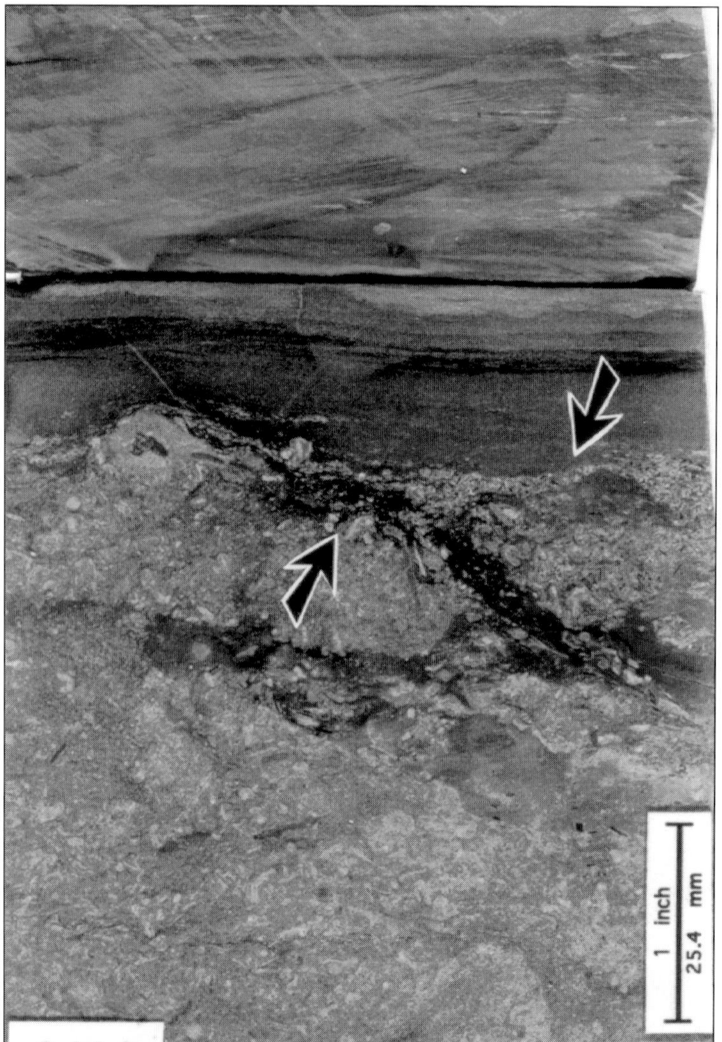

FIGURE 7 — Hardground and marine condensed section (arrows) of HST in the Neva Limestone.

FIGURE 8 — Photomicrograph of phosphatic skeletal debris (arrows) and glauconite (G) that accumulated in a marine condensed section in the Neva Limestone. Depth 3198 ft. Plane-polarized light.

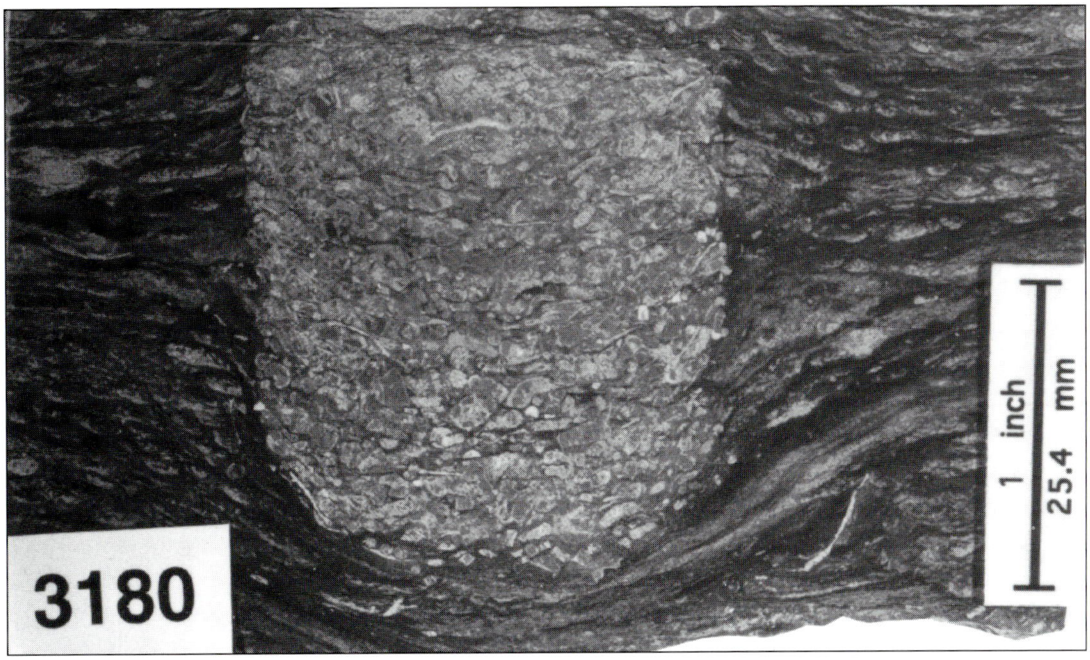

FIGURE 9 — Large horizontal burrow in the silt-rich (S-SH facies) interval of the Neva Limestone.

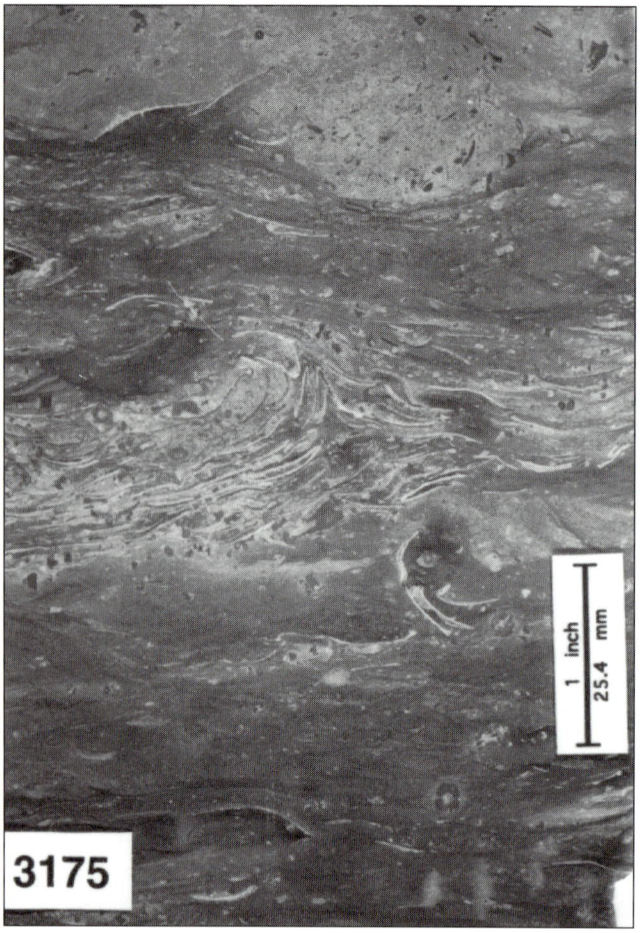

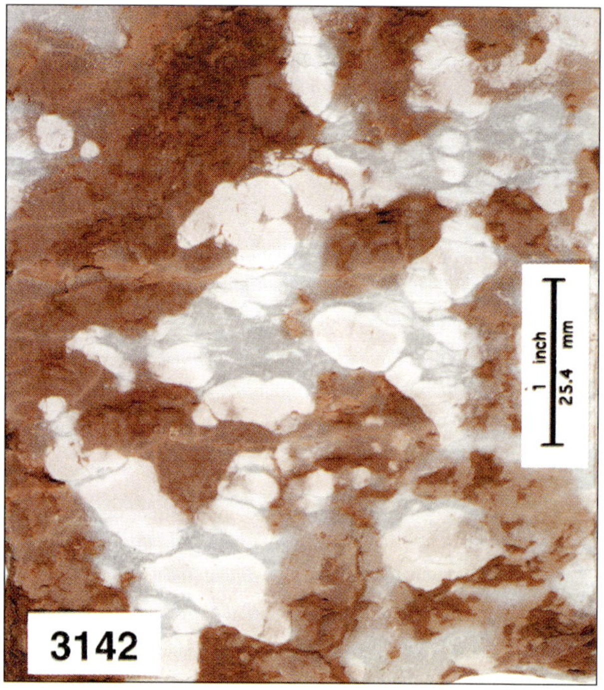

FIGURE 11 (above) — Calcareous nodules (caliche) in the nonmarine redbed (RC facies) Eskridge Shale.

FIGURE 10 (left) — Invertebrate fossils (mainly brachiopods) in silt-rich (S-SH facies) interval of the Neva Limestone.

mum flooding during the Neva is represented by a burrowed hardground (Fig. 7) and overlying marine condensed section that developed on a flooded grainstone. Phosphatic skeletal debris including conodonts and glaucony accumulated in burrows and formed a 1- to 3-cm thick layer on the hardground surface (Fig. 8).

Overlying the condensed section is a thinly laminated black calcareous shale (S-SH facies) that reflects quiescent, stratified dysoxic to anoxic marine conditions. This thin (1-ft thick) unit corresponds in part to the high gamma ray values (hot shale signature) evident on logs (Fig. 2). The forced regression in the Neva is represented by a mixed siliciclastic/carbonate interval (S-SH facies) that contains a variety of biotic indicators including large horizontal burrows (Fig. 9) and macroinvertebrate fossils (Fig. 10). The upward-shoaling limestone (OB-PG facies) succeeding the siliciclastic/carbonate interval is capped by a weathered exposure surface consisting of carbonate intraclasts in a silt- and clay-rich matrix (I-WP facies).

Caliche-rich red beds (Fig. 11) of the Eskridge Shale represent the uppermost unit in the Grenola sequence. The upper boundary for the Grenola sequence is placed at the contact between the Eskridge red claystone and the transgressive surface at the base of the overlying Cottonwood Limestone.

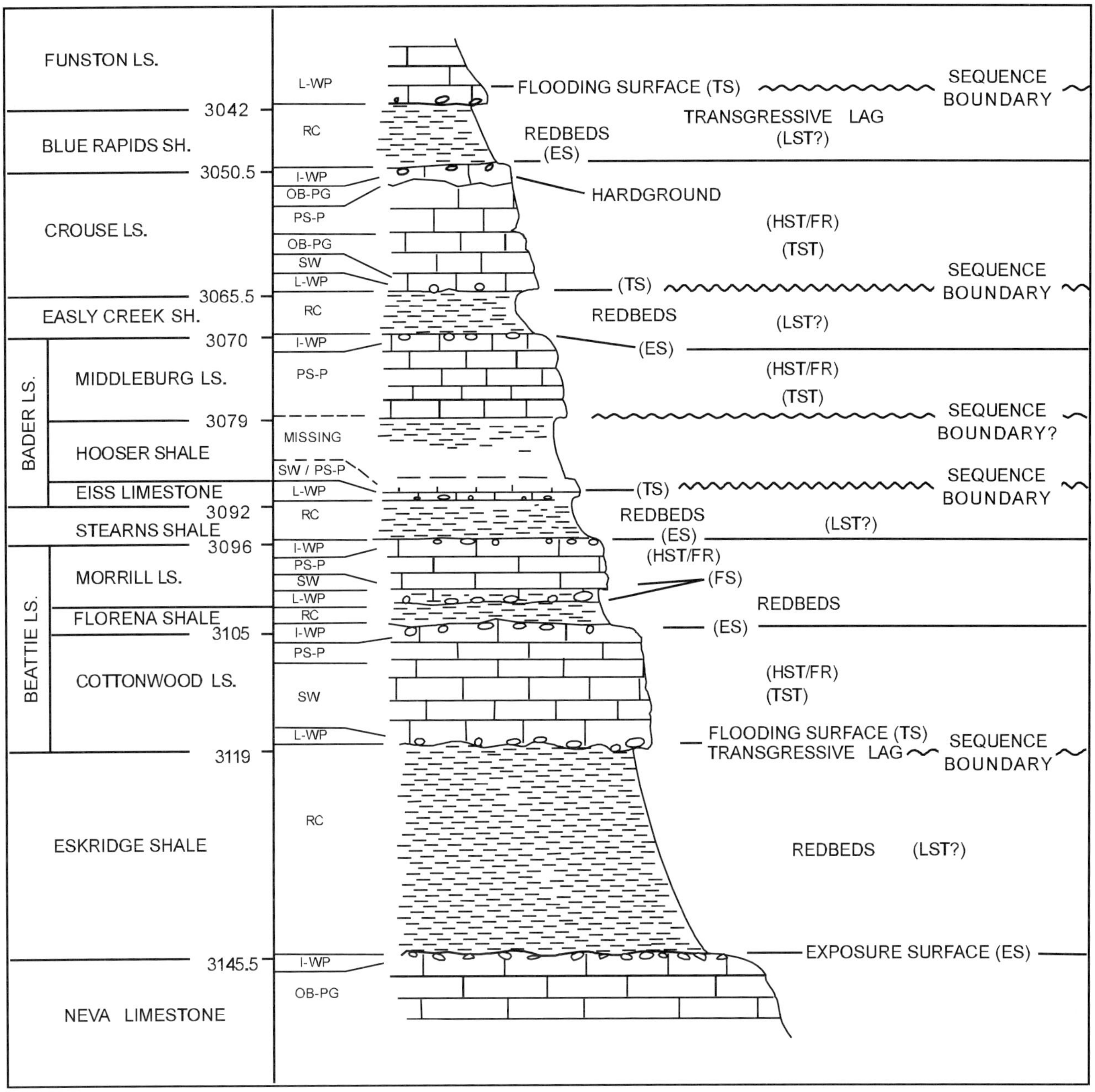

FIGURE 12 — Schematic diagram illustrating facies and depositional sequences in upper part of Council Grove Group (Beattie, Bader, and Crouse sequences).

Upper Council Grove Group Sequences: Beattie, Bader, Crouse, and Funston Limestones

Sequences in the upper part of the Council Grove apparently occurred at a higher frequency and with less water depth change than those in the lower part. They are typically thinner and lack a well-developed HST and marine condensed section.

The Beattie (Limestone) sequence contains the Cottonwood Limestone, Florena Shale, Morrill Limestone, and Stearns Shale (Fig. 12). The lower sequence boundary is a flooding surface at the

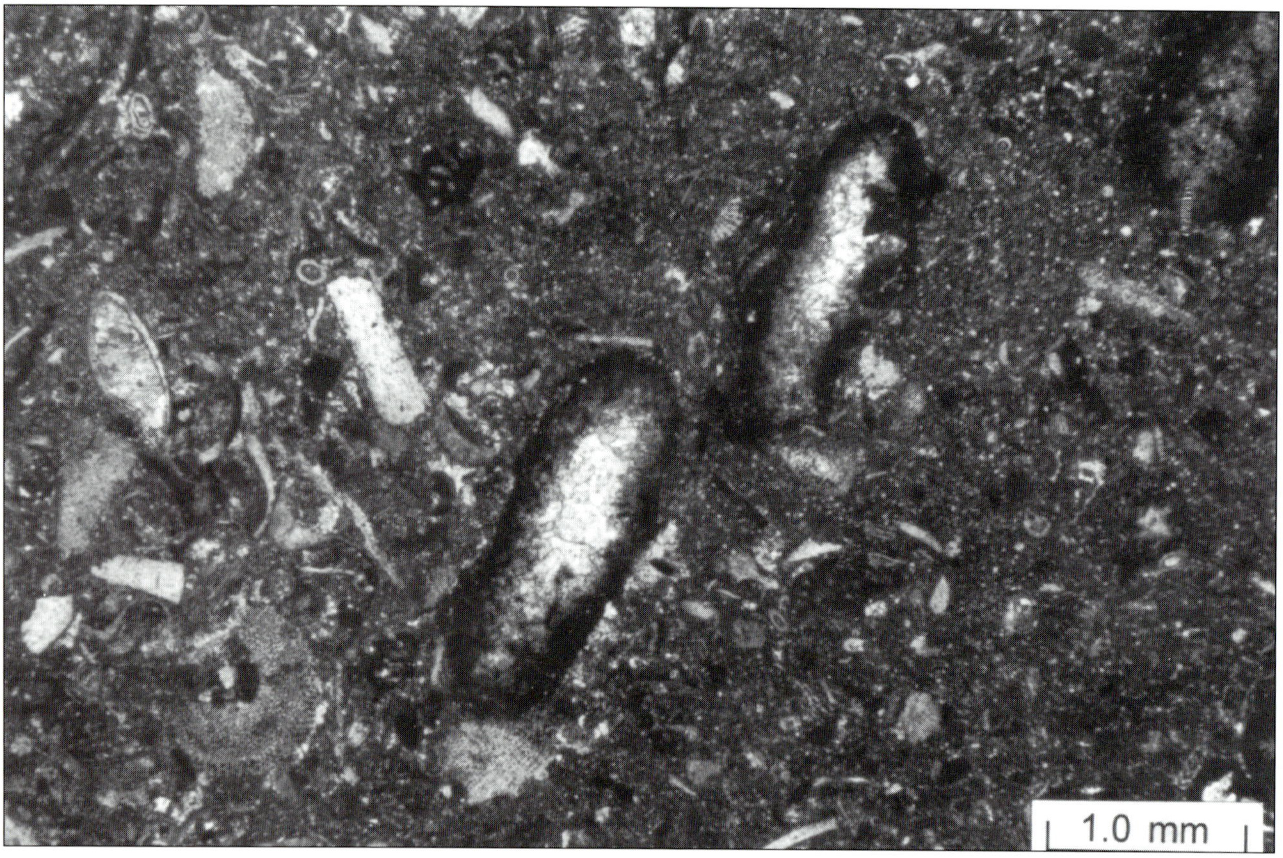

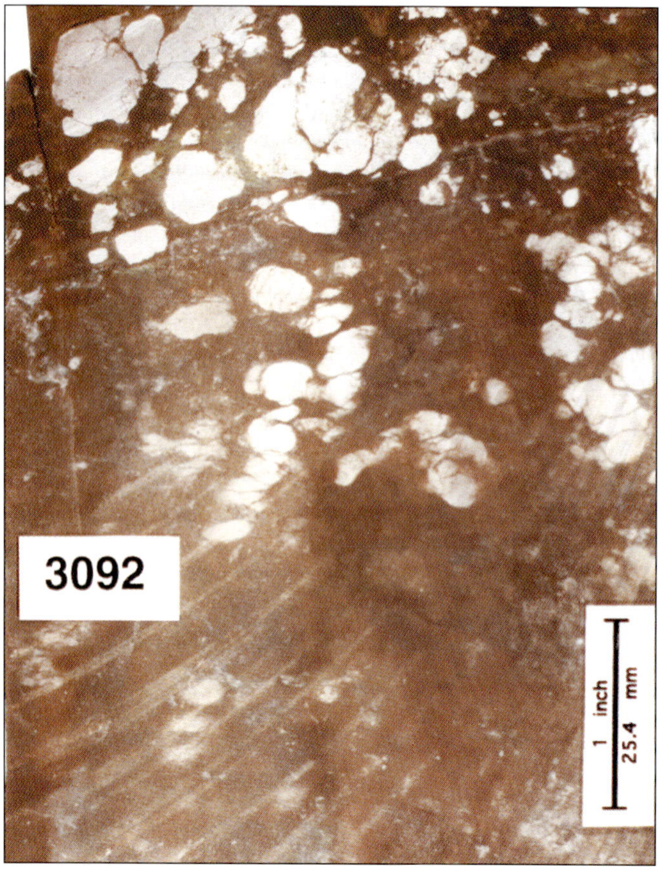

FIGURE 13 — Photomicrograph of pelmatozoan, oncoid, and ostracode wackestone (SW facies) of the TST in the Cottonwood Limestone. Depth 3118 ft. Plane-polarized light.

FIGURE 14 — Calichified red bed (RC facies) of Stearns Shale.

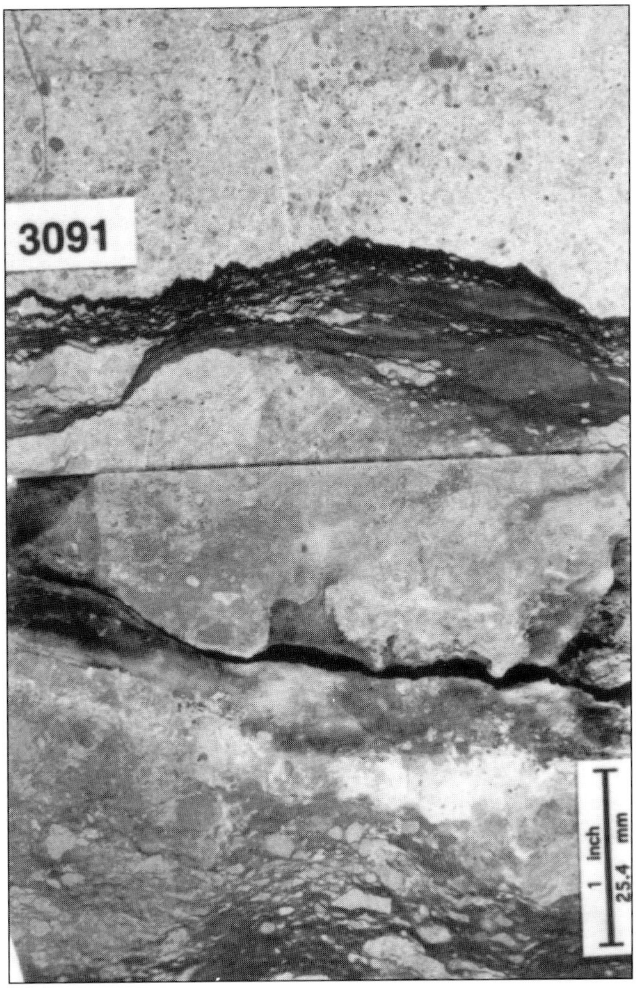

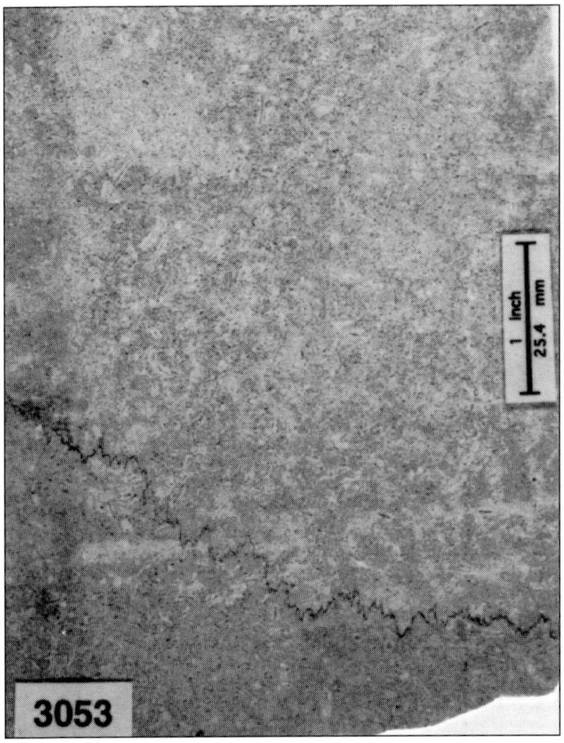

FIGURE 15 (left) — Transgressive lag (L-WP facies) marking lower boundary of Bader Limestone sequence.

FIGURE 16 (above) — Skeletal grainstone (OB-PG facies) from upward-shoaling interval in the Crouse Limestone.

base of the Cottonwood Limestone. The lower part of the Cottonwood is gray, skeletal, oncoid-rich wackestone (SW facies) of the TST (Fig. 13). It shoals upward and becomes a pelmatozoan and ostracode-rich wackestone to packstone (PS-P facies) that represents the highstand systems tract/forced regression (HST/FR). The Cottonwood was exposed to extensive surficial weathering, developed an intraclast regolith layer (I-WP facies), and contains a variety of dissolution features. The Cottonwood is overlain by red claystone (RC facies) of the Florena Shale. The transgressive Morrill Limestone overlies the Florena Shale. The SW facies in the lower part of the Morrill shoals upward to the PS-P facies and culminates in an exposure surface (ES) with intraclasts (I-WP facies). Calichified red beds (RC facies)(Fig. 14) of the Stearns Shale (LST) complete the Beattie sequence.

The Bader Limestone (Fig. 12) contains the Eiss Limestone, Hooser Shale, Middleburg Limestone, and Easly Creek Shale. The lower boundary is marked by a lag zone (Fig. 15) (L-WP facies) that formed along the transgressive surface (TS) at the base of the Eiss Limestone. Overlying this lag zone is a wackestone (SW) to packstone (PS-P) that represents the TST. The remaining parts of the Eiss Limestone and the Hooser Shale were not cored, but can be inferred from wireline logs. The caliper curve indicates washout across the Hooser similar to that observed for other redbed intervals. A sequence boundary may be placed between the Hooser red beds and the overlying Middleburg Limestone, which was partially cored. It consists of the HST/FR peloid-skeletal packstone (PS-P) and a thin (<1.5-ft thick) silty wackestone that was subaerially exposed and weathered (I-WP). The Middleburg and overlying Easly Creek Shale form the upper sequence in the Bader Limestone.

FIGURE 17 — Hardground surface (arrows) developed near top of Crouse Limestone.

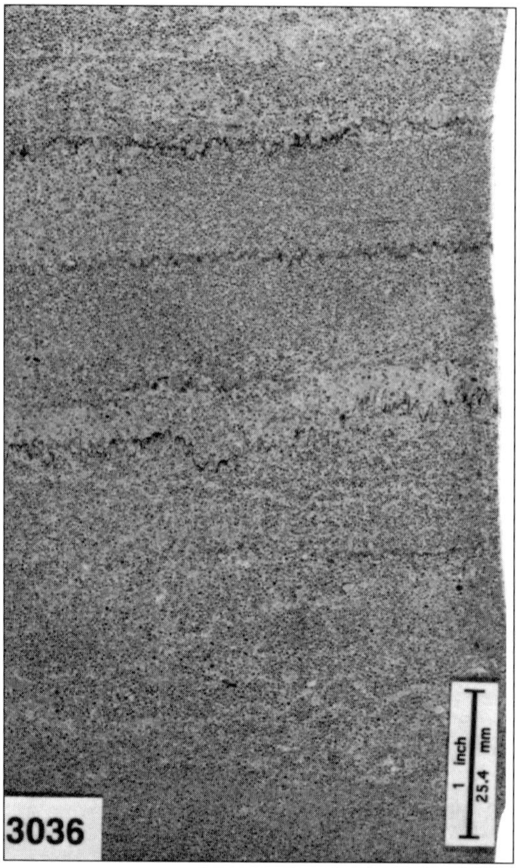

FIGURE 19 — Oolitic shoal grainstone (OB-PG facies) with low-angle crossbedding indicating high-energy environment.

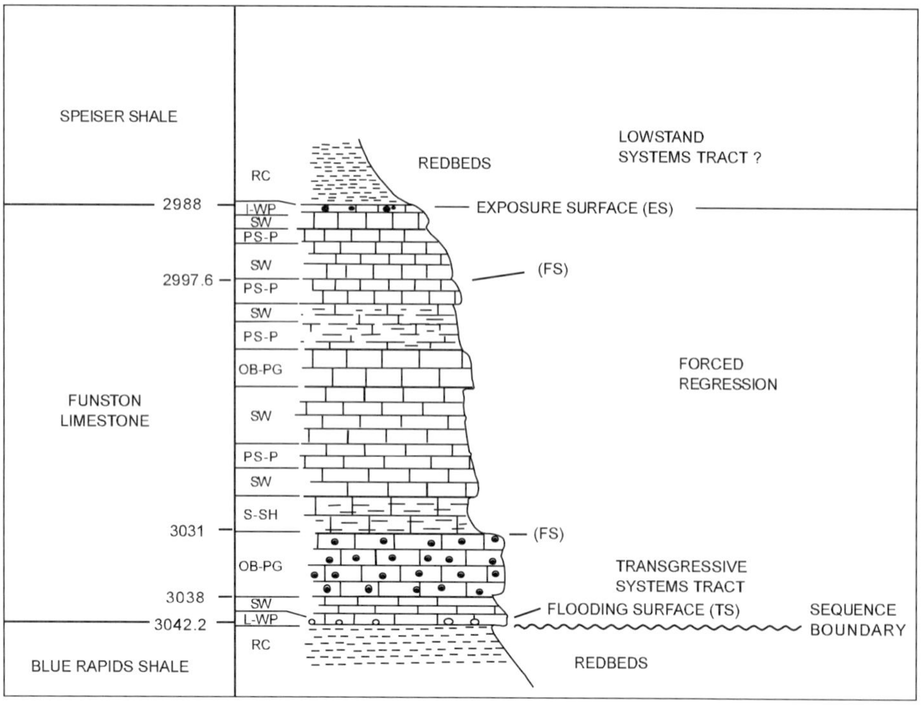

FIGURE 18 — Schematic diagram illustrating facies in cored portion of Funston sequence. Parasequence boundaries are defined by flooding surfaces (FS) within Funston interval.

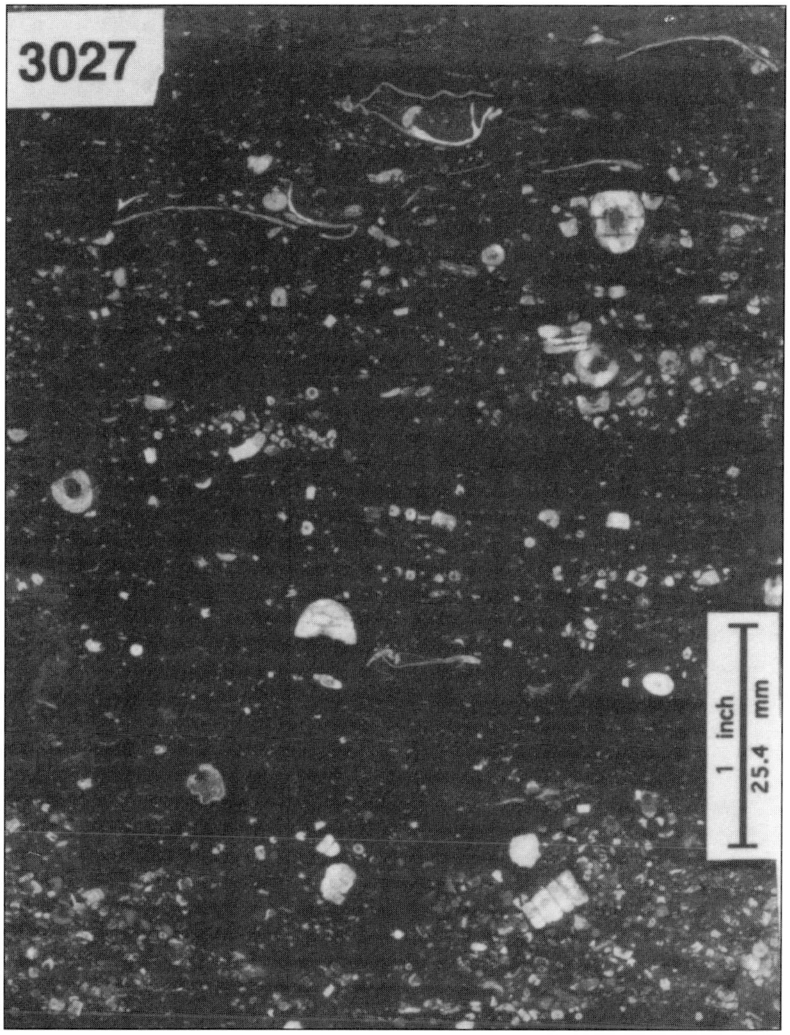

FIGURE 20 — Fossiliferous marine shale (S-SH facies) above the lower flooding surface in Funston Limestone. These flooding surfaces punctuate upward-shoaling cycles and identify parasequence boundaries.

FIGURE 21 — Carbonate regolith (I-WP facies) zone formed by subaerial weathering of upper surface of the Funston Limestone. Exposure surfaces identify the contact between marine carbonates and lowstand nonmarine red beds.

The Crouse Limestone sequence (Fig. 12) is bounded on the base by a flooding surface (FS) that contains a transgressive lag (L-WP). The TST wackestone (SW) shoals upward to the HST/FR skeletal packstone (PS-P) and grainstone (OB-PG) (Fig. 16). A hardground (Fig. 17) and minor flooding surface occur near the top of the Crouse. An exposure surface with intraclasts (I-WP) developed on the upper surface of the Crouse.

The Blue Rapids Shale (LST) overlies the Crouse Limestone. The upper boundary of the Crouse sequence occurs at the contact between the Blue Rapids red beds (RC) and the transgressive basal Funston Limestone.

The Funston Limestone sequence was initiated with a transgression on the underlying Blue Rapids Shale (Fig. 18). Skeletal wackestones (SW) of the TST rapidly shoal upward to an ooid packstone-grainstone (OB-PG). This grainstone contains crossbedding (Fig. 19) that reflects the traction transport and increased energy in the ooid shoal.

The oolite was terminated by a flooding surface and deposition of highly fossiliferous dark shale (S-SH facies)(Fig. 20). In the upper part of the Funston, deeper water wackestone (SW) succeeds shallow water packstone (PS-P) without a noticeable break in deposition. This gradual deepening event sharply contrasts with the other flooding surfaces that punctuate the forced regression and delineate the parasequences. Funston Limestone deposition was terminated by a drop in sea level and forced regression. The upper surface of the Funston was exposed and weathered, and a carbonate regolith zone (I-WP) (Fig. 21) formed on this exposure surface prior to deposition of the overlying Speiser Shale.

Vertical Facies Distribution

The depositional sequences in the Council Grove display a pattern of vertical facies succession that is relatively consistent throughout the section. From bottom to top of each sequence, the facies are L-WP, SW, PS-P, S-SH, OB-PG, I-WP, and RC. In the lower part of the Council Grove (Fig. 3), the lower section of a sequence (L-WP to SW) is interpreted to represent an upward-deepening succession from the basal transgressive lag to skeletal wackestones within the TST. Upward shallowing is recorded by the gradational change from the SW facies to the PS-P and OB-PG facies. Flooding surfaces were marked by the quick change from the PS-P or OB-PG facies to the SW facies across a bed boundary. In the Neva Limestone Member of Grenola Formation (Fig. 3), a hardground developed on the flooding surface at the top of a grainstone (OB-PG). This was followed by highstand deposition and an upward-shallowing succession of facies (S-SH) that represents forced regression. Continued regression was reflected in the succeeding OB-PG, I-WP, and RC facies.

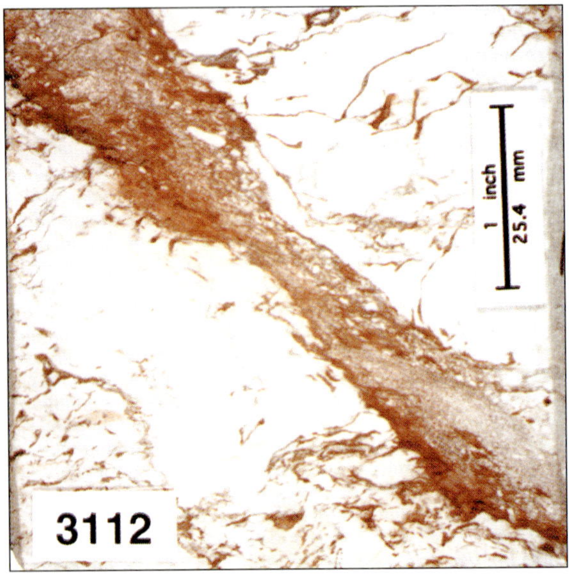

FIGURE 22 — Focused flow paleokarstic dissolution features that formed during subaerial exposure of the Cottonwood Limestone. These channels filled with redbed silt and clay.

The upper part of the Council Grove displays a vertical succession of facies consisting of L-WP, SW, PS-P, OB-PG, I-WP, and RC (Figs. 12 and 18). If the succession is punctuated by a flooding surface, it will include an S-SH facies (fossiliferous shale) in sharp contact with the underlying facies (generally PS-P and OB-PG). Gradual deepening will result in the SW facies overlying the PS-P facies. As the result of forced regressions, all four of the upper Council Grove sequences (Beattie, Bader, Crouse, and Funston) ultimately contain the I-WP and RC facies to complete the succession.

PALEOKARSTIC FEATURES

The exposure surfaces in the upper Council Grove Group are characterized by the development of a thin regolith zone of rounded carbonate intraclasts. These weathered zones identify periods of subaerial exposure prior to deposition of the overlying red beds. Paleokarstic horizons developed subjacent to some exposure surfaces. Dissolution cavities, solution-enlarged channels (Fig. 22), and vugs

FIGURE 23 — Photomicrograph of oomoldic porosity in ooid grainstone (OB-PG facies). Porosity values for this reservoir approach 30%, but permeability is <0.02 md. Depth 3035 ft. Plane-polarized light.

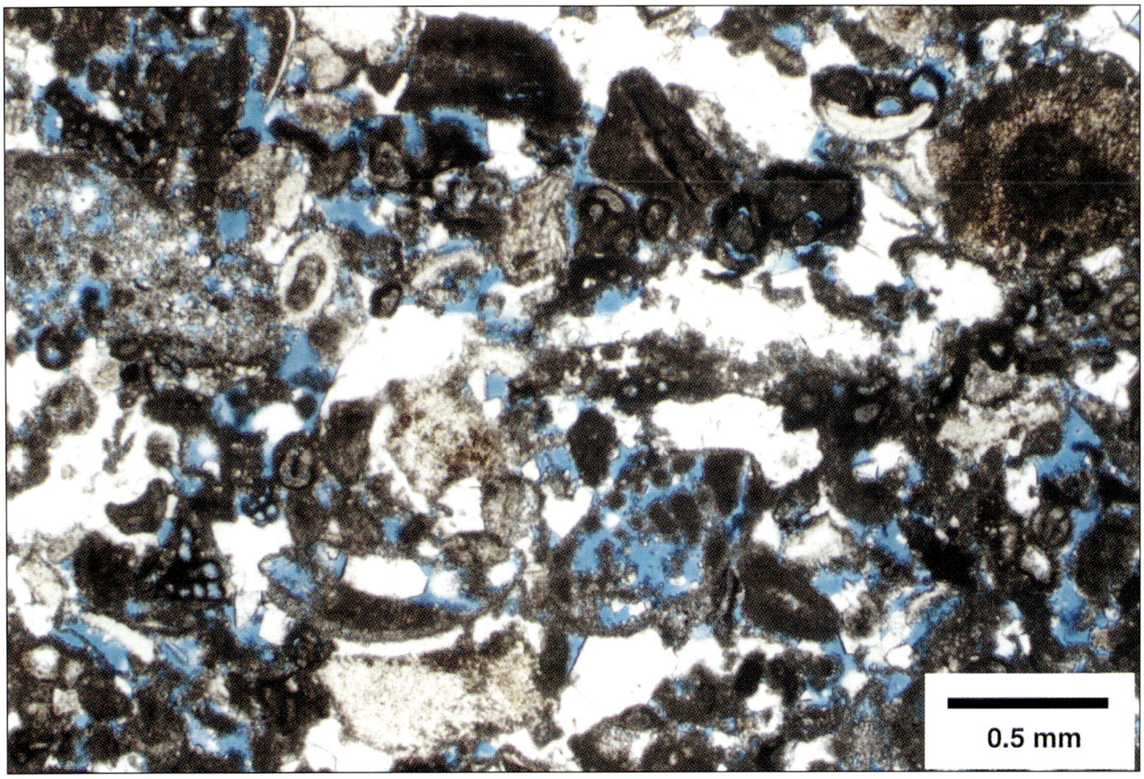

FIGURE 24 — Photomicrograph of porous skeletal packstone (PS-P facies). Depth 3099 ft. Plane-polarized light

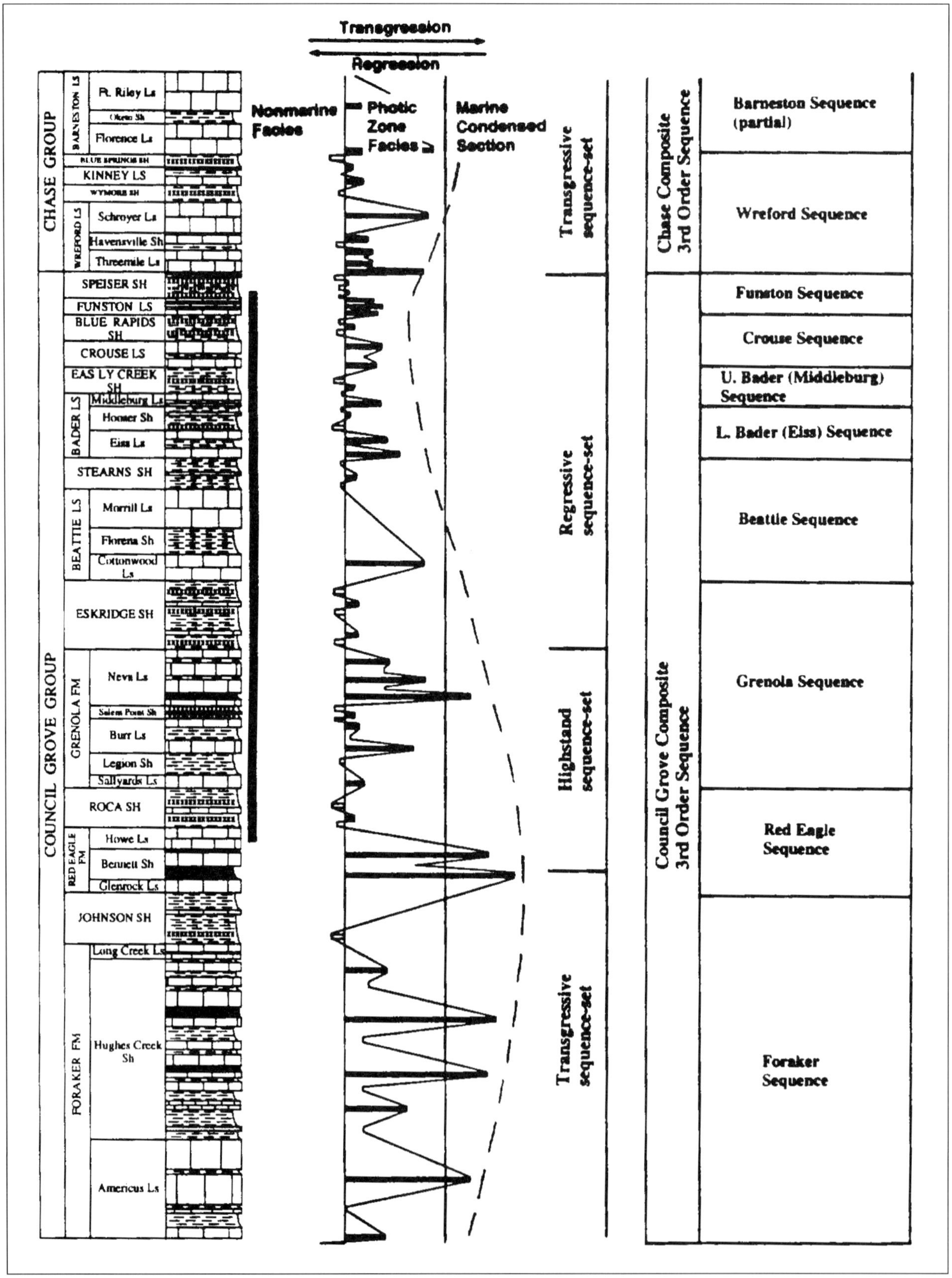

FIGURE 25 — Sequence stratigraphy of the Council Grove Group. Dashed line represents relative sea-level curve of a third-order composite sequence composed of transgressive, early highstand, and regressive sequence sets. Fourth-order sequences are listed in the right column (after Boardman and Nestell, 1994).

are evident. These features are commonly filled with red silt and clay.

RESERVOIR FACIES AND POROSITY EVOLUTION

Porosity in the Council Grove reservoirs is essentially all secondary. Porosity occurs most commonly in the grain-rich rocks and was formed by the dissolution of grains and cement. The highest porosities are in the ooid grainstone (Zone A; Fig. 2) where grains were dissolved to generate moldic pores (Fig. 23). Porosimetry-determined porosity values in the oolite approach 30%, but permeability is relatively low (0.017 md) due to the lack of interconnecting throats between the moldic pores. Porosity in the skeletal packstone and grainstone facies (PS-P and OB-PG) was formed primarily by dissolution of mollusk, bryozoan, and foraminifer grains and micrite and microspar matrix following dolomitization (Fig. 24). Porosity in these reservoirs (Zone B; Fig. 2) is considerably less than that of the oolite, but permeability is better due to the development of an intergranular pore network.

THIRD-ORDER COMPOSITE SEQUENCE

The Council Grove is interpreted as a third-order composite sequence that contains three sequence sets (Youle et al., 1994). These sequence sets reflect differences in facies trends in the fourth-order members that suggest longer term changes in the sea level. Third-order transgressive, highstand, and regressive sequence sets were identified (Fig. 25). Water depth was inferred by lithofacies and biofacies analyses (Boardman and Nestell, 1994).

The transgressive sequence set records increasing water depths associated with each third-order sea-level rise. Beginning with the Americus Limestone, the maximum water depths associated with the marine condensed section black shales increase upward to the Bennett Shale. This trend suggests deposition during a progressive long-term transgression that culminated with the Bennett.

The highstand sequence set is represented by the remainder of the Red Eagle sequence and the Neva Member of the Grenola sequence. The rate of third-order sea-level change slowed, and no marked trend toward either regression or transgression was evident.

The regressive sequence set records the falling limb of the third-order sea-level curve. Red beds became more frequent, whereas well-defined marine condensed sections were no longer developed.

The validity of third-order composite sequences is based in part on their regional correlatability. In the Council Grove Group, the vertical facies relationships in the cored interval reflect those of the outcrop 280 miles to the east (Fig. 1). Furthermore, the third-order sea-level curves for the Lower Permian in Kansas and Oklahoma are similar to those in Texas (Boardman and Nestell, 1994).

CONCLUSIONS

(1) The Council Grove Group is composed of repetitive cycles of shallow marine carbonates and siliciclastics and nonmarine red beds that document sea-level oscillations.

(2) Depositional sequences in the Council Grove display a vertical facies succession (L-WP through RC) that documents flooding and deepening, followed by a general shallowing phase that culminated in subaerial exposure, karsting, and redbed deposition. A complete succession of these genetically related strata is considered a fourth-order depositional sequence.

(3) Depositional sequences in the lower part of the Council Grove consist of (a) a basal transgressive surface, (b) a shallow marine limestone of the TST, (c) an HST with a well-developed marine condensed section, (d) shallow water carbonates deposited during forced regression that ultimately culminate with a weathered upper exposure surface, and (e) nonmarine caliche-bearing red beds that might correspond to the LST of the shelf margin.

(4) Upper Council Grove sequences are similar to those found in the lower Council Grove, but lack a well-defined marine condensed section that represents the HST.

(5) Porous reservoir rocks are generally found in ooid and skeletal packstone and grainstones of the upward-shoaling intervals within the forced regressive phase of the sequences. Large-scale karstic dissolution features (vugs and channels) developed during subaerial exposure have relatively low porosity due to sediment infilling.

(6) The differences in the character of the fourth-order sequences reflect general trends in facies that help define sequence sets. The Council Grove consists of three sequence sets (transgressive, highstand, and regressive) that compose a third-order composite sequence.

ACKNOWLEDGMENTS

The authors wish to express their appreciation to Justin Evans, Catherine Price, and Michael Sykes for their contributions to this study. The constructive criticisms and additions of Phil Heckel and Rick Erickson were valuable to the manuscript.

MIDDLE PENNSYLVANIAN SEQUENCE STRATIGRAPHY OF THE WESTERN ARKOMA BASIN AND BORDERING SHELVES, OKLAHOMA AND KANSAS: A MODEL FOR FORELAND BASINS

Allan P. Bennison
Consulting Geologist
Tulsa, OK 74119

ABSTRACT

Sequence stratigraphy for tectonically active foreland basins, such as the western Arkoma basin, differs appreciably from that for relatively passive marginal basins. Owing to its greater geographical confinement and narrowness, the basin can be completely filled by deposits. The sediments can accumulate above sea level, until an increasing but sporadic subsidence or a rise in global sea level occurs. Another important difference is that the orogenic belt bordering the basin maintains its high relief and remains a strong contributor of sediment even during eustatic sea-level rises. In contrast, the cratonic shelf sediment influx diminishes from that cratonic source, resulting in a cratonward shift of the sediment depocenter. Depocenter deposits average about four times thicker than correlative deposits on the cratonic shelf. Compressional stresses produced during episodic thrusting in the Ouachita orogenic belt were transmitted throughout much of the basin and cratonic shelf, diminishing with distance from the orogenic belt. Tectonic stress resulted in syndepositional folds, faults, local unconformities, hinge-line flexures, and occasional reactivation of ancient arches (e.g., the Chautauqua and Bourbon arches) and flanking uplifts (e.g., the Nemaha ridge and Ozark uplift).

INTRODUCTION

Modeling of stratigraphic sequences along the Middle Pennsylvanian outcrop belt (Fig. 1) in Oklahoma and Kansas has disclosed a third type of sequence quite unlike type 1 and 2 sequences described by Van Wagoner et al. (1988) and Vail et al. (1987) for passive marginal basins. This type may be fairly common for many active foreland basins and flanking shelves, such as the western Arkoma basin northward to the Chautauqua platform or arch of the Oklahoma-Kansas border area.

Although this type of sequence is commonly uncomformity bounded, the usual criterion for type 1 and 2 sequences for passive marginal basins, the initial deposits belong to the transgressive systems tract (TST) rather than to the shelf margin (SMST) or lowstand systems tract (LST) (Fig. 2). This dis-

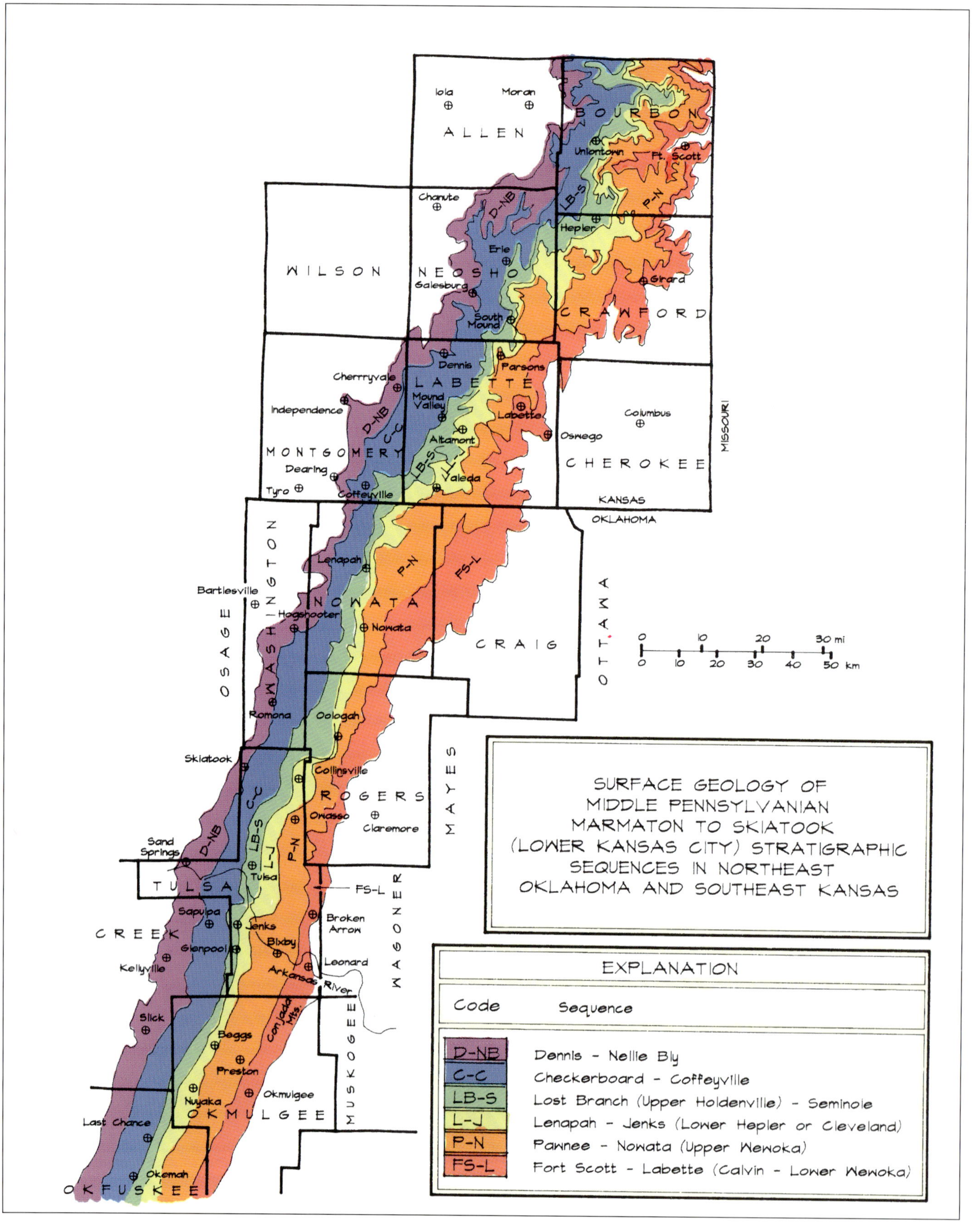

FIGURE 1 — Surface geology of Middle Pennsylvanian Marmaton to Skiatook (Lower Kansas City) stratigraphic sequences.

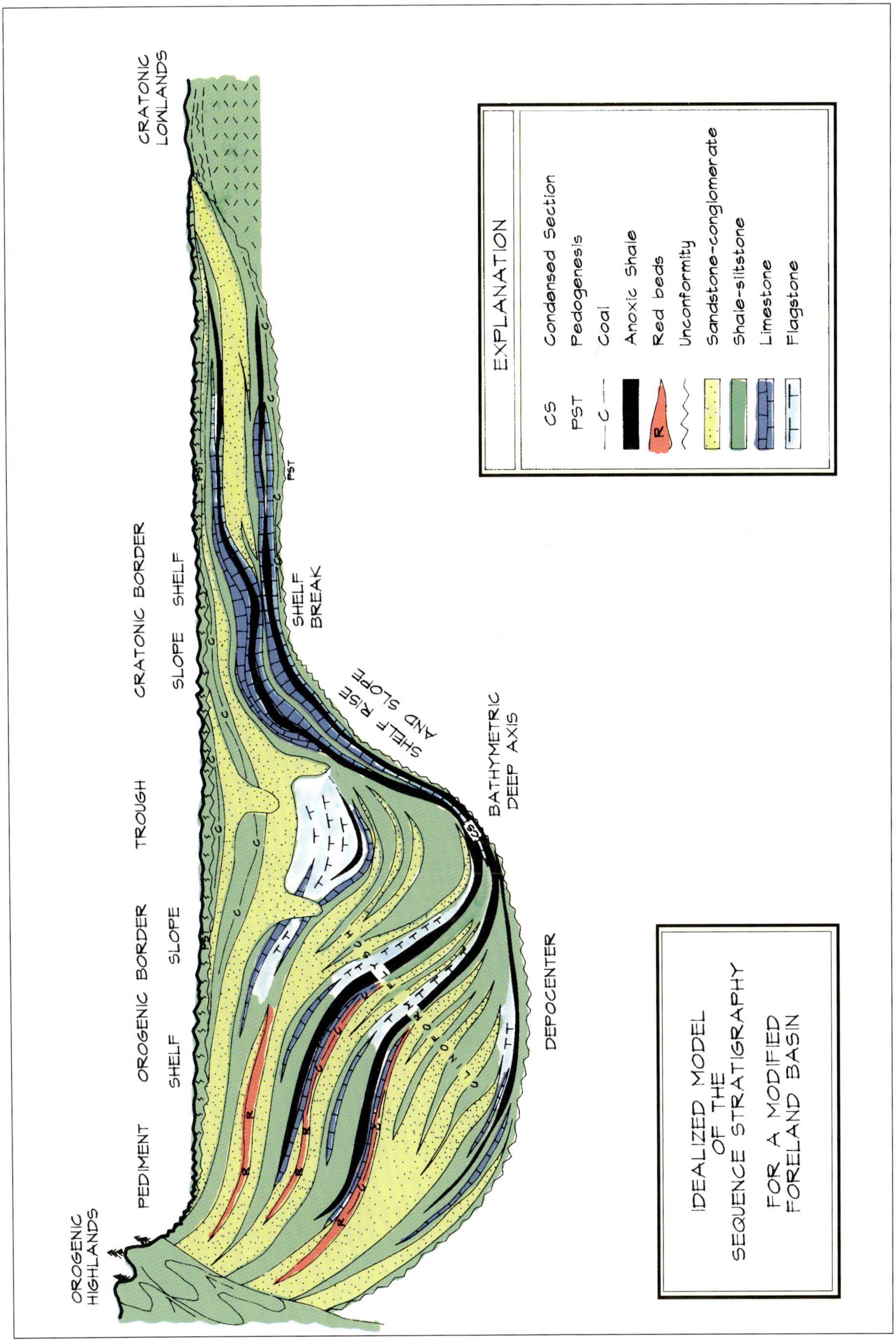

FIGURE 2 — Idealized model of the sequence stratigraphy for a modified foreland basin.

tinctive active foreland basin sequence resembles the Yoredale Series of England (Philips, 1834), in which the gross succession is limestone, shale, and sandstone, in upward order.

Therefore, the sequence described in this study may be called a modified foreland basin sequence. The succession of systems tracts remains the same, but the sequence boundary is at the base of an extremely widespread TST that had quickly spread over an eroded, commonly soil-covered, mature basinal sequence. This modified foreland basin sequence may be conceived as a third-order cycle, a larger mimic of a fourth-order cycle. Such a sequence consists of three to seven transgressive-regressive (T-R) cycles, in which each succeeding cycle usually represents an upward-shallowing trend, owing to gradual basinal filling. This trend often results in subaerial exposure and erosion. In this scenario, a limestone commonly rests discordantly on fluvial to shallow water offshore sandstones, mudrocks, and paleosols of the closing lowstand, shelf margin, or forced regression deposits of the preceding cycle. Other associated deposits such as a coal or black euxinic shale may overlie the eroded and weathered sequence. These, especially the latter, are a consequence of a rapid shelfward inundation of the marine environment characteristic of this distinctive foreland basin sequence. Such a transgression is often rapid and completed within a narrow interval of a cycle. Middle Pennsylvanian time was characterized by the advances and retreats of Gondwanan ice sheets (Heckel, 1985). These were probably of sufficient extent and volume to affect global sea level by as much as 100 m. Isotope analyses of ice cores from ice sheets in Greenland and Antarctica indicate rapid warming and meltdown intervals lasting only decades during the Pleistocene, causing precipitous rises in sea levels (Blanchon and Shaw, 1995).

Contrary to some conventional views, regional coals within this sequence commonly form initial transgressive deposits and may extend over several Mid-Continent states (Donaldson et al., 1991; Phillips et al., 1991) (Fig. 2). These coals reflect rapidly expanding wetlands, commonly with compressed fossil plants in their lowermost layers and subjacent carbonaceous mudrock, and shallow marine fossils in their uppermost, commonly pyritic, capping layers.

Black phosphatic shales, the core shales that represent maximum flooding and anoxic events, commonly separate a transgressive limestone or shale from a regressive or shoaling limestone, the well-known triad of a Kansas cyclothem (Heckel, 1977).

After a transgression, a highstand of sea level may continue throughout several minor eustatic fluctuations, resulting in a wide extent of the foreland sea and submergence of much of the bordering shelves. Many downlapping sediment plumes fail to reach the outer shelf, slope, or basin. Van Wagoner et al. (1988) illustrated several such flooding events within a highstand systems tract (HST).

Unless the preceding phase is tectonically disrupted, it is followed by basinal filling, consisting of lowstand (or forced regression) and shelf margin deposits of shallow marine to onshore fluvial deposits frequently capped by soil formation (pedogenesis). These are commonly cyclic because of eustatic fluctuations. This latter systems tract may have a much longer duration than the preceding systems tracts and may even be locally uplifted and eroded. Some resulting erosional valleys are commonly filled with clastic sediments during flooding events within the systems tract or renewed transgression of a succeeding sequence.

During the basinal enlargement and filling process, the axial deep area, depocenter, and slopes are dominated by marine mudrock or shale with several anoxic periods, prograding submarine fans of mixed turbiditic sandstone and mudrock, carbonate flagstones, and thin carbonate debris flows. During maximum marine transgression and/or highstand, deposition may lag in the paleobathymetric deep axial area, resulting in a condensed section or even nondeposition.

Slopes and shelves bordering the orogenic highlands feature landward coarsening sandstones, conglomerates, red to green shales, and thin transgressive to highstand limestone beds. Transgressive sandstones are more plentiful than on the cratonic shelf, commonly grading up from a reddish paleosol into a capping molluscan calcareous sandstone. These orogenic border sediments commonly are absent in surface outcrops, owing either to overriding thrust slabs or to later tectonic uplift and erosion.

Because of the complex interplay of tectonics, subsidence, and eustasy, the uppermost cycle may be more marine than the preceding cycle. Therefore, it may serve as a forerunner to the succeeding initial TST of the overlying sequence. Although it is usually conformable with the underlying cycle, it is commonly truncated by overlying beds of the succeeding sequence shelfward. Recognition of this distinctive type of stratigraphic sequence in an active foreland basin regime does not preclude the presence of the now classical type 1 and 2 sequences at other stratigraphic intervals in the Arkoma basin, especially in the Lower Pennsylvanian, when the Arkoma basin was much wider and received a greater volume of sediments from northern and eastern sources, owing perhaps to a less emergent Ozark uplift.

In the modeling of stratigraphic sequences, emphasis should be placed on regional coals, transgressive limestones, black phosphatic shales, and paleosols along regionally continuous surface or subsurface traverses that transect tectonic features. Gamma ray logs close to such a traverse are usually more useful than electric logs for correlations. Biostratigraphy is helpful but in some cases is less conclusive, owing to uncertain longevity of many species that prevail throughout several cycles. Exceptions are the major extinction events. The most important extinction event in this area is that between the Desmoinesian and Missourian series, which is probably a global event synchronous with that between the Westphalian and Stephanian stages in western Europe. In northern Ontario Province, Clearwater Lake is now accepted as a major impact feature dated at about 290 m.y., the approximate age of the Desmoinesian-Missourian series boundary.

TRANSGRESSIVE-REGRESSIVE CYCLES

Vail et al. (1991) claimed that transgressive-regressive facies (T-R) cycles are not reliable for making intrabasinal correlations, because the boundaries of these units may be time-transgressive. Many Mid-Continent stratigraphers, including the writer, do not accept this negative view, especially those who have worked the Pennsylvanian outcrop belt that trends across several basins and intervening arches. There are hundreds of measured sections supplemented by thousands of well logs. The transgressive phase of such a T-R cycle usually involves a transgressive interval, less than 10% of the cycle, with the remaining 90% interval composed of shoaling to regressive beds. Most of these beds show a lithic integrity over vast areas of the Mid-Continent. If these transgressions are linked to abrupt meltdowns of Gondwanan ice sheets, these may be chronostratigraphic horizons and chronostratigraphic units. They could be more useful than the discontinuities reported by Vail et al. (1991). Also, many Mid-Continent stratigraphers find much arbitrariness in defining the boundaries of systems tracts and parasequence sets. References to such boundaries in this study must be treated as strongly interpretative and far from being conclusive. Part of this problem may be due to the interplay of deposits from opposing shores in a foreland basin environment with differing tectonics. For example, the LST involving prograding seafloor fans, such as observed within the Wewoka Formation, commonly extends in a thinning classic wedge shelfward between transgressive core shales and reputed highstand deposits of marine shales and limestones to become part of the HST. Nonetheless, abrupt

FIGURE 3 — North to northwest drift of the Arkoma basin bathymetric deep axis.

regional and possibly global rises of the sea level have provided the basal transgressive deposits with a common widespread signature.

Defining bases of reputed lowstand channel fills as sequence boundaries is also unrealistic as observed in the field. Channels incised into older strata may occupy different levels within lowstand deposits. As a result, their boundary horizons often fade out laterally into interlaminated lenticular shales, siltstones, and thin-bedded sandstones within a transitional upward-coarsening succession. Such channels are probably more closely linked to a meandering stream or deltaic tidal or subtidal channel regimes than to eustatic fluctuations.

Sequences

Vail et al. (1991) defined a sequence as an unconformity-bounded sedimentary package corresponding to a third-order (0.5 to 3 m.y.) cycle. The foreland basin and shelf sequence of this study fall within the low range of this cycle. Boardman et al. (1994) and data by the writer indicate about 40 (±5) unconformity-bounded sequences compose the Pennsylvanian system in northeastern Oklahoma. According to the DNAG chart (Palmer, 1983), the Pennsylvanian period has an approximate duration of 34 m.y. This indicates an average duration of 0.8 to 0.9 m.y. for the northeastern Oklahoma sequences. To the north and northwest, in the Pennsylvanian succession of Kansas, internal unconformities and their lateral conformities appear within these sequences between T-R cycles. By definition, these should be promoted to sequences, except for their short duration of about 0.2 m.y. These fit fourth-order cycles or systems tracts lasting about 0.08 to 0.5 m.y. Throughout much of Kansas, a transgression is typically followed by a shoaling or regressive limestone overlain by forced regression clastics that coarsen upward. This clastic interval is commonly absent through either nondeposition or erosion at an unconformity surface. This is the classical cyclothem. In northeastern Oklahoma and southeastern Kansas, an average of about five cyclothems comprise a modified foreland basin sequence.

In the trough and bordering slope areas of the Arkoma basin of east-central Oklahoma, most unconformities present on the shelf fade out, but their postunconformity beds (such as black phosphatic shales and commonly associated debris flow limestones) may persist across the basin into the orogenic bordering slope and shelf facies and even into the adjacent intracratonic basins.

In this study, sequences will be named by their initial and terminal stratigraphic units, in that order. This tentative arrangement needs to be supplanted by an improved sequential nomenclature. Fortunately, the lithostratigraphic units in much of the study area coincide fairly well with its chronostratigraphic units, so that they can be used interchangeably. In other parts of the Pennsylvanian succession, such correspondence is often lacking.

Systems Tracts and Parasequence Sets

Systems tracts and parasequence sets are based on stacking patterns developed within a sequence. Owing possibly to northward thrusting and loading within the Ouachita orogenic movement and flexuring of the shelves, these are extremely difficult to define in the western Arkoma basin area of this study. The stacking of carbonate banks may be retrogradational, aggradational, or even progradational, depending on the nature of lithospheric flexuring at the shelf break or even the vigor of the organisms composing the shelf-edge carbonate banks.

In the traverse from the shelf to the depocenter or foredeep area of the western Arkoma basin, the five- to tenfold thickening of the sedimentary pile is attended by higher frequency cycles, owing largely to splaying of shelf sandstones into several clinoform-slope and basin-floor fans composed of interlaminated sandstones, siltstones, and shale (Fig. 2). These commonly show bottom marks and bioturbation, as is present in typical flysch deposits.

Parasequences

Vail et al. (1991) defined parasequences as high-frequency flooding and shoaling events comparable to fifth-order cycles (30 to 80 k.y.). These are difficult to establish within the sedimentary framework of a typical longer lasting Mid-Continent Pennsylvanian cyclothem, but such high-frequency cycles may be present in the Arkoma basin depocenter. For these additional beds composing a sequence, Vail et al. recommended the name "composite sequences." These remain for future investigation, as past interest has focused on the larger cycles and gross sedimentary divisions.

EVOLUTION OF A FORELAND BASIN SEQUENCE

Evolution of an active foreland depositional basin is more complex than that of a passive marginal basin. In the sequences from the Late Desmoinesian Pawnee Limestone to the Early Missourian Nellie Bly Formation, a northward shift of the basinal deep axis of about 40 mi is indicated. This is a relatively short interval during the much larger shift of the basinal axis, depocenter, and related features from the initial Morrowan stage position within the southern Ouachita Mountains to a Virgilian position within southern Kansas. Maps showing fixed boundaries of the Arkoma basin are not realistic. An estimate of this north-to-northwest migration is about 1 cm/yr (Fig. 3). This shift may be the result of either one or two forcing mechanisms. One is the relative northward push of the South American plate, producing a loading of thrust slabs (nappes) that depressed the adjacent foredeep. Isostatic adjustment may be a second cause. Sediment loading in the foredeep may exceed its threshold, causing the basin to sag deeper into the lithosphere.

Recent field studies by Hemish et al. (1995) revealed an approximately tenfold increase in sediment stacking of Early Desmoinesian McAlester and Savanna deposits across the Arkoma basin from the Muskogee area to Leflore County, a probable foredeep area near the Ouachita Mountains front, a distance of about 80 mi. Nevertheless, many of the lithic marker beds (such as coals, black phosphatic shales, and limestones) continue across. The considerable increase in thickness is attributed to southward-thickening wedges of turbidite clastics, especially gray silty mudrock and sandstone. Extreme basinward thickening is reported for the Atokan deposits. Roberts (1987) indicated as much as 28,000 ft of thickness for the Atoka Formation in the Ouachita Mountains. Other thickness estimates in the Arkoma basin vary from 8,000 to 15,000 ft. Tomlinson and McBee (1959) estimated a minimum of 8000 ft of Atokan beds in Atoka County. Shelfward, the Atoka Formation thins to only a few tens of feet of largely marine dark shales and thin ferruginous limestones in Mayes County, Oklahoma.

In contrast, basinward thickening for the Late Desmoinesian and Early Missourian deposits of the study seem no greater than five times. Krumme (1981) indicated about a four times basinward thickening for similar deposits, with a distinct northward shift of their depocenters. Shelf to basin stratigraphic sections in this study (Figs. 4 to 8) show a three to five times basinward thickening. This decrease in thickness of basinal fill relative to shelf deposits may be characteristic of aging basins (Beaumont et al., 1988).

The foreland sequence models in this study suggest the following tectonic cycle to account for their stratigraphic cross-sectional geometry. Either thrust load and/or sediment loading produced an initial abrupt acceleration and downwarping of previously filled to partly filled basin and shelves. This was soon followed by accelerating subsidence outpacing sediment supply and the formation of a new foredeep, flanked by opposing slopes and shelves, where subsidence rates were less. Subsidence may have been sinusoidal, at first accelerating and then decelerating to stagnation (Beaumont et al., 1987). This process was punctuated by moderate and varying amounts of eustatic rises and falls of the sea level.

As previously mentioned, transgression usually occurred within the duration of a cycle with a rapid spread of shell banks and coal-producing wetlands. During the ensuing highstand, sediment starvation often occurred at or near the basinal axial area and adjoining slopes and shelves. Deceleration of basin subsidence led to decreased sediment accommodation and shallowing with sediment infill. This longer term pattern of sequence evolution was punctuated by shorter term sea-level fluctuations.

Throughout these phases of transgression and regression, the prevailing northward progradation of thick sediments eroded off the orogenic highlands south of the basinal axis led to the shift of the depocenter toward the cratonic shelf, possibly increasing the subsidence of the basin and adjoining shelves.

The role of deltaic sedimentation in establishing the stratigraphic framework of the Arkoma basin has been overemphsized in the geological literature. The lower one-half of a sequence in the basinal and slope area is dominated by great thicknesses of marine mudrock or shale and interbedded turbidic siltstones and sandstones with sole marks, and on the cratonic shelf by carbonate banks and marine black shales. Some of the clastics from a Ouachita source onlap the cratonic slope and outer shelf as a shelfward thinning wedge. It is during the forward shoaling of the upper sequence that deltaic, fluvial, and estuarine sedimentation becomes important and widespread, largely through incised channeling extending deep into the basin and laterally landward into the inner shelf and cratonic lowlands (Krumme, 1981). Some of these channels were choked with aggrading stream loads during flooding events. Shoreline sands and muds spread out over the partly submerged shelf and then downslope into the basin, often merging with sands from the opposing shelf that borders the orogenic highland.

A narrow and highly confined foreland basin may be easily filled above sea level with fluvial and paludal sediments until base level is achieved or until subsidence and renewed marine transgression.

Sequence Boundaries

Unconformities and correlative discontinuities are designated as the optimum boundaries for stratigraphic sequences (Van Wagoner et al., 1988). Within the Arkoma basin and especially on its northern ramp and bordering Chautauqua platform, outcrop mapping of Pennsylvanian beds has revealed numerous unconformities, showing regional truncation of underlying strata. In addition to these, there are more unconformities of limited extent. Some represent incised stream channels, splay crevasses, erosion of localized syndepositional anticlines, pinnacle reefs, barrier islands, and onlap of tectonic arches. Many discontinuities exist in sediment-starved condensed sections, owing to downlapping and extremely slow deposition or nondeposition. Prime cause of regional unconformities in the Arkoma basin is probably strong orogenic uplift in the Ouachita and Arbuckle mountains and rejuvenation of tectonic movements on the interior arches. Four of the many unconformities along the Chautauqua arch in the Coffeyville area are shown in Fig. 10.

The interfaces between highstand and lowstand or shelf margin systems tracts are sequence boundaries on the widely accepted models proposed by Van Wagoner et al. (1988). An outcrop search for such boundaries in the study area was disappointing to the writer. For example, a fairly thick interval of clastic slope and submarine fans in the lower depocenter of a sequence, that would be classified by Vail et al. (1988) as an LST, commonly would persist laterally onto the shelf as a thinner HST. A channel contact commonly fades in a short distance into a succession of interlaminated sediments or into an upward-coarsening succession without definite boundaries. There is a conflict of theory and observations. The long-range unconformities are most common at the bases of widespread transgressive beds. This condition requires an objective review of foreland basin dynamics.

TECTONIC FRAMEWORK OF THE WESTERN ARKOMA BASIN AND ITS BORDERING SHELVES

Boundaries of the Arkoma basin changed in time, largely owing to the convergence of the North and South American plates (Rascoe and Adler, 1984). Its southern boundary lies deep below the Choctaw Thrust within the Ouachita Mountains area (Fig. 3). Its western boundary is arbitrarily placed along the Seminole arch. Its eastern boundary lies beyond the area of this study, and its northern boundary lies partly along the southern boundary of the Ozark uplift and a poorly defined western extension that shifted northward in time toward the Chautauqua arch or platform along the Oklahoma-Kansas border. The southern half of the Arkoma basin is marked by east-west-trending folds and normal faults, extending as far north as the Muskogee anticline. Some northeast-southwest Ozark-trending, low-amplitude folds and faults fade into its northeast corner. Its northwest quadrant contains hundreds of low-amplitude domes and plunging anticlinal noses, partly formed over buried hills of the Precambrian Tulsa Mountains. Some structures were produced by dip reversals into a series of en echelon faults, probably caused by differential north thrusting of the Ouachita orogenic front. Others may represent distant response to Ouachita thrusting. Late movements of the Ozark uplift have tilted the Pennsylvanian strata lying to the west an average of about 50 ft/mi toward the west-northwest, converting many slightly closed structures into anticlinal noses.

A small intracratonic basin, known as the Cherokee basin, is situated between the Chautauqua and Bourbon arches in southeastern Kansas. Much of its stratigraphy is a thinner version of that present south of the Chautauqua arch in northeast Oklahoma.

Regional unconformities of the Arkoma basin and adjacent areas are attributed in part to profound orogenic movements of the Ouachita orogenic front and also to episodic epeirogenic movements of a cratonic lithosphere. Therefore, they may occur at any horizon within a sedimentary sequence. However, the majority occur at the base of transgressive beds. The explanation for this may be linked to a long period of low-angled tilting, erosion, and weathering resulting in nondeposition or a soil horizon that may have lasted as long as a depositional cycle preceding renewed subsidence and transgression. Also, the orogenic pulse may have depressed the foredeep that, in turn, tilted much of the basin floor and shelf, also initiating a new transgression and a new modified foreland basin sequence. Several such sequences will now be described in upward chronological order.

Fort Scott-Labette Sequence

The Fort Scott Limestone to Labette Shale sequence is an unconformity-bounded sequence that is exposed over much of the Mid-Continent area (Fig. 4). It embraces four known cyclothems, the lower of which, the Excello cycle, transgresses and truncates various beds of the underlying Senora Formation. North of the Arkansas River, limestones of the lower parasequences (or cycles) are retrogradational phylloidal algal limestone banks typical of HSTs. These constitute the Fort Scott Limestone. South of the Arkansas River valley, in the Conjada Mountains, a condensed section indicating a paleobathymetric deep axis has been identified. Farther south in Okmulgee and Okfuskee counties,

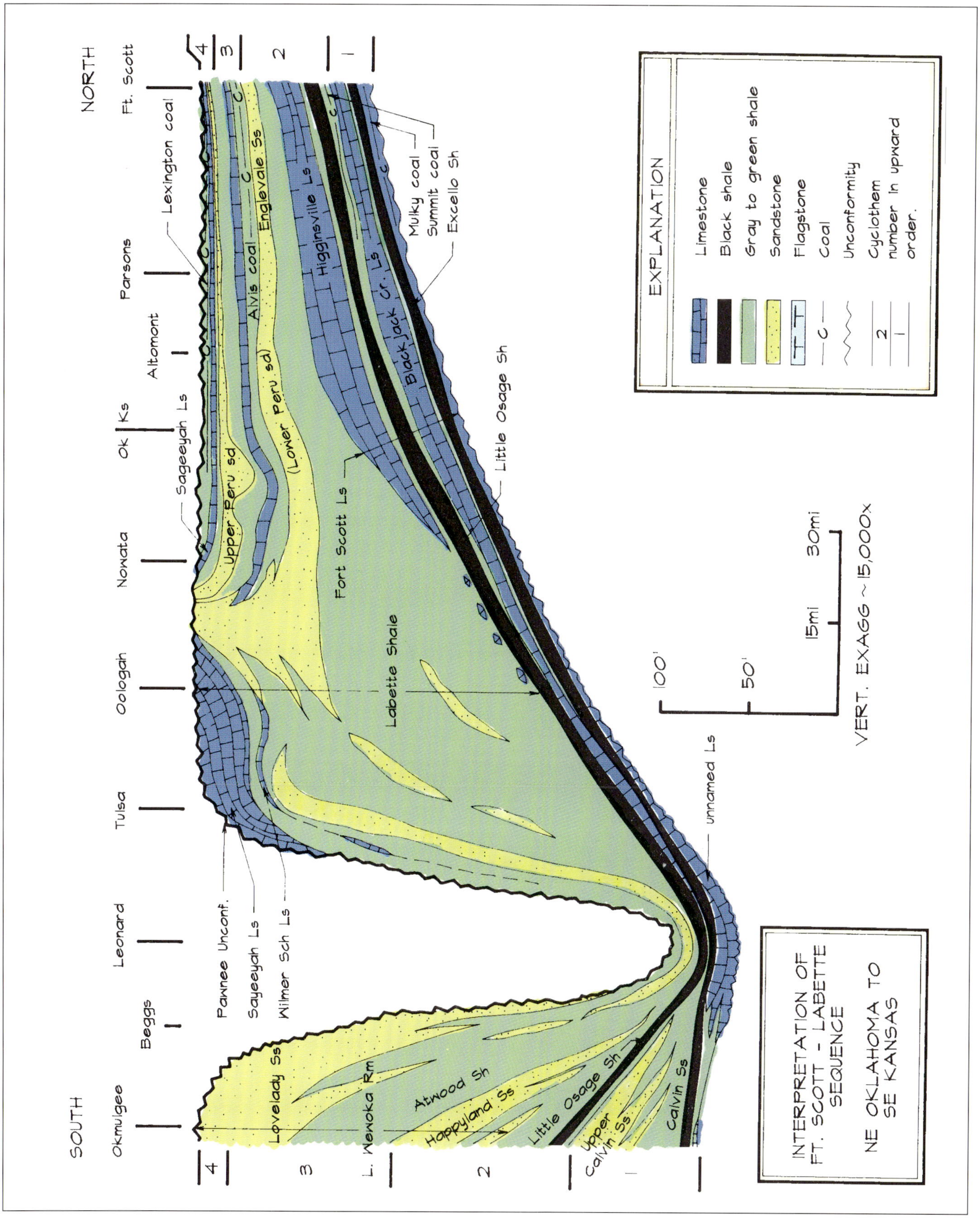

FIGURE 4 — Interpretation of Fort Scott-Labette sequence.

southeastward thickening wedges of turbiditic sandstones and mudrock have been mapped as Calvin Sandstone and overlying Wetumka Shale.

The upper two cycles are dominated by shales and sandstones of the Labette Shale, representing basinal filling episodes. Flooding events resulted in thin coals and limestones. South of its basinal axis, exposed in the northern Conjada Mountains near the Tulsa and Okmulgee county line, a southeastward thickening wedge of clinoform turbiditic sandstones and mudrocks or shales is correlated with the two lower members of the Wewoka Formation.

The Fort Scott to Labette sequence thickens from about 50 to 200 ft north of the Arkansas River and then thins from Tulsa County to about 150 ft at the Kansas border. In Kansas it gradually thins from 150 to about 100 ft at the Missouri border.

The first cyclothem is that of the incomplete Excello cycle on the cratonic shelf in northeastern Oklahoma and southeastern Kansas. The Excello Shale is a black phosphatic shale up to 5.5 ft thick (Ece, 1994) that forms the basal unconformity and core shale in much of northeastern Oklahoma. North of the Oklahoma-Kansas border area, the Mulky Coal that underlies the Excello Shale represents the very rapid cratonward transgression of the Excello cycle, which lacks a transgressive limestone. However, such a transgressive limestone is indicated in the Conjada Mountains of southeastern Tulsa County and adjoining Wagoner County where about 18 ft of a calcareous mudstone appears gradational with the overlying Excello Shale. The regressive limestone of the Excello cycle is the Blackjack Creek Limestone (Roth, 1991). It is commonly a phylloidal algal limestone bank that is as much as 40 ft thick on the Oklahoma-Kansas border. This limestone thins to less than a foot in the Conjada Mountains. Farther south in Okmulgee to Okfuskee counties, the Excello Shale grades up through dark gray fossiliferous shale into the upper Calvin Sandstone, and together they form a prominent cuesta.

The second cycle commences with the anoxic Little Osage Shale and minor associated beds. In east-central Kansas, the Summit Coal appears below the Little Osage Shale and represents the initial flooding event. A transgressive limestone of limited extent is present along U.S. Highway 166 in southern Kansas, where a thin limestone underlies the Little Osage Shale and overlies the probable underclay or paleosol of the missing Summit Coal. The Higginsville Limestone, a phylloidal algal limestone bank retrogradationally developed relative to the other Fort Scott Limestone, attains a thickness of nearly 40 ft along the Oklahoma-Kansas border, but thins southward to only a few inches of nodular marl on the east bank of Oologah Lake.

On the southern orogenic border shelf, the Little Osage Shale grades up through dark gray shale with limestone concretions and thin sideritic nodules into the Happyland Sandstone Member (Bennison, 1981), the lower sandstone member of the Wewoka Formation. On the opposing cratonic shelf, the probable correlative Englevale Sandstone is developed within the Labette Shale. These sandstones probably correlate with the subsurface lower Peru Sand.

The third cycle is based on the Wimer School Limestone that crops out from Nowata County northward across southeastern Kansas. In Kansas it overlies the thin Alvis Coal that serves as the initial transgressive bed. On the southern shelf, a possible correlative bed of the Wimer School Limestone is the basal dark-gray marly shale of the Atwood Shale in Hughes and Okfuskee counties. Northward its shales are largely concealed by overlapping and progradational turbiditic clinoform sandstones of its slope facies in Okmulgee County. Both the Labette and Atwood shales (Bennison, 1984) of the Wewoka Formation grade up into lenticular deltaic fluvial sandstones of the upper Labette Shale and into the probable correlative Lovelady Sandstone Member (Bennison, 1984) of the Wewoka Formation. This third cycle is interpreted as belonging to an upward-shallowing depositional regime.

The fourth and last cycle of the Fort Scott-Labette sequence is somewhat more complex than the preceding ones. It is dominated by the Sageeyah Limestone in northern Tulsa and Rogers counties, where it ranges up to 30 ft or more in thickness. A few miles north of Oologah, the Sageeyah Limestone is only intermittently exposed below the basal Pawnee unconformity, and still farther north (near Nowata), a thin Sageeyah Limestone — commonly less than a foot thick — is exposed with thin overlying beds of sandstone and silty to coaly shales. These beds, with only minor facies changes, persist across southeast Kansas to the Fort Scott area.

In the quarry area northeast of Tulsa and at Oologah, a thick Sageeyah Limestone directly underlies the Pawnee Limestone and both are incorporated into the Oologah Limestone. South of Tulsa in the Broken Arrow area, the Sageeyah grades into spiculitic cherty limestone and fossiliferous marly shale before disappearing below the alluvium of the Arkansas River valley. It has not been identified south of the Arkansas River. This anomalous upper member of the Labette Shale may have resulted from an abrupt subsidence of a hinge-line flexure prior to the deep and widespread marine Anna Shale transgression.

Pawnee-Nowata Sequence

The various limestones composing the Pawnee-Nowata (Fig. 5) and the underlying Sageeyah Limestone sequence are collectively called the Oologah Limestone or Big Lime in Tulsa and Rogers counties, where they are intensively mined for cement and aggregates.

Five cyclothems compose the Pawnee-Nowata sequence. In upward order, these cyclothems involve the Anna Shale, Joe Shale, Bandera Shale, Lake Neosho Shale, and Nowata Shale. The Anna Shale cycle is interpreted as the TST, the Joe Shale to Lake Neosho cycle represents the HST, with the possible exception of the semimarine Bandera Shale, and the Nowata Shale presumably passes from highstand to lowstand at the base of the Walter Johnson Sandstone and correlative horizons.

From southwest to northeast across Tulsa and Rogers counties, the Eleventh Street (Norfleet) Limestone, a transgressive member, overlaps most, if not all, of the underlying Nowata Shale including (in downward order) an upper sandstone and siltstone member (a possible Walter Johnson Sandstone correlative), the upper Nowata flagstone member, and the entire lower Nowata gray and basal black shale members. At a road cut southwest of Oologah (NW/4,NW/4 Section 4 T22N-R15E), a thin conglomeratic limestone rests discordantly on 1.5 ft of calcareous silty mudrock and 11.5 ft of the basal Nowata black shale. About 1.8 mi farther north and west of Oologah (E/2,E/2 E/2,NE/4 Section 29 T23N-R15E), a dark shale above the Altamont Limestone yielded Memorial Shale conodonts. According to P.H. Heckel (pers. comm.), this indicates complete truncation there of the Nowata Shale, about 500 ft by the unconformity underlying the Memorial Formation.

The TST of the Pawnee-Nowata sequence consists of the locally developed Lexington Coal and Childers School Limestone and the more extensive overlying Anna Shale. These rest, in turn, abruptly on onshore clastics of the Labette Shale north of the Arkansas River and on the Lovelady Sandstone Member of the Wewoka Formation south of the Arkansas River. Lexington Coal is a thin discontinuous bed in Nowata County and southeastern Kansas. In a distance of 10 mi northward from Oologah to the Winganon Road northeast of Talala, the entire thickness of the Sageeyah Limestone and some of the upper sandy beds of the underlying Labette Shale has been truncated by the unconformity underlying the Childers School Limestone, the basal transgressive limestone of the Pawnee-Nowata sequence. This represents about 30 ft of erosional downcutting.

The Childers School Limestone and Anna Shale persist to Rogers and northern Tulsa counties, where they are well exposed in quarries exploiting the thick (120 ft or more) Oologah Limestone that includes strata from the Sageeyah Limestone to the Altamont Limestone. Southward, limited outcrops indicate a facies change of the Childers School Limestone to a thin concretionary marl and downslope

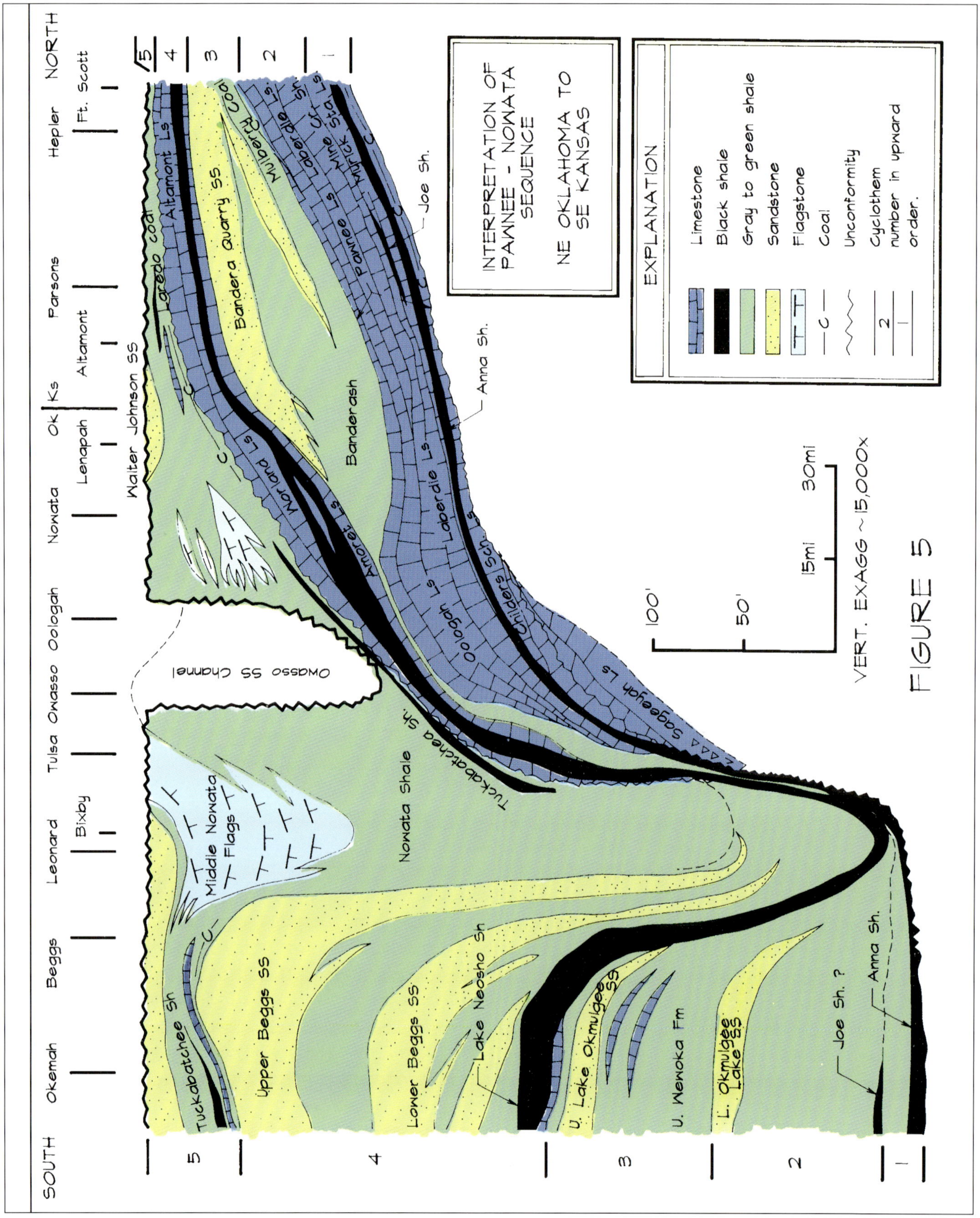

FIGURE 5 — Interpretation of Pawnee-Nowata sequence.

pinchouts of all the Oologah limestones to form a condensed section of dark gray to black shale about 6 ft thick, exposed near Leonard on the bank of the Arkansas River. This is interpreted as a paleobathymetric deep axis for the sequence. Southward in Okmulgee County, this lower cycle is almost completely exposed on the west bank of the Deep Fork River near the Oklahoma Highway 56 bridge. Here about 4 ft of the Anna black anoxic shale rests with a sharp break on the Lovelady Sandstone Member of the Wewoka Formation. The Anna Shale is overlain by almost 30 ft of dark gray shale that yields several ammonoid genera and the *Trepospira-Worthenia* assemblage.

The overlying second cycle includes the discontinuous transgressive Joe Shale and the regressive or upward-shoaling Laberdie Limestone. These beds are well exposed in road cuts along U.S. Highway 69 north of Fort Scott and along U.S. Highway 160 east of Parsons. Joe Shale usually consists of 2 to 4 ft of black anoxic shale (Price, 1981), and the overlying Laberdie Limestone consists of 20 to 30 ft of algal and crinoidal calcareous mudstone. Southward in Oklahoma in the Tulsa quarries, the Joe Shale may possibly be represented in the condensed section in the lower Pawnee Limestone, incorporating the Anna Shale, Myrick Station Limestone, and Joe Shale. The Laberdie Limestone reaches maximum thickness of about 80 ft in the quarries northeast of Tulsa, but thins abruptly southward into only a few inches of fossiliferous wackestone. It has not been observed south of the Arkansas River. Still farther south in Okmulgee County, about 1 mi northwest of the Oklahoma Highway 56 bridge at the Deep Fork River, a black phosphatic shale is present about 35 ft stratigraphically above the Anna Shale that grades up through about 210 ft of shale into the lower beds of the Okmulgee Lake Sandstone Member (Bennison, 1984) of the Wewoka Formation. Oakes and Motts (1963) illustrated these beds as progradational clinoform beds dipping north. These regressive or shoaling clastic beds are absent north of the Arkansas River, where the Bandera Shale rests directly on the Pawnee Limestone.

The third cycle is the Bandera Shale that is as much as 120 ft thick in the Oklahoma-Kansas border area east and south of Edna, Kansas. Here the cycle begins with 3 to 4 ft of a soft clayey paleosol that is overlain by a brachiopod wackestone to mudstone of similar thickness. The provisional name for this limestone is the Edna Limestone. Northward into Bourbon County, Kansas, the limestone pinches out, and the Mulberry Coal thickens to about 2 ft of minable coal. Its upper beds consist of gray clay to silt shale and the lenticular Bandera Quarry Sandstone. Southward into Oklahoma, its basal paleosol and limestone may merge into the upper layers of the Laberdie Limestone. Southward the overlying shale thins to about 2 to 3 ft in the quarry east of Nowata. Farther south near Oologah, a thin 2-in. thick coal is locally developed within the Bandera Shale. Still farther south, in the Tulsa area, it becomes a highly fossiliferous marly shale, varying from 2 to 20 ft thick. Farther south in the Broken Arrow area, the Bandera Shale apparently becomes a dark gray slightly calcareous shale about 10 ft thick, between the black anoxic Anna and Lake Neosho shales. South of the Arkansas River it is rarely exposed, except in the Preston area of Okmulgee County, where about 30 ft of an interbedded section of fossiliferous wackestone and dark gray shale is present at the approximate interval of the Bandera Shale. From east-central Kansas to Craig and Nowata counties in northern Oklahoma, thick lenticular sandstones (known as the Bandera Quarry Sandstone Member) are present in the upper beds of the Bandera Shale. Widespread extent of the Bandera Shale in the Mid-Continent indicates that it may belong to an HST in spite of its upper fluvial-deltaic nature.

The fourth cycle is a typical Kansas cyclothem. The Altamont Limestone consists of (in upward order) the transgressive Amoret Limestone, its core shale (Lake Neosho Shale), and the regressive or shoaling Worland Limestone (Schenk, 1967). These limestones thicken and thin across northeast Kansas and northeastern Oklahoma to the northern outskirts of Broken Arrow, where they become concretionary marly shales in a condensed section that includes the Bandera Shale. South of this condensed section, the Lake Neosho Shale expands in thickness from less than 5 ft to almost 30 ft in

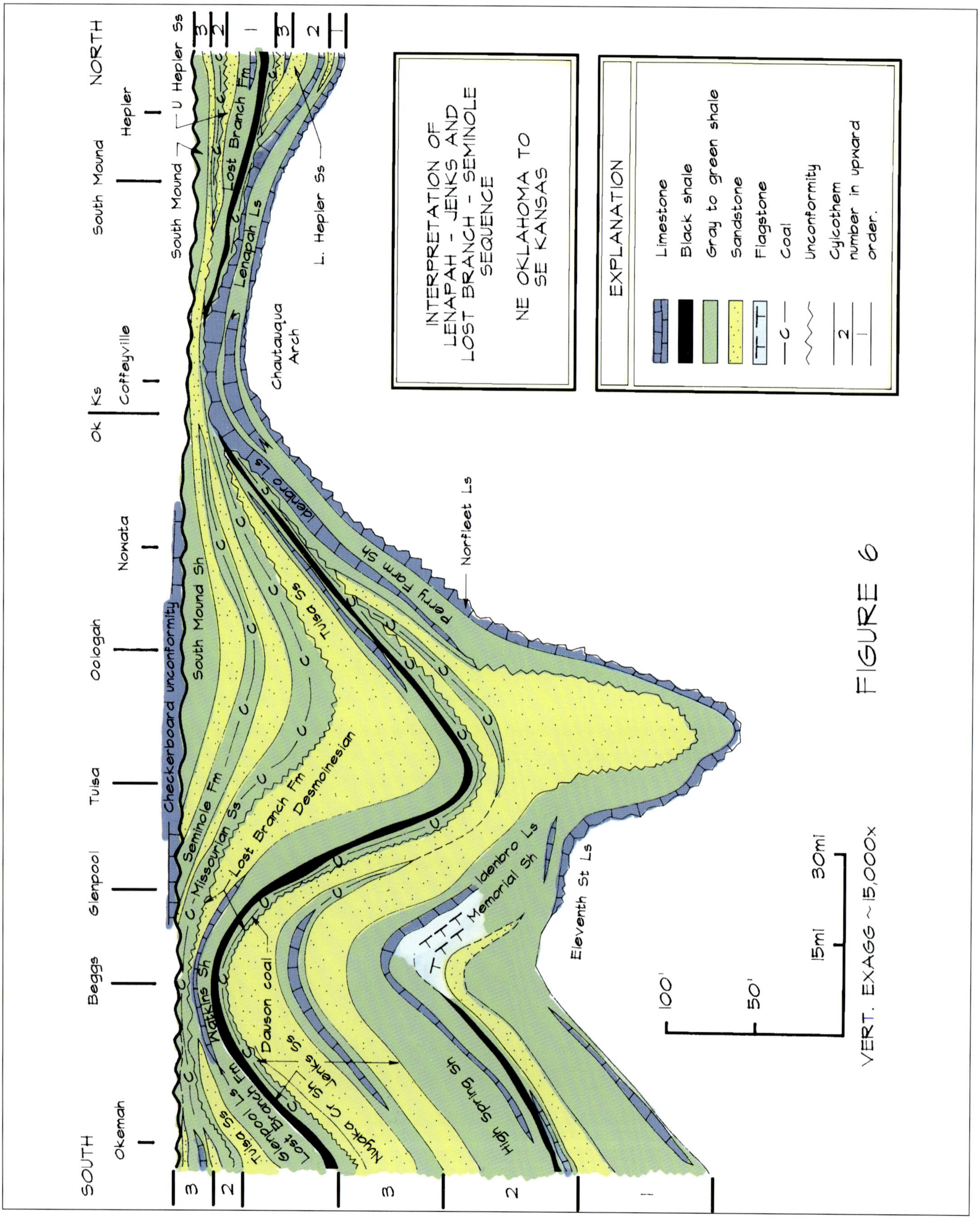

FIGURE 6 — Interpretation of Lenapah-Jenks and Lost Branch-Seminole sequences.

northern Okmulgee County. A brief shoaling interval, involving gray shale, thin siltstone, or sandstone, locally splits this anoxic black shale succession. On the east shore of Okmulgee Lake, the Lake Neosho Shale rests on a thin fossiliferous brown ferruginous wackestone that may be a correlative of the Amoret Limestone. The succeeding clastics of this cycle, lacking on the northern cratonic shelf, become very conspicuous in a clastic wedge of turbidite clinoform beds of interbedded sandstones and shales up to 350 ft thick. In Okmulgee County, these clastics are named the Beggs Sandstone Member of the Wewoka Formation (Bennison, 1984). Some of the lower clastics may correlate with the Worland Limestone that crops out north of the Arkansas River.

The fifth cycle is that of the Nowata Shale, which is incomplete throughout much of Oklahoma and Kansas, because of erosion. It is overlain by the basal unconformity of the Lenapah Limestone and correlative beds, such as the Eleventh Street Limestone.. The most complete Nowata Shale succession is in southern Tulsa County, where it is about 550 ft thick and fills a deep foredeep area over a condensed Altamont Limestone section. It is bordered on the south by a downlapping series of the Beggs clinoform beds and on the north by a series of stacked carbonate banks of the Oologah Limestone. Much of this thick Nowata Shale is composed of (in upward order) dark gray to medium gray slightly calcareous mudrock, calcareous flagstones, sandstones, and silty to clayey mudrock.

In Montgomery and Labette counties of southern Kansas, the Nowata Shale varies from 8 to 40 ft in thickness and consists of (in upward order) a paleosol on weathered Altamont Limestone, locally a coal 1 to 7 in. thick, tentatively identified with the Laredo Coal of Missouri (Howe, 1953), a thin capping bioclastic limestone, and grayish black to dark gray shale. In southern Montgomery County and adjacent Nowata County, the shale is randomly capped by the lenticular Walter Johnson Sandstone. A possible correlative of the Walter Johnson Sandstone has been mapped in southern Tulsa County. The Laredo Coal extends as far south as the gooseneck bend of the Verdigris River northeast of Lenapah, where it is a coaly shale about 13 in. thick that directly overlies a thin calcareous siltstone carrying impressions of lycopod logs and tidal zone snails. This is interpreted as a transgressive shoreface deposit. Farther south in Rogers County, flagstones within the Nowata Shale crop out from Talala Creek to the Oologah urban area. These may have been continuous with thicker flagstones in southern Tulsa County, but erosion by the overlying Lenapah Limestone and the deeply incised channel of Jenks Sandstone in the Owasso to Oologah area interrupted this continuity. The dark gray to black basal shale near the base of the formation is sheety and phosphatic on the south side of the Labette County village of Mound Valley, and is a gray black anoxic shale southwest of Oologah, west of Owasso, and at Garnett Plaza in eastern Tulsa. South of Tulsa County in Okmulgee to Hughes counties, this anoxic shale may correlate with the Tuckabatchee black shale which overlies a partly dolomitized fossiliferous limestone named the Yeager Limestone (Bennison, 1984) that is considered to be the base of the Holdenville Shale.

Lenapah-Jenks Sequence

The Lenapah Limestone to Jenks Sandstone sequence (Fig. 6) is a modified foreland basin sequence that has had conflicting interpretations due to earlier miscorrelations by Oakes (1952). What Oakes considered to be the Lower Seminole Formation is now called the Jenks Sandstone by Bennison (1984), and his Middle Seminole Shale is now the Lost Branch Formation of Heckel (1991). Both are Desmoinesian rather than Missourian age. In 1994, Heckel proposed incorporating the Lenapah Limestone and Jenks Sandstone as the Memorial Formation, an upward extension of the Memorial (Perry Farm) Shale of Dott (1941). Farther north across the Chautauqua arch, the Jenks Sandstone and correlative Hepler sandstones have been largely truncated by the Lost Branch and Seminole formations. Its correlative sequence south of the Arkansas River is tentatively placed in the High Spring Shale, and the overlying sandstones appear to be laterally continuous with the Jenks Sandstone and with the middle Holdenville sandstones of the Holdenville area. As was previously noted, its basal limestone (Eleventh Street Limestone in the Tulsa area and the Norfleet Limestone northward) overlies an unconformity which truncates much of the underlying Nowata Shale. In Rogers County near Oologah, the Lenapah Limestone rests unconformably on only a foot of the basal Nowata black shale bed.

Unlike the preceding sequences, this sequence lacks a well-defined paleobathymetric deep axis and has only a minor outcrop of black sheety shale near Mound Valley. The depocenter area along its outcrop belt is limited to southern Tulsa County. Only three cycles have been identified in this thinner than usual sequence. The basal transgressive limestone is the Norfleet-Eleventh Street Limestone that grades up into dark gray shales of the Perry Farm and Memorial shales that contain the deeper water diverse mollusk-dominated fauna of ammonoids and gastropods, such as *Worthenia* and *Trepospira*. The closing event of this parasequence is the deposition of locally thin sandstones mapped in southern Tulsa, Okmulgee, and Okfuskee counties of Oklahoma and in Bourbon County of Kansas.

The second, and middle, cycle features the transgressive Idenbro Limestone that has been partly eroded and is overlain by the Lost Branch Formation. This formation contains Idenbro pebbles in its basal paleosol. In southern Tulsa County and adjacent Okmulgee County, the Idenbro Limestone may be equivalent to the calcareous flagstones that laterally grade to shales. South of Beggs, this interval contains black phosphatic nodular shale that bears ammonoids. In Tulsa to Okfuskee counties, a thin and nameless regressive wackestone intervenes between the shale and the overlying lower Jenks Sandstone that represents the closing phase of this cycle. This sandstone is the lower member of the Jenks Sandstone (Bennison, 1984), the surface name of the subsurface lower Cleveland Sandstone (Krumme, 1981). The Jenks Sandstone is deeply channeled into both Lenapah and Nowata beds north of Owasso in northern Tulsa County and adjacent Rogers County, and is missing over the Chautauqua arch.

North of the Chautauqua arch in the Cherokee basin of southeastern Kansas, the Lost Branch Formation offlaps a sandstone that has been correlated with one of the several Hepler sandstones. However, with the revised placement of the Hepler Sandstone above the Lost Branch Creek Shale, it is now nameless, but is considered a probable correlative of the Jenks Sandstone.

The third, and upper, cycle of the Lenapah to Jenks sequence is also missing over the Chautauqua arch. This cycle comprises the nameless middle Jenks marine band and the overlying upper Jenks Sandstone. Just south of Jenks, it contains a paleosol and a 2-ft thick coal now known as the Jenks Coal (Bennison, 1984; Hemish, 1990). At Jenks it is channeled by the upper Jenks Sandstone, but, farther south, a *Myaline* bearing and nameless limestone appears above the coal. Still farther south in the Beggs to Nuyaka area, this limestone becomes a lime mudstone with a more diverse deeper marine fauna. West of Beggs and south of Okemah and Holdenville, the overlying upper Jenks Sandstone contains irregular layers of novaculite pebbles, indicating deep erosion of the Ouachita Mountains during Middle Pennsylvanian time. At Glenpool, about 3 to 5 ft of clay to silt shale, capped by a lenticular pebbly limestone containing thick shelled mollusks, may represent a minor tempestite disrupted cycle.

Lost Branch-Seminole Sequence

The Lost Branch Formation to Seminole Formation sequence is a relatively thin but widespread foreland basin sequence, involving about three T-R cycles, that straddles the Desmoinesian-Missourian stage boundary (Fig. 6). Heckel (1991) proposed the name of Lost Branch Formation for a marine shale-dominated Late Desmoinesian interval in Labette County, Kansas, that was previously correlated with either the Memorial Shale or the Holdenville Shale of Oklahoma. The overlying Seminole Formation is considered to be basal Missourian and is dominated by sandstone and coal-bearing shale beds that become conglomeratic in east-central Oklahoma. No condensed section representing a depocenter or paleobathymetric deep axis has been identified in this sequence.

The Nuyaka Creek Shale (Bennison, 1981) is a transgressive black phosphatic shale that has been traced over much of the Mid-Continent. In northeastern Oklahoma and southeastern Kansas, it is underlain by the Dawson Coal and in southern Oklahoma by the thin transgresisve Homer School Limestone. The Lost Branch Formation truncates the entire Jenks Sandstone and part of the Idenbro Limestone over the Chautauqua platform. North of the platform in the Cherokee basin, offlapping by the Nuyaka Creek Shale has exposed sandstones similar in stratigraphic position to the type Hepler Formation of Jewett (1940). They are now nameless, owing to the recent redefining of the Hepler Sandstone at a higher horizon. Other sandstones throughout Kansas and Missouri have been miscorrelated with the type Hepler Sandstone as defined by Jewett.

The HST of the sequence was of short duration and is probably represented by the dark gray marine shale above the Nuyaka Creek black phosphatic shale. The lenticular Glenpool Limestone is locally developed above the gray shale and is now interpreted to be the youngest Desmoinesian limestone in the Mid-Continent. In Bourbon County this limestone contains many carbonized wood fragments, speculated to be the remnants of a Pennsylvanian wildfire.

A shallowing episode leading to the widespread lowstand environment of the Seminole Formation is indicated by the Tulsa Sandstone of northern Oklahoma and by one of the several Hepler sandstones in Kansas. It is gradational and unconformable with the underlying Lost Branch Shale Member. This sandstone is commonly missing through truncation by the overlying Tulsa Coal and its underclay, also informally known as the Hepler Coal in Kansas. This coal horizon is now interpreted by many stratigraphers as the base of the Seminole Formation as well as that of the Missourian Stage (Hemish, 1990).

The Seminole Formation consists of two or more lowstand T-R cycles. Its lower cycle was named the Mooser Creek Member and consists of interbedded carbonaceous sandstones and coal-bearing shales. The upper cycle is known as the South Mound Shale in Kansas and consists largely of shallow marine mudrock, thin limestone, and minor lenticular sandstones and coals. Conglomerates, pebbly sandstones, and mottled red and green mudrock characterize the type Seminole Formation in Seminole County, southeastern Oklahoma. Throughout much of Oklahoma, the Seminole Formation is 40 to 80 ft thick. However, about 1.5 mi northeast of Last Chance in Okfuskee County of east-central Oklahoma, the Seminole formation consists of only 4 ft of thin smutty coal and paleosols that rests discordantly on the Holdenville Shale. According to Tomlinson and McBee (1959), the Desmoinesian-Missourian transition was a period of maximum orogenic movements within the Ouachita Mountains. These events may have resulted in some random syndepositional low-amplitude folding and faulting within the Arkoma basin and adjacent areas, such as the Chautauqua arch. The Ouachita movements may be related to the Hercynian movements of Europe and other global, middle to late Carboniferous orogenies. The widespread emergence of land areas and probably increased aridity resulted in extensive biological extinction, both on land and sea. It is reported that more than 40 floral species disappeared at that time in the eastern United States. *Lepidodendron* species were replaced by *Sigillaria* species in Oklahoma (L.R. Wilson, pers. comm., 1995). Some of the major faunal extinctions at the top of the Desmoinesian series (Boardman et al., 1994) include the brachiopod genus *Mesolobus*, the ammonoid genera *Wellerites* and *Gonioglyphioceras*, the conodont genera *Neognathodus*, and several *Idiognathodus*, *Thymospora*, and *Cappasporites* species.

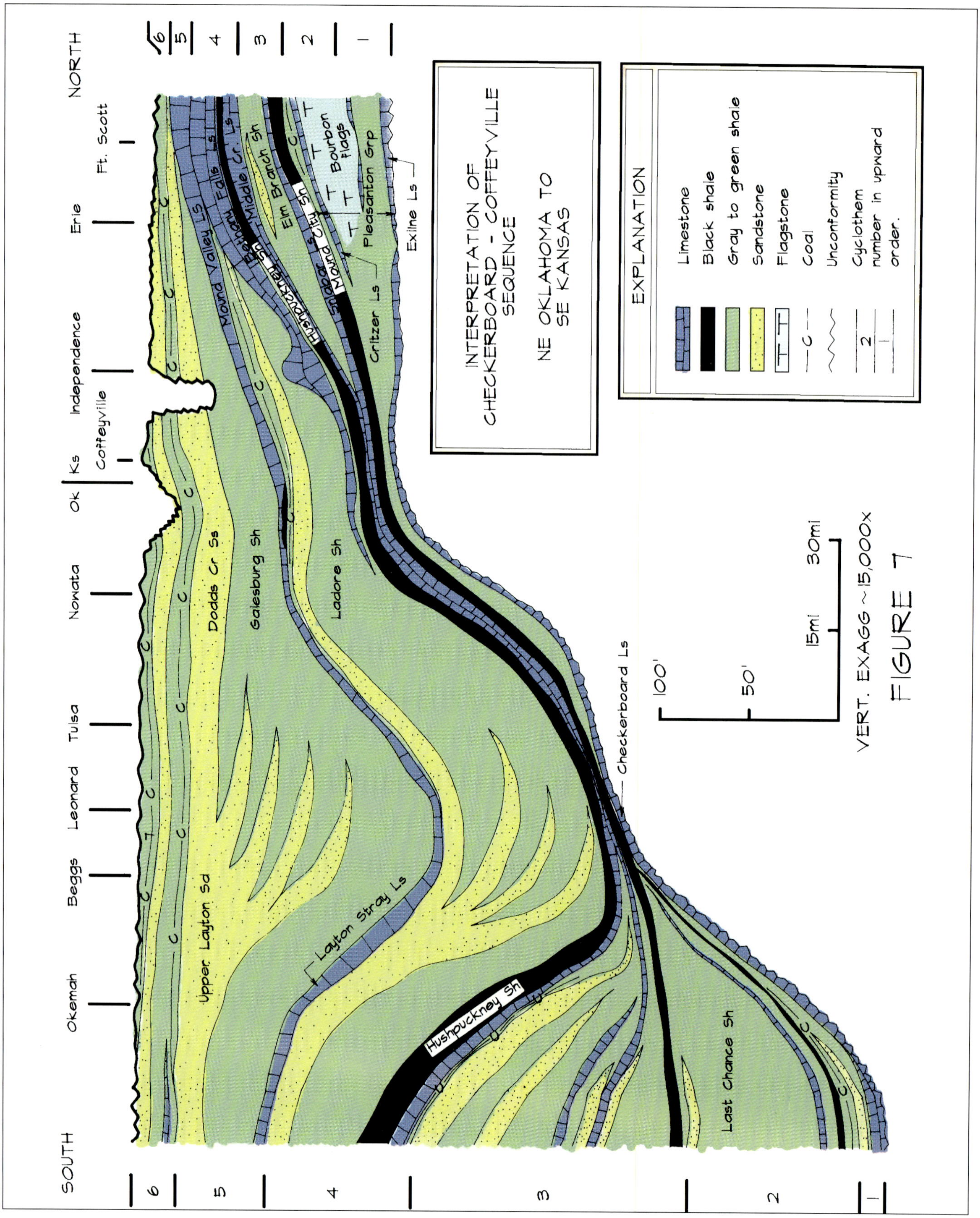

FIGURE 7 — Interpretation of Checkerboard-Coffeyville sequence.

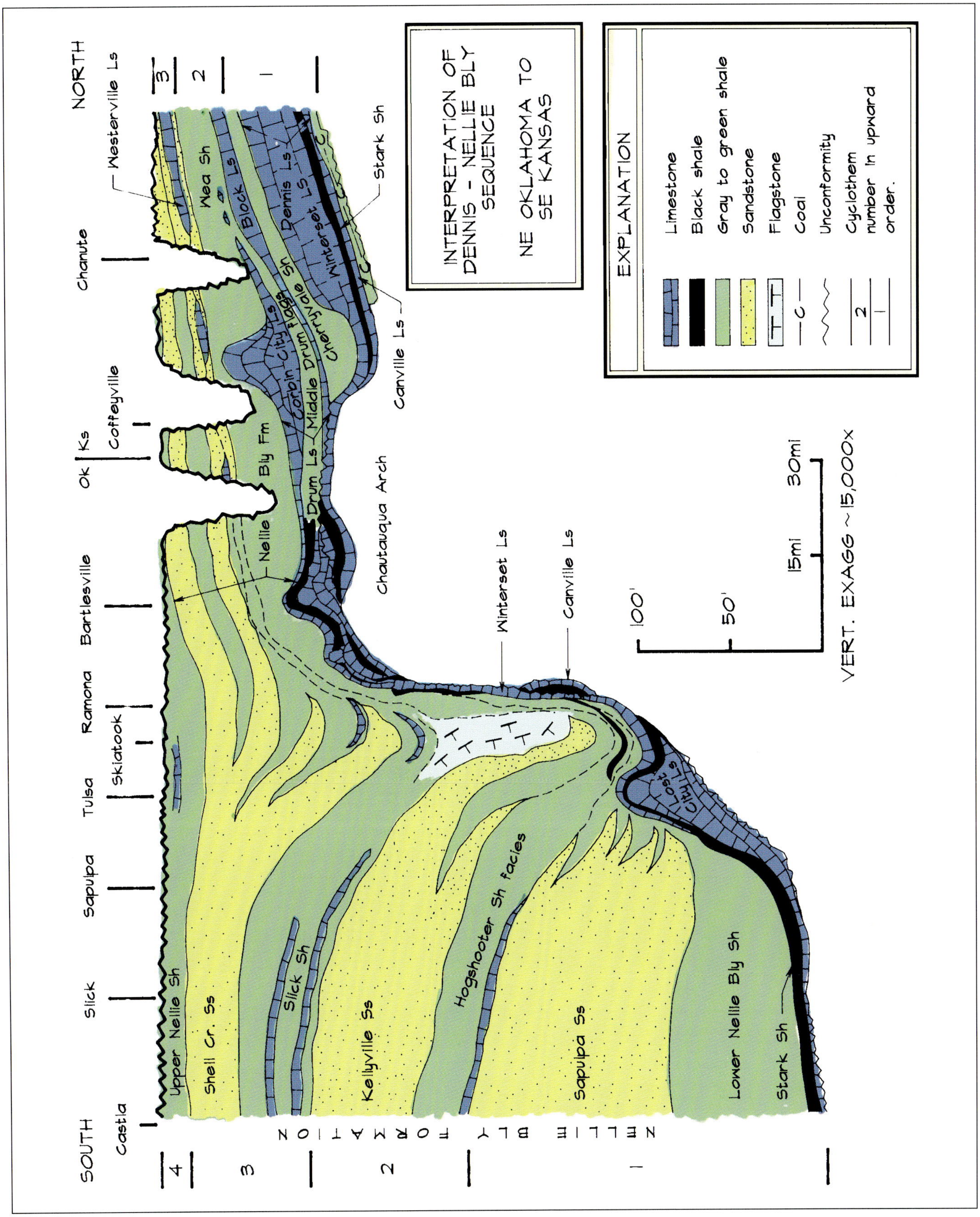

FIGURE 8 — Interpretation of Dennis-Nellie Bly sequence.

Checkerboard-Coffeyville Sequence

After the major marine withdrawal and biological crises during the Desmoinesian to Missourian transition represented by the Lost Branch-Seminole sequence (Fig. 7), the Checkerboard to Coffeyville sequence was dominated by a more marine environment and a characteristic modified foreland basin sequence involving cycles. The paleobathymetric deep axis shifted to middle Tulsa County. The transgressive Checkerboard Limestone rests unconformably on the Seminole Formation and is interpreted as the base of the sequence. However, other unconformities within the Checkerboard-Coffeyville sequence may, in the future, be used to divide it into other sequences. The succeeding highstand marine episode embraces four cycles. Lowstand to forced regression systems tracts prevailed during the last three parasequences. The highest cycle is very minor and may represent only a slight base-level fluctuation during a forced regression.

According to surface mapping south of Tulsa, the Checkerboard Limestone truncates the shallow marine South Mound Shale Member of the Seminole Formation. Although the Checkerboard Limestone rarely exceeds 3.5 ft in thickness, it is a very useful time-stratigraphic horizon for much of northern Oklahoma and southern Kansas. In east-central Kansas and adjacent Missouri, its presumed correlative bed is the Exline Limestone. In east-central Oklahoma, the Sasakwa Limestone is the probable correlative of the Checkerboard Limestone and similarly rests discordantly on the Seminole Formation (Krumme, 1981). At many localities in northern Oklahoma and southern Kansas, the Checkerboard Limestone is directly overlain by the Tacket Shale, a condensed section of both the Mound City and Hushpuckney shales, the core shale, respectively, of the Hertha Limestone and overlying Swope Limestone. South of Tulsa, the Checkerboard Limestone, Mound Valley Shale, and Hushpuckney Shale diverge and offlap to expose intervening strata to form southward-thickening wedges of mainly marine beds that are either very thin or missing in northeast Oklahoma. In the Last Chance area of Okfuskee County, these become well-developed highstand marine beds. The transgressive Checkerboard Limestone grades up into regressive shale and carbonaceous sandstone.

Northeast of Parsons, Kansas, the Pleasanton Group is offlapped by the Mound City Shale to expose as much as 140 ft of lenticular semimarine deposits along the Kansas-Missouri border (Gentile, 1976). Here, the 1- to 2-ft thick Exline Limestone Member of the Shale Hill Formation occupies the approximate position of the Checkerboard Limestone and, similarly, contains a moderate depth mollusk and brachiopod fauna overlain by dark shale bearing a similar fauna. These basal beds grade up into greenish gray to gray shales, and partly calcareous sandstones of the Unity Farm Shale that locally yield a tidal to shallow subtidal biota.

The second overlying cycle in Okfuskee County, Oklahoma, is preceded by a thin interbedded coal and carbonaceous shale that grades upward into about 1 ft of black concretionary shale. This is succeeded by about 160 ft of dark gray fossiliferous shale with thin wackestone beds (Hemish, 1987). In Bourbon County, Kansas, this interval may be occupied by the locally thick Bourbon flagstones that Underwood (1984) now consider a slope facies of the shelfward Critzer Limestone Member. A more thorough study of correlative beds of the Pleasanton Group of Kansas and Missouri and the thick post-Checkerboard Limestone and pre-Mound City Shale bed in east-central Oklahoma is needed.

The proposed third cycle is the Mound City Shale cycle. In Okfuskee County, the Mound City Shale is overlain by a highly fossiliferous limestone, formerly called the Checkerboard Limestone, but now known to be at the approximate position of the Sniabar Limestone of Kansas. The transgressive limestone bed of this cycle crops out farther south in Seminole County, where it is known as the Denay Limestone. Moderately thick clinoform marine sandstone and shale beds complete this cycle, and may correlate with the Elm Branch shale and sandstone above the Sniabar Limestone. In a recent review of the Mound City Shale, Watney and French (1994) proposed the new name, Guthrie Mountain Shale

Member for the gray shale below its black phosphatic shale bed. These limestone beds associated with the Mound City Shale have been referred to the Hertha Formation in east-central Kansas and to the lower Tacket Shale in southeastern Kansas and northeastern Oklahoma, where the associated limestones are either very thin or missing and resemble a debris flow carbonate.

The fourth cycle involves the Hushpuckney Shale, a black phosphatic core shale of the Swope Limestone of Kansas, and the overlying regressive Ladore Shale. Its basal transgressive limestone is the Middle Creek Limestone that locally exceeds 30 ft in thickness in east-central Kansas, but thins southward to less than a foot in thickness in the Oklahoma-Kansas border area. Southward in the Tulsa area, it becomes part of the thin middle marly band of the condensed Tacket Shale (Pavlicek, 1986). In Okmulgee and Okfuskee counties, the Middle Creek horizon is occupied by a thin interval of coal and a fossiliferous marly limestone that overlies thick clinoform sandstones.

The regressive or shoaling limestone of the Hushpuckney cycle in Kansas is the Bethany Falls Limestone, a mottled gray calcareous mudstone with local oolitic facies that crops out across Kansas and thins southward to the south border of Nowata County, Oklahoma, where it grades or pinches out into dark gray shale overlying the Hushpuckney Shale. These gray shales coarsen upward into the Ladore Shale and its interbedded sandstones and shales that extend across northeastern Oklahoma and southward as the middle division of the Coffeyville Formation. In Bourbon County, Kansas, the Ladore Shale is overlapped by the highly oolitic to pisolitic grainstone facies of the Mound Valley Limestone.

The fifth cycle involves the Mound Valley Limestone, which crops out almost continuously across much of southeastern Kansas and northeastern Oklahoma. West of South Coffeyville, exposures reveal a complete Kansas type cycle with a transgressive limestone, a core shale of black phosphatic sheety shale, and a regressive or shoaling limestone succeeded by shale grading up through upper Galesburg Shale into its lower sandstone bed. This apparently belongs to an LST that wedges out northward in east-central Kansas.

The sixth cycle begins with the Cedar Bluff Coal and its underclay that grades up into the upper sandstone of the Galesburg Shale. Type locality of the Cedar Bluff Coal is at Big Hill just north of Coffeyville, Kansas. Northward, this sequence wedges out in Bourbon County. Southward, it continues about halfway across Oklahoma, where thin encrinitic limestones in southern Okmulgee County indicate a deepening marine environment.

The seventh and uppermost cycle is known only in the Sand Springs area of Creek County, where it consists of a thin section of smutty coals, carbonaceous siltstone, and paleosol above the upper sandstone of the Galesburg Shale. This cycle may be interpreted as the final Coffeyville regression.

The Checkerboard-Coffeyville sequence differs in the two basins. In the Cherokee basin and Bourbon arch area of Kansas, thick carbonate banks of the Hertha prograde southward over the Pleasanton clastic wedge. In the Arkoma basin and bordering shelves, the sequence is dominated by northwestward prograding clinoform of sandstone and shale deposited during marine highstands and is succeeded by lowstand aggradational to progradational sandstones filling the basin to its brim.

Dennis-Nellie Bly Sequence

During the deposition of the Dennis to Nellie Bly sequence (Fig. 8), an abrupt shift in the exposed paleobathymetric deep axis occurred. The basinal axial area of subsidence that initiated the marine deepening event of the transgressive Lost City-Canville Limestone and overlying anoxic Stark Shale had retreated an uncertain distance south of that for the preceding Checkerboard-Coffeyville sequence. During deposition of the second and third cycle, it then advanced many miles northward to

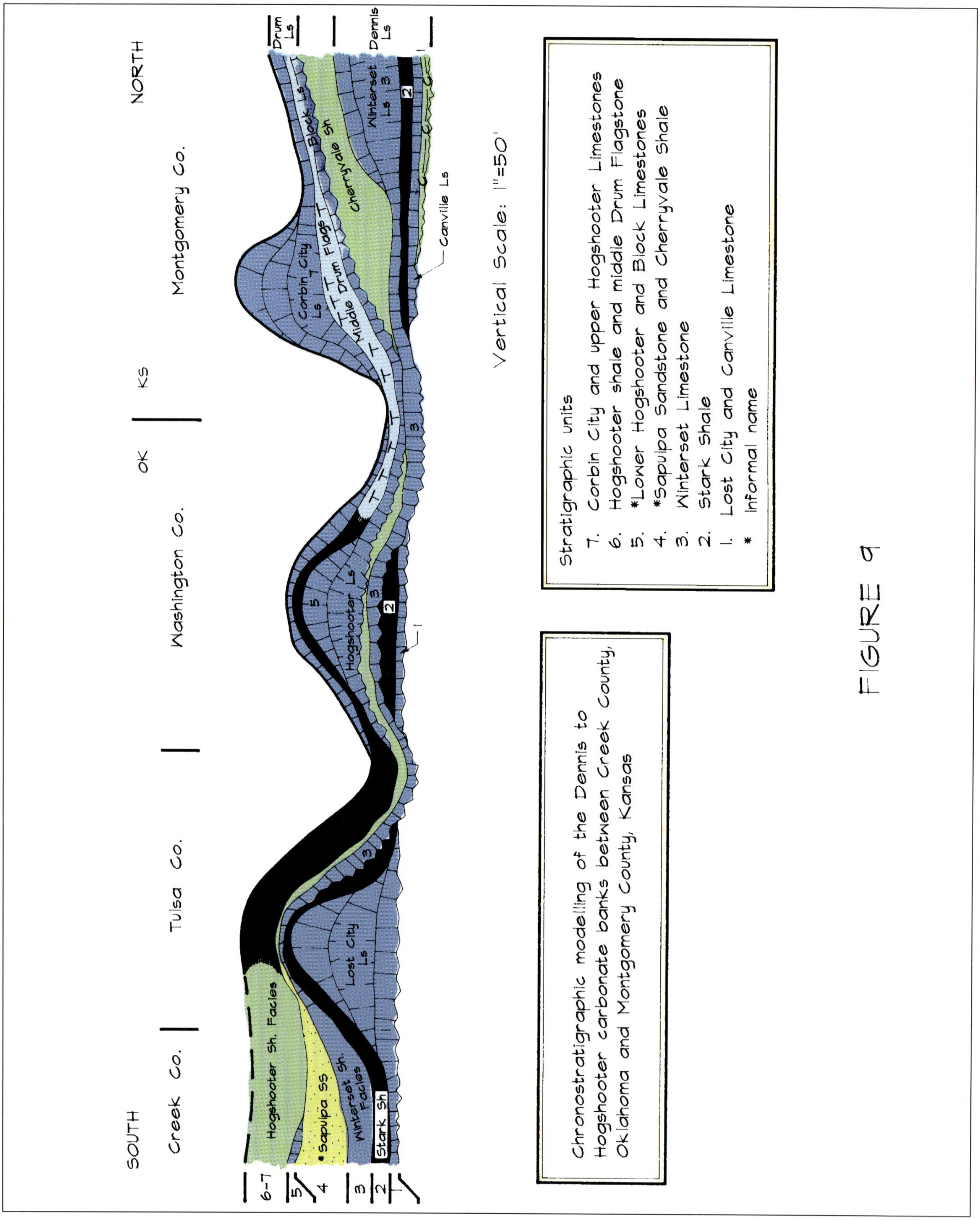

FIGURE 9 — Lithostratigraphic modeling of the Dennis to Hogshooter carbonate banks.

a position about 2 mi northwest of Sperry in northern Tulsa County. This axial migration was forced by the much greater sediment loading from the southern orogenic belt than from the northern cratonic sources. With these exceptions, the sequence resembles the geometry of the previously described foreland basin sequences.

The basal unconformity of the transgressive Lost City-Canville Limestone members of the Dennis Limestone exhibits less truncation than those of the preceding sequences. North of the Arkansas River in the Sand Springs area to Montgomery County in Kansas, Lost City-Canville Limestone is partly missing, owing to low-amplitude uplift and truncation by either the regressive Winterset Limestone of the first cycle or the overlying Hogshooter Limestone of the succeeding cycle. The unconformity capping the sequence at the base of the Dewey Limestone represents only a minor amount of erosion and, therefore, is relatively minor compared to those of previous sequences and may require reassessment as a sequence boundary. The Chanute Formation locally overlaps the Dewey Limestone in southeastern Montgomery County to directly rest on the partly eroded Nellie Bly Formation.

Moderate Early Missourian tectonic activity along the Chautauqua arch area of the Oklahoma-Kansas border area complicates the stratigraphic framework and determination of sequence boundaries. Figures 8, 9, and 10 illustrate an increase of unconformities over this arch and its ramps. For example, along the road cuts of U.S. Highway 166 west of Coffeyville (NW corner of Section 6 T35S-R16E), the uppermost Coffeyville beds (Stark Shale and Lost-City Canville Limestone) are missing, either through nondeposition or truncation, and are overlain by the upper Winterset Member of the Dennis Limestone. Still stratigraphically higher, the probably thin correlative bed of the lower Hogshooter-Drum (Block?) Limestone lies on the Winteset Limestone showing erosion of some of the uppermost beds of the Winterset Limestone and all the intervening Cherryvale Shale. In turn, the stratigraphically higher Nellie Bly Formation occurs on a truncated contact on the upper bed of the middle Drum Flagstone Member as well as its upper Corbin City Limestone Member. The upper Nellie Bly Formation and overlying Dewey Limestone are likewise truncated and are overlain by the Chanute Formation. Multiple unconformities in less than a 70-ft interval reveal that internal unconformities locally may form cyclic boundaries and even separate members and beds within a cycle.

Four cycles have been recognized within this stratigraphic framework. Preliminary tentative interpretations of its systems tracts are transgressive for the Canville Limestone to Stark Shale, highstand for the Winterset Limestone to upper Hogshooter Limestone and correlative Corbin City Limestone, and lowstand for the succeeding Nellie Bly Formation and presumed intertonguing Westerville Limestone.

The known extent of the lower cycle embracing the Dennis Limestone and overlying Cherryvale Shale (Lower Nellie Bly Formation) is northward from Spring Creek in Seminole County, Oklahoma, to Iowa. In the Sand Springs area on the Arkansas River, the shelf edge bank of Lost City Limestone is as much as 52 ft thick. In other locations where the Lost City-Canville Limestone is not truncated, it varies randomly from about 1 to 8 ft in thickness. The middle member and core shale of the Dennis Limestone (Stark Shale), typically a black phosphatic sheety shale, ranges up to 8 ft in thickness. The Winterset Limestone rarely exceeds 3 ft in thickness in Oklahoma, and commonly rests unconformably on the Lost City-Canville Limestone or on underlying Coffeyville strata. Phosphate nodules (redeposited in its basal layers from the Stark Shale) are fairly common. North of Cherryvale in Montgomery County, the Winterset expands within one mile from about 4 ft to more than a 40-ft thick algal mudstone. The Cherryvale Shale, as it was originally defined by Haworth (1898), is the regressive wedge of silty clay shales and thin lenticular sandstones that, northward, becomes the Fontana Shale of Moore (1932). In the Cherryvale area, this shale locally exceeds 60 ft in thickness. South of Independence, Kansas, the Cherryvale Shale is overlapped by the Drum Limestone (Fig.10), but it reappears for a short distance about one mile south of Hogshooter in Washington County, Oklahoma. Still farther south in the Sand Springs area a probable correlative of the Cherryvale Shale, a sandstone-dominated unit onlapping the south flank of the Lost City Limestone, thickens southward toward Sapulpa. There it has the informal name of the Sapulpa Sandstone Member of the stratigraphically enlarged Nellie Bly Formation.

The second, and overlying, cycle consists of the Hogshooter-Drum Limestone and lower beds of the Nellie Bly Formation. The thick quarried Hogshooter Limestone at Hogshooter ranges up to 28 ft of a phylloidal algal calcareous mudstone that lies unconformably on various beds of the underlying Dennis Limestone and, locally, the Cherryvale Shale (Fig. 9). Northward it becomes a thin dark gray foraminiferal mudrock, but appears to thicken toward east-central Kansas in the vicinity of Block, where it averages about 6 ft thick. Above this massive calcareous mudstone at Hogshooter is a thin, dark gray black fossiliferous clay shale which grades northward into the thicker middle Drum flagstone member of Kansas with the diverse fauna characteristic of the dysoxic facies of the core shale. Above this core shale at Hogshooter are 1 to 2 ft of a brown gray wackestone that northward apparently becomes the massive oolitic Corbin City Limestone Member of the Drum Limestone at the Heartland quarry near Independence, where it is 60 ft thick or more. In east-central Kansas, it is apparently represented by a zone of limestone concretions in marly shale in the basal Wea Shale above the Block Limestone. South of Hogshooter, at Ramona, the Hogshooter Limestone onlaps a pinnacle reef of the Dennis Limestone. South of Ramona the upper and lower Hogshooter limestones gradually thin and the middle Hogshooter core shale thickens to form more than 40 ft of dark gray shale rich in the deeper marine, largely molluscan fauna of mainly *Trepospira, Worthenia,* and ammonoid species. The provisional name for this shale is the New Harmony Shale, which is an alternate name for the Hogshooter community. Here and throughout much of southeast Kansas, the Hogshooter-Drum Limestone grades up through shale into unnamed lenticular siltstones and sandstones of the Nellie Bly Formation. These become largely thick progradational to aggradational sandstone and shale beds in Tulsa County, and are interpreted to represent an LST.

A distinct third cycle is recognizable in Creek County, Oklahoma, where it is represented by a transgressive fossiliferous ferrocarbonate bed succeeded by a dark fossiliferous shale and then a thin fossiliferous limestone. The writer has assigned the provisional name of Slick Shale for its excellent outcrops in the Slick area of Creek County. North of Creek County into much of southeastern Kansas, this part of the sequence is largely occupied by lenticular sandstone and minor encrinitic limestone lenses of questionable stratigraphic positions. In east-central Kansas, the northward-thickening Westerville Limestone probably represents the flooding event of the third cycle. Its shoaling phase in Oklahoma is tentatively selected as the Shell Creek Sandstone, the upper thick sandstone of the Nellie Bly Formation that probably extends into southeastern Kansas. Its equivalent in east-central Kansas may be the thin onshore varicolored mudrock of the lower Quivira Shale.

The fourth cycle of limited extent in Tulsa and Creek counties is represented by an unnamed member about 30 ft thick, consisting of thin marine shales and fusulinid limestones capped by sandstones that underlie the Dewey Limestone.

CONCLUSIONS

Pennsylvanian Ouachita orogenies caused by converging continental plates controlled the geodynamics of the Arkoma basin area, and the development of depositional sequences. Global eustasy from Gondwanan ice-sheet fluctuation and apparently random tectonism established the overprint of T-R cycles. Basin subsidence rate may have assumed a sinusoidal curve, probably skewed to the deceleration and stagnation phase. Basin subsidence may have resulted from a combination of thrust-

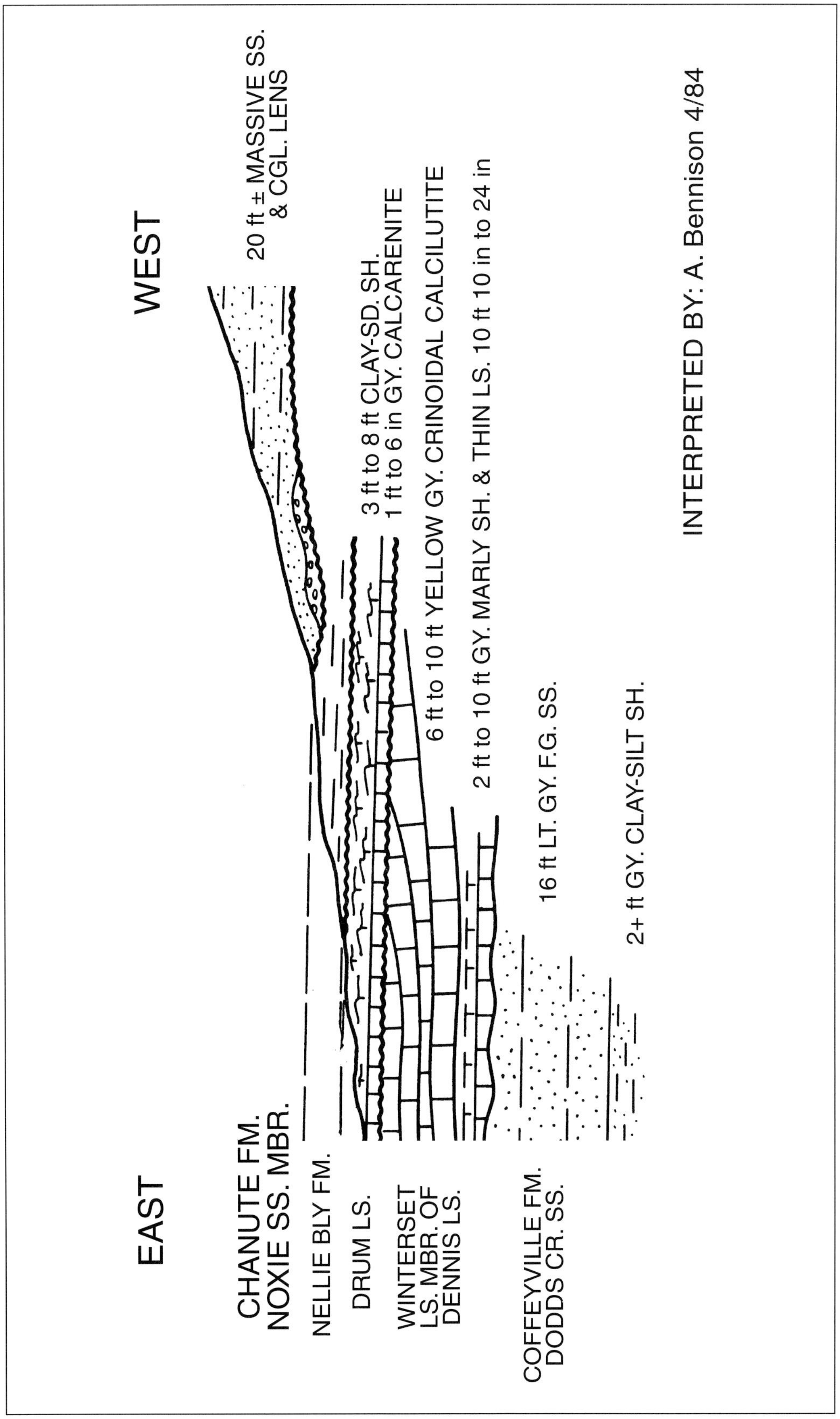

FIGURE 10 — Interpretative sketch of stratigraphy exposed along south side of U.S. Highway 166, west of Coffeyville, Kansas.

ing at the orogenic front of the Ouachita Mountains and sediment loading with isostatic compensation. A greater volume of sediments eroded from the rising Ouachita and Appalachian mountains than from the northern cratonic lowlands episodically shifted the Arkoma basin depocenter and bathymetric deep axis north to northwestward at the rate of about 1 cm/yr.

Under these conditions, transgressive processes deposited the initial systems tract (TST) of modified foreland basin sequences, usually represented by one cycle. It was then followed by one or more HSTs. Deceleration and final stagnation resulted in deposition of the lowstand and forced regression systems tracts and possibly their deep erosion and soil formation. Such final phases may have been of greater duration than those of transgression and high sea-level stands. Interplay of these various forcing mechanisms contributed to the development of this distinctive foreland basin sequence which is tentatively named a modified foreland basin stratigraphic sequence. Some unique aspects of this foreland basin sequence are that deposition can occur above sea level even in the basinal area, and interfingering of sediments from opposing sources located on either side of the basinal axis can occur.

SEQUENCE STRATIGRAPHIC MODEL FOR SIMPSON GROUP OF THE SOUTHERN MID-CONTINENT

Magell P. Candelaria[1]
Core Laboratories
Carrollton, TX 74006

C. Robertson Handford[2]
Consulting Geologist
Englewood, CO 80111

ABSTRACT

Sequence stratigraphic interpretation of Middle Ordovician Simpson Group of the southern Mid-Continent indicates significant, untapped reservoir potential remains. This reservoir potential exists within stratigraphic traps that lie off-structure, or may be too thin to be seismically resolved. Stratigraphic traps can be predicted and exploited through application of fundamental sequence model principles. Historically, exploration for Simpson sandstone reservoirs in Oklahoma has been structurally driven. As a consequence, there has been little exploratory drilling for stratigraphic traps that may exist in sparsely drilled areas off-structure. A sequence stratigraphic interpretation has led to development of a revised depositional model for strata of the Oil Creek through Bromide formations of the Simpson Group, in the southern Mid-Continent. Through this revised depositional model, widely recognized but poorly understood stratal geometries and discontinuities of the Simpson Group strata are clarified and placed within a basinwide stratigraphic framework. Within this sequence framework, each of the Oil Creek, McLish, and Bromide formations is interpreted as a third-order stratigraphic sequence. Each of the lowermost sandstones of the Oil Creek, McLish, and Bromide formations is interpreted as an areally restricted, basal lowstand to widespread early transgressive unit of a third-order stratigraphic sequence. The overlying interbedded marine shale, shallow marine carbonates, and minor sandstone are interpreted as late transgressive to highstand deposition within each sequence. Sequence boundaries may be conformable in the southern Oklahoma region, but in more updip regions, sequence boundaries are demonstrably erosional.

Stratigraphic trap prediction by the described sequence model runs counter to widely held dogma

[1] *Formerly Consulting Geologist, Midland, Texas.*

[2] *Formerly Arco Exploration and Production Technology Company, Plano, Texas.*

that Simpson sandstones produce only from closed structural highs. This sequence interpretation facilitates prediction of stratal discontinuities in stratigraphic trap geometry, and recognition of internal reservoir heterogeneity which can lead to discovery of significant new field reserves and recovery of additional reserves from bypassed pay in existing fields.

INTRODUCTION

Exploration for oil and gas fields in Oklahoma since the turn of the century has been largely focused on the search for structurally trapped reservoirs. Resolution of the stratigraphic component of structurally trapped reservoirs has been a low priority as a result of this exploration strategy. The common occurrence of structurally trapped hydrocarbons in Oklahoma has resulted in a paucity of published studies regarding stratigraphic trap relationships between structures or along the flanks of paleostructures. The purpose of this study is to establish a regional sequence depositional framework that addresses the potential, and defines the basis for a correlation rationale for identification of stratigraphic traps in the Simpson Group. This study identifies the hitherto unrecognized potential stratigraphic trap opportunities in the sparsely drilled, off-structure areas of southern Oklahoma.

Lithofacies description and depositional environment interpretations of the Simpson Group are numerous (Decker and Merritt, 1931; Schramm, 1964; Ireland, 1965; Lewis, 1978; McPherson et al., 1988), yet published literature contains few studies focused on the description of stratal geometries and their significance with respect to stratigraphic trap potential. A few studies (Dapples, 1955; Ireland, 1965; Longman, 1981; Johnson, 1991) have made reference to stratal geometries and the likelihood that Simpson sandstones, so well developed in the Ardmore basin of southern Oklahoma, are transgressive units. However, these studies have not emphasized, nor have they integrated, the described stratal relationships into an internally consistent regional sequence stratigraphic framework, as was proposed by Candelaria et al. (1994).

Working from a sequence stratigraphic perspective, Candelaria et al. (1994) proposed that stratigraphic traps may have been set up by deposition of three distinct, genetically related but temporally separated depositional systems. Each depositional system is characterized by its own set of depositional environments, stratal geometries, and lateral continuity which are a product of the interplay of variance in extant basin configuration, relative sea-level change, and sediment supply. The three depositional systems consist of (1) lowstand sandstone deposition along the margin of the southern Oklahoma aulacogen; (2) transgressive (retrogradational) shoreface sandstones which are both temporally and spatially separated from one another; and (3) aggradational carbonate shoal facies and progradational carbonate banks. Stratigraphic architecture analogous to the three depositional systems described here is likely more commonplace than is presently recognized or appreciated in the Simpson Group. Reinterpretation of stratal architecture, by use of a well log based sequence stratigraphic exploration strategy, will facilitate prediction of subtle stratigraphic traps that escape seismic detection and may lead to an era of renewed exploration for stratigraphic traps in the Simpson Group (Candelaria et al., 1994).

REGIONAL STRATIGRAPHY

Introduction

A comprehensive review of previous regional stratigraphic studies of the Simpson Group is not the purpose of this study; however, a brief review will establish the basis for the model proposed by Candelaria et al. (1994) which is elaborated on in this study. An understanding of the subsidence history of the southern Oklahoma region during Middle Ordovician time as it relates to sediment geometry and distribution is of critical importance to establishment of a sequence stratigraphic depositional model of Simpson Group strata.

The name Simpson was applied by Taff (1902) to a Middle Ordovician sequence of interbedded quartz arenites, skeletal limestones, and gray-green marine shale exposed near the town site of Simpson, presently the location of Pontotoc, in Johnston County, Oklahoma. Ulrich subdivided the Simpson sequence into formations, and later Ulrich (1929) further refined the formation nomenclature and stratigraphic boundaries. Decker and Merritt (1931) divided the succession into the five presently recognized formations of Joins, Oil Creek, McLish, Tulip Creek, and Bromide, oldest to youngest, respectively.

The fivefold subdivision of the Simpson is accepted by most regional stratigraphers. However, due to local variations in sandstone thickness and distribution, several previous studies (Cram, 1948; Ham, 1955; Gahring, 1959; Statler, 1965) have included the Tulip Creek interval with the Bromide Formation. This correlation is based on difficulty in identifying "Tulip Creek Formation" distinct from the Bromide Formation in regions outside the western southern flank of the Arbuckle Mountains, where Decker and Merritt (1931) defined Tulip Creek as a formation of the Simpson Group (Fig. 1). Cram (1948), Ham (1955), Gahring (1959), and Statler (1965) do not recognize Tulip Creek at formation status. Their correlations are based on the interpretation that Tulip Creek is actually a facies of the Bromide Formation in regions outside the type locality, as defined by Decker and Merritt (1931). In light of the ambiguity regarding the formation status of the Tulip Creek interval, the sequence model described below follows the usage of Cram (1948), Ham (1955), Gahring (1959), and Statler (1965), and omits the Tulip Creek Formation from discussions regarding Simpson Group stratigraphy.

The Middle Ordovician Simpson Group depositional basin was the remnant of the southern Oklahoma aulacogen (Schatski, 1946; Wickham, 1978). The aulacogen formed in the Cambrian Period in the region of what now comprises the Ardmore basin/Criner Hills/Marietta basin structural complex. During the Late Cambrian to Early Ordovician, regional subsidence of the aulacogen roughly equaled the sedimentation rate allowing a great thickness of Arbuckle Group shallow water carbonates to accumulate. Although the basin may not have been very deep at any one time, the subsidence rate permitted the Arbuckle Group to accumulate in excess of 6500 ft in thickness (Ireland, 1965; Ham, 1973). In the Middle Ordovician, however, differential basin subsidence permitted 2300 ft of Simpson Group sediments to accumulate in the basin relative to barely 1000 ft on the adjoining craton (Ham, 1973). Further northward, along the state boundary between Oklahoma and Kansas, the Simpson Group is only a few hundred feet in thickness, consisting of upper Simpson Group strata only (Ireland, 1965). Thus, along a basin-to-shelf traverse, the Simpson stratigraphic section exhibits dramatic southwest to northeast stratigraphic thinning as it onlaps the craton. This stratal onlap supports the interpretation of Dapples (1955) and Schramm (1964) that siliciclastic sediment transport to the basin was largely derived from cratonic sources to the north or northeast, probably from multicyclic, mature sandstones on the craton (Ireland, 1965). Differential relative subsidence between the cratonic platform and the basin was particular-

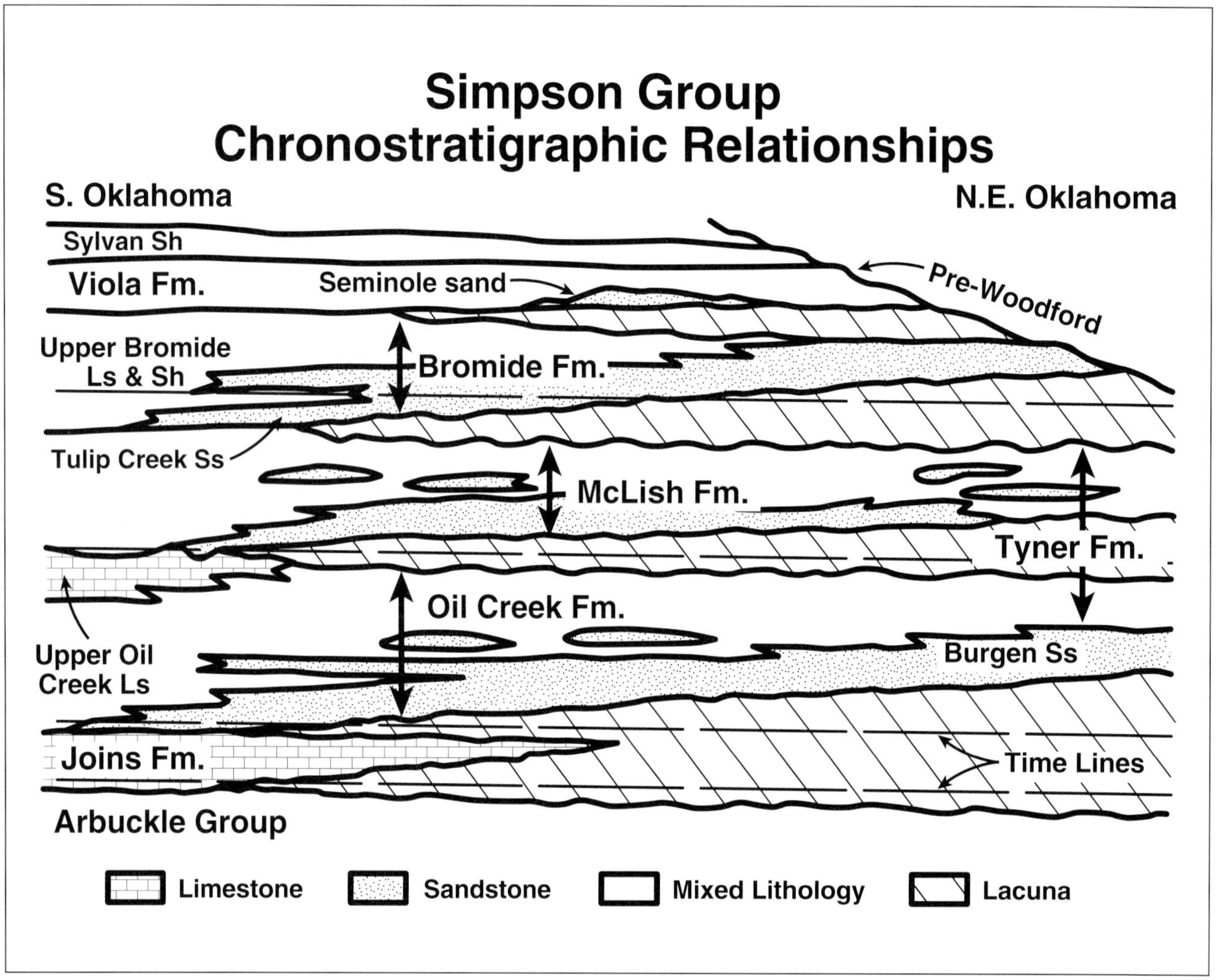

FIGURE 1 — Chronostratigraphic diagram illustrating the postulated stratigraphic relationships of Statler (1965). Note the time-transgressive nature of the Simpson Group lithostratigraphic units in addition to the isolated sand bodies as depicted by Statler. Within the proposed sequence stratigraphic interpretation, the isolated sandstones are interpreted as stranded shoreface complexes which are potential stratigraphic traps.

ly influential to sandstone distribution and stratal geometry during early Simpson deposition, as will be subsequently discussed.

Stratigraphy

Joins Formation

The stratigraphically lowest formation of the Simpson Group, the Joins Formation, overlies in apparent conformity, peritidal strata of the Arbuckle Group (Derby et al., 1991). In contrast to the other formations of the Simpson Group, the Joins Formation is more areally restricted and lacks a basal sandstone interval. The Joins predominantly consists of platform type limestones interbedded with minor shale (Ulrich, 1929) including a thin, discontinuous, basal, edgewise limestone pebble conglomerate (Schramm, 1965; Statler, 1965). The edgewise conglomerate may be indicative of a ravinement or transgressive surface developed at the Arbuckle-Joins contact. The regional significance of the edgewise pebble conglomerate remains unclear. The Joins does contain thin sandstone interbeds locally (Schramm, 1965; Statler, 1965; O'Brien and Derby, 1994), but none is developed to the extent as those occurring in the overlying Oil Creek, or younger Simpson Group formations. The more clastic-rich formations of the Simpson Group are the subject of the sequence stratigraphic interpretation described in this study.

Post-Joins Strata

As described by Decker and Merritt (1931), all the formations of the Simpson Group above the Joins, Oil Creek, McLish, and Bromide formations, oldest to youngest, respectively, are characterized by a prominent, basal quartz sandstone. The basal sandstone for each formation varies regionally in thickness, as well as on a local scale. Regionally, each basal sandstone thickens basinward, but locally the sandstones also demonstrate moderate variability in thickness between localities separated by as little as a few miles in distance (Ireland, 1965). Above the basal sandstone in each formation lies a sequence of interbedded green shale, shallow marine limestones, and minor sandstone. Each formation is capped by a carbonate-dominated succession of strata characteristic of shallow marine to peritidal depositional environments. In general, total siliciclastic content displays an antipathetic relationship to carbonate content in post-Joins strata. Siliciclastic content decreases upward in each formation, whereas the carbonate content increases upward. Ireland (1965) noted the predominance of carbonate lithologies over siliciclastic lithologies in the Bromide Formation versus formations lower in the stratigraphic section. He interpreted this occurrence as indicative of increasing tectonic stability through Simpson depositional time, during which clastic influx became less frequent and less voluminous. Ireland (1965) proposed that carbonate depositional environments became increasingly more widespread and prevalent from Oil Creek through Bromide depositional time.

Oil Creek Formation

The Oil Creek Formation attains a maximum thickness of over 800 ft in the present-day Ardmore basin where it consists of a thick basal sandstone overlain by a succession of shale, minor sandstone, and limestone, overlain by a thicker more continuous limestone with local fenestral pores. The Oil Creek thins by onlap to the north and east, although it maintains the prominent basal sandstone throughout its extent in Oklahoma. The basal Oil Creek Sandstone attains a maximum thickness of approximately 350 ft along the flank of the southern Oklahoma aulacogen (within the present Ardmore basin). Its maximum thickness is restricted to the northeast margin of the basin as it pinches out by basinward downlap to the west and southwest (Schramm, 1964). Basal Oil Creek Sandstone is anomalous in that at maximum thickness it is two to three times thicker than the maximum thickness of basal sandstones of the other Simpson Group formations. Sands derived from pre-existing, mature sandstones further northward on the craton were likely reworked and transported basinward across the cratonic shelf during Oil Creek deposition. Depositional facies of the basal Oil Creek Sandstone are interpreted as of fairly shallow marine origin. McPherson et al. (1988) interpreted sedimentary structures within the basal sandstone as indicative of shoreface environments.

Overlying the basal sandstone is a succession of interbedded greenish to dark gray marine shale, limestone, and minor sandstone. Many of the sandstones have significant lateral continuity and serve as regional subsurface correlation markers. The upper Oil Creek is dominated by a widespread shallow marine limestone interval, locally characterized by fenestral (bird's-eye) pores. Fenestral pores are characteristic of shallow marine environments, typically forming in tidal flat to supratidal flat environments (Shinn, 1968).

Oil Creek deposition ceased with development of a major exposure surface (unconformity) across the cratonic shelf (Derby et al., 1991). This unconformity marks the break between the Sauk and Tippecanoe cratonic sequences of Sloss (1963), and is the most significant exposure surface identified in the Simpson Group. During this period of exposure, mature sand (quartzarenite) was transported across the exposed shelf and accumulated as the basal sandstone of the overlying McLish Formation.

McLish Formation

Strata of the McLish Formation are much like the Oil Creek Formation in that they consist of a basal sandstone overlain by an interbedded succession of greenish marine shale, limestone, and minor sandstone. The uppermost McLish strata are dominated by a regionally thick, upper carbonate interval. The basal deposit of the McLish unconformably overlies the Oil Creek and consists of a texturally and mineralogically mature sandstone, much like the basal Oil Creek Sandstone. It does differ from the basal Oil Creek Sandstone in that at its maximum thickness, approximately 100 ft, the basal McLish Sandstone is roughly one-third of the thickness of the basal Oil Creek Sandstone. The basal McLish Sandstone onlaps the craton as it thins south to north, and is overlain by an interval of interbedded shales, thin sandstones, and minor limestones. Upper McLish strata are dominated by a mixed skeletal to nonskeletal limestone characterized by fenestral (bird's-eye) pores. The stratigraphically and geographically widespread occurrence of fenestrae in the McLish Limestone (bird's-eye) indicates much of this lithostratigraphic unit was deposited in peritidal marine environments. Widespread tidal flat-dominated environments were not uncommon along the southern cratonic margin during Ordovician time (Wilson et al., 1992). Widespread peritidal environments likely persisted in the region of the southern Oklahoma aulacogen well into Middle Ordovician (Simpson) depositional time.

Bromide Formation

McLish Limestone is abruptly overlain in apparent conformity by the basal sandstone of the Bromide Formation. The contact between the basal Bromide Sandstone and subjacent McLish Limestone is not demonstrably erosional as no biostratigraphic evidence has been found which substantiates this contact as disconformable (Derby et al., 1991). The Bromide is characterized by a basal sandstone similar to the basal McLish Sandstone in thickness and distribution. It, too, thins by shelfward onlap and is not appreciably thicker in the Ardmore basin than further shelfward. The Bromide is characterized by the presence of more sandstone intervals low in the stratigraphic succession, which demonstrate greater lateral continuity than those present in the McLish. These sandstones are interbedded with marine shale and exhibit evidence of a moderately shallow marine origin (Longman, 1981). Although characterized by multiple sandstones low in the stratigraphic section, the upper Bromide is dominated by a thick limestone unit (Bromide "dense" of informal subsurface terminology) containing skeletal grains of relatively shallow marine origin.

The Bromide is overlain by a thin shale which separates the Simpson Group from basal strata of the overlying Viola Limestone. The Viola is interpreted as a basin slope carbonate (Galvin, 1983) which implies a deepening of the environment following Bromide Formation deposition. This transgression in the southern Mid-Continent region brought Simpson Group deposition to a close (Ham, 1973).

Sequence Stratigraphy

Chronostratigraphy

The chronostratigraphic framework of the Simpson Group (Fig. 1) clearly illustrates the time-transgressive nature of lithostratigraphic units of the Simpson Group previously interpreted by numerous workers (Dapples, 1955; Ireland, 1965; Schramm, 1964, 1965; Statler, 1965; Longman, 1981; Johnson, 1991). Recognition of the time-transgressive nature of the lithostratigraphic units of the Simpson Group is critical to development of a robust sequence stratigraphic interpretation of Simpson Group strata and their geometric relations. Statler (1965) clearly illustrated these chronostratigraphic relationships in his work of 30 years ago. When viewed from a sequence stratigraphic perspective, Statler's work acquires new significance in terms of providing a regional framework for stratal geometries that are consistent with stratal geometries predicted by basic sequence stratigraphic principles. Description of these interpreted stratal geometries and their integration into a sequence stratigraphic model are presented in the following.

Each clastic-bearing formation of the Simpson Group (Oil Creek, McLish, and Bromide) is interpreted as a relatively conformable succession of genetically related strata bounded by unconformities, or their correlative conformable surfaces. As such, each formation is interpreted as a stratigraphic sequence, as defined by Mitchum (1977). The prominent basal sandstone unit of each formation is interpreted as local lowstand to widespread transgressive deposits of their respective third-order sequences. Vail et al. (1977) defined a third-order sequence as ranging from 1 to 5 m.y. in duration, from sea-level fall to the next sea-level fall.

Oil Creek Sequence

Prior to Oil Creek deposition, the southern cratonic shelf was subaerially exposed. The Oil Creek disconformably overlies Arbuckle Group strata in northern and eastern Oklahoma, although in the basin and surrounding region, the Joins Limestone overlies the Arbuckle, and is below the Oil Creek. The contact between the Oil Creek and Joins is not clearly erosional, and biostratigraphic evidence does not support a disconformity at this contact in southern Oklahoma (Derby et al., 1991). The surface over which the basal Oil Creek Sandstone lies is erosional in updip regions, but nonerosional toward the basin. Sands derived from further northward on the craton were transported, via eolian and possibly fluvial processes, across the emergent shelf to the basin margin during this period of lowstand exposure. This exposure surface is abruptly overlain by basal Oil Creek clastics which thicken dramatically into the proximal basin region and pinch out distally to the southwest. Basal Oil Creek clastics thin to the north and east as they onlap the contemporaneous shelf. Depositional facies described for the basal Oil Creek Sandstone are interpreted as of fairly shallow water origin, probably indicative of shoreface environments (McPherson et al., 1988). Based on sandstone distribution along the basin margin of the southern Oklahoma aulacogen, and basinward downlapping geometry, the basal Oil Creek Sandstone is interpreted as lowstand systems tract (LST) deposition (Fig. 2).

A basin margin accumulation within the LST occurs when base level drops to, but not below, the depositional shelfbreak. At this position, sediments transported to the basin will accumulate on the slope of the basin margin. The basal Oil Creek Sandstone is interpreted as a basin margin wedge deposited at a lowstand of sea level (lowstand wedge) during the final stages of the filling of the southern Oklahoma aulacogen. Significant water depth in the aulacogen at the time of Oil Creek lowstand deposition is not required during lowstand wedge accumulation, nor is it supported by the sedimentological evidence. On the contrary, the sedimentological data support fairly shallow marine deposi-

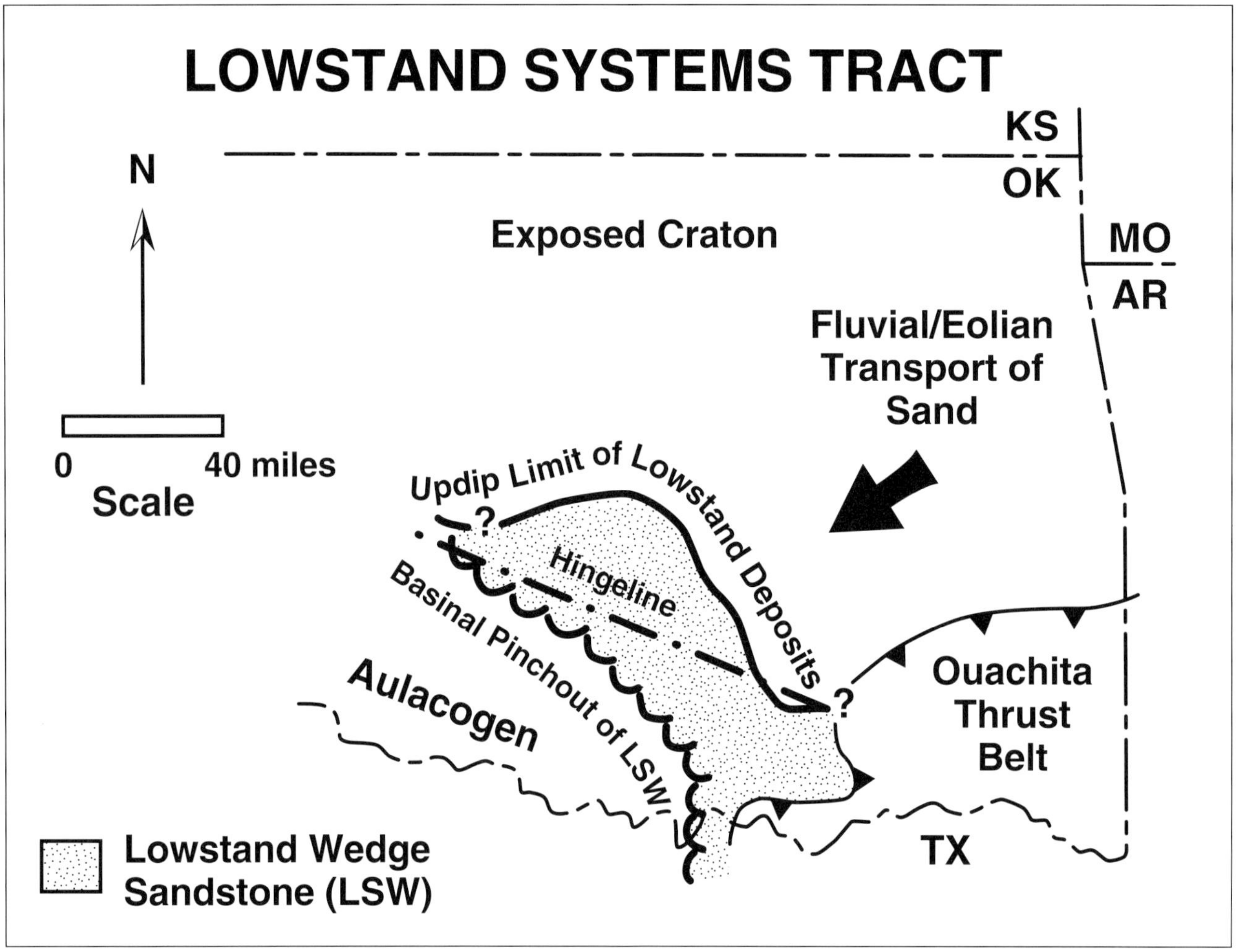

FIGURE 2 — Schematic diagram showing location of LST deposits of the basal Oil Creek Sandstone along the northeast margin of the former southern Oklahoma aulacogen. The thick basal Oil Creek Sandstone is interpreted as a lowstand wedge (lsw) accumulation of stacked shoreface sandstones.

tional environment at this time. The mechanism by which shoreface environments were maintained for the duration of Oil Creek basal sandstone deposition is explained by a sedimentation rate equal to the subsidence rate along the aulacogen basin margin, thus allowing the accumulation of some 350 ft of shallow marine clastics.

Thinner Oil Creek sands interbedded with marine shales on the shelf, further to the north, are interpreted as widespread transgressive systems tract (TST) accumulations, as suggested by Johnson (1991). These younger sands higher in the Oil Creek section exhibit depositional features indicative of shallow depositional environments (McPherson et al., 1988; Forgotson et al., *in press*). During third-order sea-level rise, the former lowstand shoreline was transgressed. At times of minor stillstand during overall transgression, shoreface complexes were established on the shelf. As the sea level continued to rise, the shoreline retrograded, stranding a series of discontinuous sand bodies along the cratonic shelf, as inferred by Statler (1965). These sandstones are interpreted as backstepping or retrogradational shoreface complexes (Fig. 3) which onlap the shelf. The basal contact between the transgressive sandstone and the shelf is both nonerosional in proximal basin localities, and erosional in more updip regions. This surface on which the shoreface complexes were established is both the transgressive surface and the third-order sequence boundary. Insufficient accommodation space

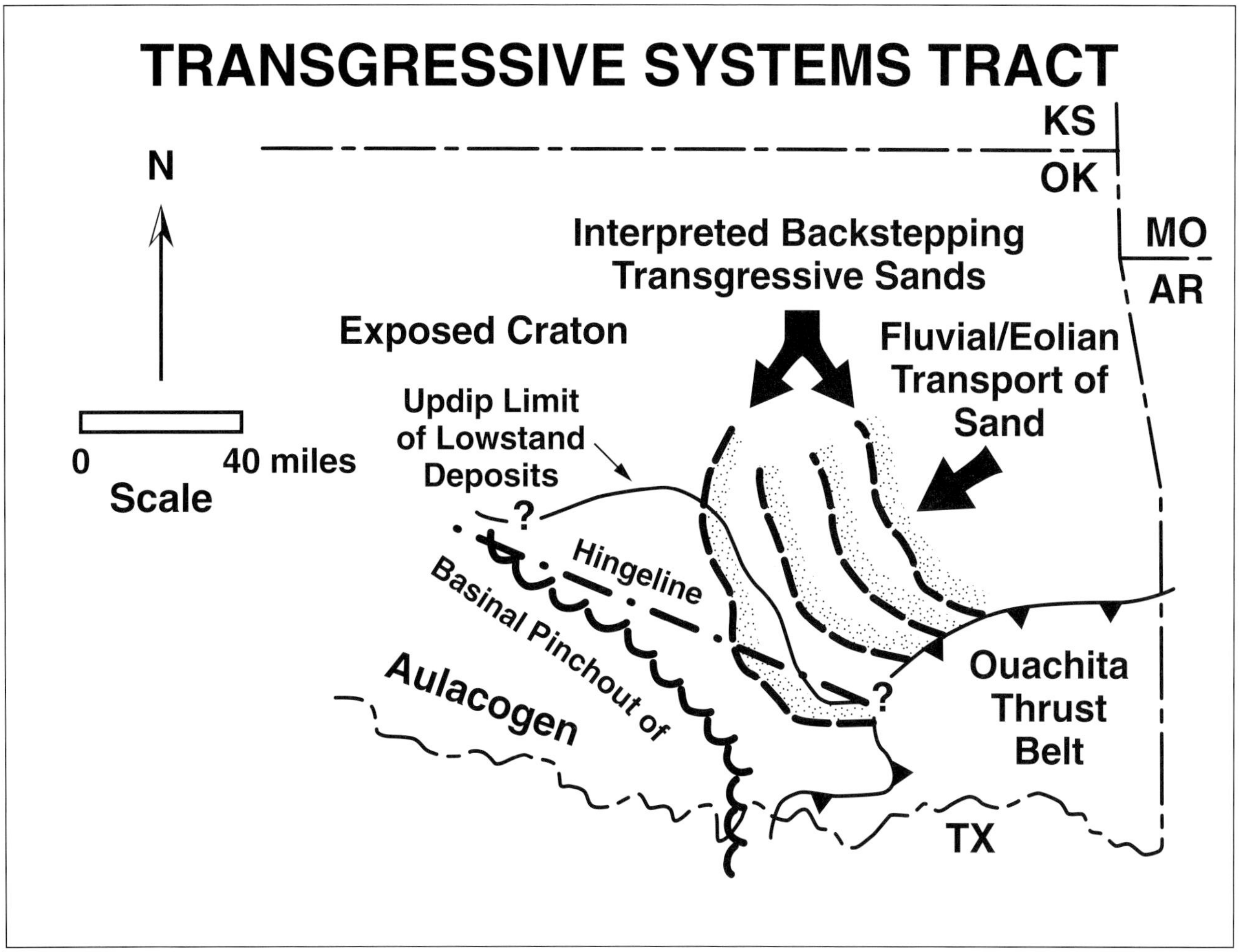

FIGURE 3 — Schematic illustration showing TST deposits consisting of retrogradational shoreface complexes which account for the widespread nature of the basal sandstone of the Oil Creek Formation on the cratonic shelf. A similar depositional environment is inferred for the basal sandstones of the McLish and Bromide formations. Based on sandstone thickness, geometry, and distribution as described in the text, basal McLish and basal Bromide sandstones are interpreted as TST deposits. No lowstand accumulations analogous to the basal Oil Creek Sandstone are known from the McLish or Bromide formations.

existed on the cratonic shelf to permit the accumulation of lowstand deposits, hence the TST overlies the subjacent sequence boundary. In the Oil Creek, the interbedded sandstone and shale deposited on the shelf are interpreted as TST deposits consisting of shoreface sandstones and offshore shale and mudstones.

A dark, organic-rich Oil Creek Shale identifiable as far northward on the craton as northeast Kansas (Watney et al., *in press*) marks the maximum transgression of the Oil Creek TST. Above the maximum transgressive or maximum flooding surface lies the carbonate-dominated upper Oil Creek stratigraphic section. The upper Oil Creek Limestone is indicative of the establishment of a shallow marine environment conducive to widespread carbonate deposition across the southern craton. In the sequence model, this carbonate environment is interpreted as the highstand systems tract (HST). The upper Oil Creek carbonate interval consists of skeletal lime wackestones, packstones, and grainstones ranging from subtidal skeletal-rich facies to supratidal algal-dominated facies (Ham, 1973). Highstand carbonates are locally interbedded with thin sandstone intervals of probable shoreface origin.

McLish and Bromide Sequences

This process of regional exposure and transgression, repeated several times throughout Simpson Group deposition, is interpreted as third-order depositional sequence cyclicity. As suggested by Ireland (1965), this inferred cyclicity was probably punctuated by many higher order, smaller scale cyclical events, as well. Third-order cyclicity is inferred to have been related to episodic changes in base level, which are interpreted as relative sea-level variations on the order of a few million years. Through this process of repeated third-order transgression and regression, each depositional sequence (formation) of the Simpson Group was deposited.

On this basis, a sequence model similar to that described for the Oil Creek is invoked to explain stratal relationships within the overlying McLish and Bromide formations. A significant difference is that during early McLish depositional time, subsidence of the former aulacogen slowed sufficiently (inadequate accommodation space), or sediment supply was less voluminous, to preclude accumulation of thick lowstand clastics along the basin margin. This difference with respect to contemporaneous basin history is significant with respect to sequence interpretation of McLish and younger strata, as well as to sandstone geometry and distribution. As determined from stratal thicknesses and distribution, no comparable siliciclastic lowstand wedge (lsw) was deposited during either McLish or Bromide depositional episodes. Within the proposed model, the basal sands of the McLish and Bromide formations consist largely of shoreface complexes deposited during transgressive episodes, as opposed to regressive deposition, as Longman (1981) suggested for the lower Bromide sandstones. At their maximum development, these basal sandstones of the McLish and Bromide formations are considerably thinner, as transgressive deposits, than the basal Oil Creek lsw sandstone (Fig. 4). The basal McLish and Bromide sandstones on the cratonic shelf are interpreted as retrogradational shoreface deposits of the TST, deposited in the same manner as TST sandstones of the Oil Creek sequence (Fig. 3). As TST deposits, the basal McLish and Bromide sandstones are progressively younger, as they thin northward, by onlap.

In the TST, a subtle maximum flooding surface (mfs) is inferred to be present within the light-colored shale and/or muddy carbonates of both the McLish and Bromide sequences. The mfs is not always characterized by dark, organic-rich marine shale, particularly in the more updip localities within well-aerated, shallow water environments. The mfs in the McLish and Bromide sequences is not currently based on physical evidence, but careful examination of well cuttings and core may provide evidence of this surface predicted by the model. As previously stated, its presence in shallow water environments may be difficult to unequivocally identify. However, seismic data may be useful in identifying the mfs in areas on the cratonic shelf. Seismic signatures may exhibit downlap onto a common surface within the TST of the McLish and Bromide sequences. This downlap surface identifies the mfs of the TST (Mitchum, 1977). Above the mfs in each sequence, deposits consisting of predominantly shallow marine skeletal carbonates, interbedded with marine shale and minor siliciclastics, are interpreted to downlap on to the mfs. The carbonate-dominated successions of strata, characteristic of the upper McLish and upper Bromide sequences, are interpreted as HST deposits, which downlap on to the mfs, analogous to the interpretation of the underlying upper Oil Creek strata.

The presence of an erosional third-order sequence boundary separating the McLish HST from the overlying TST of the Bromide sequence has not been biostratigraphically demonstrated. However, a conformable boundary is in accordance with the definition of a sequence boundary, as explained below; thus, a sequence boundary separating the McLish from the overlying Bromide sequence is interpreted in this study.

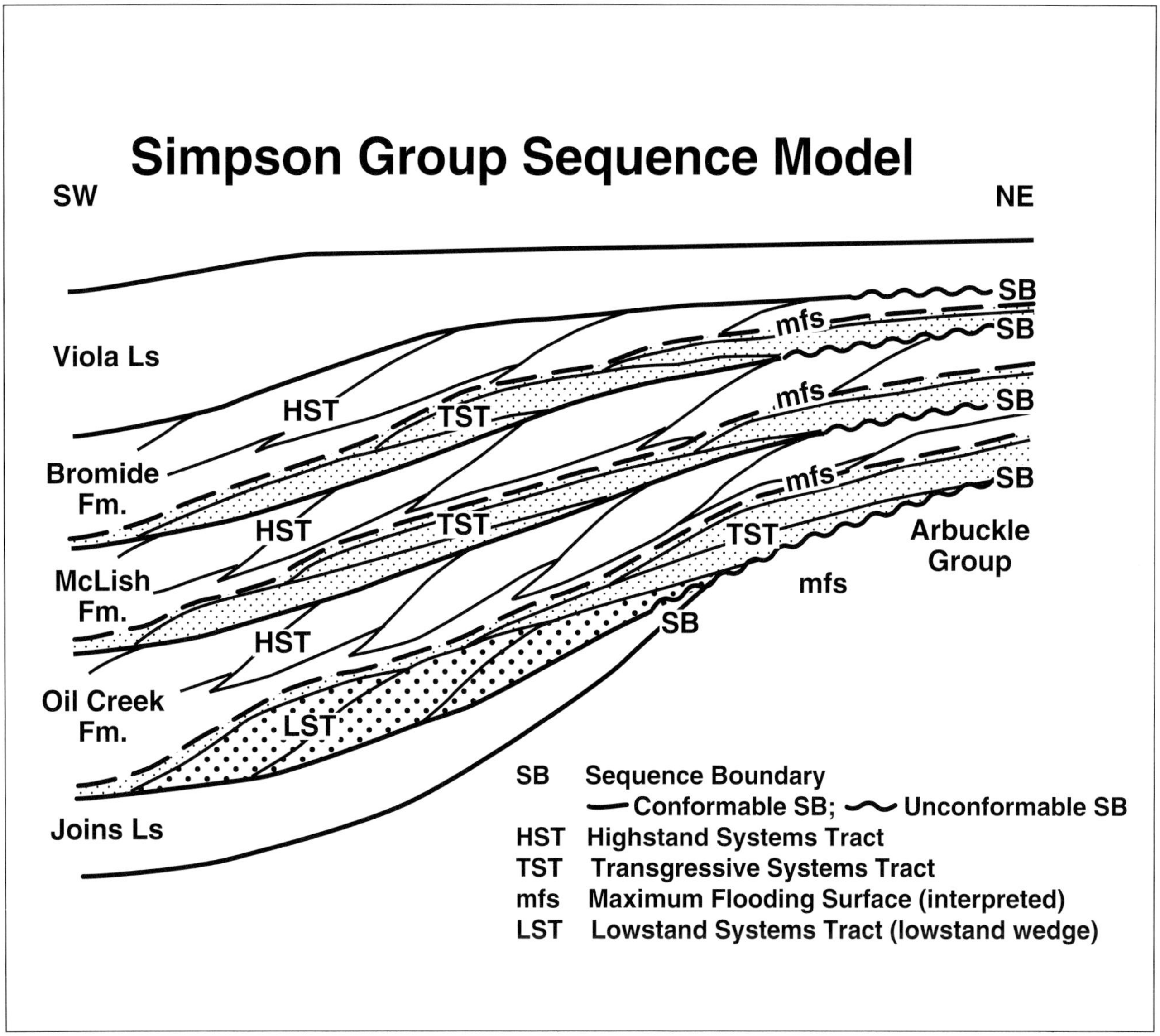

FIGURE 4 — Sequence stratigraphic framework of the Oil Creek through Bromide formations of the Simpson Group. The basal Oil Creek Sandstone represents an lsw accumulation of stacked shoreface sands. Backstepping cratonward are shoreface sands reworked as thinner TST deposits of the Oil Creek. Retrograding transgressive sandstones of the McLish and Bromide formations are also shown. On the cratonic shelf, these TST sandstones overlie third-order sequence boundaries. Each TST is overlain by an mfs which is the downlap surface onto which carbonate banks and minor shoreface sandstones prograded during the third-order, sea-level highstand for each of the Oil Creek, McLish, and Bromide sequences of the Simpson Group.

Sequence Boundaries

The credibility of the sequence interpretation originally proposed by Candelaria et al. (1994), and described above, relies heavily on adherence to the sequence stratigraphic model as it was defined (Mitchum et al., 1977; Van Wagoner et al., 1988). One of the critical issues this proposed model for the Simpson Group must address is the presence or absence of sequence boundaries between interpreted third-order sequences. Additionally, a surface identified as (or inferred to be) a sequence boundary must be consistent with the definition of such, as established by Mitchum et al. (1977), and

Van Wagoner et al. (1988). Without substantiation of all components of the basic sequence model, this sequence interpretation of the Simpson Group has no basis nor geologic significance.

A sequence boundary, by definition, is required between successive third-order sequences. As established by Mitchum et al. (1977) and Van Wagoner et al. (1988), a sequence boundary separates all strata below the boundary from all strata above the boundary, and is an unconformity across which stratal discordance is evident, and may laterally extend into a correlative conformable surface. Hence, evidence of unconformity or angular discordance between each interpreted third-order sequence of the Simpson Group is not a requirement at all localities. As a result, a third-order sequence boundary is interpreted between each formation of the Simpson Group. In addition, evidence of unconformity will not be readily apparent, and may not be discernible in a well log based interpretation of sequence stratigraphic relations. Recent studies of the correlative stratigraphic section elsewhere on the craton, e.g., portions of Kansas (Watney et al., *in press*) and Arkansas (O'Brien and Derby, *in press*) demonstrate evidence of disconformity between formations of the Simpson Group or coeval strata. Thus, unconformable Simpson Group formation boundaries have been documented on the craton with apparently conformable, correlative surfaces in more basinward areas in southern Oklahoma. These regional observations support the contention that formation boundaries in the Simpson Group are sequence boundaries, without violation of fundamental sequence stratigraphic definitions.

Implications for Stratigraphic Traps

Sandstones of the Transgressive Systems Tract

The implications for stratigraphic traps which can be predicted through the application of a sequence stratigraphic interpretation of the Simpson Group are great. The majority of the stratigraphic trap potential lies in sandstones of the TST, as these deposits are much more widespread than the areally restricted, LST basin margin wedge of the basal Oil Creek Formation. Transgressive facies of the Oil Creek, McLish, and Bromide formations consist of two distinct sand body geometries: (1) thin, continuous sheets of reworked, well-winnowed basal transgressive sandstone, or (2) multiple discontinuous, lenticular sand bodies, as envisioned by Statler (1965).

Locally, discrete transgressive sandstones in basinward localities thin and amalgamate shelfward (Ireland, 1965). Amalgamation results in one composite sandstone that is virtually indistinguishable from the two or more component sands which lie further basinward. Amalgamation arising from an onlap relationship in the shelfward direction, or downlap in the basinward direction, has been described by Ham (1955), Ireland (1965), and O'Brien and Derby (*in press*) among others. Such relationships have befuddled subsurface correlations since the initial days of the oil industry in Oklahoma, and are indicative of significant stratal discontinuity and reservoir heterogeneity from which stratigraphic traps may be identified.

Within the discontinuous transgressive sandstones, depositional facies have been interpreted as shoreface sands (Dapples, 1955; Ireland, 1965; Schramm, 1965; Lewis, 1978; Longman, 1981; Johnson, 1991). More recent study has identified the presence of upward-shallowing sedimentological features interpreted as a progression from lower shoreface to foreshore environments (Forgotson et al., *in press*). During minor stillstand or short-term, higher frequency relative sea-level variations, siliciclastics were reworked into shoreface complexes. Marine shale, mudstone, or argillaceous limestone was deposited in the offshore environment and further basinward. As transgression continued, marine shale and limestone onlapped and buried stranded retrogradational shoreface complexes. This upward-deepening succession (transgression of deeper water facies over shoreface sands) isolated potential sandstone reservoir facies within relatively impermeable strata. Shoreface sands deposited within a transgressive depositional system are commonly not in stratigraphic continuity (Posamentier et al., 1992). As a result, stratigraphic trap possibilities may exist within porous, discontinuous, transgressive shoreface complexes.

Where these TST shoreface sandstones lie on the flanks of closed structural highs, or are isolated between structures, they can be stratigraphic traps of high reservoir quality sandstone, which to date, remain largely untapped as exploration targets.

Limestones of the Highstand Systems Tract

Deposition of predominantly shallow water carbonate lithologies within the HST probably established prograding carbonate banks and aggradational grainstone shoals. These carbonate facies may be locally porous and productive off-structure, owing to updip porosity pinchout, thereby forming a stratigraphic trap. These facies provide local reservoir units as in the Oil Creek and McLish, where peritidal bird's-eye (fenestral) limestone is oil productive (Ham, 1955; Suhm, *in press*). Fenestral porosity is generally thought to be indicative of intertidal to supratidal environments (Shinn, 1968), thus these reservoirs may represent the upper range of tidal influence in the prograded carbonate bank depositional environment of the HST.

Porous packstone-grainstone shoal facies of the McLish and Bromide may be productive from stratigraphic traps in off-structure areas. These reservoir units may provide additional exploration potential when seismic interpretation (where applicable) and well log correlation techniques, developed within a sequence framework, are employed from which stratigraphic trap geometries can be predicted.

SUMMARY AND CONCLUSIONS

Use of a sequence stratigraphic model assists explanation and clarification of the various Simpson Group sand body geometries and distributions. Moreover, this sequence model assists prediction of potential stratigraphic trap origin and occurrence, as well as reservoir facies geometry. In areas such as Oklahoma where well control is highly concentrated on structural features with little off-structure well control, it becomes critical to successful exploitation of stratigraphic trap plays that the depositional system is correctly identified. Accurate interpretation of the depositional environment is critical to prediction of sand body geometry, continuity, and internal heterogeneity. In stratigraphic trap plays, it is critical that discontinuous reservoirs are not erroneously correlated as coeval lithostratigraphic units, when they may in fact represent distinct depositional sequences (Posamentier et al., 1992). The benefits of accurate knowledge of the depositional system and stratal geometries to secondary recovery efforts, and to field development, are obvious. For stratigraphic plays such as this, close integration of core depositional facies interpretation with a sequence-based depositional model and wireline log petrofacies is necessary to reduce the higher element of risk associated with stratigraphic trap exploration plays.

From this sequence stratigraphic interpretation of the Simpson Group, the following conclusions may be drawn:

- three third-order sequences — Oil Creek, McLish, and Bromide — are interpretable within Simpson Group strata of southern Oklahoma

- the highly productive Simpson Group sandstones are interpreted as both lowstand wedge accumulations at the basin margin, and as retrogradational (transgressive) shoreface complexes on the cratonic shelf

- retrogradational shoreface sands may be discontinuous bodies of high reservoir potential in local stratigraphic trap geometry, and thus constitute significant and yet under-explored potential in southern Oklahoma

- sequence interpretation affords widespread exploration potential in sparsely drilled, off-structure areas, and in areas where seismic data are difficult to acquire or are of limited resolution.

Application of this sequence stratigraphic approach to Simpson Group reservoir prediction may usher in a new era of exploration in the southern Mid-Continent. The proposed sequence model is counter to the widely held belief that Simpson reservoirs are productive only from closed structural highs. Sequence stratigraphic interpretation can explain and delineate internal reservoir heterogeneities, as well as predict stratigraphic trap geometries that lie both on- and off-structure. Stratigraphic traps within the Simpson remain largely unexplored, and lie waiting for aggressive operators willing to apply new interpretation techniques in their search for economic quantities of oil and gas from historically prolific Simpson Group reservoirs.

REFERENCES

Abbott, S.T., and R.M. Carter, 1994, The sequence architecture of mid-Pleistocene (c. 1.1-0.4 Ma) cyclothems from New Zealand: facies development during a period of orbital control on sea-level cyclicity, *in* P.L. de Boer and D.G. Smith, eds., Orbital forcing and cyclic sequences: IAS Special Publication 19, p. 367-394.

Adams, G.I., 1903, Stratigraphy and paleontology of the upper Carboniferous rocks of the Kansas section: USGS Bulletin 211, p. 15-72.

Aden, L.J., 1982, Clay mineralogy and depositional environments of Upper Cherokee (Desmoinesian) mudrocks, eastern Kansas, western Missouri, and northeastern Oklahoma: M.S. thesis, University of Iowa, 126 p.

Adey, W.H., I.G. Macintyre, R. Stuckenrath, and R.F. Dill, 1977, Relict barrier reef system of St. Croix — its implications with respect to late Cenozoic coral reef development in the western Antantic: Proceedings of Third International Coral Reef Symposium, v. 2, p. 15-21.

Adler, F.J., W.M. Caplan, M.F. Carlson, E.D. Goebel, H.T. Henslee, I.C. Hicks, T.G. Larson, M.H. McCracken, M.C. Parker, B. Rascoe, Jr., M.W. Schramm, and J.S. Wells, 1971, Future petroleum provinces of the Mid-Continent, *in* I.H. Cram, ed., Future petroleum provinces of the United States — their geology and potential: AAPG Memoir 15, v. 2, p. 985-1120.

Adlis, D.S., E.L. Grossmar, T.E. Yancey, and R.D. McLerran, 1988, Isotope stratigraphy and paleodepth changes of Pennsylvanian cyclical sedimentary deposits: Palaios, v. 3, p. 487-506.

Algeo, T.J., and B.H. Wilkinson, 1988, Periodicity of mesoscale Phanerozoic sedimentary cycles and the role of Milankovitch orbital modulation: Journal of Geology, v. 96, p. 313-322.

Allen, G.P., 1991, Sedimentary processes and facies in the Gironde estuary: a recent model for macrotidal estuarine systems, *in* D.G. Smith, G.E. Reinson, B.A. Zaitlin, and R.A. Rahmani, eds., Clastic tidal sedimentology: Canadian Society of Petroleum Geologists Memoir 16, p. 29-40.

Allen, G.P., J.C. Salomon, P. Bassoulet, Y. Dupenhoat, and C. Degrandpre, 1980, Effects of tides on mixing and suspended sediment in the Gironde estuary, France, *in* M. Wiley, ed., Estuarine processes, v. II: New York, Academic Press, p. 63-81.

Amsden, T.W., 1967, Chimneyhill Limestone sequence (Silurian) Hunton Group, Oklahoma, revised: AAPG Bulletin, v. 51, p. 942-945.

Amsden, T.W., 1975, Hunton Group (Late Ordovician, Siluran and Early Devonian) in the Anadarko basin of Oklahoma: Oklahoma Geological Survey Bulletin 121, 214 p.

Amsden, T.W., and J.E. Barrick, 1988, Late Ordovician through Early Devonian annotated correlation chart and brachiopod range charts for the southern Midcontinent region, U.S.A., with a discussion of Silurian and Devonian conodont faunas: Oklahoma Geological Survey Bulletin 143, 66 p.

Anderson, K.H., and J.S. Wells, 1968, Forest City basin of Kansas, Nebraska, and Missouri: AAPG Bulletin, v. 52, p. 264-281.

Andresen, M.J., 1961, Geology and petrology of the Trivoli Sandstone in the Illinois basin: Illinois State Geological Survey Circular 316, 31 p.

Archer, A.W., and C.G. Maples, 1984, Pennsylvanian nonmarine trace-fossil assemblages: southwestern Indiana: Journal of Paleontology, v. 58, p. 448-466.

Archer, A.W., 1991, Modeling of tidal rhythmites using modern tidal periodicities and implications for short-term sedimentation rates, *in* E.K. Franseen et al., eds., Sedimentary modeling: computer simulations for improved parameter definition: Kansas Geological Survey Bulletin 233, p. 185-194.

Archer, A.W., E.P. Kvale, and H.R. Johnson, 1991, Analysis of modern equatorial tidal periodicities as a test of information encoded in ancient tidal rhythmites, *in* D.G. Smith, G.E. Reinson, B.A. Zaitlin, and R.A. Rahmani, eds., Clastic tidal sedimentology: Canadian Society of Petroleum Geologists Memoir 16, p. 189-196.

Archer, A.W., and E.P. Kvale, 1993, Origin of gray-shale lithofacies ("clastic wedges") in U.S. Midcontinental coal measures (Pennsylvanian): an alternative explanation, *in* J.C. Cobb and B. Cecil, eds., Modern and ancient coal-forming environments: GSA Special Paper 286, p.181-192.

Archer, A.W., and R.W. West, 1993, Generalized climato-stratigraphic analyses of the exposed Permo-Pennsylvanian section of Kansas, *in* A.W. Archer, H.R. Feldman, and W.P. Lanier, eds., Incised paleovalleys of the Douglas Group in northeastern Kansas: Field Guide and Related Papers: Kansas Geological Survey Open File Report 93-24, p. 19.1-19.12.

Archer, A.W., 1994, Extraction of sedimentological information via computer-based image analyses of gray shales in Carboniferous coal-bearing sections of Indiana and Kansas, USA: Mathematical Geology, v. 236, p. 47-65.

Archer, A.W., H.R. Feldman, E.P. Kvale, and W.P. Lanier, 1994a, Pennsylvanian (Upper Carboniferous) fluvio- to tidal-estuarine coal-bearing systems: delineation of facies transitions based upon physical and biogenic sedimentary structures, *in* J.H. Calder and M.R. Gibling, eds., Palaeogeography, Palaeoclimatology, Palaeoecology, v. 106, p. 171-185.

Archer, A.W., W.P. Lanier, and H.R. Feldman, 1994b, Stratigraphic and depositional history within incised paleovalley fills and related facies, Douglas Group

(Missourian/Virgilian; Upper Carboniferous) of Kansas, U.S.A., *in* R. Boyd, B.A. Zaitlin, and R. Dalrymple, eds., Incised valley systems: origin and sedimentary sequences: SEPM Special Publication 51, p. 175-190.

Archer, A.W., G.J. Kuecher, and E.P. Kvale, 1995, The role of tidal-velocity asymmetries in the deposition of silty tidal rhythmites (Carboniferous, Eastern Interior Coal Basin): Journal of Sedimentary Research, v. A65, no. 2, p. 408-416.

Arthur, M.A., and B.B. Sageman, 1994, Marine black shales: Depositional mechanisms and environments of ancient deposits: Earth and Planetary Science Annual Reviews, v. 22, p. 499-551.

Baars, D.L., 1992, Conjugate basement rift zones in Kansas, Midcontinent, USA, *in* M.J. Ricker, H.J. Harrington, and P.R. Williams, eds., Basement tectonics 9: Boston, Kluwer Academic Publishers, p. 201-210.

Baars, D.L., S.M. Ritter, C.G. Maples, and C.A. Ross, 1994a, Redefinition of the Upper Pennsylvanian Virgilian Series in Kansas: Kansas Geological Survey Bulletin 230, p. 11-16.

Baars, D.L., S.M. Ritter, C.G. Maples, and C.A. Ross, 1994b, Proposed repositioning of the Pennsylvanian-Permian boundary in Kansas: Kansas Geological Survey Bulletin 230, p. 5-10.

Baesemann, J.F., 1973, Missourian (Upper Pennsylvanian) conodonts of northeastern Kansas: Journal of Paleontology, v. 47, p. 689-710.

Ball, D.S., 1985, The Pennsylvanian Haskell-Cass section, a perspective on controls of Midcontinent cyclothem deposition: Unpublished M.S. thesis, University of Kansas, 147 p.

Ball, S.M., 1964, Stratigraphy of the Douglas Group (Pennsylvanian, Virgilian) in the northern Midcontinent region: Unpublished Ph.D. dissertation, University of Kansas, 674 p.

Ball, S.M., 1971, The Westphalia Limestone of the northern Midcontinent: a possible storm deposit: Journal of Sedimentary Petrology, v. 41, p. 217-232.

Bally, A.W., 1987, Atlas of seismic stratigraphy: AAPG Studies in Geology 27, v. 1, 2, and 3, 125 p., 277 p., and 244 p.

Barrett, T.R., 1989, Hierarchal genetic (T-R unit) stratigraphy of the Wreford Limestone Formation (Lower Permian, Gearyan) in northeastern Kansas: Unpublished M.S. thesis, Kansas State University, 251 p.

Barrick, J.E., and D.R. Boardman, 1989, Stratigraphic distribution of morphotypes of *Idiognathodus* and *Streptognathodus* in Missourian-lower Virgilian strata, north-central Texas, *in* D.R. Boardman, J.E. Barrick, J.M. Cocke, and M.K. Nestell, eds., Middle and Late Pennsylvanian chronostratigraphic boundaries in north-central Texas glacial-eustatic events, biostratigraphy, and paleoecology, Part II: Texas Tech University Studies in Geology, p. 167-188.

Bass, N.W., 1936, Origin of the shoestring sands of Greenwood and Butler counties, Kansas: Kansas Geological Survey Bulletin 23, 1235 p.

Bates, R.L., and J.A. Jackson, 1987, Glossary of geology: Alexandria, Virginia, American Geological Institute, 788p.

Baum, G.R., and P.R. Vail, 1988, Sequence stratigraphic concepts applied to Paleogene outcrops, Gulf and Atlantic basins, *in* C.K. Wilgus et al., eds., Sea-level changes: an integrated approach: SEPM Special Publication 42, p. 309-328.

Beaumont, C., 1981, Foreland basins: Geophysics Journal Research Astronomical Society, v. 65, p. 291-329.

Beaumont, C., G. Quinlan, and J. Hamilton, 1987, The Alleghenian Orogeny and its relationship to the evolution of the Eastern Interior, North America, *in* C. Beaumont and A.J. Tankard, eds., Canadian Society of Petroleum Geologists Memoir 12, p. 425-445.

Beaumont, C.G., 1988, Orogeny and stratigraphy — numerical models of the Paleozoic in the Eastern Interior of North America: Tectonics, v. 7, p. 389-416.

Becker, L.E., and J.B. Patton, 1968, World occurrence of petroleum in pre-Silurian rocks: AAPG Bulletin, v. 52, p. 224-245.

Beckman, W.A., and L.L. Sloss, 1966, Possible pre-Springeran unconformity in southern Oklahoma: AAPG Bulletin, v. 50, p. 1342-1364.

Bennison, A.P., 1979, Mobile basin and shelf border area in northeast Oklahoma during Desmoinesian cyclic sedimentation, *in* N.J. Hyne, ed., Pennsylvanian sandstones of the Mid-Continent: Tulsa Geological Society Special Publication 1, p. 283-294.

Bennison, A.P., 1981, Type area of the Seminole and Holdenville formations, *in* R. Dott, Sr., ed., A Guidebook to the type areas of the Holdenville and Seminole formations, western Arkoma basin: AAPG Mid-Continent Regional Meeting, Oklahoma City, Field Trip 2, p. 1-10.

Bennison, A.P., 1984, Shelf to trough correlations of late Desmoinesian and early Missourian carbonate banks and related strata, northeast Oklahoma, N.J. Hyne, ed., Limestones of the Mid-Continent: Tulsa Geological Society Special Publication 2, p. 93-126.

Bennison, A.P., A.D. Buzzalini, and F.N. Murray, 1984, Upper Pennsylvanian source beds of northeastern Oklahoma and adjacent Kansas: Tulsa Geological Society Guidebook, p. 13-45.

Bennison, A.P., 1985, Trough-to-shelf sequence of the early Missourian Skiatook Group, Oklahoma and Kansas, *in* W.L. Watney, R.L. Kaesler, and K.D. Newell, eds., Recent interpretations of Late Paleozoic cyclothems: SEPM Midcontinent Section Third Annual Meeting, p. 219-244.

Benton, J.W., 1971, Subsurface stratigraphic analysis, Morrow (Pennsylvanian), north-central Texas County, Oklahoma: Unpublished M.S. thesis, University of Oklahoma, 60 p.

Berg, O.R., and D.G. Woolverton, 1985, Seismic stratigraphy II: An integrated approach to hydrocarbon exploration: AAPG Memoir 39, 276 p.

Bice, D., 1988, Synthetic stratigraphy of carbonate platform and basin systems: Geology, v. 16, p. 703-706.

Blanchon, P., and J. Shaw, 1995, Reef drowning during the last deglaciation: evidence for catastrophic sea-level rise and ice-sheet collapse: Geology, v. 23, no. 1, p. 4-8.

Boardman, D.R., and R.H. Mapes, 1984, Preliminary findings of the placement of the Desmoinesian-Missourian boundary utilizing ammonoid cephalopods, in Upper Pennsylvanian source beds of northeastern Oklahoma: Tulsa Geological Society Guidebook, p. 54-58.

Boardman, D.R., R.H. Mapes, T.E. Yancey, and J.M. Malinky, 1984, A new model for depth-related allogenic community succession within North American Pennsylvanian cyclothems and implications on the black-shale problem, *in* N.J. Hyne, ed., Limestones of the Mid-Continent, Tulsa Geological Society Special Publication 2, p. 141-182.

Boardman, D.R., and P.H. Heckel, 1989, Glacial-eustatic sea-level curve for early Late Pennsylvanian sequence in north-central Texas and biostratigraphic correlation with curve for Mid-Continent North America: Geology, v. 17, p. 802-805.

Boardman, D.R., and M.K. Nestell, 1993, Glacial-eustatic sea-level fluctuation curve for Carboniferous-Permian boundary strata based on outcrops in the North American Midcontinent and north-central Texas, *in* R.E. Crick, ed., Transactions and abstracts: AAPG Southwest Section Geological Convention, p. 15-25.

Boardman, D.R., II, and M.K. Nestell, 1994, Glacial-eustasy of the upper Carboniferous-Lower Permian boundary strata in the North American Mid-Continent: Carboniferous to Jurassic Pangea: First International Symposium of the Geological Survey of Canada, Ottawa, p. 28.

Boardman, D.R., D.A. Work, R.H. Mapes, and J.E. Barrick, 1994, Biostratigraphy of Middle and Late Desmoinesian-Virgilian ammonoites: Kansas Geological Survey Bulletin 232, 121 p.

Bond, G.C., P.A. Nickelson, and M.A. Kominz, 1984, Breakup of a supercontinent between 625 Ma and 555 Ma: new evidence and implications for continental histories: Earth and Planetary Science Letters, v. 70, p. 325-345.

Bond, G., W. Broecker, S. Johnson, J. McManus, L. Labeyrie, J. Jouzel, and G. Bonami, 1993, Correlations between climate and records from North Atlantic sediments and Greenland ice: Nature, v. 360, p. 245-249.

Bowen, D.W., L.F. Krystinik, and R.E. Grantz, 1990, Geology and reservoir characteristics of the Sorrento-Mt. Pearl field complex, Cheyenne County, Colorado, *in* S.A. Sonnenberg et al., eds., Morrow sandstones of southeast Colorado and adjacent areas: Denver, The Rocky Mountain Association of Geologists, p. 67-77.

Bowen, D.W., P. Weimer, and A.J. Scott, 1993, The relative success of siliciclastic sequence stratigraphic concepts in exploration: examples from incised valley fill and turbidite systems reservoirs, *in* P. Weimer and H.W. Posamentier, eds., Siliciclastic sequence stratigraphy: recent developments: AAPG Memoir 58, p. 15-42.

Bowsher, A.L., and J.M. Jewett, 1943, Coal resources of the Douglas Group in east-central Kansas: Kansas Geological Survey Bulletin 43, 94 p.

Boyd, R., J. Suter, and S. Penland, 1989, Relationship of sequence stratigraphy to modern sedimentary environments: Geology, v. 17, p. 926-929.

Branson, C.C., 1960, Local fossil assemblage of the Seminole Formation: Oklahoma Geology Notes, v. 20, no. 1, p.16-17.

Branson, C.C., and G.G. Huffman, 1965, Geology, oil, and gas of Craig County, Oklahoma, Part 1 — Geology of Craig County: Oklahoma Geological Survey Bulletin 99, 109 p.

Brenner, R.L., 1989, Stratigraphy, petrology, and paleogeography of the upper portions of the Cherokee Group (Middle Pennsylvanian), eastern Kansas and northern Oklahoma: Kansas Geological Survey Geology Series 3, 70 p.

Brenner, R.L., 1994, Progressive changes in siliciclastic depositional styles, middle and upper Pennsylvanian cyclothems, Mid-Continent, U.S.A.: AAPG Abstracts for 1994 Annual Meeting, Denver, Colorado, p. 110.

Brewer, J.A., R. Good, J.E. Oliver, L.D. Brown, and S. Haufman, 1983, COCORP profiling across the southern Oklahoma aulacogen — overthrusting of the Wichita Mountains and compression with the Anadarko basin: Geology, v. 11, p. 109-114.

Bristol, H.M., and R.H. Howard, 1974, Sub-Pennsylvanian valleys in the Chesterian surface of the Illinois basin and related Chesterian slump blocks: GSA Special Paper 148, p. 315-335.

Broecker, W., and J. van Donk, 1970, Insolation changes, ice volumes, and the O^{18} record in deep-sea cores: Rev. Geophys. Space Physics, v. 8, p. 169-198.

Broecker, W.S., and G.H. Denton, 1989, The role of ocean-atmospheric reorganizations in glacial cycles: Geochimica et Cosmochimica Acta, v. 53, p. 2465-2501.

Brondos, M.D., and R.L. Kaesler, 1976, Diversity of assemblages of Late Paleozoic Ostracoda, *in* R.W. Scott and R.R. West, eds., Structure and classification of paleocommunities: Stroudsburg, Pennsylvania, Dowden, Hutchinson and Ross, Inc., p. 213-234.

Brown, L.F., Jr., and W.L. Fisher, 1977, Seismic-stratigraphic interpretation of depositional systems: examples from Brazilian rift and pull-apart basins, *in* C.E. Payton, Seismic stratigraphy — applications to hydrocarbon exploration: AAPG Memoir 26, p. 213-248.

Brown, L.F., Jr., 1979, Deltaic sandstone facies of the Midcontinent, *in* N.J. Hyne, ed., Pennsylvanian sandstones of the Mid-Continent: Tulsa Geological Society Special Publication 1, p. 35-63.

Budd, D.A., A.H. Saller, and P.M. Harris, 1995, Unconformities and porosity in carbonate strata: AAPG Memoir 63, 313 p.

Busch, D.A., 1953, The significance of deltas in subsurface exploration: Tulsa Geological Society Digest, v. 21, p. 71-80.

Busch, D.A., 1971, Genetic units in delta prospecting: AAPG Bulletin, v. 55, p. 1137-1154.

Busch, D.A., 1974, Stratigraphic traps in sandstones — exploration techniques: AAPG Memoir 21, 171 p.

Busch, R.M., and H.B. Rollins, 1984, Correlation of Carboniferous strata using a hierarchy of transgressive-regressive units: Geology, v. 12, p. 471-474.

Busch, D.A., and D.A. Link, 1985, Genetic units of stratigraphy, *in* Exploration methods for sandstone reservoirs: Tulsa, OGCI Publication, p. 13-27.

Busch, R.M., R.J. West, F.J. Barrett, and T.R. Barrett, 1985, Cyclothems versus a hierarchy of transgressive-regressive units, *in* W.L. Watney, R.L. Kaesler, and K.D. Newell, convenors, Recent interpretations of Late Paleozoic cyclothems: SEPM Mid-Continent Section Field Conference Guidebook, p. 141-154.

Busch, R.M., 1988, Minor T-R units in the Lower Permian Chase Group, northeast Kansas: GSA South-Central Section Centennial Field Guide, p. 17-24.

Cade, C.M., III, 1953, The geology of the Marmaton Group of northeastern Nowata and northwestern Craig counties, Oklahoma: Tulsa Geological Society Digest, v. 21, p. 130-148.

Caldwell, C.D., 1991, Cyclic deposition of the Lower Permian, Wolfcampian, Chase Group, Western Guymon-Hugoton field, Texas County, Oklahoma, *in* Midcontinent core workshop integrated studies of petroleum reservoirs in the Midcontinent: AAPG Midcontinent Section Meeting, p. 57-80.

Calver, M.A., 1968, Distribution of Westphalian marine faunas in northern England and adjacent areas: Yorkshire Geological Society Proceedings, v. 37, p. 1-72.

Campbell, J.A., C.J. Mankin, A.B. Schwarzkopf, and J.H.

Raymer, 1988, Habitat of petroleum in Permian rocks of the Midcontinent region, *in* W.A. Morgan and J.A. Babcock, eds., Permian rocks of the Midcontinent: SEPM Midcontinent Section Special Publication 1, p. 13-35.

Candelaria, M.P., C.R. Handford, and C.L. Reed, 1994, Sequence stratigraphic model for Simpson Group of the southern Mid-Continent: the key to a new stratigraphic play (abstract for poster exhibit): AAPG National Convention Program With Abstracts, p. 114-115.

Carlisle, D., 1983, Concentration of uranium and vanadium in calcretes and gypcretes, *in* R.C.L. Wilson, ed., Residual deposits: Geological Society of London Special Publication 11, p. 185-195.

Cathles, L.M., and A. Hallam, 1991, Stress-induced changes in plate density, Vail sequences, epeirogeny, and short-lived global sea-level fluctuations: Tectonics, v. 10, p. 659-671.

Cecil, C.B., 1990, Paleoclimatic controls on stratigraphic repetition of chemical and siliciclastic rocks: Geology, v. 18, p. 533-536.

Chaplin, J.R., 1988, Lithostratigraphy of Lower Permian rocks in Kay County, north-central Oklahoma, and their stratigraphic relationships to lithic correlatives in Kansas and Nebraska, *in* W.A. Morgan and J.A. Babcock, eds., Permian rocks of the Midcontinent: SEPM Midcontinent Section Special Publication 1, p. 79-111.

Chaplin, J.R., 1994, Day 2: shoreline carbonate-to-clastic facies changes of the Chase Group in southernmost Kansas and north-central Oklahoma: Kansas Geological Society 1994 Field Trip Guidebook, Day 2, p. 2-63.

Chenoweth, P.A., 1966, Type section of the Oologah Limestone: Oklahoma Geology Notes, v. 26, no. 7, p. 193-208.

Chowns, T.M., and J.E. Elkins, 1974, The origin of quartz geodes and cauliflower cherts through the silicification of anhydrite nodules: Journal of Sedimentary Petrology, v. 44, p. 885-903.

Clement, W.A., 1977, A case history of geoseismic modeling of basal Morrow-Springer sandstones, Watonga-Chickasha Trend: Geary, Oklahoma,T13N-R10W, *in* C.E. Payton, ed., Seismic stratigraphy — applications to hydrocarbon exploration: AAPG Memoir 26, p. 451-476.

Cole, J.G., 1967, Regional stratigraphy of the Marmaton Group in northeastern Oklahoma: Shale Shaker, v. 17, no. 5, p. 112-123.

Cole, J.G., 1970, Marmaton Group east flank of the Nemaha Ridge: Shale Shaker, v. 21, p. 67-82.

Condra, G.E., and J.E. Upp, 1931, Correlation of the Big Blue Series in Nebraska: Nebraska Geological Survey Bulletin 6, 74 p.

Connolly, W.M., and R.J. Stanton, 1992, Interbasinal cyclostratigraphic correlation of Milankovitch band transgressive-regressive cycles: correlation of Desmoinesian-Missourian strata between southeastern Arizona and the Midcontinent of North America: Geology, v. 20, p. 999-1002.

Cornish, F.G., 1982, Fluvial environments and paleohydrology of the upper Morrow 'A' (Pennsylvanian) meander belt sandstone, Beaver County, Oklahoma: Shale Shaker, v. 11, p. 73-83.

Coveney, R.M., Jr., and S.P. Martin, 1983, Molybdenum and other heavy metals of the Mecca quarry and Logan quarry shales: Economic Geology, v. 78, p. 132-149.

Coveney, R.M., Jr., W.L. Watney, and C.G. Maples, 1989, Rates and durations for accumulation of Pennsylvanian black shales in the Midwestern United States, *in* E.K. Franseen and W.L. Watney, eds., Sedimentary modeling, computer simulation of depositional sequences: Kansas Geological Survey Subsurface Series, v. 12, p. 73-76.

Coveney, R.M., Jr., W.L. Watney, and C.G. Maples, 1991, Contrasting depositional models for Pennsylvanian black shale discerned from molybdenum abundances: Geology, v. 19, p. 147-150.

Cram, I.H., 1948, Cumberland oil field, Bryan and Marshall counties, Oklahoma, *in* J.V. Howell, ed., Structure of typical American oil fields: AAPG, Tulsa, Oklahoma, v. III, p. 341-358.

Creaney, S., and Q.R. Passey, 1993, Recurring patterns of total organic carbon and source rock quality within a sequence stratigraphic framework: AAPG Bulletin, v. 77, p. 386-401.

Cronoble, W.R., and C.J. Mankin, Petrology of the Hogshooter Formation: Oklahoma Geological Survey Bulletin 107, p. 1-148.

Crowell, J.C., 1978, Gondwanan glaciation, cyclothems, continental positioning, and climate change: American Journal of Science, v. 278, p. 1345-1372.

Crowell, J.C., and L.A. Frakes, 1975, The late Paleozoic glaciation, *in* K.S.W. Campbell, ed., Gondwana geology: Canberra, Australia National University Press, p. 313-331.

Crowley, T.J., J.G. Mengel, and D.A. Short, 1987, Gondwanaland's seasonal cycle: Nature, v. 329, p. 803-807.

Crowley, T.J., and S.K. Baum, 1991, Estimating Carboniferous sea-level fluctuations from Gondwana ice extent: Geology, v. 19, p. 975-977.

Cuffey, R.J., 1967, Bryozoan *Tabulipora carbonaria* in Wreford megacyclothem (Lower Permian) of Kansas: University of Kansas Paleontological Contributions, Bryozoa, article 1, 96 p.

Cunningham, C.R., H.R. Feldman, E.K. Franseen, R.A. Gastaldo, G. Mapes, C.G. Maples, and H.P. Schultze, 1994, The Upper Carboniferous (Stephanian) Hamilton Fossil-Lagerstätte (Kansas, U.S.A.): a valley-fill, tidally influenced depositional model: Lethaia, v. 26, p. 225-236.

Cushman, J.A., and J.A. Waters, 1927, Arenaceous Paleozoic Foraminifera from Texas: Cushman Foundation for Foraminiferal Research Contributions, v. 3, pt. 3, no. 46, p. 146-153, plates 26-27.

Cushman, J.A., and J.A. Waters, 1928a, Some foraminifera from the Pennsylvanian and Permian of Texas: Cushman Foundation for Foraminiferal Research Contributions, v. 4, pt. 2, no. 59, p. 31-54, plates 4-7.

Cushman, J.A., and J.A. Waters, 1928b, Additional Cisco Foraminifera from Texas: Cushman Foundation for Foraminiferal Research Contributions, v. 4, pt. 3, no. 61, p. 62-67, plate 8.

Cushman, J.A., and J.A. Waters, 1930, Foraminifera of the Cisco group of Texas (exclusive of the Fusulinidae): Texas University Bulletin 3019, p. 22-81, plates 2-12.

Dalrymple, R.W., Y. Makino, and B.A. Zaitlin, 1991, Temporal and spatial patterns of rhythmite deposition on mud flats in the macrotidal Cobequid Bay — Salmon River estuary, Bay of Fundy, Canada, *in* D.G. Smith, G.E. Reinson, B.A. Zaitlin, and R. A. Rahmani, eds., Clastic tidal sedimentology: Canadian Society of Petroleum Geologists Memoir 16, p. 137-160.

Dalrymple, R.W., B.A. Zaitlin, and R. Boyd, 1992, Estuarine facies models: conceptual basis and stratigraphic implications: Journal of Sedimentary Petrology, v. 62, p. 1130-1146.

Dalrymple, R.W., R. Boyd, and B.A. Zaitlin, 1994, History of research, valley types and internal organization of incised-valley systems: introduction to the volume, *in* R.W. Dalrymple, R. Boyd, and B. Zaitlin, eds., Incised-valley systems: orgin and sedimentary sequences: SEPM Special Publication 51, p. 3-10.

Dapples, E.C., 1955, General lithofacies relationships of St. Peter Sandstone and Simpson Group: AAPG Bulletin, v. 39, p. 444-467.

DeBoer, P.L., 1982, Cyclicity and the storage of organic matter in Middle Cretaceous pelagic sediments, *in* G. Einsele and A. Seilacher, eds., Cyclic and event stratification: New York, Springer-Verlag, p. 456-475.

DeBoer, P.L., 1991, Comment — Pennsylvanian time scales and cycle periods: Geology, v. 19, p. 408-409.

Decker, C.E., and C.A. Merritt, 1931, Stratigraphy and physical characteristics of the Simpson Group: Oklahoma Geological Survey Bulletin 55, 112 p.

Decker, C.E., 1941, Simpson Group of Arbuckle and Wichita Mountains of Oklahoma: AAPG Bulletin, v. 25, p. 650-667.

Defant, A., 1961, Physical oceanography, v. II: New York, Permagon Press, 598 p.

Demaison, G.J., and G.T. Moore, 1980, Anoxic environments and oil source bed genesis: AAPG Bulletin, v. 64, p. 1179-1209.

de Mobray, T., and M.J. Visser, 1984, Reactivation surfaces in subtidal channel deposits, Oosterschelde, southwest Netherlands: Journal of Sedimentary Petrology, v. 54, p. 811-824.

Denesen, S.L., 1985, Stratigraphy, petrography, and depositional environments of the Banzet Formation (Middle Pennsylvanian) in southeastern Kansas and northeastern Oklahoma: M.S. thesis, University of Iowa, 141 p.

Denton, G.H., and T.J. Hughes, 1983, Milankovitch theory of ice ages — hypothesis of ice-sheet linkage between regional insolation and global climate: Quaternary Research, v. 20, 125-144.

Derby, J.R., J.A. Bauer, W.B. Creath, R.I. Dresbach, R.L. Ehrington, J.D. Loch, J.H. Stitt, T.R. McHargue, J.F. Miller, M.A. Miller, J.E. Repetski, W.C. Sweet, J.F. Taylor, and M. Williams, 1991, Biostratigraphy of the Timbered Hills, Arbuckle, and Simpson groups, Cambrian and Ordovician, Oklahoma: a review of correlation tools and techniques available to the explorationist, *in* K.S. Johnson, ed., Late Cambrian-Ordovician geology of the southern Mid-continent: Oklahoma Geological Survey Circular 92, p. 15-41.

Dickinson, W.R., and H. Yarborough, 1979, Plate tectonics and hydrocarbon accumulation: AAPG Continuing Education Case Note Series 1, 150 p.

Dolson, J., D. Muller, M.J. Evetts, and J.A. Stein, 1991, Regional paleotopographic trends and production, Muddy Sandstone (Lower Cretaceous), central and northern Rocky Mountains: AAPG Bulletin, v. 75, p. 409-435.

Donaldson, A.C., and C.F. Eble, 1991, Pennsylvanian coals of central and eastern United States, *in* R.B. Taylor, D.D. Rice, and H.J. Gluskoter, eds., Economic geology, U.S., v. P-2, The geology of North America: Boulder, CO, GSA, p. 523-546.

Donovan, D.T., and E.J.W. Jones, 1979, Causes of world-wide changes in sea level: Journal of the Geological Society of London, v. 136, p. 187-192.

Dorjes, J., and J.D. Howard, 1975, Estuaries of the Georgia coast, U.S.A.: Sedimentology and biology. IV. Fluvial-marine transition indicators in an estuarine environment, Ogeeche River — Ossabaw Sound: Senckenbergia Marit., v. 7, p. 137-179.

Dott, R.H., 1941, Memorial Shale of Pennsylvanian Age, in Oklahoma: AAPG Bulletin, v. 25, no. 8, p. 1591-1597.

Doveton, J.H., 1991, Lithofacies and geochemical facies profiles from nuclear wireline logs: new subsurface templates for sedimentary modeling, *in* E.K. Franseen, W.L. Watney, C.G.St.C. Kendall, and W. Ross, eds., Sedimentary modeling: computer simulations and methods for improved parameter definition: Kansas Geological Survey Bulletin 233, p. 101-110.

Drake, D.A., 1958, Microfauna of the Cisco Series, Brown and Coleman counties, Texas: Unpublished M.S. thesis, University of Texas, 127 p.

Dubois, M.K., 1985, Application of cores in development of an exploration strategy for the Lansing-Kansas City "E" Zone, Hitchcock County, Nebraska: Kansas Geological Survey Subsurface Geology Series 6, p. 120-144.

Dunbar, C.O., and J. Rodgers, 1957, Principles of stratigraphy: New York, John Wiley and Sons, 356 p.

Early, A.K., 1951, Microfauna of the Brownwood Shale, Brown and McCulloch counties, Texas: Unpublished M.S. thesis, University of Texas, 94 p.

Ebanks, E.J., Jr., L.L. Brady, P.H. Heckel, H.A. O'Connor, G.A. Sanderson, R.R. West, and F.W. Wilson, 1979, The Mississippian and Pennsylvanian (Carboniferous) systems in the United States — Kansas: USGS Professional Paper 110-M-DD, p. Q1-Q30.

Ece, O.I., 1984, The Excello Shale of the Mid-Continent: Tulsa Geological Society Field Trip Guidebook, p. 50-53.

Ece, O.I., 1987, Stratigraphy, paleogeography and depositional environments of the upper Middle Carboniferous (Desmoinesian) cyclothemic Excello Black Shale of the Midcontinent Region, U.S.A.: Marine and Petroleum Geology, v. 4, p. 243-254.

Echols, J., 1959, Microfauna and biostratigraphy of the upper Strawn Group, Eastland and Palo Pinto counties, Texas: Unpublished M.S. thesis, University of Texas, 176 p.

Elias, M.K., 1937, Depth of deposition of the Big Blue (Late Paleozoic) sediments in Kansas: GSA Bulletin, v. 48, p. 403-432.

Elias, M.K., 1964, Depth of Late Paleozoic sea in Kansas and its megacyclic sedimentation, *in* D.F. Merriam, ed., Symposium on cyclic sedimentation: Kansas Geological Survey Bulletin 169, p. 87-106.

Enos, P., 1991, Sedimentary parameters for computer modeling, *in* E.K. Franseen, W.L. Watney, C.G. St.C. Kendall, and W. Ross, eds., Sedimentary modeling: computer simulations and methods for improved parameter definition: Kansas Geological Survey Bulletin 233, p. 63-99.

Esteban, M., 1992, Paleokarst short course notes: Redding University, England.

Esteban, M., and C.F. Klappa, 1983, Subaerial exposure, *in* P.A. Scholle, D.G. Bebout, and C.H. Moore, eds., Carbonate depositional environments: AAPG Memoir 33, p. 1-54.

Evans, G., 1979, Quaternary transgressions and regressions: Journal of Geological Society of London, v. 136, p. 125-132.

Fath, A.E., 1921, Geology of the Eldorado oil and gas field: Kansas Geological Survey Bulletin 7, 187 p.

Faucette, J.R., 1955, The geology of the Marmaton Group of southern Nowata County, Oklahoma: Tulsa Geological Society, v. 23, p. 239-259.

Feldman, H.R., and E.K. Franseen, 1991, Stratigraphy and depositional history of the Drum Limestone and associated strata (Pennsylvanian) in the Independence, Kansas, area: a field trip guidebook and road log: Kansas Geological Survey Open File Report 91-45, 29 p.

Feldman, H.R., A.W. Archer, E.P. Kvale, C. Cunningham, C.G. Maples, and R.R. West, 1993a, A tidal model of Carboniferous Konservat-Lagerstätte Formation: Palaios, v. 8, p. 485-498.

Feldman, H.R., A.W. Archer, R.R. West, C.G. Maples, C.R. Cunningham, H.P. Schultze, and W.P. Lanier, 1993b, The Garnett and Hamilton paleovalleys, and their relationship to the Douglas Group paleovalleys, *in* A.W. Archer, H.R. Feldman, and W.P. Lanier, Incised paleovalleys of the Douglas Group in northeastern Kansas: field guide and related contributions: Kansas Geological Survey Open File Report 93-24, p. 12-1-12-10.

Feldman, H.R., and A.W. Archer, 1995, Virgilian (Upper Pennsylvanian) paleosols in the upper Lawrence Formation (Douglas Group) and in the Snyderville Shale Member (Oread Formation, Shawnee Group), of the northern Midcontinent, USA: pedologic contrasts in a cyclothem sequence— discussion: Journal of Sedimentary Research, v. A65, p. 122-124.

Feldman, H.R., M.R. Gibling, A.W. Archer, W.G. Wightman, and W.P. Lanier, 1995, Stratigraphic architecture of the Tonganoxie paleovalley fill (Lower Virgilian) in northeastern Kansas: AAPG Bulletin, v. 79, p. 1019-1043.

Felton, R.M., and P.H. Heckel, 1992, GSA Abstracts With Programs, North-Central Meeting, Iowa City, v. 24, no. 4, p. 14.

Fischer, A.G., 1964, The Lofer cyclothems of the Alpine Triassic, *in* D.F. Merriam, ed., Symposium on cyclic sedimentation: Kansas Geological Survey Bulletin 169, p. 107-150.

Fischer, A.G., and D.J. Bottjer, 1991, Orbital forcing and sedimentary sequences: Journal of Sedimentary Petrology, v. 61, no. 7, p. 1063-1252.

Ford, G.W., 1975, Genesis and trend of the lowermost unit of the Vamoosa Formation (Gypsey Sandstone) in parts of northeastern and central Oklahoma: Unpublished M.S. Thesis, Oklahoma State University, 69 p.

Forgotson, J.M., 1957, Nature, usage and definition of marker-defined vertically segregated rock units: AAPG Bulletin, v. 41, p. 2108-2113.

Forgotson, J.M., A.T. Statler, and M. David, 1966, Influence of regional tectonics and local structure on deposition of Morrow Formation in western Anadarko basin: AAPG Bulletin, v. 50, p. 518-532.

Forgotson, J.M., J.T. Boyce, R.M. Knapp, H. Liu, and P.W. Smith, in press, Influence of vertical permeability barriers on the ultimate recovery from "Oil Creek" reservoirs, *in* K.S. Johnson, ed., Simpson and Viola groups in the southern Mid-Continent, 1994 Symposium: Oklahoma Geological Survey Circular.

Fouch, T.D., and E.R. Magathan, eds., 1980, Paleozoic paleogeography of the west-central United States: SEPM Rocky Mountain Section Rocky Mountain Paleogeography Symposium 1, 430 p.

Franseen, E.K., H.R. Feldman, N.L. Anderson, R.D. Miller, in press, Depositional and stratigraphic analysis of Kansas City Group strata utilizing high-resolution seismic imaging, Montgomery County, Kansas, *in* N.L. Anderson and D. Hedke, eds., Geophysical atlas of Kansas: Kansas Geological Survey Special Publication.

French, J.A., and W.L. Watney, 1990, Computer modeling of Mid-Continent cyclothems and its application in the prediction of hydrocarbon reservoirs: AAPG Bulletin, v. 74, p. 657.

French, J.A., and W.L. Watney, 1993, Integrated field, analog, and shelf-scale geologic modeling of oolitic grainstone reservoirs in the upper Pennsylvanian Kansas City Group in Kansas, *in* B. Lineville, ed., Reservoir characterization II: Tulsa, OK, PennWell Books, p. 983-993.

French, J.A., and W.L. Watney, 1993, Stratigraphy and depositional setting of the lower Missourian (Pennsylvanian) Bethany Falls and Mound Valley limestone analogues for age-equivalent and ooid-grainstone reservoirs, Kansas: Current research on Kansas geology, Summer 1993: Kansas Geological Survey Bulletin 235, p. 27-39.

French, J.A., and P.H. Heckel, 1994, Probable high-amplitude glacial-eustatic origin for Pennsylvanian cyclothems requires modification of applied sequence-stratigraphic classification and terminology: AAPG 1994 Annual Meeting Abstracts, p. 152.

Fritz, R.D., and P.L. Medlock, 1993, Sequence stratigraphy of the Hunton Group as defined by core, outcrop and log data, *in* K.S. Johnson, ed., Hunton Group core workshop and field trip: Oklahoma Geological Survey Special Publication 93-4, p. 161-180.

Gahring, R.R., 1959, History and development of North Madill field, Marshall County, Oklahoma: AAPG Bulletin, v. 43, p. 274-286.

Galloway, W.E., 1989a, Genetic stratigraphic sequences in basin analysis I: architecture and genesis of flooding-surface bounded depositional units: AAPG Bulletin, v. 73, no. 2, p. 125-142.

Galloway, W.E., 1989b, Genetic stratigraphic sequences in basin analysis II: application to northwest Gulf of Mexico Cenozoic Basin: AAPG Bulletin, v. 73, no. 2, p. 143-154.

Galloway, W.E., M.S. Yancy, and A.P. Whipple, 1977, Seismic stratigraphy model of depositional platform margin, eastern Anadarko basin, Oklahoma, *in* C.E. Payton, ed., Seismic stratigraphy — applications to hydrocarbon exploration: AAPG Memoir 26, p. 439-450.

Galvin, P.K., 1983, Deep-to-shallow carbonate ramp transition in Viola Limestone (Ordovician), southwest Arbuckle Mountains, Oklahoma [abs.]: AAPG Bulletin, v. 63, p. 466-467.

Gentile, L.F., 1982, Sedimentary and graptolite biostratigraphy of the Viola Group (Ordovician), Arbuckle Mountains and Criner Hills, Oklahoma: M.S. thesis, Oklahoma State University, 104 p.

Gentile, R.J., 1976, The geology of Bates County, Missouri: Missouri Division of Geology and Land Survey Report of Investigations 59, p. 1-112.

Gerken, L.D., 1992, Morrowan sandstones in south-central Texas County, Oklahoma: Unpublished M.S. thesis, Oklahoma State University, 140 p.

Gibling, M.R., H.R. Feldman, A.W. Archer, and W.P. Lanier, 1993, Sedimentology, stratigraphy, and paleoflow patterns of the Tonganoxie Sandstone Member and related strata in northeast Kansas and southwest Missouri, *in* A.W. Archer, H.R. Feldman, and W.P. Lanier, eds., Incised Paleovalleys of the Douglas Group in northeastern Kansas: Field guide and related contributions (SEPM Midcontinent Section 1993 Fall Field Conference): Kansas Geological Survey Open File Report 93-24, p. 3-1 - 3-39.

Gibling, M.R., H.R. Feldman, W.P. Lanier, and A.W. Archer, 1993, Fluvial and estuarine fill of the Pennsylvanian Tonganoxie paleovalley, NE Kansas and SW Missouri: GSA Abstracts With Programs 26, p. 46.

Gibson, G.R., 1965, Oil and gas in southwestern region — geologic framework: fluids in subsurface environments, a symposium: AAPG Memoir 4, p. 66-100.

Gilmer, M.H., 1993, Stratigraphy, petrology, depositional environments, and diagenesis of the Heplar Formation Pleasanton Group (Upper Pennsylvanian) in Southeastern Kansas: M.S. thesis, Iowa University, 149 p.

Ginsburg, R.M., 1971, Landward movement of carbonate mud: new model for regressive cycles in carbonates (abs.): AAPG Bulletin, v. 55, p. 340.

Glass, H.D., 1958, Petrology and sedimentation of the Pennsylvanian sediments in southern Illinois: a vertical profile: Illinois State Geological Survey Report of Investigations 204, 60 p.

Glossa, J.M., 1982, Depositional environments and diagenetic history of the Nolans Limestone (Upper Wolfcampian), Rice County, Kansas: Unpublished M.S. thesis, University of Kansas, 30 p.

Goebel, E.D., 1971, Southeastern Kansas-northeastern Oklahoma-southwestern Missouri, *in* I.H. Cram, ed., Future petroleum provinces of the United States — their geology and potential: AAPG Memoir 15, p. 1088-1097.

Goebel, K.A., E.A. Bettis, and P.H. Heckel, 1989, Upper Pennsylvanian paleosol in Stranger Shale and underlying Iatan Limestone, southwestern Iowa: Journal of Sedimentary Petrology, v. 59, no. 2, p. 224-232.

Goldhammer, R.K., 1987, Platform carbonate cycles, Middle Triassic of northern Italy: the interplay of local tectonics and eustasy: Baltimore, The Johns Hopkins University, Unpublished Ph.D. dissertation, 468 p.

Goldhammer, R.K., E.J. Oswald, and P.A. Dunn, 1989, The hierarchy of stratigraphic forcing — an example from Middle Pennsylvanian (Desmoinesian) shelf carbonates of the southwestern Paradox basin, Honaker Trail, Utah, *in* E.K. Franseen and W.L. Watney, eds., Sedimentary modeling: computer simulation of depositional sequences: Kansas Geological Survey Subsurface Geology Series 12, p. 27-30.

Goldhammer, R.K., P.A. Dunn, and L.A. Hardie, 1990, Depositional cycles, composite sea-level changes, cycle stacking patterns, and the hierarchy of stratigraphic forcing: examples from Alpine Triassic platform carbonates: GSA Bulletin, v. 102, p. 535-562.

Goldhammer, R.K., E.J. Oswald, and P.A. Dunn, 1991, Hierarchy of stratigraphic forcing: example from Middle Pennsylvanian shelf carbonates of the Paradox basin, *in* E.K. Franseen, W.L. Watney, C.G.St.C. Kendall, and W. Ross, eds., Sedimentary modeling: computer simulations and methods for improved parameter definition: Kansas Geological Survey Bulletin 233, p. 361-413.

Goldhammer, R.K., P.J. Lehman, and P.A. Dunn, 1993, The origin of high-frequency platform carbonate cycles and third-order sequences (Lower Ordovician El Paso Group, west Texas): constraints from outcrop data and stratigraphic modeling: Journal of Sedimentary Petrology, v. 63, p. 318-359.

Gonzales-Bonorino, F., and M.E. Terruggi, 1952, Lexico sedimentologico: Instituto Nacional de Investigaciones de Ciencias Naturales, Buenos Aires, Publicaciones de extension cultural y didactica, no. 6, 164 p.

Gradstein, F.M., F.P. Agterberg, J.C. Brower, and W.S. Schwarzacher, 1985, Quantitative stratigraphy: Dordrecht, D. Reidel Publishing Co., 598 p.

Greb, S.F., D.A. Williams, and A.D. Williamson, 1992, Geology and stratigraphy of the western Kentucky coal field: Kentucky Geological Survey Bulletin 2, series XI, 77 p.

Greb, S.F., 1993, Peat accumulation and sedimentation in the Livingston paleovalley (Morrowan), eastern Kentucky, *in* A.W. Archer, H.R. Feldman, and W.P. Lanier, eds., Incised paleovalleys of the Douglas Group in northeastern Kansas: field guide and related papers: Kansas Geological Survey Open File Report 93-24, p. 15.1 - 15.6.

Greb, S.F., and A.W. Archer, 1995, Rhythmic sedimentation in a mixed tide and wave deposit, Hazel Patch Sandstone (Pennsylvanian), eastern Kentucky coal field: Journal of Sedimentary Research, v. B65 p. 96-106.

Greenberg, H., 1986, Depositional and diagenetic study of the Lenapah Limestone in eastern Kansas and western Missouri: M.S. thesis, University of Iowa, p. 1-209.

Griffith, G.L., 1981, The Tonganoxie Sandstone in portions of Sedgewick, Butler, and Greenwood counties, Kansas: M.S. thesis, Wichita State University,54 p.

Guy, W.J., and J. Doveton, 1993, New interpretation strategies for predicting oil productivity using extensions of classic log analysis techniques: Reservoir Description Workshop: Application of Underutilized Technologies, Energy Research Center, Lawrence, KS, p. 51-72.

Hallock, P., and W. Schlager, 1986, Nutrient excess and the demise of coral reefs and carbonate platforms: Palaios, v. 1, p. 389-398.

Ham, W.E., 1954, Algal origin of Birds-eye Limestone in McLish Formation: Oklahoma Academy of Science Proceedings, 1952, v. 33, p. 200-203.

Ham, W.E., 1955, Regional stratigraphy and structure of the Arbuckle Mountain region, *in* Field conference on geology of the Arbuckle Mountain region: Oklahoma Geological Survey Guidebook III, p. 29-31.

Ham, W.E., and J.L. Wilson, 1967, Epeirogeny and orogeny in the central United States: American Journal of Science, v. 265, p. 332-407.

Ham, W.E., 1973, Regional geology of the Arbuckle Mountains, Oklahoma: GSA Guidebook for Field Trip 5, p. 1-18.

Handford, C.R., and R.G. Loucks, 1991, Unique signature of carbonate strata and the development of depositional sequence and systems tract models for ramps, rimmed shelves, and detached platforms: AAPG Bulletin, v. 75, p. 588.

Handford, C.R., and R.G. Loucks, 1993, Carbonate depositional sequences and systems tracts — responses of carbonate platforms to relative sea-level changes, *in* R.G. Loucks, and J.F. Sarg, eds., Carbonate sequence stratigraphy — recent developments and applications: AAPG Memoir 57, p. 3-41.

Haq, B.U., J. Hardenbol, and P.R. Vail, 1987, Chronology of fluctuating sea levels since the Triassic: Science, v. 235, p. 1156-1167.

Haq, B.U., J. Hardenbol, and P.R. Vail, 1988, Mesozoic and Cenozoic chronostratigraphy and eustatic cycles, *in* C.K. Wilgus, ed., Sea-level changes: an integrated approach: SEPM Special Publication 42, p. 71-108.

Hardie, L.A., and E.A. Shinn, 1986, Carbonate depositional environments, modern and ancient: 3 — tidal flats: Colorado School of Mines Quarterly, v. 81, no. 1, p. 1-74.

Hardie, L.A., A. Bosellini, and R.K. Goldhammer, 1986, Repeated subaerial exposure of subtidal carbonate platforms, Triassic, northern Italy: evidence for high-frequency sea-level oscillations on a 104-year scale: Paleooceanography, v. 1, p. 447-457.

Hardie, L.A., P.A. Dunn, and R.K. Goldhammer, 1991, Field and modeling studies of Cambrian carbonate cycles, Virginia Appalachians — discussion: Journal of Sedimentary Petrology, v. 61, p. 636-646.

Harland, W.B., A.V. Cox, P.G. Llewellyn, C.A.G. Pickton, A.G. Smith, and R. Walters, 1982, A geologic time scale: Cambridge University Press, 131 p.

Harlton, B.H., 1927a, Pennsylvanian Foraminifera of Oklahoma and Texas: Journal of Paleontology, v. 1, p. 305-310, plates 52-53.

Harlton, B.H., 1927b, Some Pennsylvanian Ostracoda of the Glenn and Hoxbar formations of southern Oklahoma and of the upper part of the Cisco Formation of northern Texas: Journal of Paleontology, v. 1, p. 203-212.

Harms, J.C., and R.K. Fahnestock, 1965, Stratification, bedforms and flow phenomena, *in* G. V. Middleton, ed., Primary sedimentary structures and their hydrodynamic interpretation: SEPM Special Publication 12, p. 84-115.

Harris, S.A., 1975, Hydrocarbon accumulation in "Meramec-Osage" (Mississippian) rocks, Sooner Trend, northwest-central Oklahoma, *in* B. Rascoe, Jr. and N.J. Hyne, eds., Petroleum geology of the Mid-Continent: Tulsa Geological Society Special Publication 3, p. 75-81.

Harrison, J.C., 1990, "Upper" Morrow Purdy Sandstone in parts of Texas and Cimarron counties, Oklahoma: Unpublished M.S. thesis, Oklahoma State University, 95 p.

Hatch, J.R., and J.S. Leventhal, 1992, Relationship between inferred redox potential of depositional environment and geochemistry of the Upper Pennsylvanian (Missourian) Stark Shale Member of the Dennis Limestone, Wabaunsee County, Kansas, U.S.A.: Chemical Geology, v. 99, p. 65-82.

Hattin, D.E., 1957, Depositional environment of the Wreford megacyclothem (Lower Permian) of Kansas: Kansas Geological Survey Bulletin 124, 150 p.

Haworth, E., 1898, The stratigraphy of the Kansas coal measures: Kansas University Quarterly, v. 3, p. 271-290.

Hayes, M.O., 1967, Relationship between coastal climate and bottom sediment type on the inner continental shelf: Marine Geology, v. 5, p. 111-132.

Hayes, M.O., 1975, Morphology of sand accumulation in estuaries, *in* L.E. Cronin, ed., Estuarine research, v. 2, Geology and engineering: New York, Academic Press, p. 3-22.

Heckel, P.H., and J.M. Cocke, 1969, Phylloid algal-mound complexes in outcropping Upper Pennsylvanian rocks of Midcontinent: AAPG Bulletin, v. 50, p. 1058-1074.

Heckel, P.H., and J.F. Baesemann, 1975, Environmental interpretation of conodont distribution in Upper Pennsylvanian (Missourian) megacyclothems in eastern Kansas: AAPG Bulletin, v. 59, p. 486-509.

Heckel, P.H., 1977, Origin of phosphatic black shale facies in Pennsylvanian cyclothems of Midcontinent North America: AAPG Bulletin, v. 61, 1045-1068.

Heckel, P.H., L.L. Brady, W.J. Ebanks, Jr., and R.K. Pabian, 1979, Ninth International Congress of Carboniferous stratigraphy and geology (IX-ICC), Field Trip 10: Pennsylvanian cyclic platform deposits of Kansas and Nebraska: Kansas Geological Survey, 79 p.

Heckel, P.H., 1980, Paleogeography of eustatic model for deposition of Mid-Continent Upper Pennsylvanian cyclothems, *in* T.D. Fouch and E.R. Magathan, eds., Paleozoic paleogeography of the west-central United States: SEPM Rocky Mountain Section Mountain Paleogeography Symposium 1, p. 197-216.

Heckel, P.H., 1983, Diagenetic model for carbonate rocks in Midcontinent Pennsylvanian eustatic cyclothems: Journal of Sedimentary Petrology, v. 53, p. 733-759.

Heckel, P.H., 1984, Changing concepts of Midcontinent Pennsylvanian cyclothems, North America: 9th International Carboniferous Congress, Compte Rendu, pt. 3, Southern Illinois University, p. 535-553.

Heckel, P.H., 1985, Current view of Midcontinent Pennsylvanian cyclothems: recent interpretations of Late Paleozoic cyclothems, *in* W.L. Watney, R.L. Kaesler, and K.D. Newell, compilers, SEPM Mid-Continent Section Third Annual Meeting, p. 1-69.

Heckel, P.H., 1986, Sea-level curve for Pennsylvanian eustatic marine transgressive-regressive depositional cycles along Midcontinent outcrop belt, North America: Geology, v. 14, p. 330-334.

Heckel, P.H., 1990, Evidence for global (glacial-eustatic) control over upper Carboniferous (Pennsylvanian) cyclothems in Midcontinent North America, *in* R.F.P. Hardman and J. Brooks, eds., Tectonic events responsible for Britain's oil and gas reserves: London Geological Society Special Publication 55, p. 35-47.

Heckel, P.H., 1991a, Lost Branch Formation and revision of upper Desmoinesian stratigraphy along Midcontinent Pennsylvanian outcrop belt: Kansas Geological Survey Geology Series 4, 67 p.

Heckel, P.H., 1991b, Thin widespread Pennsylvanian black shales of Midcontinent North America: a record of a cyclic succession of widespread pycnoclines in a fluctuating epeiric sea, *in* R.V. Tyson and T.H. Pearson, eds., Modern and ancient Continental Shelf Anoxia: London Geological Society Special Publication 58, p. 259-273.

Heckel, P.H., and C.P. Weibel, 1991, Current status of conodont-based biostratigraphic correlation of Upper Pennsylvanian succession between Illinois and Mid-Continent, *in* C.P. Weibel, ed., Sequence stratigraphy in mixed clastic-carbonate strata, Upper Pennsylvanian, east-central Illinois: SEPM Great Lakes Section 21st Annual Field Conference, Illinois State Geological Survey, p. 60-69.

Heckel, P.H., 1994, Evaluation of evidence for glacio-eustatic control over marine Pennsylvanian cyclothems in North America and consideration of possible tectonic effects, *in*

J.M. Dennison and F.R. Ettensohm, eds., Tectonic and eustatic controls on sedimentary cycles: SEPM Concepts in Sedimentology and Paleontology 4, p. 65-87.

Hemish, L.A., 1987, Miscorrelation of the Checkerboard Limestone in Okfuskee County proved by Oklahoma Geological Survey core-drilling: Oklahoma Geology Notes, v. 47, no. 4-1/2, p. 148-161.

Hemish, L.A., 1990, Coal geology of Tulsa, Wagoner, Creek, and Washington counties, Oklahoma: Oklahoma Geological Survey Map GM-33, 117 p.

Hemish, L.A., J. Chaplin, and N.H. Suneson, 1995, Stratigraphy and sedimentation of some selected Pennsylvanian Atokan-Desmoinsian strata in the southeastern part of the Arkoma basin, Oklahoma: Guidebook for the Spring (1955) Field Excursion, Midcontinent Pennsylvanian Stratigraphic Working Group (SEPM Midcontinent Section), Oklahoma Geological Survey.

Hentz, T.F., 1994, Sequence stratigraphy of the Upper Pennsylvanian Cleveland Formation: a major tight-gas sandstone, western Anadarko basin, Texas panhandle: AAPG Bulletin, v. 78, no. 4, p. 569- 595.

Hentz, T.F., 1994, Depositional, structural, and sequence framework of the gas-bearing Cleveland Formation (Upper Pennsylvanian), western Anadarko basin, Texas panhandle: Bureau of Economic Geology Report of Investigations 213, 73 p.

Hite, R.J., 1978, Possible genetic relationships between evaporties, phosphorites, and iron-rich sediments: Mountain Geologist, v. 14, no. 3, p. 97-107.

Hopkins, M.E., 1958, Geology and petrology of the Anvil Rock Sandstone of southern Illinois: Illinois State Geological Survey Circular 256, 49 p.

Howe, W.B., 1953, Upper Marmaton strata in western and northern Missouri: Missouri Geological Survey and Water Resources Report of Investigations 9.

Howe, W.B., 1956, Stratigraphy of Pre-Marmaton Desmonesian (Cherokee) rocks in southeastern Kansas: Kansas Geological Survey Bulletin 123, 132 p.

Howe, W.B., 1982, Stratigraphy of the Pleasanton Group, Pennsylvanian System in Missouri: Missouri Department of Natural Resources, Geology and Land Survey Division Open File Report 82-10-GI, 112 p.

Howe, W.B., 1986, Stratigraphy of exposed Pennsylvanian strata in Platte County, Missouri: Missouri Geological Survey Open File Report OFR-86-57-GI, 60 p.

Huffman, G.G., 1959, Pre-Desmoinesian isopachous and paleogeologic studies in central Mid-Continent region: AAPG Bulletin, v. 43, p. 2541-2575.

Hughes, T., 1992, Abrupt climate change related to unstable ice-sheet dynamics: toward a new paradigm: Palaeogeography, Palaeoclimatology, Palaeoecology, v. 97, p. 203-234.

Hunt, D., and M.E. Tucker, 1993, Sequence stratigraphy of carbonate shelves with an example from the mid-Cretaceous (Urgonian) of southeast France: *in* H.W. Posamentier, C.P. Summerhayes, B.U. Haq, and G.P. Allen, eds., Sequence stratigraphy and facies associations: IAS Special Publication 18, p. 307-341.

Imbrie, J., L.F. Laporte, and D.F. Merriam, 1959, Beattie Limestone facies and their bearing on cyclical sedimentation theory: Kansas Geological Society 24th Field Conference Guidebook, p. 69-78.

Imbrie, J., L.F. Laporte, and D.F. Merriam, 1964, Beattie Limestone facies (Lower Permian) of the northern Midcontinent, *in* D.F. Merriam, ed., Symposium on cyclic sedimentation: Kansas Geological Survey Bulletin 169, p. 219-238.

Ireland, H.A., 1965, Regional depositional basin and correlations of Simpson Group, *in* T. Herndon, ed., Symposium on the Simpson Group: Tulsa Geological Society Digest, v. 33, p. 74-89.

James, G.W., 1970, Stratigraphic geochemistry of a Pennsylvanian black shale (Excello) in the Midcontinent and Illinois basin: Ph.D. dissertation, Rice University, 92 p.

James, N.P., and P.W. Choquette, 1987, Paleokarst: New York, Springer-Verlag, 416 p.

Jewett, J.M., 1933, Evidence of cyclic sedimentation in Kansas during the Permian Period: Transactions of the Kansas Academy of Science, v. 36, p. 137-140.

Jewett, J.M., 1940, Oil and Gas in Linn County, Kansas: Kansas Geological Survey Bulletin 30, 29 p.

Jewett, J.M., 1941, Classifiction of the Marmaton Group, Pennsylvanian in Kansas: Kansas Geological Survey Bulletin 38, p. 285-344.

Jewett, P.W., 1995, Geologic consequences of globe-encircling equatorial currents: Geology, v. 23, no. 2, p. 117-120.

Joeckel, R.M., 1989, Geomorphology of a Pennsylvanian land surface: pedogenesis in the Rock Lake Shale Member, southeastern Nebraska: Journal of Sedimentary Petrology, v. 59, p. 469-481.

Joeckel, R. M., 1994, Virgilian (Upper Pennsylvanian) paleosols in the upper Lawrence Formation (Douglas Group) and in the Snyderville Shale Member (Oread Formation, Shawnee Group) of the northern Mid-Continent, USA: pedologic contrasts in a cyclothem sequence: Journal of Sedimentary Research, v. A64, no. 4, p. 853-866.

Johnson, J.G., G. Klapper, and C.A. Sandberg, 1985, Devonian eustatic fluctuations in Euramerica: GSA Bulletin, v. 96, p. 567-587.

Johnson, K.S., 1991, Geologic overview and economic importance of Late Cambrian and Ordovician rocks in Oklahoma: Oklahoma Geological Survey Circular 92, p. 3-14.

Jordan, D.W., and R.W. Tillman, 1987, Geologic facies analysis for enhanced oil recovery, Bartlesville Sandstone, Greenwood County, Kansas, *in* R.W. Tillman and K.J. Weber, eds., Reservoir sedimentology: SEPM Special Publication 40, p. 311-332.

Kansas City Power and Light Company, 1975, Wolf Creek Generating Station, Addendum, Standardized Nuclear Unit Power Plant System, Preliminary Safety Analysis Report, Revision 9.

Kaufman, E.G., 1984, Paleobiogeography and evolutionary response dynamic in the Cretaceous Western Interior seaway of North America: Geological Association of Canada Special Paper 27, p. 273-306.

Kendall, C.G. St. C., J. Strobel, J. Tang, P. Moore, R. Cannon, J. Bezdek, and G. Giswas, 1989, Simulation of the west Texas, New Mexican Permian Guadalupian basin margin — a response to eustatic change, an example of SEDPAK, *in* P.M. Harris and G.A. Grover, eds., Subsurface and outcrop examination of the Capitan shelf margin, northern Delaware basin: SEPM Core Workshop Note 13, p. 423-426.

Kerr, R.A., 1980, Changing global sea levels as a geological

index: Science, v. 209, p. 483-486.
Khaiwka, M.H., 1968, Geometry and depositional environments of Pennsylvanian reservoir sandstones, northwestern Oklahoma: Ph.D. dissertation, University of Oklahoma, 279 p.
Kidder, D.L., 1985, Petrology and origin of phosphate nodules from the Midcontinent Pennsylvanian epicontinental sea: Journal of Sedimentary Petrology, v. 55, p. 809-816.
King, W.E., 1988, Permian fusulinids from the Conoco Inc. 33-1 core, Kay County, Oklahoma, *in* W.A. Morgan and J.A. Babcock, eds., Permian rocks of the Midcontinent: SEPM Midcontinent Section Special Publication 1, p. 213-223.
Klein, G.deV., 1977, Clastic tidal facies: Champaign, Illinois, Continuing Education Publishing Company (CEPCO), 149 p.
Klein, G.deV., and S. Cloetingh, 1989, Tectonic subsidence during deposition of Pennsylvanian cyclothems: 28th International Geological Congress Abstracts, v. 2, p. 198-199.
Klein, G.deV., and D.A. Willard, 1989, Origin of the Pennsylvanian coal-bearing cyclothems of North America: Geology, v. 17, p. 152-155.
Klein, G. deV., 1990, Pennsylvanian time scales and cycle periods: Geology, v. 18, p. 455-457.
Klein, G. deV., 1990, Comments on sedimentary-stratigraphic verification of some geodynamic basin models: example from a cratonic and associated foreland basin, *in* A.T. Cross, ed., Quantitative dynamic stratigraphy: Englewood Cliffs, New Jersey, Prentice-Hall, p. 503-518.
Klein, G deV., and J.B. Kupperman, 1992, Pennsylvanian cyclothems: methods of distinguishing tectonically induced changes in sea level from climatically induced changes: GSA Bulletin, v. 104, p. 166-175.
Klemme, H.D., and G.F. Ulmishek, 1991, Effective petroleum source rocks of the world: stratigraphic distribution and controlling depositional factors: AAPG Bulletin, v. 75, p. 1809-1851.
Kluth, C.F., and P.J. Coney, 1981, Plate tectonics of the Ancestral Rocky Mountains: Geology, v. 9, p. 10-15.
Kluth, C.F., 1986, Plate tectonics of the ancestral Rocky Mountains, *in* J.A. Peterson, ed., Paleotectonics and sedimentation in the Rocky Mountain region, United States: AAPG Memoir 41, p. 353-369.
Knapp, R.W., and J. French, 1991, Features in Kansas cyclothems seen by high-resolution seismology: Kansas Geological Survey Bulletin 233, p. 111-121.
Knight, K.L., 1985, Stratigraphy, depositional and diagenic history of three Middle Pennsylvanian cyclothems (Breezy Hill and Fort Scott limestones), Mid-Continent North America: Ph.D. thesis, University of Iowa, 340 p.
Krumme, G.W., 1981, Stratigraphic significance of limestones of the Marmaton Group in eastern Oklahoma: Oklahoma Geological Survey Bulletin 131, p. 1-67.
Krystinik, L.F., and B.A. Blakeney, 1990, Sedimentology of the upper Morrow Formation in eastern Colorado and western Kansas, *in* S.A. Sonnenberg et al., eds., Morrow sandstones of southeast Colorado and adjacent areas: Denver, The Rocky Mountain Association of Geologists, p. 37-50.
Kuecher, G.J., B.G. Woodland, and F.M. Broadhurst, 1990, Evidence of deposition from individual tides and of tidal cycles from the Francis Creek Shale (host rock to the Mazon Creek Biota), Westphalian D (Pennsylvanian), northeastern Illinois: Sedimentary Geology, v. 68, p. 211-221.
Kumar, N., and R.M. Slatt, 1984, Submarine-fan and slope facies of Tonkawa (Missourian-Virgilian) Sandstone in deep Anadarko basin: AAPG Bulletin, v. 68, no. 12, p. 1839-1856.
Kvale, E.P., A.W. Archer, and H.R. Johnson, 1989, Daily, monthly, and yearly tidal cycles within laminated siltstones of the Mansfield Formation (Pennsylvanian) of Indiana: Geology, v. 167, p. 365-368.
Kvale, E.P., and A.W. Archer, 1990, Tidal deposits associated with low-sulfur coals, Brazil Formation (Lower Pennsylvanian), Indiana: Journal of Sedimentary Petrology, v. 60, p. 563-574.
Kvale, E.P., and A.W. Archer, 1991, Characteristics of two Pennsylvanian-age semidiurnal tidal deposits in the Illinois basin, U.S.A., *in* D.G. Smith, G.E. Reinson, B.A. Zaitlin, and R.A. Rahmani, eds., Clastic tidal sedimentology: Canadian Society Petroleum Geologists Memoir 16, p. 179-188.
Kvale, E.P., and M.L. Barnhill, 1994, Evolution of Lower Pennsylvanian estuarine facies within two adjacent paleovalleys, Illinois basin, Indiana, *in* R.W. Dalrymple, R.Boyd, and B. Zaitlin, eds., Incised-valley systems: origin and sedimentary sequences: SEPM Special Publication 51, p. 191-207.
Kvale, E.P., G.S. Fraser, A.W. Archer, A. Zawistoski, N. Kemp, and P. McGough, 1994, Evidence of seasonal precipitation in Pennsylvanian sediments of the Illinois basin: Geology, v. 22, p. 331-334.
Lanier, W.P., H.R. Feldman, and A.W. Archer, 1993, Tidally modulated sedimentation in a fluvial to estuarine transition, Douglas Group, Missourian-Virgilian, Kansas: Journal of Sedimentary Petrology, v. 63, p. 860-873.
Laporte, L.F., 1962, Paleoecology of the Cottonwood Limestone (Permian), northern Midcontinent: GSA Bulletin, v. 73, p. 521-544.
Laporte, L.F., and J. Imbrie, 1964, Phases and facies in the interpretation of cyclic deposits, *in* D.F. Merriam, ed., Symposium on cyclic sedimentation: Kansas Geological Survey Bulletin 169, p. 249-263.
Lardner, J.E., 1984, Petrology, depositional environment and diagenesis of Middle Pennsylvanian (Desmoinesian) 'Lagoda interval' Cherokee Group in east-central Kansas: M.S. thesis, University of Iowa, 156 pp.
Levorsen, A.I., 1954, Geology of petroleum: San Francisco, W.H. Freeman and Co., 703 p.
Lewis, R.D., 1978, Environments of deposition and paleoecology of the Oil Creek Formation (Middle Ordovician), Arbuckle Mountains and Criner Hills, Oklahoma (abs.): West Texas Geological Society Bulletin, v. 78-69, p. 88.
Lindsay, J.F., J.M. Kennard, and P.M. Southgate, 1993, Application of sequence stratigraphy in an intracratonic setting, Amedeus basin, central Australia, *in* H.W. Posamentier, C.P. Summerhayes, B.A. Haq, and G.P. Allen, eds., Sequence stratigraphy and facies associations: International Association of Sedimentologists Special Publication 18, p. 605-631.
Lins, T.W., 1950, Origin and environment of the Tonganoxie Sandstone in northeastern Kansas: Kansas Geological Survey Bulletin 86, pt. 5, 140 p.
Longman, M.W., 1981, Deposition of the Bromide Formation, Arbuckle Mountains, Oklahoma: Ontogency of an ancient carbonate shelf: Shale Shaker, v. 31, p. 1-18.
Loucks, R.G., and J.F. Sarg, 1993, Carbonate sequence

stratigraphy — recent developments and applications: AAPG Memoir 57, 545 p.
Loutit, T.S., J. Hardenbol, P.R. Vail, and G.R. Baum, 1988, Condensed sections, the key to age dating and correlation of continental margin sequences, *in* B.H. Lidtz, ed., Sea-level change — an integrated approach: SEPM Publication 42, p. 182-213.
Lukert, L.H., 1949, Subsurface cross sections from Marion County, Kansas, to Osage County, Oklahoma: AAPG Bulletin, v. 33, p. 131-152.
Lutz-Garihan, A.B., and R.J. Cuffey, 1979, Stratigraphy of the Lower Permian Wreford megacyclothem in southernmost Kansas and northern Oklahoma: Kansas Geological Survey Bulletin 216, 19 p.
Lynch, M., and Z. Al-Shaieb, 1991, Evidence of paleokarstic phenomena and burial diagenesis in the Ordovician Arbuckle Group of Oklahoma, *in* K.S. Johnson, ed., Late Cambrian-Ordovician geology of the southern Midcontinent: Oklahoma Geological Survey Circular 92, p. 42-60.
Maples, C.G., 1986, Enhanced paleoecological and paleoenvironmental interpretation results from analysis of early diagenetic concretions in Pennsylvanian shales: Palaios, v. 1, p. 512-516.
Maples, C.G., and A.W. Archer, 1987, Trace-fossil holotypes from the freshwater Hindostan Whetstone beds (Late Carboniferous) of Indiana: Journal of Paleontology, v. 61, p. 890-897.
Mazzullo, S.J., 1994, Permian stratigraphy and facies, Permian basin (Texas-New Mexico) and adjoining areas in the Midcontinent United States, *in* P.A. Scholle, T.M. Peryt, and D.S. Ulmer-Scholle, eds., The Permian of northern Pangea, 2: Sedimentary basins and economic resources: New York, Springer-Verlag, p. 41-60.
Mazzullo, S.J., and C.S. Teal, 1994, Road log and locality guide for Day 1: a new look at old rocks: stratigraphy, sedimentology, and structure of the Chase Group (Permian, Wolfcampian), south-central Kansas: Kansas Geological Society 1994 Field Trip Guidebook, Day 1, p. 1-60.
Mazzullo, S.J., C.S. Teal, and C.A. Burtnett, 1995a, Cycles, cyclothems, and sequences in Chase Group strata (Permian, Wolfcampian), south-central Kansas (abs.): GSA North-Central and South-Central Annual Meeting, v. 27, no. 3, p. 72-73.
Mazzullo, S.J., C.S. Teal, and C.A. Burtnett, 1995b, Wolfcampian cyclothems and sequences in Kansas: an analysis of some Permian basin correlatives, *in* P.H. Pause and M.P. Candelaria, eds., Carbonate facies and sequence stratigraphy: practical applications of carbonate models: SEPM Permian Basin Section Publication 95-36 and Permian Basin Graduate Center Joint Publication 5-95, p. 217- 244.
McCrone, A.W., 1963, Paleoecology and biostratigraphy of the Red Eagle cyclothem (Lower Permian) in Kansas: Kansas Geological Survey Bulletin 164, 114 p.
McCrone, A.W., 1964, Water depth and Midcontinent cyclothems, *in* D.F. Merriam, ed., Symposium on cyclic sedimentation: Kansas Geological Survey Bulletin 169, p. 275-281.
McKee, E.D., and E.J. Crosby, 1975, Paleotectonic investigations of the Pennsylvanian System in the United States, Part II, interpretive summary and special features of the Pennsylvanian System: USGS Professional Paper 853, 192 p.
McPherson, J.G., R.E. Denison, D.W. Kirkland, and D.M. Summers, 1988, Basal sandstones of the Oil Creek Formation in the quarry of the Pennsylvania Glass Sand Corporation, Johnson County, Oklahoma, *in* O.T. Hayward, ed., GSA South-Central Section Centennial Field Guide, p. 165-170.
Meade, R.H., 1966, Factors influencing the early stages of the compactions of clays and sands: Journal of Sedimentary Petrology, v. 36, p. 1085-1101.
Melnyk, D.H., and R.F. Maddocks, 1988, Ostracode biostratigraphy of the Permo-Carboniferous of central and north-central Texas, Part I, paleoenvironmental framework: Micropaleontology, v. 34, p. 1-20.
Melnyk, D.H., D.G. Smith, and K. Amiri-Garroussi, 1994, Filtering and frequency mapping as tools in subsurface cyclostratigraphy, with examples from the Wessex basin, UK, *in* P.L. de Boer and D.G. Smith, eds., Orbital forcing and cyclic sequences: IAS Special Publication 19, p. 35-46.
Menning, M., 1989, A synopsis of numerical time scale 1917-1986: Episodes, v. 12, no. 1, p. 3-6.
Merriam, D.F., 1963, The geologic history of Kansas: Kansas Geological Survey Bulletin 162, 311 p.
Merriam, D.F., ed., 1964, Symposium on cyclic sedimentation: Kansas Geological Survey Bulletin 169, 636 p.
Merrill, G.K., 1973, Pennsylvanian conodont paleoecology, *in* F.H.T. Rhodes, ed., Conodont paleozoology: GSA Special Paper 141, p. 239-274.
Merrill, G.K., and P.H. von Bitter, 1976, Revision of conodont biofacies nomenclature and interpretations of environmental controls in Pennsylvanian rocks of eastern and central North America: Royal Ontario Museum of Life Sciences Contributions 108, 46 p.
Merrill, G.K., and P.H. von Bitter, 1989, Conodont paleoecology in the type East Mountain Shale, north-central Texas, *in* D.R. Boardman, J.E. Barrick, J.M. Cocke, and M.K. Nestell, eds., Middle and Late Pennsylvanian chronostratigraphic boundaries in north-central Texas — glacial-eustatic events, biostratigraphy, and paleoecology, Part II: Texas Tech University Studies in Geology, p. 235-248.
Meyer, B.L., and M.H. Nederlof, 1984, Identification of source rocks on wireline logs by density/resistivity and sonic transit time/resistivity crossplots: AAPG Bulletin, v. 68, p. 121-129.
Miall, A.D., 1986, Eustatic sea level changes interpreted from the stratigraphic record: a critique of the methodology with particular reference to the North Sea Jurassic record: AAPG Bulletin v. 70, p. 131-137.
Miall, A.D., 1992, Exxon global cycle chart: an event for every occasion?: Geology, v. 20, p. 787-790.
Miller, K.G., T.J. McCahon, and R.R. West, 1992, Introduction to Lower Permian (Wolfcampian) cycles of Kansas: Project Pangea Workshop Guidebook, p. I-1 to I-47.
Miller, K.G., and R.R. West, 1993, Reevaluation of Wolfcampian cyclothems in northeastern Kansas: significance of subaerial exposure and flooding surfaces: Kansas Geological Survey Bulletin 235, p. 1-26.
Miller, F.X., 1977, The graphic correlation method in biostratigraphy, *in* E.G. Kauffman and J.E. Hazel, eds., Concepts and methods of biostratigraphy: Stroudsburg, PA, Dowden,Hutchinson and Ross, Inc., p. 165-186.
Mitchum, R.M., Jr., 1977, Seismic stratigraphy and global changes of sea level. Part 1: Glossary of terms used in

seismic stratigraphy, *in* C.E. Payton, ed., Seismic stratigraphy — applications to hydrocarbon exploration: AAPG Memoir 26, p. 205-212.

Mitchum, R.M., Jr., P.R. Vail and S. Thompson III, 1977, Seismic stratigraphy and global changes in sea level. Part 2: The depositional sequence as a basic unit for stratigraphic analysis, *in* C.E. Payton, ed., Seismic stratigraphy — applications to hydrocarbon exploration: AAPG Memoir 26, p. 53-62.

Mitchum, R.M., Jr., P.R. Vail, and J.B. Sangree, 1977, Seismic stratigraphy and global changes of sea level. Part 6: Stratigraphic interpretation of seismic reflection patterns in depositional sequences, *in* C.E. Payton, ed., Seismic stratigraphy — applications to hydrocarbon exploration: AAPG Memoir 26, p. 117-133.

Mitchum, R.M., Jr., 1977, Seismic stratigraphy and global changes of sea level. Part 11: Glossary of terms used in seismic stratigraphy, *in* C.E. Payton, ed., Seismic stratigraphy — applications to hydrocarbon exploration: AAPG Memoir 26, p. 205-212.

Mitchum, R.M., Jr., and J.C. Van Wagoner, 1991, High-frequency sequences and their stacking patterns: sequence stratigraphic evidence of high-frequency eustatic cycles: Sedimentary Geology, v. 70, p. 131-160.

Montanez, I.P., and D.A. Osleger, 1993, Parasequence stacking patterns, third-order accommodation events, and sequence stratigraphy of Middle to Upper Cambrian platform carbonates, Bonanza King Formation, southern Great basin, *in* R.G. Loucks and J.F. Sarg, eds., Carbonate sequence stratigraphy, recent developments and applications: AAPG Memoir 57, p. 305-326.

Moody, J.D., J.W. Mooney, and J. Spirak, 1970, Giant oil fields of North America, *in* M.T. Halbouty, ed., Geology of giant petroleum fields: AAPG Memoir 14, p. 8-17.

Moore, R.C., 1930, Sedimentation cycles in the Pennsylvanian of the northern Mid-Continent region: GSA Bulletin, v. 41, p. 51-52.

Moore, R.C., 1931, Pennsylvanian cycles in the northern Mid-Continent region: Illinois Geological Survey Bulletin 60, p. 247-257.

Moore, R.C., 1932, A reclassification of the Pennsylvanian system in the northern Mid-Continent region: Guidebook of the Kansas Geological Society Sixth Annual Field Conference, p. 79-97.

Moore, R.C., 1936, Stratigraphic classification of the Pennsylvanian rocks of Kansas: Kansas Geological Survey Bulletin 22, 256 p.

Moore, R.C., J.C. Frye, and J.M. Jewett, 1944, Tabular description of outcropping rocks in Kansas: Kansas Geological Survey Bulletin 52, pt. 4, 212 p.

Moore, R.C., 1949, Divisions of the Pennsylvanian System in Kansas: Kansas Geological Survey Bulletin 83, 201 p.

Moore, R.C., 1950, Late Paleozoic cyclic sedimentation in central United States: 18th International Geological Congress Report, pt. 4, p. 5-16.

Moore, R.C., J.C. Frye, J.M. Jewett, L. Wallace, and H.G. O'Connor, 1951, The Kansas rock column: Kansas Geological Survey Bulletin 89, 132 p.

Moore, R.C., 1957, Geological understanding of cyclic sedimentation represented by Pennsylvanian and Permian rocks of northern Mid-Continent region: Kansas Geological Society 21st Annual Field Conference Guidebook, p. 91-100.

Moore, R.C., 1964, Paleoecological aspects of Kansas Pennsylvanian and Permian cyclothems, *in* D.F. Merriam, ed., Symposium on cyclic sedimentation: Kansas Geological Survey Bulletin 169, v. 1, p. 287-380.

Morgan, W.A., R.E. Schneider, and J.H. Copley, 1982, Identification of subtle porosity and traps within Frisco Formation, Canadian County, Oklahoma: a geologic, seismic-waveform approach, *in* M. Halbouty, ed., The deliberate search for the subtle trap: AAPG Memoir 32, p. 93-114.

Morgan, W.A., and J.A. Babcock, eds., 1988, Permian rocks of the Midcontinent: SEPM Midcontinent Section Special Publication 1, 224 p.

Moussavi-Harami, R., and R.L. Brenner, 1984, Deltaic sedimentation on a carbonate shelf — Stanton Formation (Upper Pennsylvanian), southeastern Kansas: AAPG Bulletin, v. 68, p. 150-163.

Mudge, M.R., and E.L. Yochelson, 1962, Stratigraphy and paleontology of the uppermost Pennsylvanian and lowermost Permian rocks in Kansas: USGS Professional Paper 323, 213 p.

Mulholland, J.W., 1994, Sequence stratigraphic correlation of well log cross sections: Mountain Geologist, v. 31, p. 65-75.

Murray, J.W., 1991, Ecology and paleoecology of benthic foraminifera of Oklahoma and Texas: Harlow, England, Longman Scientific and Technical, 397 p.

Nelson, M.R., 1985, Petrology, diagenesis and depositional environment of the Lagonda interval, Cabaniss Subgroup, Cherokee Group, Middle Pennsylvanian, in northeastern Kansas: M.S. thesis, University of Iowa, 175 p.

Neumann, A.C., and I. Macintyre, 1985, Reef response to sea-level rise — keep up, catch up, or give up: Proceedings of the Fifth International Coral Reef Congress, Tahiti, v. 3, p. 105-110.

Newell, K.D., W.L. Watney, B.P. Stephens, and J.R. Hatch, 1987, Hydrocarbon potential in Forest City basin: Oil and Gas Journal v. 85, no. 42, Oct. 19, 1987, p. 58-62.

Newell, K.D., W.L. Watney, S.W.L. Cheng, and R.L. Brownrigg, 1987, Stratigraphic and spatial distribution of oil and gas production in Kansas: Kansas Geological Survey Subsurface Geology Series 9, 86 p.

Nichols, M.M., G.H. Johnson, and P.C. Peebles, 1991, Modern sediments and facies model for a microtidal coastal plain estuary, the James estuary, Virginia: Journal of Sedimentary Petrology, v. 61, p. 883-899.

Nio, S.D., and C. Yang, 1991, Diagnostic attributes of clastic tidal deposits: a review, *in* D.G. Smith, G.E. Reinson, B.A. Zaitlin, and R.A. Rahami, eds., Clastic tidal sedimentology: Canadian Society Petroleum Geologists Memoir 16, p. 3-27.

Nodine-Zeller, D.E., 1978, Radiolarian-sponge fauna from phosphatic black shales in the Middle Pennsylvanian of Kansas: Transactions of the Kansas Acadamy of Science, v. 81, p. 108-109.

Nodine-Zeller, D.E., B.K. Holdsworth, and P. Berendson, 1985, Well preserved Middle and Late Pennsylvanian radiolaria from Kansas, *in* J.T. Dutro, Jr. and H.W. Pfefferkorn, eds., 9th International Congress on Carboniferous Stratigraphy and Geology, v. 5, p. 181-186.

North American Commission on Stratigraphic Nomenclature, 1983, North American Stratigraphic Code: AAPG Bulletin,

v. 67, p. 841-875.
Oakes, M.C., 1952, Geology and mineral resources of Tulsa County, Oklahoma: Oklahoma Geological Survey Bulletin 69, p. 1-234.
Oakes, M.C., 1959, Geology and mineral resources of Creek County, Oklahoma: Oklahoma Geological Survey Bulletin 81, p. 1-134.
Oakes, M.C., and J.M. Jewett, 1943, Upper Desmoinesian and lower Missourian rocks in northeastern Oklahoma and southeastern Kansas: AAPG Bulletin, v. 27, no. 5, p. 632-640.
Oakes, M.C., and W.S. Motts, 1963, Geology and mineral resources of Okmulgee County, Oklahoma: Oklahoma Geological Survey Bulletin 91, p. 1-164.
O'Brien, J.E., and J.R. Derby, in press, Progress report on Simpson and Viola correlations from the Arbuckles to the Ozarks, with data from the Keener No. 1 Watson, 15-14S-8E, Creek County, *in* K.S. Johnson, ed., Simpson and Viola groups in the southern Mid-Continent, 1994 Symposium: Oklahoma Geological Survey Circular.
Ohlson, P.E., 1986, A 40-million-year lake record of early Mesozoic orbital climatic forcing: Science, v. 234, p. 842-848.
Palmer, A.R., 1983, The decade of North American geology: 1983 geologic time scale: Geology, v. 11, p. 503-504.
Parham, K.D., 1993, Introduction to Permian plays, *in* D.G. Bebout, W.A. White, and T.F. Hentz, eds., Atlas of major Midcontinent gas reservoirs: Gas Research Institute and Bureau of Economic Geology, University of Texas, p. 6-7.
Parham, K.D., and J.A. Campbell, 1993, Wolfcampian shallow-shelf carbonate — Hugoton embayment, Kansas and Oklahoma, *in* D.G. Bebout, W.A. White, and T.F. Hentz, eds., Atlas of major midcontinent gas reservoirs: Gas Research Institute and Bureau of Economic Geology, University of Texas, p. 9-12.
Parkinson, R.W., 1982, Petrology, conodont distribution, and depositional interpretation of the Lenapah Formation: M.S. thesis, University of Iowa, 89 p.
Parrish, J.T., 1982, Upwelling and petroleum source beds, with reference to Paleozoic: AAPG Bulletin, v. 66, no. 6, p. 750-774.
Patzkowsky, M.E., L.H. Smith, P.J. Markwick, C.J. Engherts, and E.D. Gyllenhaal, 1991, Application of the Fujita-Ziegler paleoclimate model — Early Permian and Late Cretaceous examples: Palaeogeography, Palaeoclimatology, Palaeoecology, v. 86, p. 67-85.
Pavlicek, J.A., 1986, Conodont distribution and correlation of the Tacket Shale in southeastern Kansas and eastern Oklahoma: M.S. thesis, University of Iowa, p. 1-132.
Payton, C.E., ed., 1977, Seismic stratigraphy — application to hydrocarbon exploration: AAPG Memoir 26, 516 p.
Perlmutter, M.A., and M.D. Matthews, 1993, Analysis of Carboniferous cyclothems using global cyclostratigraphy and the stratigraphic computer model SEDPAK (abs.): Annual Convention, Canadian Society of Petroleum Geologists Program and Abstracts, p. 241.
Peterson, J.A., 1980, Permian paleogeography and sedimentary provinces, west-central United States, *in* T.D. Fouch and E. R. Magathan, eds., Paleozoic paleogeography of the west-central United States: SEPM Rocky Mountain Section Rocky Mountain Paleogeography Symposium 1, p. 271-292.
Phillips, J., 1934, Illustrations of the Yorkshire II, the Mountain Limestone district: London, John Murray, p. 1-253.
Phillips, T.L., and A.T. Cross, 1991, Paleobotany and paleoecology of coal, *in* R.B. Taylor, D.D. Rice, and H.G. Gluskoter, eds., Economic geology, U.S., v. P-2, The geology of North America: Boulder, CO, GSA, p. 483-502.
Piper, D.J.W., and A.E. Aksu, 1992, Architecture of stacked Quaternary deltas correlated with global oxygen isotopic curve: Geology, v. 21, p. 415-418.
Pippin, L., 1970, Panhandle-Hugoton field, Texas, Oklahoma and Kansas — the first 50 years, *in* M.T. Halbouty, ed., Geology of giant petroleum fields: AAPG Memoir 14, p. 204-222.
Plafker, G., and J.C. Savage, 1970, Mechanism of the Chilean earthquake of May 21 and 22, 1960: GSA Bulletin, v. 81, p. 1001-1030.
Plummer, H.J., 1930, Calcareous Foraminifera in the Brownwood Shale near Bridgeport, Texas: Texas University Bulletin 3019, p. 5-21, plate 1.
Plummer, H.J., 1945, Smaller Foraminifera in the Marble Falls, Smithwick, and lower Strawn strata around the Llano uplift in Texas: Texas University Bulletin 4401, p. 209-271, plates 15-17.
Posamentier, H.W., 1988, Fluvial deposition in a sequence stratigraphic framework, *in* D.P. James and D.A. Leckie, eds., Sequences, stratigraphy, sedimentology: surface and subsurface: Canadian Society of Petroleum Geologists Memoir 15, p. 582-583.
Posamentier, H.W., and P.R. Vail, 1988, Eustatic controls on clastic deposition II — sequence and systems tracts models, *in* C.K. Wilgus et al., eds., Sea-level changes: an integrated approach: SEPM Special Publication 42, p. 125-154.
Posamentier, H.W., G.P. Allen, D.P. James, and M. Tesson, 1992a, Forced regressions in a sequence stratigraphic framework: concepts, examples, and exploration significance: AAPG Bulletin, v. 76, no. 11, p. 1687-1709.
Posamentier, H.W., G.P. Allen, and D.P. James, 1992b, High-resolution sequence stratigraphy — the East Coulee delta: Journal of Sedimentary Petrology, v. 62, p. 310-317.
Posamentier, H.W., and D.P. James, 1993, An overview of sequence-stratigraphic concepts: uses and abuses, *in* H.W. Posamentier, C.P. Summerhayes, B.A. Haq, and G.P. Allen, eds., Sequence stratigraphy and facies associations: IAS Special Publication 18, p. 3-18.
Potter, P.E., and J.A. Simon, 1961, Anvil Rock Sandstone and channel cutouts of Herrin (No. 6) coal in west-central Illinois: Illinois State Geological Survey Circular 314, 12 p.
Potter, P.E., and G.A. Desborough, 1965, Pre-Pennsylvanian Evansville Valley and Caseyville (Pennsylvanian) sedimentation in the Illinois basin: Illinois State Geological Survey Circular 384, 16 p.
Potter, P.E., 1978, Significance and origin of big rivers: Journal of Geology, v. 86, p. 13-33.
Prather, B.E., 1981, Petrology and diagenesis of the D-Zone megacyclothem of the Lansing-Kansas City groups, Hitchcock County, Nebraska: M.S. thesis, University of New Orleans, 97 p.
Prather, B.E., 1985, An Upper Pennsylvanian desert paleosol in the D-Zone of the Lansing-Kansas City groups, Hitchcock County, Nebraska: Journal of Sedimentary Petrology, v. 55, p. 213-221.

Price, R.C., 1981, Stratigraphy, petrology, and depositional environments of the Pawnee Limestone, Midcontinent North America: Ph.D. dissertation, University of Iowa, p. 1-279.

Prosser, C.S., 1895, The classification of the Upper Paleozoic rocks of Central Kansas: Journal of Geology, v. 3, p. 682-705 and 764-800.

Puckette, J., 1993, The petrography and diagenetic history of an upper Morrowan valley-fill sandstone: Oklahoma State University, unpublished report.

Radke, B.M., and R.L. Mathis, 1980, On the formation and occurrence of saddle dolomite: Journal of Sedimentary Petrology, v. 50, p. 1149-1204.

Rahmani, R.A., 1989, Cretaceous tidal estuarine and deltaic deposits, Drumheller, Alberta: Field Trip Guide to Second International Research Symposium on Clastic Tidal Deposits, Canadian Society of Petroleum Geologists, Calgary, Canada, 55 p.

Rascoe, B., Jr., 1962, Regional stratigraphic analysis of Pennsylvanian and Permian rocks in western Mid-Continent, Colorado, Kansas, Oklahoma, and Texas: AAPG Bulletin, v. 46, p. 1345-1363.

Rascoe, B., Jr., 1978, Sedimentary cycles in the Virgilian Series (Upper Pennsylvanian) of the Anadarko basin, Parts 1 and 2: Shale Shaker, v. 28, p. 123-131, 144-149.

Rascoe, B., Jr., and F.J. Adler, 1983, Permo-Carboniferous hydrocarbon accumulations, Mid-Continent, USA: AAPG Bulletin, v. 67, p. 979-1001.

Rascoe, B., Jr., and N.J. Hyne, eds., 1987, Petroleum geology of the Mid-Continent: Tulsa Geological Society Special Publication 3, p. 75-81.

Rascoe, B., Jr., and K. S. Johnson, 1988, Anadarko basin and Hugoton embayment, *in* L.L. Sloss, ed., Sedimentary cover — North American craton: U.S.:The geology of North America: Boulder, CO, GSA, v. D-2, p. 318-321.

Rascoe, B., Jr., 1988, Permian System in western Midcontinent, *in* W.A. Morgan and J.A. Babcock, eds., Permian rocks of the Midcontinent: SEPM Midcontinent Section Special Publication 1, p. 3-12.

Ravn, R.L., J.W. Swade, M.R. Howes, J.L. Gregory, R.R. Anderson, and P.E. Van Dorpe, 1984, Stratigraphy of the Cherokee Group and revision of Pennsylvanian stratigraphic nomenclature in Iowa: Iowa Geological Survey Technical Information Series 12, 76 p.

Read, J.F., 1989, Controls on evolution of Cambro-Ordovician passive margins, U.S. Appalachian, *in* P.D. Crevello, J.L. Wilson, J.L. Sarg, and J.F. Read, eds., Controls on carbonate platform and basin development: SEPM Special Publication 44, p. 147-165.

Read, J.F., and R.K. Goldhammer, 1988, Use of Fischer plots to define third-order sea-level curves in Ordovician peritidal cyclic carbonates, Appalachians: Geology, v. 16, p. 895-899.

Read, J.F., W.F. Koerschner II, D.A. Osleger, G.A. Bollinger, and C. Coruth, 1991, Field and modeling studies of Cambrian carbonate cycles, Virginia Appalachians — Reply: Journal of Sedimentary Petrology, v. 61, p. 647-652.

Reike, H.H., and G.V. Chilingarian, 1974, Compaction of argillaceous sediments: Developments in Sedimentology 16, 424 p.

Reineck, H.E., and F. Wunderlich, 1968, Classification and origin of flaser and lenticular bedding: Sedimentology, v. 11, p. 99-104.

Reinholtz, P.N., 1982, Distribution, petrology, and depositional environment of 'Bush City Shoestring Sandstone' and 'Centerville Lagonda Sandstone' in Cherokee Group (Middle Pennsylvanian), southeastern Kansas: M.S. thesis, University of Iowa, 180 p.

Reinson, G.E., ed., 1989, Modern and ancient examples of clastic tidal deposits — a core and peel workshop: Canadian Society of Petroleum Geologists, Second International Research Symposium on Clastic Tidal Deposits, August 22-25, Calgary, Alberta, 126 p.

Reisz, R.R., M.J. Heaton, and B.R. Pynn, 1982, Vertebrate fauna of Late Pennsylvanian Rock Lake Shale near Garnett, Kansas: Journal of Paleontology, v. 56, p. 741.

Retallack, G.J., 1988, Field recognition of paleosols, *in* J. Reinhardt and W.R. Sigleo, eds., Paleosols and weathering through geologic time — principles and applications: GSA Special Paper 216, p. 1-20.

Rich, J.L., 1951, Three critical environments of deposition, and criteria for recognition of rocks deposited in each of them: GSA Bulletin, v. 66, p. 1-22.

Ries, E.R., 1954, Geology and mineral resources of Okfuskee County, Oklahoma: Oklahoma Geological Survey Bulletin 74, p. 1-175.

Ritter, S.M., 1987, Biofacies-based refinement of Early Permian biostratigraphy, in central and western USA, *in* R.L. Austin, ed., Conodonts, investigative techniques and applications: Ellis Harwood Ltd. for British Micropaleontological Society, Chichester, West Sussex, England, p. 382-403.

Roberts, M.T., 1987, Carboniferous shelf and basin facies of eastern Oklahoma: Tulsa Geological Society Guidebook, March 1987, p. 1-65.

Ross, C.A., 1963, Standard Wolfcampian Series (Permian), Glass Mountains, Texas: GSA Memoir 88, 205 p.

Ross, C.A., and J.R.P. Ross, 1985, Late Paleozoic depositional sequences are synchronous and worldwide: Geology, v. 13, p. 27-30.

Ross, C.A., 1986, Paleozoic evolution of southern margin of Permian basin: GSA Bulletin, v. 97, p. 536-554.

Ross, C.A., and J.R.P. Ross, 1987, Late Paleozoic sea levels and depositional sequences, *in* C.A. Ross and D. Haman, eds., Timing and depositional history of eustatic sequences: constraints on seismic stratigraphy: Cushman Foundation for Foraminiferal Research Special Publication 24, p. 137-168.

Ross, C.A., and J.R.P. Ross, 1988, Late Paleozoic transgressive deposition, *in* D.K. Wilgus et al., eds., Sea-level changes: an integrated approach: SEPM Special Publication 42, p. 227-247.

Rossignol-Strick, 1982, Petroleum origin — heavy rain, river plume, ocean stratificiation: AAPG Bulletin, v. 66, p. 137-168.

Rossinski, V. Jr., H.R. Wanless, and P.K. Stewart, 1992, Penetrative calcretes and their stratigraphic implications: Geology, v. 20, no. 4, p. 331-334.

Roth, S.M., 1991, Regional stratigraphic analysis of the Blackjack Creek Limestone (Desmoinesian, Middle Pennsylvanian) in southeast Kansas and northeast Oklahoma: M.S. thesis, Kansas State University, 243 p..

Rutan, D., 1980, The environment of deposition of the Ireland Sandstone Member and related strata: Unpublished M.S. thesis, University of Kansas, 120 p.

Ryland, J.S., 1967, Polyzoa, *in* H. Barnes, ed., Oceanography and Marine Biology Annual Review, v. 5, p. 343-359.

Saller, A.H., and K. Yaremko, 1994, Dolomitization and porosity development in the middle and upper Wabamun Group, southeast Peace River arch, Alberta, Canada: AAPG Bulletin, v. 78, p. 1406-1430.

Sanders, D.T., 1959, Sandstones of the Douglas and Pedee groups in northeastern Kansas: Kansas Geological Survey Bulletin 134, p. 125-159.

Sanderson, G.A., and G.J. Verville, 1970, Morphologic variability of the genus *Schwagerina* in the Lower Permian Wreford Limestone of Kansas: Paleontology, v. 13, p. 175-183.

Sanderson, G.A., and G.J. Verville, 1988, Permian fusulinids from the Amoco Production Company No. 1 Eldon Hargrave core, Riley County, Kansas, *in* W.A. Morgan and J.A. Babcock, eds., Permian rocks of the Midcontinent: SEPM Midcontinent Section Special Publication 1, p. 205-212.

Sangree, J.B., P.R. Vail, and R.M. Sneider, 1988, Evolution of facies interpretation of the shelf-slope: application of the new eustatic framework in the Gulf of Mexico: Proceedings of the 20th Annual Offshore Technology Conference, v. 2, OTC 5695, p. 133-144.

Sangree, J.G., P.R. Vail, and R.M. Mitchum, Jr., 1990, A summary of exploration applications of sequence stratigraphy, *in* J.M. Armentrout and B.F. Perkins, eds., Sequence stratigraphy as an exploration tool: concepts and practices in the Gulf Coast: SEPM Gulf Coast Section Foundation 11th Annual Research Conference Program and Abstracts, p. 53-58.

Sarg, J.F., 1988, Carbonate sequence stratigraphy, *in* C.K. Wilgus et al., eds., Sea-level changes: an integrated approach: SEPM Special Publication 42, p. 155-181.

Schatski, N.S., 1946, The Great Donetz basin and the Wichita System, comparative tectonics of ancient platforms: Izv. Akad. Nauk SSSR, Ser. Geol., no. 6, p. 57-90.

Schenk, P.E., 1967, Facies and phases of the Altamont Limestone, Iowa to Oklahoma: GSA Bulletin, v. 78, 1369-1384.

Schlager, W., 1992, Sedimentology and sequence stratigraphy of reefs and carbonate platforms: a short course: AAPG Continuing Education Course Note Series 34, 71 p.

Schramm, W.W., Jr., 1964, Paleogeologic and quantitative lithofacies analysis of Simpson Group, Oklahoma: AAPG Bulletin, v. 48, p. 1164-1195.

Schramm, W.W., Jr., 1965, Resume of Simpson (Ordovician) Stratigraphy *in* T. Herndon, ed., Symposium on the Simpson Group: Tulsa Geological Society Digest, v. 33, p. 26-34.

Schulte, G.S., 1959, The Cottage Grove and Noxie sandstones ("Layton") in south-central Kansas: Kansas Geological Society 24th Field Conference Guidebook, p. 101-106.

Schumm, S.A., 1993, River response to base level change: implications for sequence stratigraphy: Journal of Geology, v. 101, p. 279-294.

Schutter, S.R., and P.H. Heckel, 1985, Missourian (Early Late Pennsylvanian) climate in Midcontinent North America: International Journal of Coal Geology, v. 5, p. 111-140.

Schwarzacher, W., and A.G. Fischer, 1982, Limestone-shale bedding and perturbations of the earth's orbit, *in* G. Einsele and A. Seilacher, eds., Cyclic and event stratification: New York, Springer-Verlag, p. 72-95.

Scotese, C.R., and W.S. McKerrow, 1990, Revised world maps and introduction, *in* W.S. McKerrow and C.R. Scotese, eds., Palaeozoic palaeogeography and biogeography: Geological Society of London Memoir 12, p. 1-21.

Scott, R.W., D. Fee, R. Magee, and H. Laali, 1978, Epeiric depositional models for the Lower Cretaceous Washita Group, north-central Texas: University of Texas Bureau of Economic Geology Report of Investigation 94, 23 p.

Scott, R.W., S.H. Frost, and B.L. Shaffer, 1988, Early Cretaceous sea-level curves, Gulf Coast and southeastern Arabia, *in* C.K. Wilgus et al., eds., Sea-level changes: an integrated approach: SEPM Special Publication 42, p. 275-284.

Scott, R.W., T.J.A. Simo, and J.P. Masse, 1993, Economic resources in Cretaceous carbonate platforms: an overview: *in* J.A. Simo et al., eds., Cretaceous carbonate platforms: AAPG Memoir 56, p. 15-23.

Scott, R.W., P.C. Franks, J.A. Stein, J.A. Bergen, and M.J. Evetts, 1994, Graphic correlation tests the synchronous Mid-Cretaceous depositional cycles: Western Interior and Gulf Coast: Rocky Mountain Association of Geologists Unconformity Control Symposium, p. 89-96.

Seddon, G., and W.C. Sweet, 1971, An ecologic model for conodonts: Journal of Paleontology, v. 45, p. 869-880.

Shanley, K.W., and P.J. McCabe, 1994, Perspectives on the sequence stratigraphy of continental strata: AAPG Bulletin, v. 78, p. 544-568.

Shaw, A.B., 1964, Time in stratigraphy: New York, McGraw Hill, 365 p.

Shelby, J.M., 1980, Geologic and economic significance of the upper Morrow chert conglomerate reservoir of the Anadarko basin: Journal of Petroleum Technology, v. 32, p. 489-495.

Shelton, J.W., 1973, Models of sand and sandstone deposits: a methodology for determining sand genesis and trend: Oklahoma Geological Survey Bulletin 118, 122 p.

Shinn, E.A., 1968, Practical significance of birdseye structures in carbonate rocks: Journal of Sedimentary Petrology, v 38, p. 215-223.

Shumard, C.B., 1991, Stockholm Southwest field — USA, *in* N.H. Foster and E.A. Beaumont, eds., Stratigraphic Traps II: AAPG Treatise of Petroleum Geology, Atlas of Oil and Gas Fields, p. 269-304.

Siever, R., 1951, The Mississippian-Pennsylvanian unconformity in southern Illinois: AAPG Bulletin, v. 35, no. 3, p. 542-581.

Sloss, L.L., 1963, Sequences in cratonic interior of North America: GSA Bulletin, v. 74, p. 93-114.

Sloss, L.L., and R.C. Speed, 1974, Relationships of cratonic and continental margin tectonic episodes, *in* W.R. Dickinson, ed., Tectonics and sedimentation: SEPM Special Publication 22, p. 98-119.

Sloss, L.L., 1984, Comparative anatomy of cratonic unconformities, *in* L.L. Sloss, ed., Interregional unconformities and hydrocarbon accumulation: AAPG Memoir 36, p. 1-7.

Sloss, L.L., 1988, Forty years of sequence stratigraphy: GSA Bulletin, v. 100, p. 1661-1665.

Smith, D.G., 1988, Tidal bundles and mud couplets in the McMurray Formation, northeastern Alberta, Canada: Canadian Petroleum Geologists Bulletin, v. 36, p. 216-219.

Sohn, I.G., 1982, Biostratigraphic evaluation of the Carboniferous and Permian ostracodes of Texas, *in* R.F.

Maddocks, ed., Texas Ostracoda: Guidebook of excursions and related papers for the 8th International Symposium on Ostracoda, Deptartment of Geosciences, University of Houston, p. 103-109.

Sonnenberg, S.A., 1985, Tectonic and sedimentation model for Morrow Sandstone deposition, Sorrento field area, Denver basin, Colorado: Mountain Geologist, v. 22, p. 180-191.

Sonnenberg, S.A., 1990, Upper Morrow Purdy Sandstone, Interstate field, Morton County, Kansas, *in* S.A. Sonnenberg et al., eds., Morrow sandstones of southeast Colorado and adjacent areas: Denver, The Rocky Mountain Association of Geologists, p. 207-216.

Sonnenberg, S.A., L.T. Shannon, K. Rader, and W.F. Von Drehle, 1990, Regional structure and stratigraphy of the Morrowan Series, southeast Colorado and adjacent areas, *in* S.A. Sonnenberg et al., eds., Morrow sandstones of southeast Colorado and adjacent areas: Denver, The Rocky Mountain Association of Geologists, p. 1-8.

Spriggs, J.L., 1989, Depositional environments of the Tonganoxie Sandstone Member of the Stranger Formation (Upper Pennsylvanian) in northeastern Kansas: Unpublished M.S. thesis, Kansas State University, 169 p.

Statler, A.T., 1965, Stratigraphy of the Simpson Group in Oklahoma: Tulsa Geological Society Digest, v. 33, p. 162-209.

Steeples, D.W., 1987, Earthquakes: Lawrence, Kansas Geological Survey, 7 p.

Stitt, J.H., 1983, Trilobites, biostratigraphy, and lithostratigraphy of the McKenzie Hill Limestone (Lower Ordovician), Wichita and Arbuckle Mountains, Oklahoma: Oklahoma Geological Survey Bulletin 134, 54 p.

Stone, W.P., Jr., 1984, Origin and evolution of oolite in the Drum Limestone, Montgomery County, Kansas, *in* Limestones of the Mid-Continent, N.J. Hyne, ed., Tulsa Geological Society Special Publication 2, p. 51-74.

Suhm, R.W., in press, Simpson stratigraphy of the southern Midcontinent, *in* K.S. Johnson, ed., Simpson and Viola groups in the southern Mid-Continent, 1994 Symposium: Oklahoma Geological Survey Circular.

Sutherland, P.K., 1988, Late Mississippian and Pennsylvanian depositional history in the Arkoma basin area, Oklahoma and Kansas: GSA Bulletin, v. 100, p. 1787-1802.

Swade, J.W., 1985, Conodont distribution, paleoecology, and preliminary biostratigraphy of the upper Cherokee and Marmaton groups (upper Desmoinesian, Middle Pennsylvanian) from two cores in south-central Iowa: Iowa Geological Survey Technical Information Series 14, 71 p.

Swallow, G.C., 1866, Preliminary report of the Kansas Geological Survey: Lawrence, Kansas, John Speer, Printer, 94 p.

Swanson, D.C., 1979, Deltaic deposits in the Pennsylvanian upper Morrow Formation of the Anadarko basin, *in* N.J. Hyne, ed., Pennsylvanian sandstones of the Mid-Continent: Tulsa Geological Society Special Publication 1, p. 115-168.

Sweet, W.C., 1984, Graphic correlation of upper Middle and Upper Ordovician rocks, North American Midcontinent Province, U.S.A., *in* D.L. Bruton, ed., Aspects of the Ordovician System: University of Oslo Paleontological Contributions 295, p. 23-35.

Sweet, W.C., 1992, Middle and Late Ordovician conodonts from southwestern Kansas — their biostratigraphic significance, *in* J.R. Chaplin and J.E. Barrick, eds., Special papers in paleontology and stratigraphy: a tribute to Thomas W. Amsden: Oklahoma Geological Survey Bulletin 145, p. 181-191.

Taff, J.A., 1901, Description of the Coalgate Quadrangle: USGS Geologic Atlas and Folio 74.

Taff, J.A., 1902, Description of the Atoka quadrangle: USGS Atlas and Folio 79, 8 p.

Tanner, W.F., 1956, Geology of Seminole County, Oklahoma: Oklahoma Geological Survey Bulletin 74, p. 1-170.

Tasch, P., 1957, Fauna and paleoecology of the Pennsylvanian Dry Shale of Kansas, *in* Treatise on marine ecology and paleoecology: GSA Memoir 67, v. 2, p. 356-406.

Tessier, B., 1993, Upper intertidal rhythmites in the Mont-Saint-Michel Bay (NW France): perspectives for paleoreconstruction: Marine Geology, v. 110, p. 355-367.

Tomlinson, C.W., and M. McBee, Jr., 1959, Pennsylvanian sediments and orogenies of Ardmore district, Oklahoma, *in* C.C. Branson, ed., Petroleum geology of southern Oklahoma, v. II: AAPG, Tulsa, OK, p. 3-52.

Tomlinson, C.W., and M. McBee, Jr., 1962, Pennsylvanian sediments and orogenies of Ardmore district, Oklahoma, *in* C.C. Branson, ed., Pennsylvanian systems in the United States: AAPG, Tulsa, OK, p. 461-500.

Toomey, D.F., and R. Mitchell, 1986, Facies relationships and paleodepositional settings of the Herington to Winfield stratigraphic interval (Lower Permian), southern Kansas-northern Oklahoma: SEPM Midcontinent Section Field Guidebook,173 p.

Toomey, D.F., 1992, Microfacies correlation of the Early Permian Barneston Limestone, Conoco Test Facility to Vap's Pass, Kay County, northern Oklahoma, *in* J.R. Chaplin and J.E. Barrick, eds., Special papers in paleontology and stratigraphy: a tribute to Thomas W. Amsden: Oklahoma Geological Survey Bulletin 145, p. 193-219.

Twiss, P.C., 1991a, Chase Group from the near-surface Amoco No. 1 Hargrave core, Riley County, Kansas: AAPG Midcontinent Section Core Workshop (Integrated studies of petroleum reservoirs in the Midcontinent), p. 123-141.

Twiss, P.C., 1991b, Nodular and bedded chert in the Florence Limestone (Lower Permian) of east-central Kansas: GSA Abstracts With Program, v. 23, p. 101.

Udden, J.A., 1912, Geology and mineral resources of the Peoria quadrangle, Illinois: USGS Bulletin 506, p. 1-103.

Ulmer-Scholle, D.S., and P.A. Scholle, 1994, Replacement of evaporites within the Park City Formation, Bighorn basin, Wyoming, USA: Sedimentology, v. 41, p. 1203-1222.

Ulrich, E.O., 1929, New classifiction of the Paleozoic deposits in Oklahoma: GSA Bulletin, v. 40, no. 1, p. 85-86.

Underwood, M.R., 1984, Stratigraphy and depositional history of the Bourbon Flags (Upper Pennsylvanian), enigmatic unit, Linn and Bourbon counties in east-central Kansas: M.S. thesis, University of Iowa, 111 p.

Vail, P.R., R.M. Mitchum, Jr., and S. Thompson, III, 1977, Seismic stratigraphy and global changes of sea level. Part 3: Relative changes of sea level from coastal onlap, *in* C.E. Payton, ed., Seismic stratigraphy — applications to hydrocarbon exploration: AAPG Memoir 26, p. 63-81.

Vail, P.R., R.M. Mitchum, Jr., and S. Thompson III, 1977a, Seismic stratigraphy and global changes of sea level. Part 4: Global cycles of relative changes in sea level, *in* C.E. Payton, ed., Seismic stratigraphy — applications to hydrocarbon exploration: AAPG Memoir 26, p. 83-98.

Vail, P.R., R.G. Todd, and J.B. Sangree, 1977b, Seismic stratigraphy and global changes of sea level. Part 5: Chronostratigraphic significance of seismic reflections, *in* Seismic stratigraphy — applications to hydrocarbon exploration: AAPG Memoir 26, p. 99-116.

Vail, P.R., and R.G. Todd, 1981, Northern North Sea Jurassic unconformities, chronostratigraphy and sea-level changes from seismic stratigraphy, *in* L.V. Illing and G.D. Hobson, eds., Petroleum geology of the Continental Shelf of northwest Europe: London, Institute of Petroleum, p. 216-235.

Vail, P.R., J. Hardenbol, and R.G. Todd, 1984, Jurassic unconformities, chronostratigraphy, and sea-level changes from seismic stratigraphy and biostratigraphy, *in* J.S. Schlee, ed., Interregional unconformities and hydrocarbon accumulation: AAPG Memoir 36, p. 129-144.

Vail, P.R., 1987, Seismic stratigraphy interpretation using sequence stratigraphy. Part 1: Seismic stratigraphy interpretation procedure, *in* A.W. Bally, ed., Atlas of seismic stratigraphy, v. 1: AAPG Studies in Geology 27, p. 1-10.

Vail, P.R., F. Audemard, S.A. Bowman, P.N. Eisner, and C. Perez-Cruz, 1991, The stratigraphic signatures of tectonics, eustasy and sedimentology — an overview, *in* G. Einsele, W. Ricken, and A. Seilacher, eds., Cycles and events in stratigraphy: New York, Springer-Verlag, p. 617-659.

Valderrama, M.H., K.C. Nielsen, G.A. McMechan, and H. Hunter, 1994, Three-dimensional seismic interpretation of the triangle zone of the frontal Ouachita Mountains and Arkoma basin, Pittsburg County, Oklahoma, *in* N.H. Suneson and L.A. Hemish, eds., 1991, Geology and resources of the eastern Ouachita Mountains frontal belt and southeastern Arkoma basin, Oklahoma: Oklahoma Geological Survey Guidebook 29, p. 225-241.

Van Siclen, D.C., 1958, Depositional topography — examples and theory: AAPG Bulletin, v. 42, p. 1897-1913.

Van Wagoner, J.C., 1985, Reservoir facies distribution as controlled by sea level change: Abstracts With Programs: SEPM Midyear Meeting, Golden, Colorado, p. 91-92.

Van Wagoner, J.C., R.M. Mitchum, Jr., H.W. Posamentier, and P.R. Vail, 1987, Seismic stratigraphy interpretation using sequence stratigraphy. Part 2: Key definitions of sequence stratigraphy, *in* A.W. Bally, ed., Atlas of seismic stratigraphy, v. 1: AAPG Studies in Geology 27, p. 11-14.

Van Wagoner, J.C., H.W. Posamentier, R.M. Mitchum, Jr., P.R. Vail, J.F. Sarg, T.S. Loutit, and J. Hardenbol, 1988, An overview of the fundamentals of sequence stratigraphy and key definitions, *in* C.K. Wilgus et al., eds., Sea level change: an integrated approach: SEPM Special Publication 42, p. 39-45.

Van Wagoner, J.C., R.M. Mitchum, Jr., K.M. Campion, and V.D. Rahmanian, 1990, Siliciclastic sequence stratigraphy in well logs, cores, and outcrops: concepts for high-resolution correlation of time and facies: AAPG Methods in Exploration 7, 55 p.

Van Wagoner, J.C., 1995, Sequence stratigraphy and marine to nonmarine facies architecture of foreland basin strata, Book Cliffs, Utah, USA, *in* J.C. Van Wagoner and G.T. Betram, eds., Sequence stratigraphy of foreland basin deposits — outcrop to subsurface examples from the Cretaceous of North America: AAPG Memoir 64, p. 137-224.

Veevers, J.J., and C. McA. Powell, 1987, Late Paleozoic glacial episodes in Gondwanaland reflected in transgressive-regressive depositional sequences in Eurameria: GSA Bulletin, v. 98, p. 475-487.

Visher, G.S., S. Saitta B., and R. Phares S., 1971, Pennsylvanian delta pattern and petroleum occurrences in eastern Oklahoma: AAPG Bulletin, v. 55, p. 1206-1230.

Visher, G.S., and J.M. Coleman, 1977, The Coffeyville Format (Pennsylvanian) of northern Oklahoma, a model for an epeiric sea delta: Short Course in Advanced Petroleum Geology Field Trip, University of Tulsa, 25 p.

von Bitter, P.H., 1972, Environmental control of conodont distribution in the Shawnee Group (Upper Pennsylvanian) of eastern Kansas: University of Kansas Paleontological Contributions, Art. 59, 105 p.

Waite, L.E., 1993, Upper Pennsylvanian seismic sequences and facies of the eastern and southern Horseshoe Atoll, Midland basin, west Texas, *in* R.G. Loucks and J.F. Sarg, eds., Carbonate sequence stratigraphy: recent developments and applications: AAPG Memoir 57, p. 213-240.

Walker, R.G., 1990, Facies modeling and sequence stratigraphy: Journal of Sedimentary Petrology, v. 60, p. 777-786.

Walters, R.F., 1958, Differential entrapment of oil and gas in Arbuckle dolomite of central Kansas, *in* B. Rascoe, Jr. and N.J. Hyne, eds., Petroleum geology of the Mid-Continent: Tulsa Geological Society Special Publication 3, p. 29-34.

Walton, A.W., and G. Griffith, 1985, Deltaic deposition in the Tonganoxie or "Stalnaker" Sandstone (Stranger Formation, Virgilian): TXO Robison C-1, Harper County, Kansas, *in* W.L. Watney, A.W. Walton, and J.H. Doveton, compilers, Core studies in Kansas: Kansas Geological Survey Subsurface Geology Series 6, p. 145-160.

Wanless, H.R., 1931, Pennsylvanian cycles in western Illinois: Illinois Geological Survey Bulletin 60, p. 179-193.

Wanless, H.R., and J.M. Weller, 1932, Correlation and extent of Pennsylvanian cyclothems: GSA Bulletin, v. 43, p. 1003-1016.

Wanless, H.R., and F.P. Shepard, 1936, Sea level and climatic changes related to Late Paleozoic cycles: GSA Bulletin, v. 47, p. 1177-1206.

Wanless, H.R., 1950, Late Paleozoic cycles of sedimentation in the United States: 18th International Geological Congress, London, 1948, rept. 4, p. 17-28.

Wanless, H.R., 1952, Studies of field relations of coal beds: Second Conference on the Origin and Constitution of Coal, Crystal Cliffs, Nova Scotia, p. 148-175.

Wanless, H.R., 1955, Pennsylvanian rocks of Eastern Interior basin: AAPG Bulletin, v. 39, no. 9, p. 1753-1820.

Wanless, H.R., 1964, Local regional factors in Pennsylvanian cyclic sedimentation, *in* D.F. Merriam, ed., Symposium on cyclic sedimentation: Kansas Geological Survey Bulletin 169, v. 2, p. 593-606.

Wanless, H.R., J.R. Baroffio, J.C. Gamble, J.C. Horner, D.R. Orlopp, A. Rocha-Campos, J.E. Souther, P.C. Trescott, R.S. Vail, and C.R. Wright, 1970, Late Paleozoic deltas in the central and eastern United States, *in* Deltaic sedimentation, modern and ancient: SEPM Special Publication 15, p. 215-245.

Wanless, H.R., and C.R. Wright, 1978, Paleoenvironmental maps of Pennsylvanian rocks, Illinois basin and northern Midcontinent region: GSA, MC-23.

Waters, J.A., 1927, A group of Foraminifera from the Dornick Hills Formation of the Ardmore basin: Journal of

Paleontology, v. 1, p. 129-133, plate 22.
Watney, W.L., and W.J. Ebanks, Jr., 1978, Early subaerial exposure and freshwater diagenesis of the Upper Pennsylvanian cyclic sediments in northern Kansas and southern Nebraska (abs.): AAPG Bulletin, v. 62, p. 570-571.
Watney, W.L., 1980, Cyclic sedimentation of the Lansing and Kansas City groups (Missourian) in northwestern Kansas and southwestern Nebraska — a guide for petroleum exploration: Kansas Geological Survey Bulletin 220, 72 p.
Watney, W.L., 1984, Favorable reservoir trends in Upper Pennsylvanian Lansing-Kansas City cyclic carbonates in western Kansas, *in* N.J. Hyne, ed., Limestones of the Mid-Continent: Tulsa Geological Society Special Publication 2, p. 201-248.
Watney, W.L., 1985a, Evaluation of the significance of tectonic, sedimentary control versus eustatic control of Upper Pennsylvanian cyclothems in the western Midcontinent, *in* Recent interpretations of Late Paleozoic cylclothems: SEPM Midcontinent Section Proceedings of Third Annual Meeting and Field Conference, p. 105-140.
Watney, W.L., 1985b, Resolving controls on epeiric sedimentation using trend-surface analysis: Mathematical Geology, v. 17, p. 427-454.
Watney, W.L., J.A. French, and E.K. Franseen, 1989, Sequence stratigraphic interpretations and modeling of cyclothems in the Upper Pennsylvanian (Missourian) Lansing and Kansas City groups in eastern Kansas: Kansas Geological Society 41st Annual Field Conference Guidebook, 211 p.
Watney, W.L., J.H. Doveton, J.A. French, and D.R. Collins, 1990, Petrophysical characterization of Upper Pennsylvanian depositional sequences in western Kansas: GSA South-Central Section Abstracts With Programs, v. 22, no. 1, p. 35.
Watney, W.L., J.A. French, J.C. Wong, and R. Black, 1992, Time series analysis of natural gamma ray logs from the Mid-Continent — tectonic or eustatic signal? (abs.): Society of Sedimentary Geology Mid-Year Meeting, p. 65.
Watney, W.L., 1993, Regional geologic perspective for Virgilian Douglas Group in Kansas, *in* A.W. Archer, H.R. Feldman, and W.P. Lanier, eds., Incised paleovalleys of the Douglas Group in northeastern Kansas. field guide and related papers: Kansas Geological Survey OFR 93-24, p. 9.1 - 9.16.
Watney, W.L., J. Harff, J.C. Davis, G. Bohling, and J.C. Wong, 1993, Controls on petroleum accumulation in Upper Pennsylvanian cyclic shelf carbonates in western Kansas, USA, interpreted by space modeling and multivariate techniques, *in* J. Harff and J.F. Merriam, eds., Computer basin analysis: New York, Plenum Press, p. 215-237.
Watney, W.L., B. Stephens, and D. Newell, in press, Shallowing-upward events and their implications for internal correlations and depositional environment of the St. Peter Sandstone in the Forest City basin, NE Kansas, *in* K.S. Johnson, ed., Simpson and Viola groups in the southern Mid-Continent, 1994 Symposium: Oklahoma Geological Survey Circular.
Watney, W.L., G.C. Bohling, and J.H. Doveton, 1994, Systematic wireline log predictions of sequence stratigraphic units: a case study from the Upper Pennsylvanian of Kansas: AAPG Annual Convention Abstracts, p. 279.
Watney, W.L., and J.A. French, 1994, Revisions of the stratigraphic nomenclature and classification of the Pleasanton, Marmaton, and Kansas City groups of Kansas: Kansas Geological Survey Open File Report 1994-34, p. 1-18.
Watney, W.L., J.A. French, and W.J. Guy, in review, Modeling petroleum resevoirs in Pennsylvanian strata of the Midcontinent, U.S.A.
Watts, A.B., E.D. Karner, and M.S. Steckler, 1982, Lithospheric flexure and the evolution of sedimentary basins: Philosophical Transactions of the Royal Society of London, ser. A, v. 305, p. 249-281.
Weaver, G.D., 1954, Geology and mineral resources of Hughes County, Oklahoma: Oklahoma Geological Survey Bulletin 70, p. 1-150.
Weibel, C.P., 1991, Application of cyclothemic-based sequence stratigraphy to Upper Pennsylvanian strata, east-central Illinois, *in* C.P. Weibel, ed., Sequence stratigraphy in mixed clastic-carbonate strata, Upper Pennsylvanian, east-central Illinois: SEPM Great Lakes Section 21st Annual Field Conference, Illinois Geological Survey, p. 1-25.
Weimer, R.J., 1984, Relation of unconformities, tectonics, and sea-level changes, Cretaceous of Western Interior, USA, *in* J.S. Schlee, ed., Interregional unconformities and hydrocarbon accumulation: AAPG Memoir 36, p. 7-35.
Weimer, R.J., 1988, Record of sea level changes, Cretaceous of Western Interior, U.S.A., *in* C.K. Wilgus et al., eds., Sea level changes: an integrated approach: SEPM Special Publication 42, p. 285-288.
Weimer, R.J., S.A. Sonnenberg, and L.T. Shannon, 1988, Production from valley-fill deposits, Morrow Sandstone, southeast Colorado: new exploration challenges and rewards (abs.): AAPG Bulletin, v. 72, p. 884.
Weimer, R.J., 1992, Sequence stratigraphy: challenges and problems in future exploration and production: Rocky Mountain Association of Geologists Outcrop, p. 3-4.
Weimer, R.J., 1992, Developments in sequence stratigraphy: foreland and cratonic basins: AAPG Bulletin, v. 76, no. 7, p. 965-982.
Weimer, R.J., and S.A. Sonnenberg, 1994, Sequence stratigraphy and petroleum geology of the central Denver basin: AAPG Annual Meeting Field Guide 8, 75 p.
Weirich, T.E., 1953, Shelf principle of oil origin, migration, and accumulation: AAPG Bulletin, v. 37, p. 2027-2045.
Weller, J.M., 1930, Cyclical sedimentation of the Pennsylvanian period and its significance: Journal of Geology, v. 38, p. 97-135.
Weller, J.M., 1931, The conception of cyclical sedimentation during the Pennsylvanian Period: Illinois Geological Survey Bulletin 60, p. 163-177.
Weller, J.M., 1956, Argument for diastrophic control of late Paleozoic cycles: AAPG Bulletin, v. 40, p. 17-50.
Weller, J.M., 1958, Cyclothems and larger sedimentary cycles of the Pennsylvanian: Journal of Geology, v. 66, p. 195-207.
Wells, J.S., 1971, Forest City basin, *in* I.H. Cram, ed., Future petroleum provinces of the United States — their geology and potential: AAPG Memoir 15, p. 1098-1103.
Wescott, W.A., 1993, Geomorphic thresholds and complex response of fluvial systems — some implications for sequence stratigraphy: AAPG Bulletin, v. 77, p. 1208-1218.
West, R.R., 1972, Relationship between community analysis and depositional environments: an example from the North American Carboniferous: 24th International Geological Congress, Section 7, p. 130-146.
West, R.R., and R.M. Busch, 1985, Sixth-order transgressive-

regressive units in the Wewoka Formation (Anna Shale Member) of Oklahoma: Proceedings of the Third Annual Meeting and Field Conference, SEPM Mid-Continent Section, p. 155-171.

Wheeler, D.M., A.J. Scott, V.J. Coringrato, and P.E. Devine, 1990, Stratigraphy and depositional history of the Morrow Formation, southeast Colorado and southwest Kansas, *in* S.A. Sonnenberg et al., eds., Morrow sandstones of southeast Colorado and adjacent areas: Denver, The Rocky Mountain Association of Geologists, p. 9-35.

Wheeler, H.E., and H.H. Murray, 1957, Base-level control patterns in cyclothemic sedimentation: AAPG Bulletin, v. 41, p. 1985-2011.

Wickham, J., 1978, The southern Oklahoma aulacogen, in structural style of the Arbuckle region: GSA South-Central Region Field Trip 3, p. 8-41.

Wilson, F.S., 1954, Foraminifera from the Pennsylvanian Wolf Mountain Shale near Graford, Texas: Unpublished M.S. thesis, University of Tulsa, 47 p.

Wilson, J.L., P.L. Medlock, R.D. Fritz, K.L. Canter, and R.G. Geesaman, 1992, A review of Cambro-Ordovician breccias in North America, *in* M.P. Candelaria and C.L. Reed, eds., Paleokarst, karst-related diagenesis, and reservoir development: Examples from the Ordovician-Devonian Age strata of west Texas and the Mid-Continent: SEPM Permian Basin Section Publication 92-33, p. 19-29.

Wilson, L.R., 1972, Fossil plants of the Seminole Formation in Tulsa County: Tulsa Geological Digest, v. 37, p. 151-161.

Wilson, L.R., 1984, Evidence for a new Desmoinesian-Missourian boundary in Tulsa County, Oklahoma, *in* A.K. Sharma et al., eds., Proceedings of the Symposium on Evolutionary Botany and Biostratigraphy: New Delhi, A.K. Ghosh Commemorative Volume, p. 251-265.

Winchell, R.L., 1957, Relationship of the Lansing Group and the Tongonoxie ("Stalnaker") Sandstone in south-central Kansas: Kansas Geological Survey Bulletin 127, pt. 4, p. 123-152.

Woodruff, J.M., 1984, The depositional environment of the Garnett fossil locality, Rock Lake Shale Member of the Pennsylvanian Staton Formation [sic], Anderson County, Kansas: Unpublished M.S. thesis, Duke University, 179 p.

Youle, J.C., 1992, Sequence stratigraphy of the lower Middle Pennsylvanian and distribution of selected sandstones, eastern Hugoton embayment, southwestern Kansas: M.S. thesis, University of Kansas, 205 p.

Youle, J.C., W.L. Watney, and L.L. Lambert, 1994, Stratal hierarchy and sequence stratigraphy — Middle Pennsylvanian, southwestern Kansas, U.S.A., *in* G.D. Klein, ed., Pangea: paleoclimate, tectonics, and sedimentation during accretion, zenith, and breakup of a supercontinent: GSA Special Paper 288, p. 267-285.

Zangerl, R., and E.S. Richardson, 1963, The paleoecologic history of two Pennsylvanian black shales: Chicago, Field Museum of Natural History, Fieldiana Geological Memoirs, v. 4, 352 p.

Zaitlin, B.A., R.W. Dalrymple, and R. Boyd, 1994, The stratigraphic organization of incised-valley systems associated with relative sea-level changes, *in* R.W. Dalrymple, R. Boyd, and B. Zaitlin, eds., Incised-valley systems: orgin and sedimentary sequences: SEPM Special Publication 51, p. 45-60.

Zeller, D.E., 1968, The stratigraphic succession in Kansas: Kansas Geological Survey Bulletin 189, 81 p.